Photopolarimetry in Remote Sensing

NATO Science Series

A Series presenting the results of scientific meetings supported under the NATO Science Programme.

The Series is published by IOS Press, Amsterdam, and Kluwer Academic Publishers in conjunction with the NATO Scientific Affairs Division

Sub-Series

I. Life and Behavioural Sciences	IOS Press
II. Mathematics, Physics and Chemistry	Kluwer Academic Publishers
III. Computer and Systems Science	IOS Press
IV. Earth and Environmental Sciences	Kluwer Academic Publishers
V. Science and Technology Policy	IOS Press

The NATO Science Series continues the series of books published formerly as the NATO ASI Series.

The NATO Science Programme offers support for collaboration in civil science between scientists of countries of the Euro-Atlantic Partnership Council. The types of scientific meeting generally supported are "Advanced Study Institutes" and "Advanced Research Workshops", although other types of meeting are supported from time to time. The NATO Science Series collects together the results of these meetings. The meetings are co-organized bij scientists from NATO countries and scientists from NATO's Partner countries – countries of the CIS and Central and Eastern Europe.

Advanced Study Institutes are high-level tutorial courses offering in-depth study of latest advances in a field.
Advanced Research Workshops are expert meetings aimed at critical assessment of a field, and identification of directions for future action.

As a consequence of the restructuring of the NATO Science Programme in 1999, the NATO Science Series has been re-organised and there are currently Five Sub-series as noted above. Please consult the following web sites for information on previous volumes published in the Series, as well as details of earlier Sub-series.

http://www.nato.int/science
http://www.wkap.nl
http://www.iospress.nl
http://www.wtv-books.de/nato-pco.htm

Series II: Mathematics, Physics and Chemistry – Vol. 161

Photopolarimetry in Remote Sensing

edited by

Gorden Videen
Army Research Laboratory,
Adelphi, Maryland, U.S.A.

Yaroslav Yatskiv
Main Astronomical Observatory of the National Academy of Sciences of Ukraine,
Kiev, Ukraine

and

Michael Mishchenko
NASA Goddard Institute for Space Studies,
New York, U.S.A.

Kluwer Academic Publishers

Dordrecht / Boston / London

Published in cooperation with NATO Scientific Affairs Division

Proceedings of the NATO Advanced Study Institute on
Photopolarimetry in Remote Sensing
Yalta, Ukraine
20 September–4 October 2003

A C.I.P. Catalogue record for this book is available from the Library of Congress.

ISBN 1-4020-2366-9
ISBN 1-4020-2368-5 (e-book)

Published by Kluwer Academic Publishers,
P.O. Box 17, 3300 AA Dordrecht, The Netherlands.

Sold and distributed in North, Central and South America
by Kluwer Academic Publishers,
101 Philip Drive, Norwell, MA 02061, U.S.A.

In all other countries, sold and distributed
by Kluwer Academic Publishers,
P.O. Box 322, 3300 AH Dordrecht, The Netherlands.

Printed on acid-free paper

Printed in the Netherlands.

TABLE OF CONTENTS

Backscatter Polarization

Biological Systems

Astrophysical Phenomena

Comets

Photopolarimetry Instrumentation

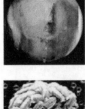
Advertisement appearing in Nature (May 22, 2003) outlining the ASI.

Preface

The NATO Advanced Study Institute on *"Photopolarimetry in Remote Sensing"* met on the outskirts of the Ukrainian Black-sea city of Yalta, 20 September – 3 October, 2003. The focus of the meeting was recent advances in polarimetric methodologies used in remote sensing, including, but not limited to terrestrial and astrophysical characterization, medical diagnostics, environmental and military monitoring. Following the ASI, some participants attended an accompanying workshop focusing on international cooperation in Kyiv, 4 – 10 October.

The concept for the ASI was put forth at the NATO Advanced Research Workshop on *"Optics of Cosmic Dust"* (held in Bratislava, Slovakia, chaired by Gorden Videen and Miroslav Kocifaj) and was proposed by Nikolai Voshchinnikov, Michael Mishchenko and Vera Rosenbush. It should be noted that none of these plotters actually co-chaired either of these events. Like most good things, the initial planning was facilitated by many bottles of vodka and brandy, too many to count actually. Because of the pioneering observational efforts and support infrastructure of the Main Astronomical Observatory of Ukraine, it was decided that the ASI would take place in Ukraine on the Black-sea coast. Preparation of the ASI began in earnest in the autumn of 2002 with formal acceptance of our NATO proposal.

While Co-chairs Yaroslav Yatskiv and Gorden Videen busied themselves with bureaucratic necessities, the actual preparations were made by the Local Organizing Committee. Special recognition is owed to Vera Rosenbush

Chairman Yaroslav Yatskiv rallies the troops at the first conference dinner.

who is responsible for all the good things that happened. In addition to the LOC, we thank the management and staff of the Sanatorium Druzhba for extending the warmest of welcomes to the participants. Sasha Krysyuk participated in all the conference activities and insisted that we receive the best of everything. We could not have wished for a better host.

ASI and Sanatorium Druzhba management and staff stand in front of flags of the NATO ASI participant countries. Back row (left to right): Anatoliy Vid'machenko, Sasha Krysyuk, Klaus Jockers and Michael Mishchenko. Front row (left to right): Valentin Babenko, Tamara Bulba, Ivan Andronov, Irina Kulyk, Yuriy Shkuratov, Nikolai Kiselev, Anny-Chantal Levasseur-Regourd, Ted Kostiuk, Vera Rosenbush, Gorden Videen, Alexander Fedorov, Zhanna Platonova, and Yaroslav Yatskiv.

Many fields have made contributions to the art of polarimetry, and our goal was to select representative lecturers who have contributed to the various aspects. We were lucky to have a group of outstanding lecturers willing to invest the time to provide illuminating and entertaining lectures on the fundamental research in their fields. This book is a compilation of significant contributions taken primarily from these key lectures. We are grateful to those who were able to devote the significant time and effort necessary to document this work.

The Swallow's Nest, symbol of the Crimea, is a short distance from the Druzhba.

While the lectures represent the key component of a NATO ASI, another critical component is providing opportunities for interactions. Not only is this important for lecturers to elucidate key points and

to provide details on a more personal level, but it is also critical to provide the time for the communications that ultimately will lead to advances and collaborations that will drive the field into the future. The LOC organized these activities that included an opening reception, two formal conference dinners and field trips to the Crimean Astrophysical Observatory (with a visit to the Uspensky Monastery and Chufut Kale), and tours of Livadia and Alupka

Picnic at CrAO was a carnivorous affair. Shown (left to right): Ivan Mishchenko, Matt Easley, Olga Kalashnikova, James McDonald, Marina Prokopjeva, Daria Dubkova and Pavel Litvinov.

palaces, Ai Petra, and Sevastopol. In sum, we did our best to provide the necessary elements to commence fruitful collaborations, i.e., food, drink, and discussion, and hope our participant colleagues are able to turn this opportunity to their advantage.

Gorden Videen
Yaroslav Yatskiv
Michael Mishchenko

April, 2004

Organizational Structure

CHAIRS

Yaroslav Yatskiv	Main Astronomical Observatory of the National Academy of Sciences of Ukraine
Gorden Videen	United States Army Research Laboratory

ORGANIZING COMMITTEE

Vladimir Grinin	Crimean Astrophysical Observatory, Ukraine
Vsevolod Ivanov	St. Petersburg University, Russia
Theodore Kostiuk	NASA Goddard Space Flight Center, USA
Michael Mishchenko	NASA Goddard Institute for Space Studies, New York, USA
Alexander Morozhenko	Main Astronomical Observatory of the National Academy of Sciences of Ukraine

LECTURERS

Oleg Dubovik	NASA Goddard Space Flight Center, USA
Francisco Gonzalez	University of Cantabria, Spain
Vladimir Grinin	Crimean Astrophysical Observatory, Ukraine
Keith Hopcraft	University of Nottingham, UK
James Hough	University of Hertfordshire, UK
Vsevolod V. Ivanov	St. Petersburg University, Russia
Theodore Kostiuk	NASA Goddard Space Flight Center, USA
A. Chantal Levasseur-Regourd	University of Paris, France
Andreas Macke	University of Kiel, Germany
Michael Mishchenko	NASA Goddard Institute for Space Studies, USA
Yuriy Shkuratov	Kharkov University, Ukraine
Nikolai Voshchinnikov	St. Petersburg University, Russia
Dmitry Zimnyakov	Saratov State University, Russia

LOCAL ORGANIZING COMMITTEE

Tamara Bulba	Alla Rostopchina- Shakhovskaya
Zhanna Dlugach	Dmitry Shakhovskoy
Irina Kulyk	Anatoliy Vid'machenko
Vera Rosenbush	

LOCAL CONTACT ADDRESS:

POLAR-2003
Main Astronomical Observatory
National Academy of Sciences of Ukraine
Zabolotnoho Str. 27
Kiev 03680
UKRAINE
Fax: (380) + 44 + 266 21 47
Tel: (380) + 44 + 266 08 69
E-mail : polar2003@mao.kiev.ua

CONFERENCE LOCALE:

Sanatorium KURPATY, Druzhba ("Friendship")
Alupka highway 12
Yalta 98659
Crimea
UKRAINE
Fax: (380) + 0654 + 24 83 09
Tel: (380) + 0654 + 31 47 93

ACKNOWLEDGMENT OF SUPPORT:

Primary support for the NATO ASI was provided by the NATO Science Committee. Additional funding was provided by the US OFFICE OF NAVAL RESEARCH INTERNATIONAL FIELD OFFICE and the US National Science Foundation. Some travel fellowships were provided by the US National Science Foundation and the Scientific and Technical Research Council of Turkey. Additional support was provided by the US Army Research Laboratory. Support for the Workshop on Remote Sensing Techniques and Instrumentation was provided by the European Research Office of the US Army and Science of Technology Center of Ukraine. Any opinions, findings and conclusions or recommendations expressed in this material are those of the authors and do not necessarily reflect the views of the European Research Office of the US Army, the Office of Naval Research International Field Office, the NATO Science Committee, the US Army Research Laboratory, the US National Science Foundation, the Technical Research Council of Turkey, the participants, authors, or the editors.

Participants of NATO ASI pose for a group photo in front of Sanatorium Druzhba.

MAXWELL'S EQUATIONS, ELECTROMAGNETIC WAVES, AND STOKES PARAMETERS

MICHAEL I. MISHCHENKO AND LARRY D. TRAVIS

NASA Goddard Institute for Space Studies, 2880 Broadway, New York, NY 10025, USA

1. Introduction

The theoretical basis for describing elastic scattering of light by particles and surfaces is formed by classical electromagnetics. In order to make this volume sufficiently self-contained, this introductory chapter provides a summary of those concepts and equations of electromagnetic theory that will be used extensively in later chapters and introduces the necessary notation.

We start by formulating the macroscopic Maxwell equations and constitutive relations and discussing the fundamental time-harmonic plane-wave solution that underlies the basic optical idea of a monochromatic parallel beam of light. This is followed by the introduction of the Stokes parameters and a discussion of their ellipsometric content. Then we consider the concept of a quasi-monochromatic beam of light and its implications and briefly discuss how the Stokes parameters of monochromatic and quasi-monochromatic light can be measured in practice. In the final two sections, we discuss another fundamental solution of Maxwell's equations in the form of a time-harmonic outgoing spherical wave and introduce the concept of the coherency dyad, which plays a vital role in the theory of multiple light scattering by random particle ensembles.

2. Maxwell's equations and constitutive relations

The theory of classical optics phenomena is based on the set of four Maxwell's equations for the macroscopic electromagnetic field at interior points in matter, which in SI units read:

$$\nabla \cdot \mathcal{D}(\mathbf{r}, t) = \rho(\mathbf{r}, t), \tag{2.1}$$

$$\nabla \times \mathcal{E}(\mathbf{r}, t) = -\frac{\partial \mathcal{B}(\mathbf{r}, t)}{\partial t}, \tag{2.2}$$

$$\nabla \cdot \mathcal{B}(\mathbf{r}, t) = 0, \tag{2.3}$$

G. Videen et al. (eds.), Photopolarimetry in Remote Sensing, 1-44.
© 2004 *Kluwer Academic Publishers. Printed in the Netherlands.*

$$\nabla \times \mathcal{H}(\mathbf{r}, t) = \mathcal{J}(\mathbf{r}, t) + \frac{\partial \mathcal{D}(\mathbf{r}, t)}{\partial t}, \tag{2.4}$$

where $\mathcal{E}$ is the electric and $\mathcal{H}$ the magnetic field, $\mathcal{B}$ the magnetic induction, $\mathcal{D}$ the electric displacement, and ρ and $\mathcal{J}$ the macroscopic (free) charge density and current density, respectively. All quantities entering Eqs. (2.1)–(2.4) are functions of time, t, and spatial coordinates, $\mathbf{r}$. Implicit in the Maxwell equations is the continuity equation

$$\frac{\partial \rho(\mathbf{r}, t)}{\partial t} + \nabla \cdot \mathcal{J}(\mathbf{r}, t) = 0, \tag{2.5}$$

which is obtained by combining the time derivative of Eq. (2.1) with the divergence of Eq. (2.4) and taking into account that $\nabla \cdot (\nabla \times \mathbf{a}) = 0$. The vector fields entering Eqs. (2.1)–(2.4) are related by

$$\mathcal{D}(\mathbf{r}, t) = \epsilon_0 \mathcal{E}(\mathbf{r}, t) + \mathcal{P}(\mathbf{r}, t), \tag{2.6}$$

$$\mathcal{H}(\mathbf{r}, t) = \frac{1}{\mu_0} \mathcal{B}(\mathbf{r}, t) - \mathcal{M}(\mathbf{r}, t), \tag{2.7}$$

where $\mathcal{P}$ is the electric polarization (average electric dipole moment per unit volume), $\mathcal{M}$ is the magnetization (average magnetic dipole moment per unit volume), and ϵ_0 and μ_0 are the electric permittivity and the magnetic permeability of free space, respectively.

Equations (2.1)–(2.7) are insufficient for a unique determination of the electric and magnetic fields from a given distribution of charges and currents and must be supplemented with so-called constitutive relations:

$$\mathcal{P}(\mathbf{r}, t) = \epsilon_0 \chi(\mathbf{r}) \mathcal{E}(\mathbf{r}, t), \tag{2.8}$$

$$\mathcal{B}(\mathbf{r}, t) = \mu(\mathbf{r}) \mathcal{H}(\mathbf{r}, t), \tag{2.9}$$

$$\mathcal{J}(\mathbf{r}, t) = \sigma(\mathbf{r}) \mathcal{E}(\mathbf{r}, t), \tag{2.10}$$

where χ is the electric susceptibility, μ the magnetic permeability, and σ the conductivity. Equations (2.6) and (2.8) yield

$$\mathcal{D}(\mathbf{r}, t) = \epsilon(\mathbf{r}) \mathcal{E}(\mathbf{r}, t), \tag{2.11}$$

where

$$\epsilon(\mathbf{r}) = \epsilon_0 [1 + \chi(\mathbf{r})] \tag{2.12}$$

is the electric permittivity. For linear and isotropic media, χ, μ, σ, and ϵ are scalars independent of the fields. The microphysical derivation and the range of validity of the macroscopic Maxwell equations are discussed in detail by Jackson [1].

The constitutive relations (2.9)–(2.11) connect the field vectors at the same moment of time t and are valid for electromagnetic fields in a vacuum and also for electromagnetic fields in macroscopic material media provided that the

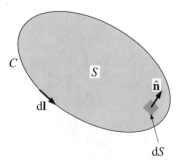

Figure 2.1: A finite surface S bounded by a closed contour C.

fields are constant or vary in time rather slowly. For a rapidly varying field in a material medium, the state of the medium depends not only on the current value of the field but also on the values of the field at all previous times. Therefore, for a linear, time-invariant medium, the constitutive relations (2.9)–(2.11) must be replaced by the following general causal relations that take into account the effect of the prior history on the electromagnetic properties of the medium:

$$\mathcal{D}(\mathbf{r},t) = \int_{-\infty}^{t} dt' \widetilde{\epsilon}(\mathbf{r},t-t')\mathcal{E}(\mathbf{r},t'), \tag{2.13}$$

$$\mathcal{B}(\mathbf{r},t) = \int_{-\infty}^{t} dt' \widetilde{\mu}(\mathbf{r},t-t')\mathcal{H}(\mathbf{r},t'), \tag{2.14}$$

$$\mathcal{J}(\mathbf{r},t) = \int_{-\infty}^{t} dt' \widetilde{\sigma}(\mathbf{r},t-t')\mathcal{E}(\mathbf{r},t'). \tag{2.15}$$

The medium characterized by the constitutive relations (2.13)–(2.15) is called time-dispersive.

It is straightforward to rewrite the Maxwell equations and the continuity equation in an integral form. Specifically, integrating Eqs. (2.2) and (2.4) over a surface S bounded by a closed contour C (see Fig. 2.1) and applying the Stokes theorem,

$$\int_{S} dS(\nabla \times \mathbf{A}) \cdot \hat{\mathbf{n}} = \oint_{C} d\mathbf{l} \cdot \mathbf{A}, \tag{2.16}$$

yield

$$\oint_{C} d\mathbf{l} \cdot \mathcal{E} = -\frac{\partial}{\partial t} \int_{S} dS \mathcal{B} \cdot \hat{\mathbf{n}}, \tag{2.17}$$

$$\oint_{C} d\mathbf{l} \cdot \mathcal{H} = \int_{S} dS \mathcal{J} \cdot \hat{\mathbf{n}} + \frac{\partial}{\partial t} \int_{S} dS \mathcal{D} \cdot \hat{\mathbf{n}}, \tag{2.18}$$

where we employ the usual convention that the direction of the differential

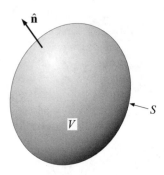

Figure 2.2: A finite volume V bounded by a closed surface S.

length vector **dl** is related to the direction of the unit vector along the local normal to the surface $\hat{\mathbf{n}}$ according to the right-hand rule.

Similarly, integrating Eqs. (2.1), (2.3), and (2.5) over a finite volume V bounded by a closed surface S (see Fig. 2.2) and using the Gauss theorem,

$$\int_V d\mathbf{r} \nabla \cdot \mathbf{A} = \oint_S dS \mathbf{A} \cdot \hat{\mathbf{n}}, \qquad (2.19)$$

we derive

$$\oint_S dS \boldsymbol{\mathcal{D}} \cdot \hat{\mathbf{n}} = \int_V d\mathbf{r}\rho, \qquad (2.20)$$

$$\oint_S dS \boldsymbol{\mathcal{B}} \cdot \hat{\mathbf{n}} = 0, \qquad (2.21)$$

$$\oint_S dS \boldsymbol{\mathcal{J}} \cdot \hat{\mathbf{n}} = -\frac{\partial}{\partial t} \int_V d\mathbf{r}\rho, \qquad (2.22)$$

where the unit vector $\hat{\mathbf{n}}$ is directed along the outward local normal to the surface.

3. Boundary conditions

The Maxwell equations are strictly valid only for points in whose neighborhood the physical properties of the medium, as characterized by the constitutive parameters χ, μ, and, σ vary continuously. However, across an interface separating one medium from another the constitutive parameters may change abruptly, and one may expect similar discontinuous behavior of the field vectors $\boldsymbol{\mathcal{E}}$, $\boldsymbol{\mathcal{D}}$, $\boldsymbol{\mathcal{B}}$, and $\boldsymbol{\mathcal{H}}$. The boundary conditions at such an interface can be derived from the integral form of the Maxwell equations as follows. Consider two different continuous media separated by an interface S as shown in Fig. 3.1. Let $\hat{\mathbf{n}}$ be a unit vector along the local normal to the interface, pointing from medium 1 toward medium 2. Let us take the integral in Eq.

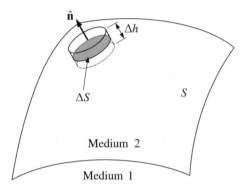

Figure 3.1: Pillbox used in the derivation of boundary conditions for the $\mathcal{B}$ and $\mathcal{D}$.

(2.21) over the closed surface of a small cylinder with bases parallel to a small surface element ΔS such that half of the cylinder is in medium 1 and half in medium 2. The contribution from the curved surface of the cylinder vanishes in the limit $\Delta h \rightarrow 0$, and we thus obtain

$$(\mathcal{B}_2 - \mathcal{B}_1) \cdot \hat{\mathbf{n}} = 0, \tag{3.1}$$

which means that the normal component of the magnetic induction is continuous across the interface.

Similarly, evaluating the integrals on the left- and right-hand sides of Eq. (2.20) over the surface and volume of the cylinder, respectively, we derive

$$(\mathcal{D}_2 - \mathcal{D}_1) \cdot \hat{\mathbf{n}} = \lim_{\Delta h \to 0} \Delta h \rho = \rho_S, \tag{3.2}$$

where ρ_S is the surface charge density (charge per unit area) measured in coulombs per square meter. Thus, there is a discontinuity in the normal component of $\mathcal{D}$ if the interface carries a layer of surface charge density.

Let us now consider a small rectangular loop of area ΔA formed by sides of length Δl perpendicular to the local normal and ends of length Δh parallel to the local normal, as shown in Fig. 3.2. The surface integral on the right-hand side of Eq. (2.17) vanishes in the limit $\Delta h \rightarrow 0$,

$$\lim_{\Delta h \to 0} \int_{\Delta A} dS \mathcal{B} \cdot (\hat{\mathbf{n}} \times \hat{\mathbf{l}}) = \lim_{\Delta h \to 0} \Delta l \Delta h \mathcal{B} \cdot (\hat{\mathbf{n}} \times \hat{\mathbf{l}}) = 0,$$

so that

$$\hat{\mathbf{l}} \cdot (\mathcal{E}_2 - \mathcal{E}_1) = 0. \tag{3.3}$$

Since the orientation of the rectangle – and hence also of $\hat{\mathbf{l}}$ – is arbitrary, Eq. (3.3) means that the vector $\mathcal{E}_2 - \mathcal{E}_1$ must be perpendicular to the interface. Thus,

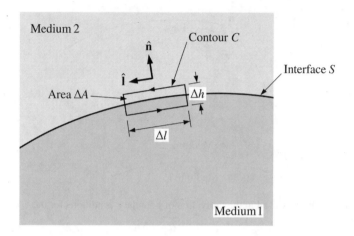

Figure 3.2: Rectangular loop used in the derivation of boundary conditions for the $\mathcal{E}$ and $\mathcal{H}$.

$$\hat{\mathbf{n}} \times (\mathcal{E}_2 - \mathcal{E}_1) = \mathbf{0}, \tag{3.4}$$

where $\mathbf{0}$ is a zero vector. This implies that the tangential component of $\mathcal{E}$ is continuous across the interface.

Similarly, Eq. (2.18) yields

$$\hat{\mathbf{l}} \cdot (\mathcal{H}_2 - \mathcal{H}_1) = \lim_{\Delta h \to 0} \Delta h (\hat{\mathbf{n}} \times \hat{\mathbf{l}}) \cdot \mathcal{J} = (\hat{\mathbf{n}} \times \hat{\mathbf{l}}) \cdot \mathcal{J}_S, \tag{3.5}$$

where $\mathcal{J}_S$ is the surface current density measured in amperes per meter. Since

$$\hat{\mathbf{l}} = (\hat{\mathbf{n}} \times \hat{\mathbf{l}}) \times \hat{\mathbf{n}}, \tag{3.6}$$

we can use the vector identity

$$(\mathbf{a} \times \mathbf{b}) \cdot \mathbf{c} = \mathbf{a} \cdot (\mathbf{b} \times \mathbf{c}) \tag{3.7}$$

to derive

$$[(\hat{\mathbf{n}} \times \hat{\mathbf{l}}) \times \hat{\mathbf{n}}] \cdot (\mathcal{H}_2 - \mathcal{H}_1) = (\hat{\mathbf{n}} \times \hat{\mathbf{l}}) \cdot [\hat{\mathbf{n}} \times (\mathcal{H}_2 - \mathcal{H}_1)] = (\hat{\mathbf{n}} \times \hat{\mathbf{l}}) \cdot \mathcal{J}_S. \tag{3.8}$$

Since this equality must be valid for any orientation of the rectangle and, thus, of the tangent unit vector $\hat{\mathbf{l}}$, we finally have

$$\hat{\mathbf{n}} \times (\mathcal{H}_2 - \mathcal{H}_1) = \mathcal{J}_S, \tag{3.9}$$

which means that there is a discontinuity in the tangential component of $\mathcal{H}$ if the interface can carry a surface current. Media with finite conductivity cannot support surface currents so that

$$\hat{\mathbf{n}} \times (\mathcal{H}_2 - \mathcal{H}_1) = 0 \qquad \text{(finite conductivity)}. \qquad (3.10)$$

The boundary conditions (3.1), (3.2), (3.4), (3.9), and (3.10) are useful in solving the differential Maxwell equations in different adjacent regions with continuous physical properties and then linking the partial solutions to determine the fields throughout all space.

4. Time-harmonic fields

Let us now assume that all fields and sources are time harmonic (or monochromatic), which means that their time dependence can be fully described by expressing them as sums of terms proportional to either $\cos \omega t$ or $\sin \omega t$, where ω is the angular frequency. It is standard practice to represent real monochromatic fields as real parts of the respective complex time-harmonic fields, e.g.,

$$\mathcal{E}(\mathbf{r}, t) = \mathrm{Re}\, \mathbf{E}(\mathbf{r}, t) = \mathrm{Re}[\mathbf{E}(\mathbf{r})\exp(-i\omega t)]$$

$$= \tfrac{1}{2}[\mathbf{E}(\mathbf{r})\exp(-i\omega t) + \mathbf{E}^*(\mathbf{r})\exp(i\omega t)] \qquad (4.1)$$

and analogously for $\mathcal{D}$, $\mathcal{H}$, $\mathcal{B}$, $\mathcal{J}$, ρ, $\mathcal{P}$, and $\mathcal{M}$, where $i = \sqrt{-1}$, $\mathbf{E}(\mathbf{r})$ is complex, and the asterisk denotes a complex-conjugate value. Equations (2.1)–(2.5) then yield the following frequency-domain Maxwell equations and continuity equation for the time-independent components of the complex fields:

$$\nabla \cdot \mathbf{D}(\mathbf{r}) = \rho(\mathbf{r}), \qquad (4.2)$$

$$\nabla \times \mathbf{E}(\mathbf{r}) = i\omega \mathbf{B}(\mathbf{r}), \qquad (4.3)$$

$$\nabla \cdot \mathbf{B}(\mathbf{r}) = 0, \qquad (4.4)$$

$$\nabla \times \mathbf{H}(\mathbf{r}) = \mathbf{J}(\mathbf{r}) - i\omega \mathbf{D}(\mathbf{r}), \qquad (4.5)$$

$$-i\omega \rho(\mathbf{r}) + \nabla \cdot \mathbf{J}(\mathbf{r}) = 0, \qquad (4.6)$$

where we emphasize the typographical distinction between the real quantities $\mathcal{E}$, $\mathcal{D}$, $\mathcal{H}$, $\mathcal{B}$, $\mathcal{J}$, and ρ and their complex counterparts $\mathbf{E}$, $\mathbf{D}$, $\mathbf{H}$, $\mathbf{B}$, $\mathbf{J}$, and ρ.

The constitutive relations remain unchanged in the frequency domain for a non-dispersive medium:

$$\mathbf{D}(\mathbf{r}) = \epsilon(\mathbf{r})\mathbf{E}(\mathbf{r}), \qquad (4.7)$$

$$\mathbf{B}(\mathbf{r}) = \mu(\mathbf{r})\mathbf{H}(\mathbf{r}), \qquad (4.8)$$

$$\mathbf{J}(\mathbf{r}) = \sigma(\mathbf{r})\mathbf{E}(\mathbf{r}). \qquad (4.9)$$

For a time-dispersive medium, we can substitute the monochromatic fields of the form (4.1) into Eqs. (2.13)–(2.15), which yields

$$\mathbf{D}(\mathbf{r}) = \epsilon(\mathbf{r}, \omega)\mathbf{E}(\mathbf{r}), \qquad (4.10)$$

$$\mathbf{B}(\mathbf{r}) = \mu(\mathbf{r}, \omega)\mathbf{H}(\mathbf{r}), \qquad (4.11)$$

$$\mathbf{J}(\mathbf{r}) = \sigma(\mathbf{r}, \omega)\mathbf{E}(\mathbf{r}), \qquad (4.12)$$

where

$$\epsilon(\mathbf{r}, \omega) = \int_0^\infty dt \tilde{\epsilon}(\mathbf{r}, t) \exp(i\omega t), \qquad (4.13)$$

$$\mu(\mathbf{r}, \omega) = \int_0^\infty dt \tilde{\mu}(\mathbf{r}, t) \exp(i\omega t), \qquad (4.14)$$

$$\sigma(\mathbf{r}, \omega) = \int_0^\infty dt \tilde{\sigma}(\mathbf{r}, t) \exp(i\omega t) \qquad (4.15)$$

are complex functions of the angular frequency. Note that we use sloping Greek letters in Eqs. (4.7)–(4.9) and upright Greek letters in Eqs. (4.10)–(4.12) to differentiate between the frequency-independent and the frequency-dependent constitutive parameters, respectively. Equations (4.2) and (4.5) can be rewritten in the form

$$\nabla \cdot [\varepsilon(\mathbf{r}, \omega) \mathbf{E}(\mathbf{r})] = 0, \qquad (4.16)$$
$$\nabla \times \mathbf{H}(\mathbf{r}) = -i\omega\varepsilon(\mathbf{r}, \omega) \mathbf{E}(\mathbf{r}), \qquad (4.17)$$

where

$$\varepsilon(\mathbf{r}, \omega) = \epsilon(\mathbf{r}, \omega) + i \frac{\sigma(\mathbf{r}, \omega)}{\omega} \qquad (4.18)$$

is the so-called complex permittivity. Again, the reader should note the typographical distinction between the frequency-dependent electric permittivity ϵ (which can, in principle, be complex-valued for a dispersive medium) and the complex permittivity ε. We will show later that a direct consequence of a complex-valued ε and/or μ is a non-zero imaginary part of the refractive index (Eq. (6.19)), which causes absorption of electromagnetic energy (Eq. (6.20)) by converting it into other forms of energy, e.g., heat.

The scalar or the vector product of two real vector fields is not equal to the real part of the respective product of the corresponding complex vector fields. Instead,

$$\begin{aligned} c(\mathbf{r}, t) &= \boldsymbol{a}(\mathbf{r}, t) \cdot \boldsymbol{b}(\mathbf{r}, t) \\ &= \tfrac{1}{4}[\mathbf{a}(\mathbf{r})\exp(-i\omega t) + \mathbf{a}^*(\mathbf{r})\exp(i\omega t)] \\ &\quad \cdot [\mathbf{b}(\mathbf{r})\exp(-i\omega t) + \mathbf{b}^*(\mathbf{r})\exp(i\omega t)] \\ &= \tfrac{1}{2}\mathrm{Re}[\mathbf{a}(\mathbf{r}) \cdot \mathbf{b}^*(\mathbf{r}) + \mathbf{a}(\mathbf{r}) \cdot \mathbf{b}(\mathbf{r})\exp(-2i\omega t)], \end{aligned} \qquad (4.19)$$

and similarly for a vector product. Usually the angular frequency ω is so high that traditional optical measuring devices are not capable of following the rapid oscillations of the instantaneous product values but rather respond to a time average

$$\langle c(\mathbf{r}, t) \rangle_t = \frac{1}{T} \int_t^{t+T} d\tau c(\mathbf{r}, \tau), \qquad (4.20)$$

where T is a time interval long compared with $1/\omega$. Therefore, Eqs. (4.19) and (4.20) imply that the time average of a product of two real fields is equal to one half of the real part of the respective product of one complex field with the complex conjugate of the other, e.g.,

$$\langle c(\mathbf{r},t) \rangle_t = \tfrac{1}{2} \mathrm{Re}[\mathbf{a}(\mathbf{r}) \cdot \mathbf{b}^*(\mathbf{r})]. \tag{4.21}$$

5. The Poynting vector

Both the value and the direction of the electromagnetic energy flow are described by the so-called Poynting vector $\mathcal{S}$. The expression for $\mathcal{S}$ can be derived by considering conservation of energy and taking into account that the magnetic field does no work and that for a local charge q the rate of doing work by the electric field is $q(\mathbf{r},t)\mathbf{v}(\mathbf{r},t)\cdot\mathcal{E}(\mathbf{r},t)$, where $\mathbf{v}$ is the velocity of the charge. Indeed, the total rate of doing work by the electromagnetic field in a finite volume V is given by

$$\int_V d\mathbf{r} \mathcal{J}(\mathbf{r},t) \cdot \mathcal{E}(\mathbf{r},t) \tag{5.1}$$

and represents the rate of conversion of electromagnetic energy into mechanical or thermal energy. This power must be balanced by the corresponding rate of decrease of the electromagnetic field energy within the volume V. Using Eqs. (2.2) and (2.4) and the vector identity

$$\nabla \cdot (\mathbf{a} \times \mathbf{b}) = \mathbf{b} \cdot (\nabla \times \mathbf{a}) - \mathbf{a} \cdot (\nabla \times \mathbf{b}), \tag{5.2}$$

we derive

$$\int_V d\mathbf{r} \mathcal{J} \cdot \mathcal{E} = \int_V d\mathbf{r} \mathcal{E} \cdot \left(\nabla \times \mathcal{H} - \frac{\partial \mathcal{D}}{\partial t} \right)$$

$$= - \int_V d\mathbf{r} \left[\nabla \cdot (\mathcal{E} \times \mathcal{H}) + \mathcal{E} \cdot \frac{\partial \mathcal{D}}{\partial t} + \mathcal{H} \cdot \frac{\partial \mathcal{B}}{\partial t} \right]. \tag{5.3}$$

Let us first consider a linear medium without dispersion and introduce the total electromagnetic energy density,

$$u(\mathbf{r},t) = \tfrac{1}{2}[\mathcal{E}(\mathbf{r},t) \cdot \mathcal{D}(\mathbf{r},t) + \mathcal{B}(\mathbf{r},t) \cdot \mathcal{H}(\mathbf{r},t)], \tag{5.4}$$

and the so-called Poynting vector,

$$\mathcal{S}(\mathbf{r},t) = \mathcal{E}(\mathbf{r},t) \times \mathcal{H}(\mathbf{r},t). \tag{5.5}$$

The latter represents electromagnetic energy flow and has the dimension [energy/(area × time)]. Using also the Gauss theorem (2.19), we finally obtain

$$\int_V d\mathbf{r} \mathcal{J} \cdot \mathcal{E} + \int_V d\mathbf{r} \frac{\partial u}{\partial t} + \oint_S d\mathcal{S} \mathcal{S} \cdot \hat{\mathbf{n}} = 0, \tag{5.6}$$

where the closed surface S bounds the volume V and $\hat{\mathbf{n}}$ is a unit vector in the direction of the local outward normal to the surface. Equation (5.6) manifests the conservation of energy by requiring that the rate of the total work done by the fields on the sources within the volume, the time rate of change of electromagnetic energy within the volume, and the electromagnetic energy flowing out through the volume boundary per unit time add up to zero. Since the volume V is arbitrary, Eq. (5.3) also can be written in the form of a differential continuity equation:

$$\frac{\partial u}{\partial t} + \nabla \cdot \mathbf{S} = -\boldsymbol{\mathcal{J}} \cdot \boldsymbol{\mathcal{E}}. \tag{5.7}$$

Since $\nabla \cdot (\nabla \times \mathbf{a}) = 0$, it is clear from Eq. (5.7) that adding the curl of a vector field to the Poynting vector will not change the energy balance, which seems to suggest that there is a degree of arbitrariness in the definition of the Poynting vector. However, relativistic considerations discussed in section 12.10 of Jackson [1] show that the definition (5.5) is, in fact, unique.

Let us now allow the medium to be dispersive. Instead of Eq. (5.1), we now consider the integral

$$\frac{1}{2} \int_V d\mathbf{r} \mathbf{J}^*(\mathbf{r}) \cdot \mathbf{E}(\mathbf{r}) \tag{5.8}$$

whose real part gives the time-averaged rate of work done by the electromagnetic field (cf. Eq. (4.21)). Using Eqs. (4.3), (4.5), and (5.2), we derive

$$\frac{1}{2} \int_V d\mathbf{r} \mathbf{J}^*(\mathbf{r}) \cdot \mathbf{E}(\mathbf{r}) = \frac{1}{2} \int_V d\mathbf{r} \mathbf{E}(\mathbf{r}) \cdot [\nabla \times \mathbf{H}^*(\mathbf{r}) - i\omega \mathbf{D}^*(\mathbf{r})]$$

$$= -\frac{1}{2} \int_V d\mathbf{r} \{ \nabla \cdot [\mathbf{E}(\mathbf{r}) \times \mathbf{H}^*(\mathbf{r})]$$

$$+ i\omega [\mathbf{E}(\mathbf{r}) \cdot \mathbf{D}^*(\mathbf{r}) - \mathbf{B}(\mathbf{r}) \cdot \mathbf{H}^*(\mathbf{r})] \}. \tag{5.9}$$

If we now define the complex Poynting vector by

$$\mathbf{S}(\mathbf{r}) = \tfrac{1}{2}[\mathbf{E}(\mathbf{r}) \times \mathbf{H}^*(\mathbf{r})] \tag{5.10}$$

and the complex electric and magnetic energy densities by

$$w_e(\mathbf{r}) = \tfrac{1}{4}[\mathbf{E}(\mathbf{r}) \cdot \mathbf{D}^*(\mathbf{r})], \tag{5.11}$$

$$w_m(\mathbf{r}) = \tfrac{1}{4}[\mathbf{B}(\mathbf{r}) \cdot \mathbf{H}^*(\mathbf{r})], \tag{5.12}$$

respectively, and apply the Gauss theorem, we then have

$$\frac{1}{2} \int_V d\mathbf{r} \mathbf{J}^*(\mathbf{r}) \cdot \mathbf{E}(\mathbf{r}) + \oint_S dS \mathbf{S}(\mathbf{r}) \cdot \hat{\mathbf{n}} + 2i\omega \int_V d\mathbf{r}[w_e(\mathbf{r}) - w_m(\mathbf{r})] = 0. \tag{5.13}$$

Obviously, the real part of Eq. (5.13) manifests the conservation of energy for the corresponding time-averaged quantities. In particular, the time-averaged Poynting vector $\langle \mathbf{S}(\mathbf{r},t)\rangle_t$ is equal to the real part of the complex Poynting vector,

$$\langle \mathbf{S}(\mathbf{r},t)\rangle_t = \mathrm{Re}[\mathbf{S}(\mathbf{r})]. \tag{5.14}$$

The net rate W at which the electromagnetic energy crosses the surface S is given by

$$W = -\oint_S \mathrm{d}S\langle \mathbf{S}(\mathbf{r},t)\rangle_t \cdot \hat{\mathbf{n}}. \tag{5.15}$$

The rate is defined such that it is positive if there is a net transfer of electromagnetic energy into the volume V and is negative otherwise.

6. Plane-wave solution

Consider an infinite homogeneous medium. The use of the formulas

$$\nabla \cdot (f\mathbf{a}) = f \nabla \cdot \mathbf{a} + (\nabla f) \cdot \mathbf{a}, \tag{6.1}$$
$$\nabla \times (f\mathbf{a}) = f \nabla \times \mathbf{a} + (\nabla f) \times \mathbf{a}, \tag{6.2}$$
$$\nabla \exp(\mathrm{i}\mathbf{k} \cdot \mathbf{r}) = \mathrm{i}\mathbf{k} \exp(\mathrm{i}\mathbf{k} \cdot \mathbf{r}) \tag{6.3}$$

in Eqs. (4.3), (4.4), (4.16), and (4.17) shows that the complex field vectors

$$\mathbf{E}(\mathbf{r},t) = \mathbf{E}_0 \exp(\mathrm{i}\mathbf{k} \cdot \mathbf{r} - \mathrm{i}\omega t), \tag{6.4}$$
$$\mathbf{H}(\mathbf{r},t) = \mathbf{H}_0 \exp(\mathrm{i}\mathbf{k} \cdot \mathbf{r} - \mathrm{i}\omega t), \tag{6.5}$$

where $\mathbf{E}_0$, $\mathbf{H}_0$, and $\mathbf{k}$ are constant complex vectors, are a solution of the Maxwell equations provided that

$$\mathbf{k} \cdot \mathbf{E}_0 = 0, \tag{6.6}$$

$$\mathbf{k} \cdot \mathbf{H}_0 = 0, \tag{6.7}$$

$$\mathbf{k} \times \mathbf{E}_0 = \omega\mu\mathbf{H}_0, \tag{6.8}$$

$$\mathbf{k} \times \mathbf{H}_0 = -\omega\varepsilon\mathbf{E}_0. \tag{6.9}$$

The so-called wave vector $\mathbf{k}$ is usually expressed as

$$\mathbf{k} = \mathbf{k}_R + \mathrm{i}\mathbf{k}_I, \tag{6.10}$$

where $\mathbf{k}_R$ and $\mathbf{k}_I$ are real vectors. Thus

$$\mathbf{E}(\mathbf{r},t) = \mathbf{E}_0 \exp(-\mathbf{k}_I \cdot \mathbf{r})\exp(\mathrm{i}\mathbf{k}_R \cdot \mathbf{r} - \mathrm{i}\omega t), \tag{6.11}$$
$$\mathbf{H}(\mathbf{r},t) = \mathbf{H}_0 \exp(-\mathbf{k}_I \cdot \mathbf{r})\exp(\mathrm{i}\mathbf{k}_R \cdot \mathbf{r} - \mathrm{i}\omega t). \tag{6.12}$$

The $\mathbf{E}_0 \exp(-\mathbf{k}_I \cdot \mathbf{r})$ and $\mathbf{H}_0 \exp(-\mathbf{k}_I \cdot \mathbf{r})$ are the complex amplitudes of the electric and magnetic fields, respectively, and $\phi = \mathbf{k}_R \cdot \mathbf{r} - \omega t$ is their phase.

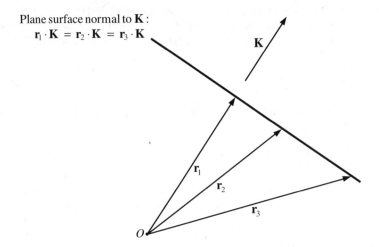

Figure 6.1: Plane surface normal to a real vector $\mathbf{K}$.

The vector $\mathbf{k}_R$ is normal to the surfaces of constant phase, whereas $\mathbf{k}_I$ is normal to the surfaces of constant amplitude. Indeed, a plane surface normal to a real vector $\mathbf{K}$ is described by $\mathbf{r} \cdot \mathbf{K} = \text{constant}$, where $\mathbf{r}$ is the radius vector drawn from the origin of the reference frame to any point in the plane (see Fig.

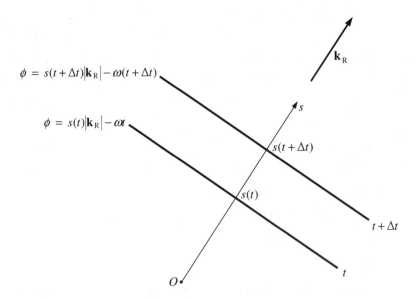

Figure 6.2: The plane of constant phase $\phi = \text{constant}$ travels a distance Δs over the time period Δt. The s-axis is drawn from the origin of the coordinate system along the vector $\mathbf{k}_R$.

6.1). Also, it is easy to see that surfaces of constant phase propagate in the direction of $\mathbf{k}_R$ with the phase velocity

$$v = \omega / |\mathbf{k}_R|. \tag{6.13}$$

Indeed, the planes corresponding to the instantaneous times t and $t + \Delta t$ are separated by the distance $\Delta s = \omega \Delta t / |\mathbf{k}_R|$ (see Fig. 6.2), which gives Eq. (6.13). Thus Eqs. (6.4) and (6.5) describe a plane electromagnetic wave propagating in a homogeneous medium without sources. This is a very important solution of the Maxwell equations because it embodies the concept of a perfectly monochromatic parallel beam of light of infinite lateral extent and represents the transport of electromagnetic energy from one point to another.

Equations (6.4) and (6.8) yield

$$\mathbf{H}(\mathbf{r},t) = \frac{1}{\omega\mu} \mathbf{k} \times \mathbf{E}(\mathbf{r},t). \tag{6.14}$$

Therefore, a plane electromagnetic wave always can be considered in terms of only the electric (or only the magnetic) field.

The electromagnetic wave is called homogeneous if $\mathbf{k}_R$ and $\mathbf{k}_I$ are parallel (including the case $\mathbf{k}_I = \mathbf{0}$); otherwise it is called inhomogeneous. When $\mathbf{k}_R \parallel \mathbf{k}_I$, the complex wave vector can be expressed as $\mathbf{k} = (k_R + ik_I)\hat{\mathbf{n}}$, where $\hat{\mathbf{n}}$ is a real unit vector in the direction of propagation and both k_R and k_I are real and nonnegative.

According to Eqs. (6.6) and (6.7), the plane electromagnetic wave is transverse: both $\mathbf{E}_0$ and $\mathbf{H}_0$ are perpendicular to $\mathbf{k}$. Furthermore, it is evident from either Eq. (6.8) or Eq. (6.9) that $\mathbf{E}_0$ and $\mathbf{H}_0$ are mutually perpendicular: $\mathbf{E}_0 \cdot \mathbf{H}_0 = 0$. Since $\mathbf{E}_0$, $\mathbf{H}_0$, and $\mathbf{k}$ are, in general, complex vectors, the physical interpretation of these facts can be far from obvious. It becomes most transparent when both ε, μ, and $\mathbf{k}$ are real. The reader can easily verify that in this case the real field vectors $\boldsymbol{\mathcal{E}}$ and $\boldsymbol{\mathcal{H}}$ are mutually perpendicular and lie in a plane normal to the direction of wave propagation $\hat{\mathbf{n}}$ (see Fig. 6.3).

Taking the vector product of $\mathbf{k}$ with the left-hand side and the right-hand side of Eq. (6.8) and using Eq. (6.9) and the vector identity

$$\mathbf{a} \times (\mathbf{b} \times \mathbf{c}) = \mathbf{b}(\mathbf{a} \cdot \mathbf{c}) - \mathbf{c}(\mathbf{a} \cdot \mathbf{b}) \tag{6.15}$$

together with Eq. (6.6) yield

$$\mathbf{k} \cdot \mathbf{k} = \omega^2 \varepsilon \mu. \tag{6.16}$$

In the practically important case of a homogeneous plane wave, we obtain from Eq. (6.16)

$$k = k_R + ik_I = \omega\sqrt{\varepsilon\mu} = \frac{\omega m}{c}, \tag{6.17}$$

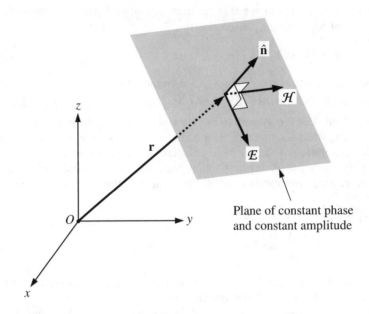

Figure 6.3: Plane wave propagating in a homogeneous medium with no dispersion and losses.

where k is the wave number,

$$c = \frac{1}{\sqrt{\epsilon_0 \mu_0}} \tag{6.18}$$

is the speed of light in a vacuum (cf. Eq. (6.13)), and

$$m = \frac{ck}{\omega} = m_R + i m_I = \sqrt{\frac{\epsilon \mu}{\epsilon_0 \mu_0}} = c\sqrt{\epsilon \mu} \tag{6.19}$$

is the complex refractive index with a non-negative real part m_R and a non-negative imaginary part m_I. Thus, the complex electric field vector of the homogeneous plane wave has the form

$$\mathbf{E}(\mathbf{r}, t) = \mathbf{E}_0 \exp\left(-\frac{\omega}{c} m_I \hat{\mathbf{n}} \cdot \mathbf{r}\right) \exp\left(i\frac{\omega}{c} m_R \hat{\mathbf{n}} \cdot \mathbf{r} - i\omega t\right). \tag{6.20}$$

If the imaginary part of the refractive index is non-zero, then it determines the decay of the amplitude of the wave as it propagates through the medium, which is thus absorbing. On the other hand, a medium is nonabsorbing if it is non-dispersive ($\epsilon = \epsilon$ and $\mu = \mu$) and lossless ($\sigma = 0$), which causes the refractive index $m = m_R = c\sqrt{\epsilon \mu}$ to be real-valued. The real part of the refractive index determines the phase velocity of the wave:

$$v = \frac{c}{m_R}. \tag{6.21}$$

In a vacuum, $m = m_R = 1$ and $v = c$.

As follows from Eqs. (5.10), (5.14), (6.4), (6.5), (6.8), and (6.15), the time-averaged Poynting vector of a plane wave is

$$\langle S(\mathbf{r},t)\rangle_t = \mathrm{Re}\left(\frac{\mathbf{k}^*[\mathbf{E}(\mathbf{r})\cdot\mathbf{E}^*(\mathbf{r})] - \mathbf{E}^*(\mathbf{r})[\mathbf{k}^*\cdot\mathbf{E}(\mathbf{r})]}{2\omega\mu^*}\right). \tag{6.22}$$

If the wave is homogeneous, $\mathbf{k}\cdot\mathbf{E}(\mathbf{r}) = 0$ causes $\mathbf{k}^*\cdot\mathbf{E}(\mathbf{r}) = 0$. Therefore,

$$\langle S(\mathbf{r},t)\rangle_t = \tfrac{1}{2}\mathrm{Re}\left(\sqrt{\frac{\varepsilon}{\mu}}\right)|\mathbf{E}_0|^2 \exp\left(-2\frac{\omega}{c}m_I\hat{\mathbf{n}}\cdot\mathbf{r}\right)\hat{\mathbf{n}}. \tag{6.23}$$

Thus, $\langle S(\mathbf{r},t)\rangle_t$ is in the direction of propagation and its absolute value, called intensity, is attenuated exponentially provided that the medium is absorbing:

$$I(\mathbf{r}) = |\langle S(\mathbf{r},t)\rangle_t| = I_0\exp(-\alpha\hat{\mathbf{m}}\cdot\mathbf{r}), \tag{6.24}$$

where I_0 is the intensity at $\mathbf{r} = \mathbf{0}$. The absorption coefficient α is

$$\alpha = 2\frac{\omega}{c}m_I = \frac{4\pi m_I}{\lambda_0}, \tag{6.25}$$

where

$$\lambda_0 = \frac{2\pi c}{\omega} \tag{6.26}$$

is the free-space wavelength. The intensity has the dimension of monochromatic energy flux, [energy/(area $\times$ time)], and is equal to the amount of electromagnetic energy crossing a unit surface element normal to $\hat{\mathbf{n}}$ per unit time.

The expression for the time-averaged energy density of a plane wave propagating in a medium without dispersion follows from Eqs. (4.7), (4.8), (4.21) and (5.4):

$$\langle u(\mathbf{r},t)\rangle_t = \tfrac{1}{4}[\varepsilon\mathbf{E}(\mathbf{r})\cdot\mathbf{E}^*(\mathbf{r}) + \mu\mathbf{H}(\mathbf{r})\cdot\mathbf{H}^*(\mathbf{r})]. \tag{6.27}$$

Assuming further that the medium is lossless and recalling Eqs. (4.3), (6.8) and (6.16) and the vector identity

$$(\mathbf{a}\times\mathbf{b})\cdot(\mathbf{c}\times\mathbf{d}) = (\mathbf{a}\cdot\mathbf{c})(\mathbf{b}\cdot\mathbf{d}) - (\mathbf{a}\cdot\mathbf{d})(\mathbf{b}\cdot\mathbf{c}), \tag{6.28}$$

we derive

$$\langle u(\mathbf{r},t)\rangle_t = \tfrac{1}{2}\varepsilon|\mathbf{E}_0|^2. \tag{6.29}$$

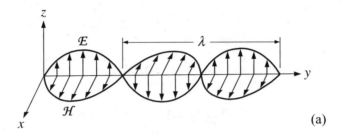

(a)

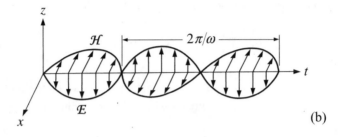

(b)

Figure 6.4: Plane electromagnetic wave described by Eqs. (6.31) and (6.32).

Comparison of Eqs. (6.23), (6.24), and (6.29) yields

$$I(\mathbf{r}) = \frac{1}{\sqrt{\epsilon\mu}} \langle u(\mathbf{r},t)\rangle_t = v\langle u(\mathbf{r},t)\rangle_t, \tag{6.30}$$

where v is the speed of light in the nonabsorbing material medium. The physical interpretation of this result is quite clear: the amount of electromagnetic energy crossing a surface element of unit area normal to the direction of propagation per unit time is equal to the product of the speed of light and the amount of electromagnetic energy per unit volume.

Figure 6.4 gives a simple example of a plane electromagnetic wave propagating in a nonabsorbing homogeneous medium and described by the following real electric and magnetic field vectors:

$$\mathcal{E}(\mathbf{r},t) = \mathcal{E}\cos(ky - \omega t - \pi/2)\,\hat{\mathbf{z}}, \tag{6.31}$$
$$\mathcal{H}(\mathbf{r},t) = \mathcal{H}\cos(ky - \omega t - \pi/2)\,\hat{\mathbf{x}}, \tag{6.32}$$

where $\mathcal{E}$, $\mathcal{H}$, and k are real and $\hat{\mathbf{x}}$ and $\hat{\mathbf{z}}$ are the unit vectors along the x-axis and the z-axis, respectively. Panel (a) shows the electric and magnetic fields as a function of y at the moment $t = 0$, while panel (b) depicts the fields as a function of time at any point in the plane $y = 0$. The period of the sinusoids in panel (a) is given by

$$\lambda = \frac{2\pi}{k} \tag{6.33}$$

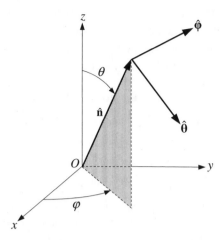

Figure 7.1: Local coordinate system used to describe the direction of propagation and polarization state of a plane electromagnetic wave at the observation point O.

and defines the wavelength of light in the nonabsorbing material medium, whereas the period of the sinusoids in panel (b) is equal to $2\pi/\omega$.

It is straightforward to verify that the choice of the $\exp(i\omega t)$ rather than $\exp(-i\omega t)$ time dependence in the complex representation of time-harmonic fields in Eq. (4.1) would have led to $m = m_R - im_I$ with a non-negative m_I. The $\exp(-i\omega t)$ time factor convention adopted here was used in many books on optics and light scattering (e.g., [2–6]), electromagnetics (e.g., [1,7–9]), and solid-state physics. On the other hand, van de Hulst [10], Kerker [11], and Hovenier and van der Mee [12] use the time factor $\exp(i\omega t)$, which implies a non-positive imaginary part of the complex refractive index. It does not matter in the final analysis which convention is chosen because all measurable quantities of practical interest are always real. However, it is important to remember that once a choice of the time factor has been made, its consistent use throughout all derivations is imperative.

7. Coherency matrix and Stokes parameters

Traditional optical devices cannot measure the electric and magnetic fields associated with a beam of light but rather measure quantities that are time averages of real-valued linear combinations of products of field vector components and have the dimension of the intensity. In order to define these quantities, we use polar spherical coordinates associated with the local right-handed Cartesian coordinate system with origin at the observation point, as shown in Fig. 7.1. Assuming that the medium is homogeneous and has no dispersion and losses, we specify the direction of propagation of a plane electromagnetic wave by a unit vector $\hat{\mathbf{n}}$ or, equivalently, by a couple $\{\theta, \varphi\}$, where $\theta \in [0, \pi]$ is the polar (zenith) angle measured from the positive z-axis and $\varphi \in [0, 2\pi)$ is the azimuth angle measured from the positive x-axis in the clockwise direction when looking in the direction of the positive z-axis. Since the component of the electric field vector along the direction of propagation $\hat{\mathbf{n}}$ is equal to zero, the electric field at the observation point can be expressed as $\mathbf{E} = \mathbf{E}_\theta + \mathbf{E}_\varphi$, where $\mathbf{E}_\theta$ and $\mathbf{E}_\varphi$ are the θ- and φ-components of the electric field vector, respectively. The component $\mathbf{E}_\theta = E_\theta \hat{\boldsymbol{\theta}}$ lies in the meridional plane (i.e., the plane through $\hat{\mathbf{n}}$ and the z-axis), whereas the component $\mathbf{E}_\varphi = E_\varphi \hat{\boldsymbol{\varphi}}$ is perpendicular to this plane. $\hat{\boldsymbol{\theta}}$ and $\hat{\boldsymbol{\varphi}}$ are the corresponding unit vectors such that $\hat{\mathbf{n}} = \hat{\boldsymbol{\theta}} \times \hat{\boldsymbol{\varphi}}$.

The specification of a unit vector $\hat{\mathbf{n}}$ uniquely determines the meridional plane of the propagation direction except when $\hat{\mathbf{n}}$ is oriented along the positive or negative direction of the z axis. Although it may seem redundant to specify φ in addition to θ when $\theta = 0$ or π, the unit $\hat{\boldsymbol{\theta}}$ and $\hat{\boldsymbol{\varphi}}$ vectors and, thus, the electric field vector components $\mathbf{E}_\theta$ and $\mathbf{E}_\varphi$ still depend on the orientation of the meridional plane. Therefore, we always assume that the specification of $\hat{\mathbf{n}}$ implicitly includes the specification of the appropriate meridional plane in cases when $\hat{\mathbf{n}}$ is parallel to the z axis.

Consider a plane electromagnetic wave propagating in a homogeneous medium without dispersion and losses and given by

$$\mathbf{E}(\mathbf{r}, t) = \mathbf{E}_0 \exp(ik\hat{\mathbf{n}} \cdot \mathbf{r} - i\omega t) \qquad (7.1)$$

with a real k. The simplest complete set of linearly independent quadratic combinations of the electric field vector components with non-zero time averages consists of the following four quantities:

$$E_\theta(\mathbf{r}, t) E_\theta^*(\mathbf{r}, t) = E_{0\theta} E_{0\theta}^*, \qquad E_\theta(\mathbf{r}, t) E_\varphi^*(\mathbf{r}, t) = E_{0\theta} E_{0\varphi}^*,$$

$$E_\varphi(\mathbf{r}, t) E_\theta^*(\mathbf{r}, t) = E_{0\varphi} E_{0\theta}^*, \qquad E_\varphi(\mathbf{r}, t) E_\varphi^*(\mathbf{r}, t) = E_{0\varphi} E_{0\varphi}^*.$$

The products of these quantities and $\frac{1}{2}\sqrt{\epsilon/\mu}$ have the dimension of monochromatic energy flux and form the 2×2 coherency (or density) matrix $\boldsymbol{\rho}$ [2]:

$$\boldsymbol{\rho} = \begin{bmatrix} \rho_{11} & \rho_{12} \\ \rho_{21} & \rho_{22} \end{bmatrix} = \frac{1}{2}\sqrt{\frac{\epsilon}{\mu}} \begin{bmatrix} E_{0\theta} E_{0\theta}^* & E_{0\theta} E_{0\varphi}^* \\ E_{0\varphi} E_{0\theta}^* & E_{0\varphi} E_{0\varphi}^* \end{bmatrix}. \qquad (7.2)$$

The completeness of the set of the four coherency matrix elements means that any plane wave characteristic directly observable with a traditional optical instrument is a real-valued linear combination of these quantities.

Since ρ_{12} and ρ_{21} are, in general, complex, it is convenient to introduce an alternative complete set of four real, linearly independent quantities called Stokes parameters [13]. We first group the elements of the 2×2 coherency matrix into a 4×1 coherency column vector:

$$\mathbf{J} = \begin{bmatrix} \rho_{11} \\ \rho_{12} \\ \rho_{21} \\ \rho_{22} \end{bmatrix} = \frac{1}{2}\sqrt{\frac{\epsilon}{\mu}} \begin{bmatrix} E_{0\theta}E_{0\theta}^* \\ E_{0\theta}E_{0\varphi}^* \\ E_{0\varphi}E_{0\theta}^* \\ E_{0\varphi}E_{0\varphi}^* \end{bmatrix}. \tag{7.3}$$

The Stokes parameters I, Q, U, and V are then defined as the elements of a 4×1 column Stokes vector $\mathbf{I}$ as follows:

$$\mathbf{I} = \begin{bmatrix} I \\ Q \\ U \\ V \end{bmatrix} = \mathbf{DJ} = \frac{1}{2}\sqrt{\frac{\epsilon}{\mu}} \begin{bmatrix} E_{0\theta}E_{0\theta}^* + E_{0\varphi}E_{0\varphi}^* \\ E_{0\theta}E_{0\theta}^* - E_{0\varphi}E_{0\varphi}^* \\ -E_{0\theta}E_{0\varphi}^* - E_{0\varphi}E_{0\theta}^* \\ i(E_{0\varphi}E_{0\theta}^* - E_{0\theta}E_{0\varphi}^*) \end{bmatrix}, \tag{7.4}$$

where

$$\mathbf{D} = \begin{bmatrix} 1 & 0 & 0 & 1 \\ 1 & 0 & 0 & -1 \\ 0 & -1 & -1 & 0 \\ 0 & -i & i & 0 \end{bmatrix}. \tag{7.5}$$

Conversely,

$$\mathbf{J} = \mathbf{D}^{-1}\mathbf{I}, \tag{7.6}$$

where

$$\mathbf{D}^{-1} = \frac{1}{2}\begin{bmatrix} 1 & 1 & 0 & 0 \\ 0 & 0 & -1 & i \\ 0 & 0 & -1 & -i \\ 1 & -1 & 0 & 0 \end{bmatrix}. \tag{7.7}$$

By virtue of being real-valued quantities and having the dimension of energy flux, the Stokes parameters form a complete set of quantities that are needed to characterize a plane electromagnetic wave, inasmuch as it is subject to practical analysis. This means that (i) any other observable quantity is a linear combination of the four Stokes parameters, and (ii) it is impossible to distinguish between two plane waves with the same values of the Stokes parameters using a traditional optical device (the so-called principle of optical equivalence). Indeed, the two complex amplitudes $E_{0\theta} = a_\theta \exp(i\Delta_\theta)$ and $E_{0\varphi} = a_\varphi \exp(i\Delta_\varphi)$ are characterized by four real numbers: the non-negative amplitudes a_θ and a_φ and the phases Δ_θ and $\Delta_\varphi = \Delta_\theta - \Delta$. The Stokes parameters carry information about the amplitudes and the phase difference Δ, but not about Δ_θ. The latter is the only quantity that could be used to distinguish different waves with the same a_θ, a_φ, and Δ (and thus the same Stokes parameters), but it vanishes when a field vector component is multiplied by the complex conjugate value of the same or another field vector component.

The first Stokes parameter, I, is the intensity introduced in the previous section, with the explicit definition here applicable to a homogeneous, nonabsorbing medium. The Stokes parameters Q, U, and V describe the polarization state of the wave. The ellipsometric interpretation of the Stokes parameters will be the subject of the next section. It is easy to verify that the Stokes parameters of a plane monochromatic wave are not completely independent but rather are related by the quadratic identity

$$I^2 = Q^2 + U^2 + V^2. \tag{7.8}$$

We will see later, however, that this identity may not hold for a quasi-monochromatic beam of light.

The coherency matrix and the Stokes vector are not the only representations of polarization and not always the most convenient ones. Two other frequently used representations are the real so-called modified Stokes column vector given by

$$\mathbf{I}^{\mathrm{MS}} = \begin{bmatrix} I_{\mathrm{v}} \\ I_{\mathrm{h}} \\ U \\ V \end{bmatrix} = \mathbf{BI} = \begin{bmatrix} \frac{1}{2}(I + Q) \\ \frac{1}{2}(I - Q) \\ U \\ V \end{bmatrix} \tag{7.9}$$

and the complex circular-polarization column vector defined as

$$\mathbf{I}^{\mathrm{CP}} = \begin{bmatrix} I_2 \\ I_0 \\ I_{-0} \\ I_{-2} \end{bmatrix} = \mathbf{AI} = \frac{1}{2} \begin{bmatrix} Q + iU \\ I + V \\ I - V \\ Q - iU \end{bmatrix}, \tag{7.10}$$

where

$$\mathbf{B} = \begin{bmatrix} 1/2 & 1/2 & 0 & 0 \\ 1/2 & -1/2 & 0 & 0 \\ 0 & 0 & 1 & 0 \\ 0 & 0 & 0 & 1 \end{bmatrix}, \tag{7.11}$$

$$\mathbf{A} = \frac{1}{2} \begin{bmatrix} 0 & 1 & i & 0 \\ 1 & 0 & 0 & 1 \\ 1 & 0 & 0 & -1 \\ 0 & 1 & -i & 0 \end{bmatrix}. \tag{7.12}$$

It is easy to verify that

$$\mathbf{I} = \mathbf{B}^{-1} \mathbf{I}^{MS} \tag{7.13}$$

and

$$\mathbf{I} = \mathbf{A}^{-1} \mathbf{I}^{CP}, \tag{7.14}$$

where

$$\mathbf{B}^{-1} = \begin{bmatrix} 1 & 1 & 0 & 0 \\ 1 & -1 & 0 & 0 \\ 0 & 0 & 1 & 0 \\ 0 & 0 & 0 & 1 \end{bmatrix} \tag{7.15}$$

and

$$\mathbf{A}^{-1} = \begin{bmatrix} 0 & 1 & 1 & 0 \\ 1 & 0 & 0 & 1 \\ -i & 0 & 0 & i \\ 0 & 1 & -1 & 0 \end{bmatrix}. \tag{7.16}$$

The usefulness of the modified and circular-polarization Stokes vectors will be illustrated in the following section.

We conclude this section with a caution. It is important to remember that whereas the Poynting vector can be defined for an arbitrary electromagnetic field, the Stokes parameters can only be defined for transverse fields such as plane waves discussed in the previous section or spherical waves discussed in Section 12. Quite often the electromagnetic field at an observation point is not a well-defined transverse electromagnetic wave, in which case the Stokes vector formalism cannot be applied directly.

8.   Ellipsometric interpretation of the Stokes parameters

In this section we show how the Stokes parameters can be used to derive the ellipsometric characteristics of the plane electromagnetic wave given by Eq. (7.1). Writing

$$E_{0\theta} = a_\theta \exp(i\Delta_\theta), \tag{8.1}$$
$$E_{0\varphi} = a_\varphi \exp(i\Delta_\varphi) \tag{8.2}$$

with real nonnegative amplitudes a_θ and a_φ and real phases Δ_θ and Δ_φ and recalling the definition (7.4), we obtain for the Stokes parameters

$$I = \frac{1}{2} \sqrt{\frac{\epsilon}{\mu}} \, (a_\theta^2 + a_\varphi^2), \tag{8.3}$$

$$Q = \frac{1}{2} \sqrt{\frac{\epsilon}{\mu}} \, (a_\theta^2 - a_\varphi^2), \tag{8.4}$$

$$U = -\sqrt{\frac{\epsilon}{\mu}}\, a_\theta a_\varphi \cos\Delta, \tag{8.5}$$

$$V = \sqrt{\frac{\epsilon}{\mu}}\, a_\theta a_\varphi \sin\Delta, \tag{8.6}$$

where

$$\Delta = \Delta_\theta - \Delta_\varphi. \tag{8.7}$$

Substituting Eqs. (8.1) and (8.2) into Eq. (7.1), we have for the real electric vector

$$\mathcal{E}_\theta(\mathbf{r},t) = a_\theta \cos(\delta_\theta - \omega t), \tag{8.8}$$

$$\mathcal{E}_\varphi(\mathbf{r},t) = a_\varphi \cos(\delta_\varphi - \omega t), \tag{8.9}$$

where

$$\delta_\theta = \Delta_\theta + k\hat{\mathbf{n}}\cdot\mathbf{r}, \tag{8.10}$$

$$\delta_\varphi = \Delta_\varphi + k\hat{\mathbf{n}}\cdot\mathbf{r}. \tag{8.11}$$

At any fixed point O in space, the endpoint of the real electric vector given by Eqs. (8.8)–(8.11) describes an ellipse with specific major and minor axes and orientation (see the top panel of Fig. 8.1). The major axis of the ellipse makes an angle ζ with the positive direction of the φ-axis such that $\zeta \in [0,\pi)$. By definition, this orientation angle is obtained by rotating the φ-axis in the *clockwise* direction when looking in the direction of propagation, until it is directed along the major axis of the ellipse. The ellipticity is defined as the ratio of the minor to the major axes of the ellipse and is usually expressed as $|\tan\beta|$, where $\beta \in [-\pi/4, \pi/4]$. By definition, β is positive when the real electric vector at O rotates clockwise, as viewed by an observer looking in the direction of propagation (Fig. 8.1(a)). The polarization for positive β is called right-handed, as opposed to the left-handed polarization corresponding to the anti-clockwise rotation of the electric vector.

To express the orientation ζ of the ellipse and the ellipticity $|\tan\beta|$ in terms of the Stokes parameters, we first write the equations representing the rotation of the real electric vector at O in the form

$$\mathcal{E}_q(\mathbf{r},t) = a\sin\beta\sin(\delta - \omega t), \tag{8.12}$$

$$\mathcal{E}_p(\mathbf{r},t) = a\cos\beta\cos(\delta - \omega t), \tag{8.13}$$

where $\mathcal{E}_p$ and $\mathcal{E}_q$ are the electric field components along the major and minor axes of the ellipse, respectively (Fig. 8.1). One easily verifies that a positive (negative) β indeed corresponds to the right-handed (left-handed) polarization. The connection between Eqs. (8.8)–(8.11) and Eqs. (8.12)–(8.13) can be established by using the simple transformation rule for rotation of a two-dimensional coordinate system:

(a) Polarization ellipse

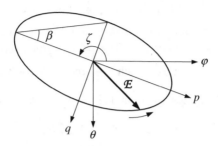

(b) Elliptical polarization $(V \neq 0)$

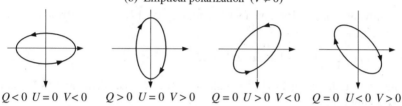

$Q < 0$ $U = 0$ $V < 0$ $Q > 0$ $U = 0$ $V > 0$ $Q = 0$ $U > 0$ $V < 0$ $Q = 0$ $U < 0$ $V > 0$

(c) Linear polarization $(V = 0)$

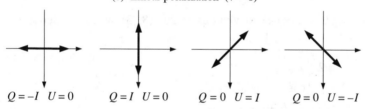

$Q = -I$ $U = 0$ $Q = I$ $U = 0$ $Q = 0$ $U = I$ $Q = 0$ $U = -I$

(d) Circular polarization $(Q = U = 0)$

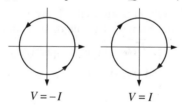

$V = -I$ $V = I$

Figure 8.1: Ellipse described by the tip of the real electric vector at a fixed point O in space (top panel) and particular cases of elliptical, linear, and circular polarization. The plane electromagnetic wave propagates in the direction $\hat{\boldsymbol{\theta}} \times \hat{\boldsymbol{\varphi}}$ (i.e., towards the reader).

$$\mathcal{E}_\theta(\mathbf{r},t) = -\mathcal{E}_q(\mathbf{r},t)\cos\zeta + \mathcal{E}_p(\mathbf{r},t)\sin\zeta, \qquad (8.14)$$

$$\mathcal{E}_\varphi(\mathbf{r},t) = -\mathcal{E}_q(\mathbf{r},t)\sin\zeta - \mathcal{E}_p(\mathbf{r},t)\cos\zeta. \qquad (8.15)$$

By equating the coefficients of $\cos\omega t$ and $\sin\omega t$ in the expanded Eqs. (8.8) and (8.9) with those in (8.14) and (8.15), we obtain

$$a_\theta \cos\delta_\theta = -a\sin\beta\sin\delta\cos\zeta + a\cos\beta\cos\delta\sin\zeta, \tag{8.16}$$

$$a_\theta \sin\delta_\theta = a\sin\beta\cos\delta\cos\zeta + a\cos\beta\sin\delta\sin\zeta, \tag{8.17}$$

$$a_\varphi \cos\delta_\varphi = -a\sin\beta\sin\delta\sin\zeta - a\cos\beta\cos\delta\cos\zeta, \tag{8.18}$$

$$a_\varphi \sin\delta_\varphi = a\sin\beta\cos\delta\sin\zeta - a\cos\beta\sin\delta\cos\zeta. \tag{8.19}$$

Squaring and adding Eqs. (8.16) and (8.17) and Eqs. (8.18) and (8.19) gives

$$a_\theta^2 = a^2(\sin^2\beta\cos^2\zeta + \cos^2\beta\sin^2\zeta), \tag{8.20}$$

$$a_\varphi^2 = a^2(\sin^2\beta\sin^2\zeta + \cos^2\beta\cos^2\zeta). \tag{8.21}$$

Multiplying Eqs. (8.16) and (8.18) and Eqs. (8.17) and (8.19) and adding yields

$$a_\theta a_\varphi \cos\Delta = -\tfrac{1}{2}a^2\cos2\beta\sin2\zeta. \tag{8.22}$$

Similarly, multiplying Eqs. (8.17) and (8.18) and Eqs. (8.16) and (8.19) and subtracting gives

$$a_\theta a_\varphi \sin\Delta = -\tfrac{1}{2}a^2\sin2\beta. \tag{8.23}$$

Comparing Eqs. (8.3)–(8.6) with Eqs. (8.20)–(8.23), we finally derive

$$I = \frac{1}{2}\sqrt{\frac{\epsilon}{\mu}}\,a^2, \tag{8.24}$$

$$Q = -I\cos2\beta\cos2\zeta, \tag{8.25}$$

$$U = I\cos2\beta\sin2\zeta, \tag{8.26}$$

$$V = -I\sin2\beta. \tag{8.27}$$

The parameters of the polarization ellipse are thus expressed in terms of the Stokes parameters as follows. The major and minor axes are given by $\sqrt{2I\sqrt{\mu/\epsilon}}\cos\beta$ and $\sqrt{2I\sqrt{\mu/\epsilon}}\,|\sin\beta|$, respectively (cf. Eqs. (8.12) and (8.13)). Equations (8.25) and (8.26) yield

$$\tan2\zeta = -\frac{U}{Q}. \tag{8.28}$$

Because $|\beta| \le \pi/4$, we have $\cos2\beta \ge 0$ so that $\cos2\zeta$ has the same sign as $-Q$. Therefore, from the different values of ζ that satisfy Eq. (8.28) but differ by $\pi/2$, we must choose the one that makes the sign of $\cos2\zeta$ to be the same as that of $-Q$. The ellipticity and handedness follow from

$$\tan2\beta = -\frac{V}{\sqrt{Q^2 + U^2}}. \tag{8.29}$$

Thus, the polarization is left-handed if V is positive and is right-handed if V is negative (Fig. 8.1(b)).

The electromagnetic wave is linearly polarized when $\beta = 0$; then the electric vector vibrates along the line making the angle ζ with the φ-axis (cf. Fig. 8.1) and $V = 0$. Furthermore, if $\zeta = 0$ or $\zeta = \pi/2$ then U vanishes as well. This explains the usefulness of the modified Stokes representation of polarization given by Eq. (7.9) in situations involving linearly polarized light as follows. The modified Stokes vector has only one non-zero element and is equal to $[I\ 0\ 0\ 0]^T$ if $\zeta = \pi/2$ (the electric vector vibrates along the θ-axis, i.e., in the meridional plane) or $[0\ I\ 0\ 0]^T$ if $\zeta = 0$ (the electric vector vibrates along the φ-axis, i.e., in the plane perpendicular to the meridional plane), where T indicates the transpose of a matrix (see Fig. 8.1(c)).

If, however, $\beta = \pm\pi/4$, then both Q and U vanish, and the electric vector describes a circle in the clockwise ($\beta = \pi/4$, $V = -I$) or anti-clockwise ($\beta = -\pi/4$, $V = I$) direction, as viewed by an observer looking in the direction of propagation (Fig. 8.1(d)). In this case the electromagnetic wave is circularly polarized; the circular-polarization vector $\mathbf{I}^{CP}$ has only one non-zero element and takes the values $[0\ 0\ I\ 0]^T$ and $[0\ I\ 0\ 0]^T$, respectively (see Eq. (7.10)).

The polarization ellipse along with a designation of the rotation direction (right- or left-handed) fully describes the temporal evolution of the real electric vector at a fixed point in space. This evolution can also be visualized by plotting the curve in (θ, φ, t) coordinates described by the tip of the electric vector as a function of time. For example, in the case of an elliptically polarized plane wave with right-handed polarization, the curve is a right-handed helix with an elliptical projection onto the $\theta\varphi$-plane centered around the t-axis (cf. Fig. 8.2(a)). The pitch of the helix is simply $2\pi/\omega$, where ω is the angular frequency of the wave. Another way to visualize a plane wave is to fix a moment in time and draw a three-dimensional curve in (θ, φ, s) coordinates described by the tip of the electric vector as a function of a spatial coordinate $s = \hat{\mathbf{n}} \cdot \mathbf{r}$ oriented along the direction of propagation $\hat{\mathbf{n}}$. According to Eqs. (8.8)–(8.11), the electric field is the same for all position-time combinations with constant $ks - \omega t$. Therefore, at any instant of time (say, $t = 0$) the locus of the points described by the tip of the electric vector originating at different points on the s axis is also a helix with the same projection onto the $\theta\varphi$-plane as the respective helix in the (θ, φ, t) coordinates, but with opposite handedness. For example, for the wave with right-handed elliptical polarization shown in Fig. 8.2(a), the respective curve in the (θ, φ, s) coordinates is a left-handed elliptical helix shown in Fig. 8.2(b). The pitch of this helix is the wavelength λ. It is now clear that the propagation of the wave in time and space can be represented by progressive movement in time of the helix shown in Fig. 8.2(b) in the direction of $\hat{\mathbf{n}}$ with the speed of light. With

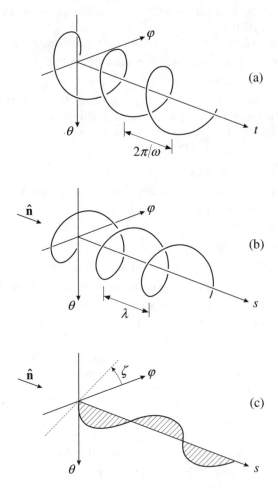

Figure 8.2: (a) The helix described by the tip of the real electric vector of a plane electromagnetic wave with right-handed polarization in (θ, φ, t) coordinates at a fixed point in space. (b) As in (a), but in (θ, φ, s) coordinates at a fixed moment in time. (c) As in (b), but for a linearly polarized wave.

increasing time, the intersection of the helix with any plane $s = $ constant describes a right-handed vibration ellipse. In the case of a circularly polarized wave, the elliptical helix becomes a helix with a circular projection onto the $\theta\varphi$-plane. If the wave is linearly polarized, then the helix degenerates into a simple sinusoidal curve in the plane making the angle ζ with the φ-axis (Fig. 8.2(c)).

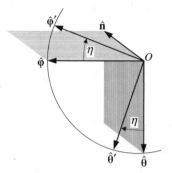

Figure 9.1: Rotation of the θ- and φ- axes through an angle $\eta \geq 0$ around $\hat{\mathbf{n}}$ in the clockwise direction when looking in the direction of propagation.

9. Rotation transformation rule for Stokes parameters

The Stokes parameters of a plane electromagnetic wave are always defined with respect to a reference plane containing the direction of wave propagation. If the reference plane is rotated about the direction of propagation, then the Stokes parameters are modified according to a rotation transformation rule, which can be derived as follows. Consider a rotation of the coordinate axes θ and φ through an angle $0 \leq \eta < 2\pi$ in the *clockwise* direction when looking in the direction of propagation (Fig. 9.1). The transformation rule for rotation of a two-dimensional coordinate system yields

$$E'_{0\theta} = E_{0\theta} \cos\eta + E_{0\varphi} \sin\eta, \tag{9.1}$$

$$E'_{0\varphi} = -E_{0\theta} \sin\eta + E_{0\varphi} \cos\eta, \tag{9.2}$$

where the primes denote the electric field vector components with respect to the new reference frame. It then follows from Eq. (7.4) that the rotation transformation rule for the Stokes column vector is

$$\mathbf{I'} = \mathbf{L}(\eta)\mathbf{I}, \tag{9.3}$$

where

$$\mathbf{L}(\eta) = \begin{bmatrix} 1 & 0 & 0 & 0 \\ 0 & \cos 2\eta & -\sin 2\eta & 0 \\ 0 & \sin 2\eta & \cos 2\eta & 0 \\ 0 & 0 & 0 & 1 \end{bmatrix} \tag{9.4}$$

is the so-called Stokes rotation matrix for angle η. It is obvious that a $\eta = \pi$ rotation does not change the Stokes parameters.

Because

$$(\mathbf{I}^{MS})' = \mathbf{BI'} = \mathbf{BL}(\eta)\mathbf{I} = \mathbf{BL}(\eta)\mathbf{B}^{-1}\mathbf{I}^{MS}, \tag{9.5}$$

the rotation matrix for the modified Stokes vector is given by

$$\mathbf{L}^{MS}(\eta) = \mathbf{BL}(\eta)\mathbf{B}^{-1} = \begin{bmatrix} \cos^2\eta & \sin^2\eta & -\frac{1}{2}\sin 2\eta & 0 \\ \sin^2\eta & \cos^2\eta & \frac{1}{2}\sin 2\eta & 0 \\ \sin 2\eta & -\sin 2\eta & \cos 2\eta & 0 \\ 0 & 0 & 0 & 1 \end{bmatrix}. \qquad (9.6)$$

Similarly, for the circular polarization representation,

$$(\mathbf{I}^{CP})' = \mathbf{AI}' = \mathbf{AL}(\eta)\mathbf{I} = \mathbf{AL}(\eta)\mathbf{A}^{-1}\mathbf{I}^{CP}, \qquad (9.7)$$

and the corresponding rotation matrix is diagonal [12]:

$$\mathbf{L}^{CP}(\eta) = \mathbf{AL}(\eta)\mathbf{A}^{-1} = \begin{bmatrix} \exp(i2\eta) & 0 & 0 & 0 \\ 0 & 1 & 0 & 0 \\ 0 & 0 & 1 & 0 \\ 0 & 0 & 0 & \exp(-i2\eta) \end{bmatrix}. \qquad (9.8)$$

10. Quasi-monochromatic light

The definition of a monochromatic plane electromagnetic wave given by Eq. (7.1) implies that the complex amplitude $\mathbf{E}_0$ is constant. In reality, the complex amplitude often fluctuates in time, albeit much slower than the time factor $\exp(-i\omega t)$. The fluctuations of the complex amplitude include, in general, fluctuations of both the amplitude and the phase of the real electric vector.

It is straightforward to verify that the electromagnetic field given by

$$\mathbf{E}(\mathbf{r}, t) = \mathbf{E}_0(t)\exp(i\mathbf{k}\cdot\mathbf{r} - i\omega t), \qquad (10.1)$$

$$\mathbf{H}(\mathbf{r}, t) = \mathbf{H}_0(t)\exp(i\mathbf{k}\cdot\mathbf{r} - i\omega t) \qquad (10.2)$$

still satisfies the Maxwell equations (2.1)–(2.4) at any moment in time provided that the medium is homogeneous and that

$$\mathbf{k}\cdot\mathbf{E}_0(t) = 0, \qquad (10.3)$$

$$\mathbf{k}\cdot\mathbf{H}_0(t) = 0, \qquad (10.4)$$

$$\mathbf{k}\times\mathbf{E}_0(t) = \omega\mu\mathbf{H}_0(t), \qquad (10.5)$$

$$\mathbf{k}\times\mathbf{H}_0(t) = -\omega\varepsilon\mathbf{E}_0(t), \qquad (10.6)$$

$$\left|\frac{\partial\mathbf{E}_0(t)}{\partial t}\right| \ll \omega|\mathbf{E}_0(t)|, \qquad (10.7)$$

$$\left|\frac{\partial\mathbf{H}_0(t)}{\partial t}\right| \ll \omega|\mathbf{H}_0(t)|. \qquad (10.8)$$

Equations (10.1)–(10.8) collectively define a *quasi-monochromatic* beam of light. Although the typical frequency of the fluctuations of the complex electric and magnetic field amplitudes is much smaller than the angular frequency ω, it is still so high that most optical instruments are incapable of tracing the instantaneous values of the Stokes parameters but rather respond to an average of the Stokes parameters over a relatively long period of time. Therefore, the definition of the Stokes parameters for a quasi-monochromatic beam of light propagating in a homogeneous nonabsorbing medium must be modified as follows:

$$I = \frac{1}{2}\sqrt{\frac{\epsilon}{\mu}}\,[\langle E_{0\theta}(t)E_{0\theta}^*(t)\rangle_t + \langle E_{0\varphi}(t)E_{0\varphi}^*(t)\rangle_t], \tag{10.9}$$

$$Q = \frac{1}{2}\sqrt{\frac{\epsilon}{\mu}}\,[\langle E_{0\theta}(t)E_{0\theta}^*(t)\rangle_t - \langle E_{0\varphi}(t)E_{0\varphi}^*(t)\rangle_t], \tag{10.10}$$

$$U = -\frac{1}{2}\sqrt{\frac{\epsilon}{\mu}}\,[\langle E_{0\theta}(t)E_{0\varphi}^*(t)\rangle_t + \langle E_{0\varphi}(t)E_{0\theta}^*(t)\rangle_t], \tag{10.11}$$

$$V = i\frac{1}{2}\sqrt{\frac{\epsilon}{\mu}}\,[\langle E_{0\varphi}(t)E_{0\theta}^*(t)\rangle_t - \langle E_{0\theta}(t)E_{0\varphi}^*(t)\rangle_t], \tag{10.12}$$

where

$$\langle f(t)\rangle_t = \frac{1}{T}\int_t^{t+T} d\tau\, f(\tau) \tag{10.13}$$

denotes the average over a time interval T long compared with the typical period of fluctuation.

The Stokes identity (7.8) is not, in general, valid for a quasi-monochromatic beam. Indeed, now we have

$$I^2 - Q^2 - U^2 - V^2$$

$$= \frac{\epsilon}{\mu}[\langle a_\theta^2\rangle_t\langle a_\varphi^2\rangle_t - \langle a_\theta a_\varphi \cos\Delta\rangle_t^2 - \langle a_\theta a_\varphi \sin\Delta\rangle_t^2]$$

$$= \frac{\epsilon}{\mu}\frac{1}{T^2}\int_t^{t+T} dt'\int_t^{t+T} dt''\,\{[a_\theta(t')]^2[a_\varphi(t'')]^2$$
$$- a_\theta(t')a_\varphi(t')\cos[\Delta(t')]a_\theta(t'')a_\varphi(t'')\cos[\Delta(t'')]$$
$$- a_\theta(t')a_\varphi(t')\sin[\Delta(t')]a_\theta(t'')a_\varphi(t'')\sin[\Delta(t'')]\}$$

$$= \frac{\epsilon}{\mu}\frac{1}{T^2}\int_t^{t+T} dt'\int_t^{t+T} dt''\,\{[a_\theta(t')]^2[a_\varphi(t'')]^2$$
$$- a_\theta(t')a_\varphi(t')a_\theta(t'')a_\varphi(t'')\cos[\Delta(t') - \Delta(t'')]\}$$

$$= \frac{\epsilon}{\mu}\frac{1}{2T^2}\int_t^{t+T} dt'\int_t^{t+T} dt''\,\{[a_\theta(t')]^2[a_\varphi(t'')]^2 + [a_\theta(t'')]^2[a_\varphi(t')]^2$$
$$- 2a_\theta(t')a_\varphi(t')a_\theta(t'')a_\varphi(t'')\cos[\Delta(t') - \Delta(t'')]\}$$

$$\geq \frac{\epsilon}{\mu} \frac{1}{2T^2} \int_t^{t+T} dt' \int_t^{t+T} dt'' \{[a_\theta(t')]^2 [a_\varphi(t'')]^2 + [a_\theta(t'')]^2 [a_\varphi(t')]^2$$
$$- 2a_\vartheta(t')a_\varphi(t')a_\vartheta(t'')a_\varphi(t'')\}$$

$$= \frac{\epsilon}{\mu} \frac{1}{2T^2} \int_t^{t+T} dt' \int_t^{t+T} dt'' [a_\theta(t')a_\varphi(t'') - a_\theta(t'')a_\varphi(t')]^2$$

$$\geq 0, \qquad\qquad\qquad (10.14)$$

thereby yielding

$$I^2 \geq Q^2 + U^2 + V^2. \qquad\qquad (10.15)$$

The equality holds only if the ratio $a_\theta(t)/a_\varphi(t)$ of the real amplitudes and the phase difference $\Delta(t)$ are independent of time, which means that $E_{0\theta}(t)$ and $E_{0\varphi}(t)$ are completely correlated. In this case the beam is said to be fully (or completely) polarized. This definition includes a monochromatic plane wave, but is, of course, more general. On the other hand, if $a_\theta(t)$, $a_\varphi(t)$, $\Delta_\theta(t)$, and $\Delta_\varphi(t)$ are totally uncorrelated and $\langle a_\theta^2 \rangle_t = \langle a_\varphi^2 \rangle_t$ then $Q = U = V = 0$, and the quasi-monochromatic beam of light is said to be unpolarized (or natural). This means that the parameters of the vibration ellipse traced by the endpoint of the electric vector fluctuate in such a way that there is no preferred vibration ellipse.

When two or more quasi-monochromatic beams propagating in the same direction are mixed incoherently, which means that there is no permanent phase relation between the separate beams, then the Stokes vector of the mixture is equal to the sum of the Stokes vectors of the individual beams:

$$\mathbf{I} = \sum_n \mathbf{I}_n, \qquad\qquad (10.16)$$

where n numbers the beams. Indeed, inserting Eqs. (8.1) and (8.2) in Eq. (10.9), we obtain for the total intensity

$$I = \frac{1}{2}\sqrt{\frac{\epsilon}{\mu}} \sum_n \sum_m \langle a_{\theta n} a_{\theta m} \exp[i(\Delta_{\theta n} - \Delta_{\theta m})]$$
$$+ a_{\varphi n} a_{\varphi m} \exp[i(\Delta_{\varphi n} - \Delta_{\varphi m})] \rangle_t$$

$$= \frac{1}{2}\sqrt{\frac{\epsilon}{\mu}} \left\{ \sum_n I_n + \sum_n \sum_{m \neq n} \langle a_{\theta n} a_{\theta m} \exp[i(\Delta_{\theta n} - \Delta_{\theta m})]\right.$$
$$\left. + a_{\varphi n} a_{\varphi m} \exp[i(\Delta_{\varphi n} - \Delta_{\varphi m})] \rangle_t \right\}.$$
$$(10.17)$$

Since the phases of different beams are uncorrelated, the second term on the right-hand side of the relation above vanishes. Hence

$$I = \sum_n I_n, \tag{10.18}$$

and similarly for Q, U, and V. Of course, this additivity rule also applies to the coherency matrix $\boldsymbol{\rho}$, the modified Stokes vector $\mathbf{I}^{MS}$, and the circular-polarization vector $\mathbf{I}^{CP}$.

The additivity of the Stokes parameters allows us to generalize the principle of optical equivalence (Section 7) to quasi-monochromatic light as follows: it is impossible by means of a traditional optical instrument to distinguish between various incoherent mixtures of quasi-monochromatic beams that form a beam with the same Stokes parameters (I, Q, U, V). For example, there is only one kind of unpolarized light, although it can be composed of quasi-monochromatic beams in an infinite variety of optically indistinguishable ways.

According to Eqs. (10.15) and (10.16), it always is possible *mathematically* to decompose any quasi-monochromatic beam into two incoherent parts, one unpolarized with a Stokes vector

$$[I - \sqrt{Q^2 + U^2 + V^2} \quad 0 \quad 0 \quad 0]^T,$$

and one fully polarized, with a Stokes vector

$$[\sqrt{Q^2 + U^2 + V^2} \quad Q \quad U \quad V]^T.$$

Thus, the intensity of the fully polarized component is $\sqrt{Q^2 + U^2 + V^2}$, so that the degree of (elliptical) polarization of the quasi-monochromatic beam is

$$P = \frac{\sqrt{Q^2 + U^2 + V^2}}{I}. \tag{10.19}$$

The degree of linear polarization is defined as

$$P_L = \sqrt{Q^2 + U^2} \Big/ I \tag{10.20}$$

and the degree of circular polarization as

$$P_C = \frac{V}{I}. \tag{10.21}$$

P vanishes for unpolarized light and is equal to unity for fully polarized light. For a partially polarized beam $(0 < P < 1)$ with $V \neq 0$, the sign of V indicates the preferential handedness of the vibration ellipses described by the endpoint of the electric vector. Specifically, a positive V indicates left-handed polarization and a negative V indicates right-handed polarization. By analogy with Eqs. (8.28) and (8.29), the quantities $-U/Q$ and $|V| \big/ \sqrt{Q^2 + U^2}$ can be interpreted as specifying the preferential orientation and ellipticity of the

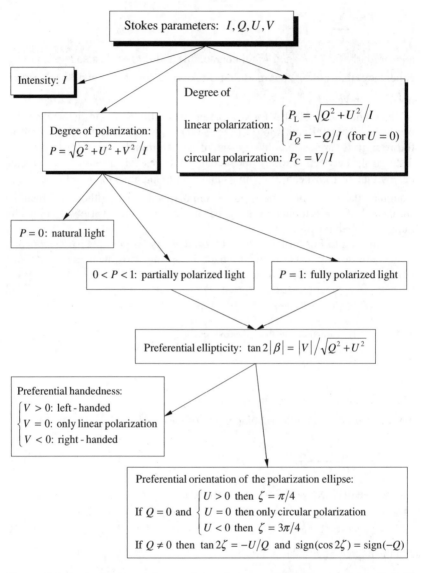

Figure 10.1: Analysis of a quasi-monochromatic beam with Stokes parameters I, Q, U, and V.

vibration ellipse. Unlike the Stokes parameters, these quantities are not additive. According to Eqs. (9.3) and (9.4), the P, P_L, and P_C are invariant with respect to rotations of the reference frame around the direction of propagation.

When $U = 0$, the ratio

$$P_Q = -\frac{Q}{I} \qquad (10.22)$$

is also called the degree of linear polarization (or the *signed* degree of linear polarization). P_Q is positive when the vibrations of the electric vector in the φ-direction (i.e., in the direction perpendicular to the meridional plane of the beam) dominate those in the θ-direction, and is negative otherwise.

The standard polarimetric analysis of a general quasi-monochromatic beam with Stokes parameters I, Q, U, and V is summarized in Fig. 10.1.

11. Measurement of the Stokes parameters

Most detectors of electromagnetic radiation, especially those in the visible and infrared spectral range, are insensitive to the polarization state of the beam impinging on the detector surface and can measure only the first Stokes parameter of the beam, viz., the intensity. Therefore, to measure the entire Stokes vector of the beam, one has to insert between the source of light and the detector one or several optical elements that modify the first Stokes parameter of the radiation reaching the detector in such a way that it contains information about the second, third, and fourth Stokes parameters of the original beam. This is usually done with so called polarizers and retarders.

A polarizer is an optical element that attenuates the orthogonal components of the electric field vector of an electromagnetic wave unevenly. Let us denote the corresponding attenuation coefficients as p_θ and p_φ and consider first the situation when the two orthogonal transmission axes of a polarizer coincide with the θ- and φ-axes of the laboratory coordinate system (see Fig. 11.1). This means that after the electromagnetic wave goes through the polarizer, the orthogonal components of the electric field change as follows:

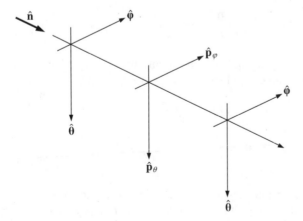

Figure 11.1: The transmission axes of the polarizer coincide with those of the laboratory reference frame.

$$E'_\theta = p_\theta E_\theta, \quad 0 \leq p_\theta \leq 1, \tag{11.1}$$

$$E'_\varphi = p_\varphi E_\varphi, \quad 0 \leq p_\varphi \leq 1. \tag{11.2}$$

It then follows from the definition of the Stokes parameters that the Stokes vector of the wave modifies according to

$$\mathbf{I'} = \mathbf{PI}, \tag{11.3}$$

where

$$\mathbf{P} = \frac{1}{2} \begin{bmatrix} p_\theta^2 + p_\varphi^2 & p_\theta^2 - p_\varphi^2 & 0 & 0 \\ p_\theta^2 - p_\varphi^2 & p_\theta^2 + p_\varphi^2 & 0 & 0 \\ 0 & 0 & 2p_\theta p_\varphi & 0 \\ 0 & 0 & 0 & 2p_\theta p_\varphi \end{bmatrix} \tag{11.4}$$

is the so-called Mueller matrix of the polarizer.

An important example of a polarizer is a neutral filter with $p_\theta = p_\varphi = p$, which equally attenuates the orthogonal components of the electric field vector and does not change the polarization state of the wave:

$$\mathbf{P} = p^2 \begin{bmatrix} 1 & 0 & 0 & 0 \\ 0 & 1 & 0 & 0 \\ 0 & 0 & 1 & 0 \\ 0 & 0 & 0 & 1 \end{bmatrix}. \tag{11.5}$$

In contrast, an ideal linear polarizer transmits only one orthogonal component of the wave (say, the θ component) and completely blocks the other one $(p_\varphi = 0)$:

$$\mathbf{P} = \frac{p_\theta^2}{2} \begin{bmatrix} 1 & 1 & 0 & 0 \\ 1 & 1 & 0 & 0 \\ 0 & 0 & 0 & 0 \\ 0 & 0 & 0 & 0 \end{bmatrix}. \tag{11.6}$$

An ideal perfect linear polarizer does not change one orthogonal component $(p_\theta = 1)$ and completely blocks the other one $(p_\varphi = 0)$:

$$\mathbf{P} = \frac{1}{2} \begin{bmatrix} 1 & 1 & 0 & 0 \\ 1 & 1 & 0 & 0 \\ 0 & 0 & 0 & 0 \\ 0 & 0 & 0 & 0 \end{bmatrix}. \tag{11.7}$$

If the transmission axes of a polarizer are rotated relative to the laboratory coordinate system (Fig. 11.2) then its Mueller matrix with respect to the laboratory coordinate system also changes. To obtain the resulting Stokes vector with respect to the laboratory coordinate system, we need to

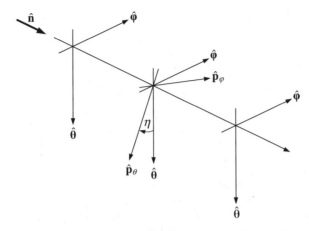

Figure 11.2: The polarizer transmission axes are rotated through an angle $\eta \geq 0$ around $\hat{\mathbf{n}}$ in the clockwise direction when looking in the direction of propagation.

1. "rotate" the initial Stokes vector through the angle η in the clockwise direction in order to obtain the Stokes parameters of the original beam with respect to the polarizer axes;
2. multiply the "rotated" Stokes vector by the original (non-rotated) polarizer Mueller matrix; and finally
3. "rotate" the Stokes vector thus obtained through the angle $-\eta$ in order to calculate the Stokes parameters of the resulting beam with respect to the laboratory coordinate system.

The final result is as follows:

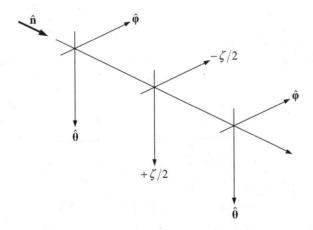

Figure 11.3: Propagation of a beam through a retarder.

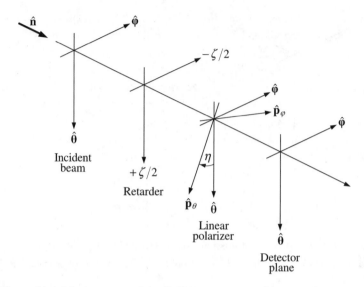

Figure 11.4: Measurement of the Stokes parameters with a retarder and an ideal perfect linear polarizer rotated with respect to the laboratory reference frame.

$$\mathbf{I'} = \mathbf{L}(-\eta)\mathbf{PL}(\eta)\mathbf{I}. \tag{11.8}$$

Hence the Mueller matrix of the rotated polarizer computed with respect to the laboratory coordinate system is given by

$$\mathbf{P}(\eta) = \mathbf{L}(-\eta)\mathbf{PL}(\eta) \tag{11.9}$$

with $\mathbf{P}(0) = \mathbf{P}$.

A retarder is an optical element that changes the phase of the beam by causing a phase shift of $+\zeta/2$ along the θ-axis and a phase shift of $-\zeta/2$ along the φ-axis (Fig. 11.3). We thus have

$$E'_\theta = \exp(+i\zeta/2)E_\theta, \tag{11.10}$$
$$E'_\varphi = \exp(-i\zeta/2)E_\varphi, \tag{11.11}$$

which yields

$$\mathbf{I'} = \mathbf{R}(\zeta)\mathbf{I}, \tag{11.12}$$

where

$$\mathbf{R}(\zeta) = \begin{bmatrix} 1 & 0 & 0 & 0 \\ 0 & 1 & 0 & 0 \\ 0 & 0 & \cos\zeta & \sin\zeta \\ 0 & 0 & -\sin\zeta & \cos\zeta \end{bmatrix} \tag{11.13}$$

is the Mueller matrix of the retarder.

Consider now the optical path shown in Fig. 11.4. The beam of light goes through a retarder and a rotated ideal perfect linear polarizer and then impinges on the surface of a polarization-insensitive detector. The Stokes vector of the resulting beam impinging on the detector surface is given by

$$\mathbf{I'} = \mathbf{P}(\eta)\mathbf{R}(\zeta)\mathbf{I}, \tag{11.14}$$

where the polarizer Mueller matrix is

$$\mathbf{P}(\eta) = \frac{1}{2}\begin{bmatrix} 1 & \cos 2\eta & -\sin 2\eta & 0 \\ \cos 2\eta & \cos^2 2\eta & -\cos 2\eta \sin 2\eta & 0 \\ -\sin 2\eta & -\cos 2\eta \sin 2\eta & \sin^2 2\eta & 0 \\ 0 & 0 & 0 & 0 \end{bmatrix} \tag{11.15}$$

(cf. Eqs. (11.7) and (11.9)). Hence the intensity of the resulting beam as a function of η and ζ is given by

$$I'(\eta, \zeta) = \tfrac{1}{2}(I + Q\cos 2\eta - U\sin 2\eta \cos\zeta - V\sin 2\eta \sin\zeta). \tag{11.16}$$

This formula suggests a simple way to determine the Stokes parameters of the original beam by measuring the intensity of the resulting beam using four different combinations of η and ζ:

$$I = I'(0°, 0°) + I'(90°, 0°), \tag{11.17}$$

$$Q = I'(0°, 0°) - I'(90°, 0°), \tag{11.18}$$

$$U = -2I'(45°, 0°) + I, \tag{11.19}$$

$$V = I - 2I'(45°, 90°). \tag{11.20}$$

Other methods for measuring the Stokes parameters and practical aspects of polarimetry are discussed in detail in [14–17].

12. Spherical wave solution

As we have seen, plane electromagnetic waves represent a fundamental solution of the Maxwell equations underlying the concept of a monochromatic parallel beam of light. Another fundamental solution representing the outward propagation of electromagnetic energy from a point-like source is a transverse spherical wave. To derive this solution, we need Eqs. (4.3), (4.4), (4.11), (4.16), (4.17), (6.1), and (6.2) as well as the following formulas:

$$\nabla \frac{\exp(ikr)}{r} = \left(ik - \frac{1}{r}\right)\frac{\exp(ikr)}{r}\hat{\mathbf{r}}, \tag{12.1}$$

$$\nabla \cdot \mathbf{a} = \frac{\partial a_r}{\partial r} + \frac{2a_r}{r} + \frac{1}{r}\frac{\partial a_\theta}{\partial \theta} + \frac{a_\theta}{r\tan\theta} + \frac{1}{r\sin\theta}\frac{\partial a_\varphi}{\partial \varphi}, \tag{12.2}$$

$$\nabla \times \mathbf{a} = \left(\frac{1}{r} \frac{\partial a_\varphi}{\partial \theta} + \frac{a_\varphi}{r \tan \theta} - \frac{1}{r \sin \theta} \frac{\partial a_\theta}{\partial \varphi} \right) \hat{\mathbf{r}}$$

$$+ \left(\frac{1}{r \sin \theta} \frac{\partial a_r}{\partial \varphi} - \frac{\partial a_\varphi}{\partial r} - \frac{a_\varphi}{r} \right) \hat{\mathbf{\theta}}$$

$$+ \left(\frac{\partial a_\theta}{\partial r} + \frac{a_\theta}{r} - \frac{1}{r} \frac{\partial a_r}{\partial \theta} \right) \hat{\mathbf{\varphi}}. \tag{12.3}$$

It is then straightforward to verify that the complex field vectors

$$\mathbf{E}(\mathbf{r}, t) = \frac{\exp(ikr)}{r} \mathbf{E}_1(\hat{\mathbf{r}}) \exp(-i\omega t), \tag{12.4}$$

$$\mathbf{H}(\mathbf{r}, t) = \frac{\exp(ikr)}{r} \mathbf{H}_1(\hat{\mathbf{r}}) \exp(-i\omega t) \tag{12.5}$$

are a solution of the Maxwell equations in the limit $kr \to \infty$ provided that the medium is homogeneous and that

$$\hat{\mathbf{r}} \cdot \mathbf{E}_1(\hat{\mathbf{r}}) = 0, \tag{12.6}$$

$$\hat{\mathbf{r}} \cdot \mathbf{H}_1(\hat{\mathbf{r}}) = 0, \tag{12.7}$$

$$k\hat{\mathbf{r}} \times \mathbf{E}_1(\hat{\mathbf{r}}) = \omega\mu \mathbf{H}_1(\hat{\mathbf{r}}), \tag{12.8}$$

$$k\hat{\mathbf{r}} \times \mathbf{H}_1(\hat{\mathbf{r}}) = -\omega\varepsilon \mathbf{E}_1(\hat{\mathbf{r}}), \tag{12.9}$$

where the wavenumber $k = k_R + ik_I = \omega\sqrt{\varepsilon\mu} = \omega m/c$ may be complex and the $\mathbf{E}_1(\hat{\mathbf{r}})$ and $\mathbf{H}_1(\hat{\mathbf{r}})$ are independent of the distance r from the origin.

Equations (12.4)–(12.9) describe an outgoing transverse spherical wave propagating radially with the phase velocity $v = \omega/k_R$ and having mutually perpendicular complex electric and magnetic field vectors. The wave is homogeneous in that the real and imaginary parts of the complex wave vector $k\hat{\mathbf{r}}$ are parallel. The surfaces of constant phase coincide with the surfaces of constant amplitude and are spherical. Obviously,

$$\mathbf{H}(\mathbf{r}, t) = \frac{k}{\omega\mu} \hat{\mathbf{r}} \times \mathbf{E}(\mathbf{r}, t), \tag{12.10}$$

which allows one to consider the spherical wave in terms of the electric (or magnetic) field only. The time-averaged Poynting vector of the wave is given by

$$\langle \mathbf{S}(\mathbf{r}, t) \rangle_t = \tfrac{1}{2} \mathrm{Re} \left(\sqrt{\frac{\varepsilon}{\mu}} \right) \frac{|\mathbf{E}_1(\hat{\mathbf{r}})|^2}{r^2} \exp\left(-2 \frac{\omega}{c} m_I r \right) \hat{\mathbf{r}}, \tag{12.11}$$

where, as before, $m_I = ck_I/\omega$. The intensity of the spherical wave is defined

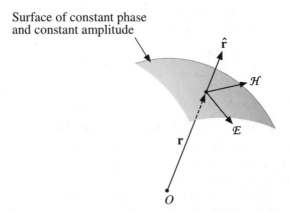

Figure 12.1: Spherical electromagnetic wave propagating in a homogeneous medium with no dispersion and losses.

as the absolute value of the time-averaged Poynting vector,

$$I(\mathbf{r}) = \left|\langle S(\mathbf{r},t)\rangle_t\right| = \tfrac{1}{2}\mathrm{Re}\left(\sqrt{\frac{\varepsilon}{\mu}}\right)\frac{|\mathbf{E}_1(\hat{\mathbf{r}})|^2}{r^2}\exp\left(-2\frac{\omega}{c}m_\mathrm{I}r\right). \qquad (12.12)$$

The intensity has the dimension of monochromatic energy flux and specifies the amount of electromagnetic energy crossing a unit surface element normal to $\hat{\mathbf{r}}$ per unit time. The intensity is attenuated exponentially by absorption and in addition decreases as the inverse square of the distance from the origin.

In the case of a medium with no dispersion and losses, the real electric and magnetic field vectors are mutually orthogonal and are normal to the direction of propagation $\hat{\mathbf{r}}$ (Fig. 12.1). The energy conservation law takes the form

$$\oint_S \mathrm{d}S\langle S(\mathbf{r},t)\rangle_t \cdot \hat{\mathbf{r}} = \oint_S \mathrm{d}S I(\mathbf{r})$$

$$= \frac{1}{2}\sqrt{\frac{\epsilon}{\mu}}\frac{1}{r^2}\oint_S \mathrm{d}S|\mathbf{E}_1(\hat{\mathbf{r}})|^2$$

$$= \frac{1}{2}\sqrt{\frac{\epsilon}{\mu}}\int_{4\pi} \mathrm{d}\hat{\mathbf{r}}|\mathbf{E}_1(\hat{\mathbf{r}})|^2$$

$$= \text{constant}, \qquad (12.13)$$

where S is the sphere of radius r and

$$\mathrm{d}\hat{\mathbf{r}} = \frac{\mathrm{d}S}{r^2} = \sin\theta\,\mathrm{d}\theta\,\mathrm{d}\varphi \qquad (12.14)$$

is an infinitesimal solid angle element around the direction $\hat{\mathbf{r}}$. It is also easy to show that in the case of a nonabsorbing medium, the time-averaged energy density of a spherical wave is given by

$$\langle u(\mathbf{r}, t) \rangle_t = \tfrac{1}{2} \epsilon \frac{|\mathbf{E}_1(\hat{\mathbf{r}})|^2}{r^2}. \tag{12.15}$$

Equations (12.12) and (12.15) show that $I(\mathbf{r}) = v \langle u(\mathbf{r}, t) \rangle_t$, where $v = 1/\sqrt{\epsilon \mu}$ is the speed of light in the material medium. This is the same result as that obtained previously for a plane wave propagating in a nonabsorbing medium (cf. Eq. (6.30)).

In complete analogy with the case of a plane wave, the coherency matrix, the coherency column vector, and the Stokes column vector of a spherical wave propagating in a homogeneous medium with no dispersion and losses can be defined as

$$\boldsymbol{\rho}(\mathbf{r}) = \begin{bmatrix} \rho_{11}(\mathbf{r}) & \rho_{12}(\mathbf{r}) \\ \rho_{21}(\mathbf{r}) & \rho_{22}(\mathbf{r}) \end{bmatrix} = \frac{1}{2} \sqrt{\frac{\epsilon}{\mu}} \frac{1}{r^2} \begin{bmatrix} E_{1\theta}(\hat{\mathbf{r}}) E_{1\theta}^*(\hat{\mathbf{r}}) & E_{1\theta}(\hat{\mathbf{r}}) E_{1\varphi}^*(\hat{\mathbf{r}}) \\ E_{1\varphi}(\hat{\mathbf{r}}) E_{1\theta}^*(\hat{\mathbf{r}}) & E_{1\varphi}(\hat{\mathbf{r}}) E_{1\varphi}^*(\hat{\mathbf{r}}) \end{bmatrix},$$

$$\tag{12.16}$$

$$\mathbf{J}(\mathbf{r}) = \begin{bmatrix} \rho_{11}(\mathbf{r}) \\ \rho_{12}(\mathbf{r}) \\ \rho_{21}(\mathbf{r}) \\ \rho_{22}(\mathbf{r}) \end{bmatrix} = \frac{1}{2} \sqrt{\frac{\epsilon}{\mu}} \frac{1}{r^2} \begin{bmatrix} E_{1\theta}(\hat{\mathbf{r}}) E_{1\theta}^*(\hat{\mathbf{r}}) \\ E_{1\theta}(\hat{\mathbf{r}}) E_{1\varphi}^*(\hat{\mathbf{r}}) \\ E_{1\varphi}(\hat{\mathbf{r}}) E_{1\theta}^*(\hat{\mathbf{r}}) \\ E_{1\varphi}(\hat{\mathbf{r}}) E_{1\varphi}^*(\hat{\mathbf{r}}) \end{bmatrix}, \tag{12.17}$$

$$\mathbf{I}(\mathbf{r}) = \begin{bmatrix} I(\mathbf{r}) \\ Q(\mathbf{r}) \\ U(\mathbf{r}) \\ V(\mathbf{r}) \end{bmatrix} = \mathbf{D}\mathbf{J}(\mathbf{r}) = \frac{1}{2} \sqrt{\frac{\epsilon}{\mu}} \frac{1}{r^2} \begin{bmatrix} E_{1\theta}(\hat{\mathbf{r}}) E_{1\theta}^*(\hat{\mathbf{r}}) + E_{1\varphi}(\hat{\mathbf{r}}) E_{1\varphi}^*(\hat{\mathbf{r}}) \\ E_{1\theta}(\hat{\mathbf{r}}) E_{1\theta}^*(\hat{\mathbf{r}}) - E_{1\varphi}(\hat{\mathbf{r}}) E_{1\varphi}^*(\hat{\mathbf{r}}) \\ - E_{1\theta}(\hat{\mathbf{r}}) E_{1\varphi}^*(\hat{\mathbf{r}}) - E_{1\varphi}(\hat{\mathbf{r}}) E_{1\theta}^*(\hat{\mathbf{r}}) \\ i[E_{1\varphi}(\hat{\mathbf{r}}) E_{1\theta}^*(\hat{\mathbf{r}}) - E_{1\theta}(\hat{\mathbf{r}}) E_{1\varphi}^*(\hat{\mathbf{r}})] \end{bmatrix},$$

$$\tag{12.18}$$

respectively. All these quantities have the dimension of monochromatic energy flux. As before, the first Stokes parameter is the intensity (defined this time by Eq. (12.12)).

13. Coherency dyad

The definition of the coherency and Stokes vectors explicitly exploits the transverse character of an electromagnetic wave and requires the use of a local spherical coordinate system. However, in some cases it is convenient to introduce an alternative quantity, which also provides a complete optical specification of a transverse electromagnetic wave, but is defined without explicit use of a coordinate system. This quantity is called the coherency dyad and, in the general case of an arbitrary electromagnetic field, is given by

$$\vec{\rho}(\mathbf{r},t) = \mathbf{E}(\mathbf{r},t) \otimes \mathbf{E}^*(\mathbf{r},t), \qquad (13.1)$$

where $\otimes$ denotes the dyadic product of two vectors. It is then clear that the coherency and Stokes vectors of a transverse time-harmonic electromagnetic wave can be expressed in terms of the coherency dyad as follows:

$$\mathbf{J} = \frac{1}{2}\sqrt{\frac{\varepsilon}{\mu}}
\begin{bmatrix}
\hat{\boldsymbol{\theta}} \cdot \vec{\rho} \cdot \hat{\boldsymbol{\theta}} \\
\hat{\boldsymbol{\theta}} \cdot \vec{\rho} \cdot \hat{\boldsymbol{\varphi}} \\
\hat{\boldsymbol{\varphi}} \cdot \vec{\rho} \cdot \hat{\boldsymbol{\theta}} \\
\hat{\boldsymbol{\varphi}} \cdot \vec{\rho} \cdot \hat{\boldsymbol{\varphi}})
\end{bmatrix}, \qquad (13.2)$$

$$\mathbf{I} = \frac{1}{2}\sqrt{\frac{\varepsilon}{\mu}}
\begin{bmatrix}
\hat{\boldsymbol{\theta}} \cdot \vec{\rho} \cdot \hat{\boldsymbol{\theta}} + \hat{\boldsymbol{\varphi}} \cdot \vec{\rho} \cdot \hat{\boldsymbol{\varphi}} \\
\hat{\boldsymbol{\theta}} \cdot \vec{\rho} \cdot \hat{\boldsymbol{\theta}} - \hat{\boldsymbol{\varphi}} \cdot \vec{\rho} \cdot \hat{\boldsymbol{\varphi}} \\
-\hat{\boldsymbol{\theta}} \cdot \vec{\rho} \cdot \hat{\boldsymbol{\varphi}} - \hat{\boldsymbol{\varphi}} \cdot \vec{\rho} \cdot \hat{\boldsymbol{\theta}} \\
i(\hat{\boldsymbol{\varphi}} \cdot \vec{\rho} \cdot \hat{\boldsymbol{\theta}} - \hat{\boldsymbol{\theta}} \cdot \vec{\rho} \cdot \hat{\boldsymbol{\varphi}})
\end{bmatrix}, \qquad (13.3)$$

whereas the products $\vec{\rho} \cdot \hat{\mathbf{n}}$ and $\hat{\mathbf{n}} \cdot \vec{\rho}$ vanish. It follows from the definition of the coherency dyad that it is Hermitian:

$$(\vec{\rho})^{\mathrm{T}} = \rho^*. \qquad (13.4)$$

The coherency dyad is a more general quantity than the coherency and Stokes vectors because it can be applied to any electromagnetic field and not just to a transverse electromagnetic wave. The simplest example of a situation in which the coherency dyad can be introduced, whereas the Stokes vector cannot involves the superposition of two plane electromagnetic waves propagating in different directions. The more general nature of the coherency dyad makes the latter very convenient in studies of random electromagnetic fields created by large stochastic groups of scatterers. For example, the additivity of the Stokes parameters (Section 10) is a concept that can be applied only to transverse waves propagating in exactly the same direction, whereas the average coherency dyad of a random electromagnetic field at an observation point can sometimes be reduced to an incoherent sum of coherency dyads of transverse waves propagating in various directions [18,19].

It is important to remember, however, that when the coherency dyad is applied to an arbitrary electromagnetic field, it may not always have as definite a physical meaning as, for example, the Poynting vector. The relationship between the coherency dyad and the actual physical observables may change depending on the problem at hand and must be established carefully whenever this quantity is used in a theoretical analysis of a specific measurement procedure. For example, the right-hand sides of Eqs. (13.2) and (13.3) may become rather meaningless if the products $\vec{\rho} \cdot \hat{\mathbf{n}}$ and $\hat{\mathbf{n}} \cdot \vec{\rho}$ do not vanish.

14. Historical notes and further reading

The equations of classical electromagnetics were written originally by James Clerk Maxwell (1831–1979) in Cartesian component form [20] and were cast in the modern vector form by Oliver Heaviside (1850–1925). The subsequent experimental verification of Maxwell's theory by Heinrich Rudolf Hertz (1857–1894) made it a well-established discipline. Since then classical electromagnetics has been a cornerstone of physics and has played a critical role in the development of a great variety of scientific, engineering, and biomedical disciplines. The fundamental nature of Maxwell's electromagnetics was ultimately asserted by the development of the special theory of relativity by Jules Henri Poincaré (1854–1912) and Hendrik Antoon Lorentz (1853–1928).

Sir George Gabriel Stokes (1819–1903) was the first to discover that four parameters, now known as the Stokes parameters, could characterize the polarization state of any light beam, including partially polarized and unpolarized light [13]. Furthermore, he noted that unlike the quantities entering the amplitude formulation of the optical field, these parameters could be directly measured by a suitable optical instrument. The fascinating subject of polarization had attracted the attention of many other great scientists before and after Stokes, including Augustin Jean Fresnel (1788–1827), Dominique François Arago (1786–1853), Thomas Young (1773–1829), Subrahmanyan Chandrasekhar (1910–1995), and Hendrik van de Hulst (1918–2000). Even Poincaré, who is rightfully considered to be one the greatest geniuses of all time, could not help but contribute to this discipline by developing a useful polarization analysis tool now known as the Poincaré sphere [21].

The two-volume monograph by Sir Edmund Whittaker [22] remains by far the most complete and balanced account of the history of electromagnetism from the time of William Gilbert (1544–1603) and René Descartes (1596–1650) to the relativity theory. This magnificent work should be read by everyone interested in a masterful and meticulously documented recreation of the sequence of events and publications that shaped physics as we know it. In this era of endless attempts to rewrite the history of modern physics by ignorant and/or biased authors (including those of the numerous guides for "dummies"), many of whom could not be bothered to study the original sources and barely could understand the subject matter, the monumental treatise by Whittaker is an ideal starting point for those individuals who want to form their own opinions as it provides a presentation of facts rather gossip, prejudices, and intentional distortions.

Comprehensive modern accounts of classical electromagnetics and optics can be found in [1,2,7]. Extensive treatments of theoretical and experimental polarimetry were provided by Shurcliff [14], Azzam and Bashra [15], Kliger *et al.* [16], and Collett [17]. Pye [23] describes numerous manifestations of polarization in science and nature.

Acknowledgments

We thank Joop Hovenier and Gorden Videen for many useful discussions and Nadia Zakharova for help with graphics. This research was funded by the NASA Radiation Sciences Program managed by Donald Anderson.

References

1. J. D. Jackson (1998). *Classical Electrodynamics* (John Wiley & Sons, New York).
2. M. Born and E. Wolf (1999). *Principles of Optics* (Cambridge University Press, Cambridge).
3. C. F. Bohren and D. R. Huffman (1983). *Absorption and Scattering of Light by Small Particles* (John Wiley & Sons, New York).
4. P. W. Barber and S. C. Hill (1990). *Light Scattering by Particles: Computational Methods* (World Scientific, Singapore).
5. M. I. Mishchenko, J. W. Hovenier, and L. D. Travis, eds. (2000). *Light Scattering by Nonspherical Particles: Theory, Measurements, and Applications* (Academic Press, San Diego).
6. M. I. Mishchenko, L. D. Travis, and A. A. Lacis (2002). *Scattering, Absorption, and Emission of Light by Small Particles* (Cambridge University Press, Cambridge).
7. J. A. Stratton (1941). *Electromagnetic Theory* (McGraw Hill, New York).
8. L. Tsang, J. A. Kong, and K.-H. Ding (2000). *Scattering of Electromagnetic Waves: Theories and Applications* (John Wiley & Sons, New York).
9. J. A. Kong (2000). *Electromagnetic Wave Theory* (EMW Publishing, Cambridge, MA).
10. H. C. van de Hulst (1957). *Light Scattering by Small Particles* (John Wiley & Sons, New York).
11. M. Kerker (1969). *The Scattering of Light and Other Electromagnetic Radiation* (Academic Press, San Diego).
12. J. W. Hovenier and C. V. M. van der Mee (1983). Fundamental relationships relevant to the transfer of polarized light in a scattering atmosphere. *Astron. Astrophys.* **128**, 1–16.
13. G. G. Stokes (1852). On the composition and resolution of streams of polarized light from different sources. *Trans. Cambridge Philos. Soc.* **9**, 399–416.
14. W. A. Shurcliff (1962). *Polarized Light: Production and Use* (Harvard University Press, Cambridge, MA).
15. R. M. A. Azzam and N. M. Bashara (1977). *Ellipsometry and Polarized Light* (North Holland, Amsterdam).
16. D. S. Kliger, J. W. Lewis, and C. E. Randall (1990). *Polarized Light in Optics and Spectroscopy* (Academic Press, San Diego).
17. E. Collett (1992). *Polarized Light: Fundamentals and Applications* (Marcel Dekker, New York).
18. M. I. Mishchenko (2002). Vector radiative transfer equation for arbitrarily shaped and arbitrarily oriented particles: a microphysical derivation from statistical electromagnetics. *Appl. Opt.* **41**, 7114–7134.
19. M. I. Mishchenko (2003). Microphysical approach to polarized radiative transfer: extension to the case of an external observation point. *Appl. Opt.* **42**, 4963–4967.
20. J. Clerk Maxwell (1891). *A Treatise on Electricity and Magnetism* (Clarendon Press, Oxford) (reprinted by Dover, New York, 1954).
21. H. Poincaré (1892). *Théorie Mathématique de la Lumière*, Vol. 2 (Georges Carré, Paris).

22. E. Whittaker (1987). *A History of the Theories of Aether and Electricity*, Vols. I and II (American Institute of Physics, New York).
23. D. Pye (2001). *Polarised Light in Science and Nature* (Institute of Physics Publishing, Bristol, UK).

Evgeniy Zubko, Mark Thoreson, Vsevolod Ivanov, Pavel Litvinov, Oleg Dubovik, Valery Loiko, Matthew Easley, Olga Kalashnikova, Rosario Vilaplana, Ben Veihelmann, James McDonald, Hal Maring, Karine Chamaillard, Michael Mishchenko, and Tracy Smith (from left to right). Photo courtesy Ivan Mishchenko.

POLARIZED LIGHT SCATTERING BY LARGE NONSPHERICAL PARTICLES

A. MACKE[1] AND K. MUINONEN[2]

[1] *IFM-GEOMAR, Leibniz-Institut für Meereswissenschaften an der Universität Kiel,*
Düsternbrooker Weg 20, D-24105 Kiel, Germany
[2] *Observatory, P.O.Box 14, Kopernikuksentie 1, FI-00014 University of Helsinki, Finland*

Abstract. Methodology and applications of the Ray Optics Method (ROM) are shown for polarized light scattering by large nonspherical particles. Particle geometries under consideration are hexagonal columns, various types of dendrites, fractal polycrystals, and Gaussian random spheres. The two latter particle shapes serve as examples for highly irregularly shaped scatterers as they can be found in atmospheric ice clouds and interplanetary dust particles. After briefly discussing the applicability of ROM in terms of size parameter and absorption strength, the effects of particle shape and orientation on the phase matrix elements P_{11}, P_{22}, P_{33}, P_{44}, P_{12}, and P_{34} are discussed. In general it is shown that the polarization information from single scattering can be used to distinguish between hexagonal columns, plates, and irregular particles as well as between 3D randomly and horizontally oriented particles. The scattering characteristics of Gaussian random particles are illustrated through ray-tracing computations using the refractive index of ice. Similarities to scattering by the fractal crystal particle are found in the case of highly irregular, fractal Gaussian particles with power law correlation functions.

1. Introduction

Naturally occurring particles come in a large variety of sizes and shapes. The particle geometries basically are determined by the molecular lattice structure and by the growth conditions. The present chapter summarizes

45

G. Videen et al. (eds.), Photopolarimetry in Remote Sensing, 45-64.

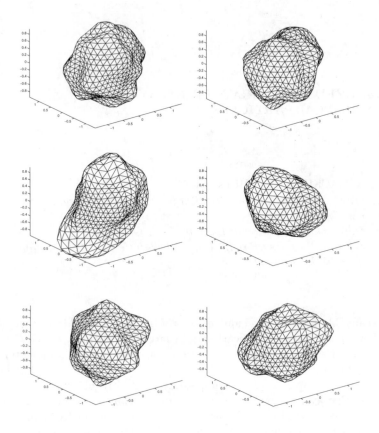

Figure 1.   Sample shapes of random Gaussian spheres with $\nu = 4$ and $\sigma = 0.12$. The maximum degree of spherical harmonics included was $l_{\max} = 10$.

recent work on light scattering by large nonspherical particles where the Ray Optics Method (geometric optics and forward diffraction) provides a reliable way of accounting for arbitrary particle geometries. Examples are various dust particles as they occur in our atmosphere, in interplanetary space, and on the surfaces of atmosphereless planetary bodies, as well as ice crystals in high altitude cirrus clouds on Earth, Mars, and Jupiter.

The Gaussian random sphere (G-sphere; [23, 18]) offers a few-parameter approach to model the shapes of irregular particles. The size and shape of the G-sphere are specified by the mean radial distance from the origin to the surface and the covariance function of the logarithmic radial distance. The covariance function is given as a series of Legendre polynomials with non-negative coefficients. For each degree, these coefficients provide the spectral weights of the corresponding spherical harmonics components in the G-sphere. Weighting the spectrum toward higher-degree harmonics results in

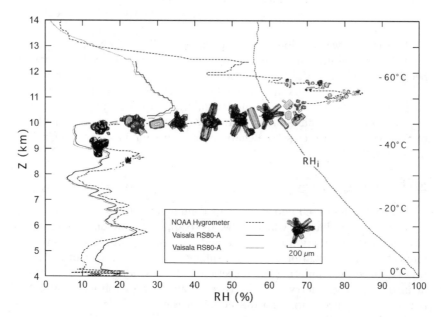

Figure 2. Schematic illustration of the development of ice particle size and shape. Ice particle replicates are plotted along a vertical profile of relative humidity with respect to water.

sample spheres with larger numbers of hills and valleys per solid angle. Increasing the variance of the logarithmic radius enhances the hills and valleys radially.

There is evidence for a power-law-type covariance function for natural irregular shapes ([19, 24]), in which case the shape of the G-sphere is fully described by two parameters, the standard deviation of the radial distance σ and the power-law index ν. Figure 1 shows some sample shapes of the G-sphere with $\nu = 4$ and $\sigma = 0.12$. Gaussian shape statistics are particularly relevant due to the Central Limit Theorem stating the tendency toward Gaussian statistics for complex systems.

The temperature and pressure conditions on our planet only allow for the hexagonal lattice of ice (I_h). Accordingly, many ice particles in cirrus clouds show a basic hexagonal structure ranging from pristine columns to plates to bullets to dendrites. In addition, highly irregularly shaped particles exist owing to rapidly changing growth conditions, collisions and splitting processes. Figure 2 from [12] illustrates the changes in shape and size of atmospheric ice crystals as they grow from micrometer sized "quasi-spheres" to millimeter sized dendrites of hexagonal columns (bullet rosettes) to melting lumps of irregularly shaped ice particles. Since the particle sizes vary along three orders of magnitude, the average scattering and absorption properties also depend significantly on the specific size distribution.

In Section 2 we describe the theoretical aspects of the ray-optics approximation, divided into the geometric optics and forward diffraction parts. In Section 3 we address the accuracy of the approximation via comparisons to more exact computations. In Section 4 we outline certain scattering characteristics of pristine ice crystals and irregularly shaped ice particles, including fractal crystals and Gaussian particles. Particle orientation is the topic of Section 5, and we close the chapter through summary and conclusions in Section 6.

2. Scattering theory

The particles described above are large compared to the wavelength of the incoming solar radiation, which allows us to apply the ray-optics method to calculate their extinction (scattering and absorption) properties. Here, the intensity and polarization state of a sufficiently large number of incoming rays is traced by simulating the reflection and refraction processes at a predefined particle geometry. By collecting all rays that are refracted out of the particle and by accounting for diffraction at the particles geometrical cross section, it is possible to obtain the scattering and absorption properties of that particle.

In contrast to the Mie-theory, ROM is applicable to arbitrary particle shapes as long as the smallest dimension of the particle is large compared to the wavelength of the incoming radiation. For ROM to be valid, the curvature radii R_c on the particle surface must be much larger than the wavelength of incident light λ and the central phase shifts across the surface irregularities must be much larger than one (e.g., [22]),

$$kR_c >> 1, \qquad 2kR_c|m - 1| >> 1, \tag{1}$$

where $k = 2\pi/\lambda$.

The historical development of ROM is described in [7]. With the strong increase in computer power, ROM has been applied to a huge field of problems in many natural sciences and technologies. See also [2] for a derivation of ROM and its application to various problems in the branch of optics. The present chapter refers only to the application of ROM to light scattering problems as described in 2.1. A list of relevant publications in this field can be found in chapter 10, 11, and 15 of [16]. Among recent works on classical geometric optics is that in [3] for random particles generated using a Gaussian random process in three-dimensional space [25]. As for near-future extensions toward physical optics, the reader is referred to [20] and references therein.

2.1. GEOMETRIC OPTICS METHOD

The scattering matrix $\mathbf{P}$ relates the Stokes vectors of the incident and scattered light $\mathbf{I}_i$ and $\mathbf{I}_s$; denoting the directions of incidence and scattering by Ω_i and Ω_s, respectively, and the distance between the particle center and the observer by r,

$$\mathbf{I}_s(\Omega_s, \Omega_i) = \frac{\sigma_s(\Omega_i)}{4\pi r^2} \mathbf{P}(\Omega_s, \Omega_i) \cdot \mathbf{I}_i(\Omega_i),$$

$$\int_{4\pi} \frac{d\Omega_s}{4\pi} P_{11}(\Omega_s, \Omega_i) = 1, \qquad (2)$$

where σ_s is the scattering cross section and P_{11} is the scattering phase function, both depending in general on the incident direction (Ω_i). Note that the incident direction $\Omega_i = (0,0)$ in the incident ray coordinate system.

For particles large compared to the wavelength, the scattering cross section and scattering matrix can be divided into the forward diffraction and geometric optics parts (denoted by superscripts D and G),

$$\sigma_s(\Omega_i) = \sigma_s^D(\Omega_i) + \sigma_s^G(\Omega_i),$$

$$\mathbf{P}(\Omega_s, \Omega_i) = \frac{1}{\sigma_s(\Omega_i)} \left[\sigma_s^D(\Omega_i)\mathbf{P}^D(\Omega_s, \Omega_i) + \sigma_s^G(\Omega_i)\mathbf{P}^G(\Omega_s, \Omega_i) \right],$$

$$\int_{4\pi} \frac{d\Omega_s}{4\pi} P_{11}^D(\Omega_s, \Omega_i) = \int_{4\pi} \frac{d\Omega_s}{4\pi} P_{11}^G(\Omega_s, \Omega_i) = 1. \qquad (3)$$

In ROM, we strictly require

$$\sigma_s^D(\Omega_i) = A_\perp(\Omega_i),$$

$$\sigma_e(\Omega_i) = \sigma_a(\Omega_i) + \sigma_s(\Omega_i) = 2A_\perp(\Omega_i), \qquad (4)$$

where σ_e and σ_a are the extinction and absorption cross sections, and $A_\perp$ is the cross-sectional area. The absorption cross section is due solely to geometric optics: $\sigma_a = \sigma_a^G$. The asymmetry parameter can be divided into the forward diffraction and geometric optics parts as in Eq. (3) for the scattering matrix.

For the geometric optics contribution, for external incidence, Snell's law is applied in the form

$$\sin \iota = \mathrm{Re}(m) \sin \tau, \qquad (5)$$

where ι and τ are the angles of incidence and refraction, respectively. When determining τ, we thus make use of $\mathrm{Re}(m)$ only and assume that $\mathrm{Im}(m)$ has either negligible influence on τ or is large enough to entirely eliminate internal ray propagation. To be more accurate, Snell's law can be generalized rigorously to complex refractive indices (e.g., [17]).

For external incidence, the Mueller matrices of the reflected and re-fracted rays (denoted by subscripts r and t) can be obtained from

$$\mathbf{M}_r = \mathbf{R} \cdot \mathbf{L} \cdot \mathbf{M}_i$$
$$\mathbf{M}_t = \mathbf{T} \cdot \mathbf{L} \cdot \mathbf{M}_i, \tag{6}$$

where $\mathbf{L}$ is the rotation to the plane of incidence and $\mathbf{R}$ and $\mathbf{T}$ are the Fresnel reflection and transmission matrices,

$$\mathbf{L} = \begin{bmatrix} 1 & 0 & 0 & 0 \\ 0 & \cos 2\eta & \sin 2\eta & 0 \\ 0 & -\sin 2\eta & \cos 2\eta & 0 \\ 0 & 0 & 0 & 1 \end{bmatrix},$$

$$\mathbf{R} = \frac{1}{2} \begin{bmatrix} r_\| r_\|^* + r_\perp r_\perp^* & r_\| r_\|^* - r_\perp r_\perp^* & 0 & 0 \\ r_\| r_\|^* - r_\perp r_\perp^* & r_\| r_\|^* + r_\perp r_\perp^* & 0 & 0 \\ 0 & 0 & 2\mathrm{Re}(r_\| r_\perp^*) & 2\mathrm{Im}(r_\| r_\perp^*) \\ 0 & 0 & -2\mathrm{Im}(r_\| r_\perp^*) & 2\mathrm{Re}(r_\| r_\perp^*) \end{bmatrix},$$

$$\mathbf{T} = \frac{1}{2} \begin{bmatrix} t_\| t_\|^* + t_\perp t_\perp^* & t_\| t_\|^* - t_\perp t_\perp^* & 0 & 0 \\ t_\| t_\|^* - t_\perp t_\perp^* & t_\| t_\|^* + t_\perp t_\perp^* & 0 & 0 \\ 0 & 0 & 2\mathrm{Re}(t_\| t_\perp^*) & 2\mathrm{Im}(t_\| t_\perp^*) \\ 0 & 0 & -2\mathrm{Im}(t_\| t_\perp^*) & 2\mathrm{Re}(t_\| t_\perp^*) \end{bmatrix}, \tag{7}$$

where η is the rotation angle and $r_\|$, $r_\perp$, $t_\|$, and $t_\perp$ are Fresnel's coefficients:

$$r_\| = \frac{m \cos \iota - \cos \tau}{m \cos \iota + \cos \tau}, \qquad r_\perp = \frac{\cos \iota - m \cos \tau}{\cos \iota + m \cos \tau},$$
$$t_\| = \frac{2 \cos \iota}{m \cos \iota + \cos \tau}, \qquad t_\perp = \frac{2 \cos \iota}{\cos \iota + m \cos \tau}. \tag{8}$$

Since the Mueller matrices in Eq. (7) interrelate flux densities that are not conserved, in practical ray tracing, energy conservation is established by renormalizing the refraction coefficients in $\mathbf{T}$ so that

$$|r_\||^2 + \frac{\mathrm{Re}(m^* \cos \tau)}{\cos \iota} |t_\||^2 = 1, \qquad |r_\perp|^2 + \frac{\mathrm{Re}(m \cos \tau)}{\cos \iota} |t_\perp|^2 = 1. \tag{9}$$

The internal incidence is treated analogously except that inside the particle, rays can be totally reflected and attenuated due to absorption. The condition for total internal reflection is $\sin \iota > 1/\mathrm{Re}(m)$, again depending on $\mathrm{Re}(m)$ only. Exponential absorption $\exp[-2\mathrm{Im}(m)k\Delta r]$ is assumed along the ray path (Δr is the path length).

To summarize, rays are traced until the flux decreases below a specific cutoff value or the ray has undergone a specific number of internal or exter-nal reflections. Scattered rays carry six Stokes vectors and the six separate sets of results are utilized to construct the geometric optics scattering phase matrix.

Note that the concept of localized rays allows for the application of radiative transfer theory to account for multiple-scattering processes inside the otherwise homogeneous host particle. This way, internal inhomogenieties like inclusions can be taken into account, as in [10] which introduce a geometric optics / Monte Carlo radiative transfer technique of light scattering by large particles with multiple internal inclusions. Applications of this concept can be found in [13] for planetary regolith particles and in [6] for cirrus clouds. Recently, [24] introduced diffuse Lambertian surface elements and internal screens to obtain a match between the geometric optics computations and the experimental data by [27] using G-spheres with power-law covariance functions.

2.2. DIFFRACTION

Owing to the wave nature of light, diffraction at the particles aperture must be accounted for and added to the geometrical part described above. If $P_{11}^{G}(\Theta)$ denotes the geometric optics part and $P_{11}^{D}(\Theta)$ the diffraction part, the total phase function including reflection/refraction and diffraction processes is given by

$$P_{11}(\Theta) = \frac{1}{\omega_0}\left[(2\omega_0 - 1)P_{11}^{G}(\Theta) + P_{11}^{D}(\Theta)\right] \qquad (10)$$

For light scattering studies, the far field approximation (Fraunhofer diffraction) can be applied to the general vector diffraction equation [2]. For the diffracted electric field it follows that

$$\underline{E}(\underline{r}) = \frac{ie^{ikr}}{2\pi r}\underline{k} \times \int_S \underline{n} \times \underline{E}(\underline{r}')e^{i\underline{k}\cdot\underline{r}}ds \qquad (11)$$

Here, $\underline{k}$ is the wave vector and $\underline{n}$ the normal vector of the surface S along which the integration has to be performed (see Fig. 3).

If we further assume a plane incoming wave that propagates perpendicularly to the surface, the diffraction contribution to the scattering phase functions can be obtained from Eq. (11) as

$$P(\theta, \phi) = \frac{k^2}{4\pi^2 S}\frac{\cos^2\theta(E_x^2 + E_y^2) + \sin^2\theta(E_x\cos\phi + E_y\sin\phi)^2}{E_x^2 + E_y^2} \cdot$$
$$\left|\int_S e^{ik\sin\theta(x\cos\phi + y\sin\phi)}dxdy\right|^2 \qquad (12)$$

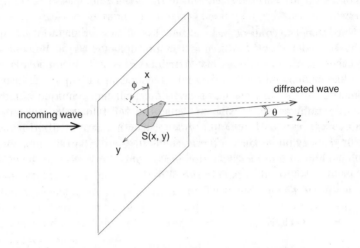

Figure 3.   Geometry of the diffraction problem.

E_x and E_y are the component of the electric field amplitude of the incoming wave. S denotes the surface value of the particle projection perpendicular to the direction of the incoming radiation.

The surface integral $F(S, k; \theta > 0, \phi)$ in eq. (12) can be evaluated by means of the Gaussian integral equation for plane surfaces

$$F(S, k; \theta, \phi) = \int_{\delta S} \frac{e^{ik \sin \theta (x \cos \phi + y \sin \phi)}}{2ik \sin \theta \cos \phi} dy - \int_{\delta S} \frac{e^{ik \sin \theta (x \cos \phi + y \sin \phi)}}{2ik \sin \theta \sin \phi} dx,$$
(13)

where the one dimensional integration is performed along the boundary of the crystal surface projection. In particular for pristine ice crystals, this boundary is described by a piecewise linear closed polygon along the horizontal plane perpendicular to the direction of the incident radiation, defined by some points $p^1, p^2, ..., p^{n+1}$ with $p^{n+1} = p^1$ with components p_x^i, p_y^i. For this situation the integral can be split into several integrals along linear paths that can be solved analytically:

$$F(S, k; \theta, \phi) = \frac{1}{2} \frac{1}{k^2 \sin^2 \theta} \sum_{j=1}^{n} \left(e^{ik \sin \theta (p_x^{j+1} \cos \phi + p_y^{j+1} \sin \phi)} - e^{ik \sin \theta (p_x^j \cos \phi + p_y^j \sin \phi)} \right).$$

$$\frac{(p_y^{j+1} - p_y^j)/\cos \phi - (p_x^{j+1} - p_x^j)/\sin \phi}{(p_x^{j+1} - p_x^j)\cos \phi + (p_y^{j+1} - p_y^j)\sin \phi},$$

$$\phi \neq 0, \tfrac{\pi}{2}, \pi, \tfrac{3\pi}{2}, 2\pi \qquad (14)$$

$$F(S, k; \theta) = \frac{1}{k^2 \sin^2 \theta} \cdot \left\{ \begin{array}{l} \sum_{j=1}^{n} \frac{(p_y^{j+1} - p_y^j)}{(p_x^{j+1} - p_x^j)} \left(e^{\pm ik \sin \theta p_x^{j+1}} - e^{\pm ik \sin \theta p_x^j} \right), \\ \phi = 0(+), \pi(-), 2\pi(+) \\ \sum_{j=1}^{n} \frac{(p_x^{j+1} - p_x^j)}{(p_y^{j+1} - p_y^j)} \left(e^{\pm ik \sin \theta p_y^{j+1}} - e^{\pm ik \sin \theta p_y^j} \right), \\ \phi = \frac{\pi}{2}(+), \frac{3\pi}{2}(-) \end{array} \right\}$$

(15)

Again by applying the Gaussian integral equation, the surface area of the crystal projection, i.e. the geometrical cross section A_g can be calculated analytically as

$$A_g = \frac{1}{2} \sum_{j=1}^{n} (p_x^j p_y^{j+1} - p_x^{j+1} p_y^j)$$

(16)

3. Accuracy of Geometric Optics

In the following, results are shown for the non-zero scattering phase matrix elements $(P_{11}(= a_1), p_{22} = P_{22}/P_{11}(= a_2/a_1), p_{33} = P_{33}/P_{11}(= a_3/a_1), p_{44} = P_{44}/P_{11}(= a_4/a_1), p_{12} = P_{12}/P_{11}(= -b_1/a_1), p_{34} = P_{34}/P_{11}(= b_2 a_1))$ for 3D randomly oriented particles. The discussion of the remaining phase matrix elements for particles with prefered 2D orientation is left for future work.

Since ROM becomes less valid as the size parameter (ratio of particle size and wavelength) decreases (cf. Eq. (1)), its application must be regarded as problematic for the smallest [almost all] ice crystals in the solar [thermal] spectral range. The error of ROM when applied to non-spherical particles is quantified in [9]. By comparing results from ROM and the exact T-matrix method [14], it was found that the approximation of ROM produces smaller errors in the case of non-spherical particles than in the case of surface- or volume-equivalent spheres. For moderately absorbing particles, the scattering properties are in good agreement for size parameter above 60, and the single scattering albedo (1 - absorptivity) for size parameter above 10. Later, [28] and [15] showed that non-absorbing particles reach a satisfying agreement above size parameter of about 120. As an example, Fig. 4 from [28] shows the scattering matrix elements resulting from T-Matrix and from ROM calculations for a randomly oriented non-absorbing circular ice cylinder at visible wavelengths with diameter-to-length ratio 1, surface-equivalent sphere size parameter 180, and refractive index 1.311. The agreement is very good except for the p_{22} and the p_{33} element in the backscattering region. Here, the geometrical optics reflection/refraction processes at the particles edges may be in error due to the small distances

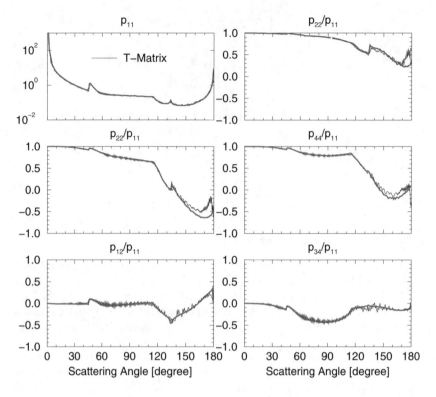

Figure 4.   Scattering matrix elements for randomly oriented circular cylinders with diameter-to-length ratio 1, surface-equivalent sphere size parameter 180, and refractive index 1.311. Thin curves show T-Matrix computations, thick curves represent ray-tracing results.

between neighbored particle edges. However, the scattering intensity (p_{11}) is hardly affected by this.

4. Particle geometries

4.1. PRISTINE CRYSTALS

As mentioned in section 1, ice crystals often obey a certain hexagonal macroscopic structure. Still, the hexagonal symmetry allows for a large variety of particle shapes. As an example Fig. 5 shows the phase matrix elements for 3D randomly oriented hexagonal columns, plates, and dendrites at a wavelength of 550 nm. The notation in parantheses is used in several books on polarized light scattering (e.g. [16])

In addition, the same matrix elements for randomly horizontally oriented ice columns are shown. Such preferred orientations are frequently

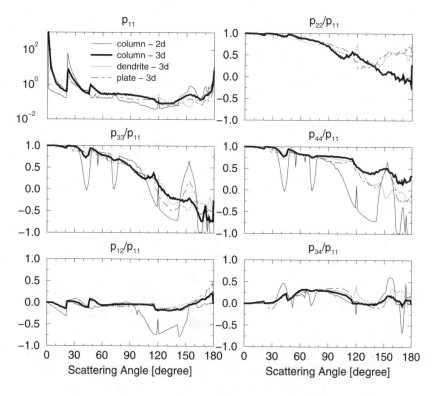

Figure 5. Scattering phase matrix elements for 3D randomly oriented hexagonal columns, plates, dendrites, and for hexagonal columns randomly oriented in the horizontal plane.

observed from Radar and Lidar measurements (specular reflections). The ice columns are illuminated perpendicular to the columnar axis (symmetry axis). This idealized case also is shown to demonstrate the differences between 3D and 2D random orientation. In fact, the latter allows for a larger number of characteristic ray paths that show up as halo features both in the scattering phase function and in the polarization properties. Examples are the 120°peak that arises from total internal reflections, and the much stronger pronounced 22°halo. In 3D random orientation, these features are superimposed by the results of various other ray paths and washed out, becoming less pronounced, if not completely averaged out.

4.2. IRREGULARLY SHAPED PARTICLES

Because many light-scattering particles are irregular in shape, the question arises as to how to represent the geometry of such particles. Here we dis-

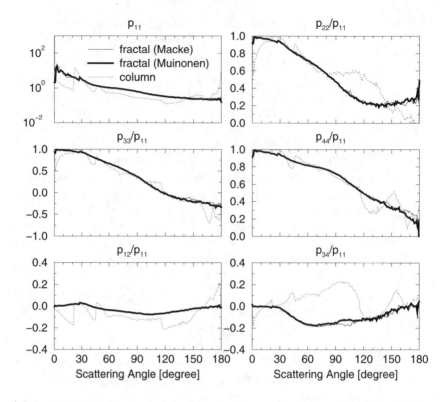

Figure 6. ROM results for the scattering phase matrix (excluding diffraction) of a 3D randomly oriented Koch fractals and hexagonal columns at a wavelength of 0.5 μm. The results for the very same fractal have been calculated with the ray-tracing codes by Macke and by Muinonen.

cuss two approaches, the Gaussian random spheres introduced in [23] and the fractal polycrystal or Koch fractal suggested in [11]. Figure 6 shows a comparison between ROM results (excluding diffraction) from Macke and from Muinonen, i.e. by two almost independent programs. The excellent agreement confirms the correct realisations of the ray-tracing procedure in both codes.

A comparison between Figs. 5 and 6, i.e. between symmetric and irregularly shaped particles shows that the latter obey much smoother angular dependencies due to the absence of characteristic ray paths like halo features. From a polarimetric point of view, some scattering angle ranges exist, were a clear distinction between the two particle geometries can be achieved. This is pronounced most strongly for the p_{22} and the p_{34} phase matrix elements.

The degree of polarization of unpolarized incoming radiation (as it is the case for solar irradiation) is equal to $-p_{12}$. Regularly shaped particles like hexagonal columns allow for strong polarizing predominant ray paths like halos and retroreflection. These features are absent for the fractal polycrystal particle, of course.

The phase matrix element p_{22} is equal to 1 for spherical particles. Therefore, any deviations from 1 are a clear indication of particle nonsphericity. Interestingly, the deviations from unity are largest at backscattering (scattering angle = 180°). Also the differences between columns and Koch fractals are almost largest here. Thus, lidar measurements of the p_{22} element not only allows for a distinction between spheres and nonspherical particle, but also between regularly and irregularly shaped particles.

The p_{34} element takes small negative values for the irregular particle and small positive values for the hexagonal particle at most scattering angles. This element describes the efficiency of the transformation of linearly polarized light to a circularly polarized mode ([5]). Obviously, such a transformation is less likely for the irregularly shaped particle. Further discussions of the phase matrix elements for various particle shapes and refractive indices can be found in [5].

In order to compare the scattering characteristics of Gaussian particles to those of the pristine crystals and fractal polycrystals, we compute scattering phase matrix elements for Gaussian random particles with standard deviations $\sigma = 0.05$, 0.1, and 0.2 and the power-law indices $\nu = 2$ and $\nu = 4$ of the correlation function using refractive index $m = 1.311$. The maximum degree of the spherical harmonics and Legendre expansions was set to $l_{\max} = 10$. For infinite series expansions, the index $\nu = 4$ corresponds to fractal dimension $D = (8 - \nu)/2 = 2$, that is, a smooth particle, whereas $\nu = 2$ corresponds to fractal dimension $D = 3$ and thus to an extremely irregular, non-differentiably Gaussian surface.

The results are shown in Fig. 7 for the power-law index $\nu = 2$. Those for $\nu = 4$ reveal similar overall features that are closer to the results for spherical particles. All of the scattering phase matrix elements have a fluctuating angular dependence due to the differing proportions of externally reflected rays and rays that have undergone two refractions and a varying number of internal reflections. Note, in particular, how the minima of P_{11} correspond accurately to the maxima of $-P_{21}/P_{11}$.

It is slightly surprising that the overall characteristics of the scattering phase matrix elements are in gross general agreement with the characteristics for the pristine crystals and the fractal polycrystal. As the single common parameter is the refractive index, it would be tempting to conclude that the overall characteristics can provide hints about the refractive index independently from the detailed nonspherical particle shapes. Finally, the results for the Gaussian particles with $\sigma = 0.2$ and $\nu = 2$ are particularly

A. MACKE[1] AND K. MUINONEN[2]

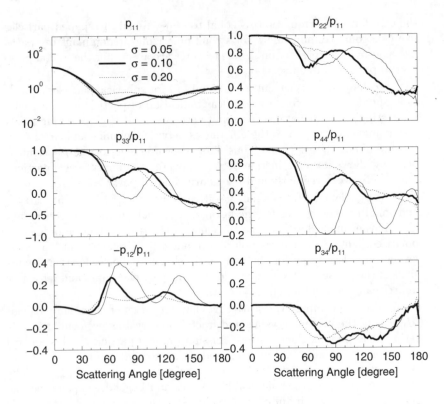

Figure 7. Phase matrix elements for the random Gaussian spheres.

similar to the results for the fractal polycrystal.

5. Particle orientations

Ice particles in cirrus clouds may show certain preferred orientations depending on the particle geometry, size, and environmental conditions. In fact, a number of fascinating halo phenomena like subsuns and various arcs can be explained only by a predominant horizontal particle orientation. Of course, the orientation is seldom perfectly horizontal. Even in calm air, the falling crystals are subject to turbulent flow around the particle that causes tumbling. Figure 8 shows the scattering phase matrix elements of such tumbling particles with degrees of maximum tilt angle around the horizontal direction equal to $10°$, $20°$, and $40°$.

The single scattering properties of particles that are not 3D randomly oriented become bidirectional, i.e. the phase matrix depends on both the direction of illumination and the direction of scattering. For the results shown

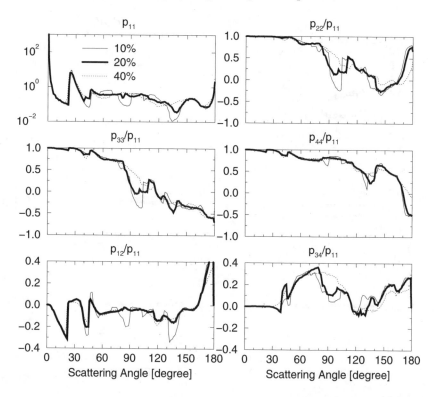

Figure 8. Phase matrix elements for a hexagonal ice column with 10, 20, and 40°random deviation from perfect horizontal orientation.

in Fig. 8 a solar zenith angle of 60°has been selected. The dependency on the direction of incident light is very pronounced. However, a full discussion of this bidirectionality is beyond the scope of the present chapter.

Figure 8 clearly shows that particle orientation has a strong effect on the scattered intensity and polarization pattern, at least for scattering angles larger than about 50°. The P_{22} and the P_{33} elements show a strong signal at about 100°scattering angles that changes from negative to positive values as the orientation changes from 10°to 40°, i.e. as the particle loses its preferential orientation. Thus, satellite observation of the polarized reflected solar radiation at specific viewing angles may provide a passive detection of the degree of particle orientation.

As mentioned above, a more direct way of measuring particle orientation results from active, i.e. from lidar and radar based backscattering measurements. Figure 9 shows the size-distribution-averaged phase matrix elements at a visible non-absorbing wavelength in the backscattering direction as a

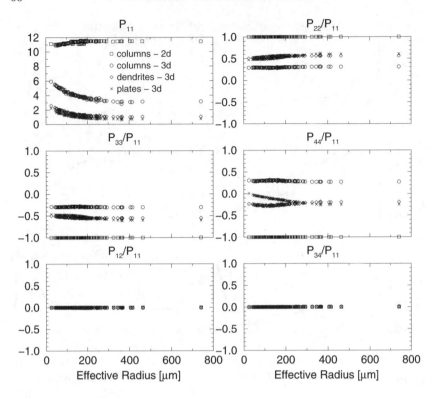

Figure 9.   Backscattering phase matrix elements for 3D randomly oriented hexagonal columns, plates, dendrites, and for hexagonal columns randomly oriented in the horizontal plane.

function of the effective radius (see [4] for a definition of effective particle radius and application in radiative transfer) that characterizes the individual size distributions under consideration. The same particle shapes have been used as in Fig. 5, i.e. 2D and 3D oriented columns and 3D oriented plates and dendrites. The size distributions used for this study were taken from aircraft measurements during several field campaigns of the European Cloud and Radiation Experiment EUCREX [26].

It must be noted that classical geometric optics does not provide scattering matrix elements at discrete directions but only for finite-sized scattering-angle bins. The backscattering phase matrix elements shown in Fig. 9 have been calculated for the angular bin [179.95°, 180°]. i.e. for a bin size of 0.05°.

The backscattering intensity of the horizontally oriented particle is much stronger than that of its 3D oriented counterparts. This is due to the stronger likelihood for retroreflection for this illumination condition and particle geometry. The direction of the incident light is perpendicular to the symmetry axis of the hexagonal cylinder, so that the light rays are bound to the hexagonal plane. In this geometry multiple total internal reflection at the 60°prisms can lead to strong backscattering at exactly 180°scattering angle.

Perfectly horizontally oriented plates provide a δ-function type of phase function that peaks at exactly 0°and 180°scattering angle and is zero everywhere else. For this reason, results for 2D oriented plates are not shown.

As the size of the ice particle increases, the ratio of the longest to the smallest dimension (aspect ratio) also becomes larger [1]. This in turn decreases the probability for retroreflection at 90°prisms for the 3D randomly oriented particles. Accordingly, the backscattering intensity decreases with effective radius for these particle orientations. For the horizontally oriented columns, the light rays do not interact with the hexagonal base planes, and the intensity increases with size.

Different particle geometries show up in the other diagonal elements of the backscattering phase matrix, but not in the two off-diagonal elements p_{12} and p_{34}.

A combination of measuring intensity and polarization of backscattered laser light may allow us to distinguish between particle orientation and particle types.

6. Summary and Conclusion

The Ray Optics Method offers a conceptually easy and widely applicable method for studying the problem of polarized light scattering by nonspherical particles large compared to the wavelength of the incident radiation. This method can be extended easily to account for randomly tilted particle facets, internal scattering and absorbing inhomogeneities. There is free software available for ray-optics computations [8, 21].

The scattering phase matrices highlighted in the present chapter are strongly different from those for spherical particles. However, there are hints about overall similarities for Gaussian, hexagonal, and fractal polycrystal particles that can be due to a single physical parameter, the refractive index of the particles. We close with a question amplified by the present work: for a given refractive index, is there an "asymptotic attractor" scattering phase matrix to which the ray optics computations converge for extreme shape irregularity, independent of the detailed modeling of the irregularity?

7. Acknowledgments

AM thanks Ingo Schlimme and Susann Klotzsche for performing ROM calculations for pristine ice crystals. The work of Ingo Schlimme was supported by ESA/ESTEC in the frame of the EarthCare algorithm study and by GKSS Research Center. KM is grateful to the Academy of Finland for partial funding of his work.

References

1. Auer, A. H. and D. L. Veal, 1970: The dimensions of ice crystals in natural clouds. *J. Atmos. Sci*, **27**, 919–926.
2. Born, M. and E. Wolf, 1999: *Principles of Optics*. Cambridge University Press.
3. Grynko, Y. and S. Yu, 2003: Scattering matrix calculated in geometric optics approximation for semitransparent particles faceted with various shapes. *J. Quant. Spectros. Radiat. Transfer*, **78**, 319–340.
4. Hansen, J. E. and L. D. Travis, 1974: Light scattering in planetary atmospheres. *Space Sci. Rev.*, **16**, 527–610.
5. Kokhanovsky, A. A., 2003: *Polarization Optics of Random Media*. Springer.
6. Labonnote, L., G. Brogniez, J. Buriez, M. Doutriaux-Bouchert, J. Gayet, and A. Macke, 2001: Polarized light scattering by inhomogeneous hexagonal monocrystals. validation with ADEOS-POLDER measurements. *J. Geophys. Res.*, **106**(D11), 12,139–12,153.
7. Mach, E., 1953: *The Principles of Physical Optics. A Historical and Philosophical Treatment*. Dover Publications, New York (reprint).
8. Macke, A., 1998: free ray optics code. http://www.ifm.uni-kiel.de/fb/fb1/me/research/Projekte/RemSens/SourceCodes/codes.html. 24.02.2004.
9. Macke, A., M. I. Michshenko, K. Miunonen, and B. E. Carlson, 1995: Scattering of light by large nonspherical particles: ray tracing approximation versus T-matrix method. *Optics Letters*, **20**, 1934–1936.
10. Macke, A., M. I. Mishchenko, and B. Cairns, 1996a: The influence of inclusions on light scattering by large ice particles. *J. Geophys. Res.*, **101**, 23,311–23,316.
11. Macke, A., J. Müller, and E. Raschke, 1996b: Single scattering properties of atmospheric ice crystals. *J. Atmos. Sci*, **53**, 2813–2825.
12. Miloshevich, L., H. Voemel, A. Paukkunen, A. Heymsfield, and S. Oltmann, 2001: Characterization and correction of relative humidity measurements from Vaisala RS80-A radiosondes at cold temperatures. *JAOT*, **18**, 135–156.
13. Mishchenko, M. and A. Macke, 1997: Asymmetry parameters of the phase function for isolated and densely packed spherical particles with multiple internal inclusions in the geometrics optics limit. *J. Quant. Spectros. Radiat. Transfer*, **57**, 767–794.
14. Mishchenko, M. I., 1993: Light scattering by size-shape distribution of randomly oriented axially symmetric particles of a size comparable to a wavelength. *Appl. Opt.*, **32**, 4652–4666.
15. Mishchenko, M. I. and A. Macke, 1999: How big should ice crystals be to produce haloes. *Appl. Opt.*, **38**(9), 1626–1629.
16. Mishhchenko, M. I., J. W. Hovenier, and L. D. Travis, eds., 2000: *Light Scattering by Nonspherical Particles. Theory, Measurements and Applications*. Academic Press.
17. Modest, M. F., 1993: *Radiative Heat Transfer*. McGraw-Hill, New York.
18. Muinonen, K., 1998: Introducing the gaussian shape hypothesis for asteroids and comets. *Astronomy and Astrophysics*, **332**, 1087–1098.
19. Muinonen, K., 2002: Light-scattering approximations for small irregular particles. pp. 219–222. Army Research Laboratory, Adelphi, Maryland, U.S.A.
20. Muinonen, K., 2003: Light scattering by tetrahedral particles in the kirchhoff approximation. In: T. Wriedt, ed., *Electromagnetic and Light Scattering — Theory and Applications VII*, pp. 251–254.
21. Muinonen, K., 2004: free ray optics code. http://www.astro.helsinki.fi/~psr/ (free software Siris). 24.02.2004.

22. Muinonen, K., L. Lamberg, P. Fast, and K. Lumme, 1997: Ray optics regime for gaussian random spheres. *J. Quant. Spectros. Radiat. Transfer*, **57**, 197–205.
23. Muinonen, K., T. Nousiainen, P. Fast, K. Lumme, and J. I. Peltoniemi, 1996: Light scattering by Gaussian random particles: ray optics approximation. *J. Quant. Spectros. Radiat. Transfer*, **55**, 577–601.
24. Nousiainen, T., K. Muinonen, and P. Räisänen, 2003: Scattering of light by large saharan dust particles in a modified ray-optics approximation. *J. Geophys. Res.*, **108**(D1), 4025, doi:10.1029/2001JD001277.
25. Peltoniemi, J., 1993: Radiative transfer in stochastically inhomogeneous media. *J. Quant. Spectros. Radiat. Transfer*, **50**, 655–671.
26. Raschke, E. and Co-authors, 1996: European Cloud and Radiation Experiment (EU-CREX). Final report on the project EV5V-CT-0130 EUCREX-2. Available at GKSS Research Center, Postfach 1160, D-21494 Geesthacht. Pp154.
27. Volten, H., O. Muñoz, J. F. de Haan, W. Vassen, J. Hovenier, K. Muinonen, and T. Nousiainen, 2001: Scattering matrices of mineral aerosol particles at 441.6 nm and 632.8 nm. *J. Geophys. Res.*, **106**, 17375–17401.
28. Wielaard, D. J., M. I. Mishchenko, A. Macke, and B. E. Carlson, 1997: Improved T-matrix computations for large, nonabsorbing and weakly absorbing nonspherical particles and comparison with geometric optics approximation. *Appl. Opt.*, **36**, 4305–4313.

Andreas Macke while preparing parts of the manuscript.

Midnight photo-shoot at the Swallow's Nest

Tracy Smith, Hester Volten
and Karri Muinonen in Livadia Palace

Marina Prokopjeva
Daria Dubkova
in Sevastopol.

OPTIMIZATION OF NUMERICAL INVERSION IN PHOTOPOLARIMETRIC REMOTE SENSING

OLEG DUBOVIK[1,2]

[1]*Laboratory for Terrestrial Physics, Code 923, NASA Goddard Space Flight Center, Greenbelt, Maryland 20771 USA*

[2]*Also at Goddard Earth Science and Technology Center, University of Maryland Baltimore County, 1000 Hilltop Circle, Baltimore, Maryland 21250 USA*

Abstract. Remote sensing is one primary tool for studying the interactions of solar radiation with the atmosphere and surface and their influence on the Earth radiation balance. During the past three decades the radiation measured from satellite, aircraft and ground have been employed successfully for characterizing radiative properties of land, ocean, atmospheric gases, aerosols, clouds, etc. One of the challenges in implementing remote sensing is the development of a reliable inversion procedure required for deriving information about the atmospheric or surface component interaction with the measured radiation. The inversion is particularly crucial and demanding for interpreting high complexity measurements where many unknowns should be derived simultaneously. Therefore the deployment of remote-sensing sensors of the next generation with diverse observational capabilities inevitably would be coupled with significant investments into inverse-algorithm development. Numerous publications offer a wide diversity of inversion methodologies suggesting somewhat different inversion methods. Such uncertainty in methodological guidance leads to excessive dependence of inversion algorithms on the personalized input and preferences of the developer. This study is an attempt to outline unified principles addressing such important aspects of inversion optimization as accounting for errors in the data used, inverting multi-source data with different levels of accuracy, accounting for *a priori* and ancillary information, estimating retrieval errors, clarifying potential of employing different mathematical inverse operations   (e.g. comparing iterative versus matrix inversion), accelerating iterative convergence, etc. The described concept uses the principles of statistical estimation and suggests a generalized multi-term least-square-type formulation

G. Videen et al. (eds.), Photopolarimetry in Remote Sensing, 65-106.

that complementarily unites advantages of a variety of practical inversion approaches, such as *Phillips-Tikhonov-Twomey* constrained inversion, *Kalman filters*, *Gauss-Newton* and *Levenberg-Marquardt* iterations, etc. The proposed methodology has resulted from the multi-year efforts of developing inversion algorithms for retrieving comprehensive aerosol properties from ground-based remote sensing observations.

1. Introduction

For the last few decades remote sensing has provided the scientific community with the global distribution of climatically important information about radiative properties of the Earth atmosphere and surface. Future expectations are increasingly high, because remote sensing still has significant potential in improving the volume and accuracy of retrieved information. The indirect nature of observations is an inherent feature of remote-sensing measurements. Indeed, the atmospheric radiation measured from space, ground, etc. is a result of complex interactions of incident solar light with atmospheric components and surface scattering and absorbing radiation. Retrieving optical and radiative properties of natural objects from radiation measurements demands two types of development efforts. First, a capability of modeling atmospheric characteristics is required. That capability is vital for building a so-called *"forward model"* retrieval algorithm that adequately simulates the measured atmospheric radiation coming from the atmospheric or surface objects with known properties. The second necessary component of retrieval is the so-called *"inversion"* procedure that utilizes an inverse transformation by recovering unknown *input* parameters of the *forward model* from known *output* of the *forward model*. Investing in a particular atmospheric remote-sensing approach is motivated by the achievements in atmospheric radiation modeling. Therefore, in remote sensing applications the *forward-model* development usually is feasible and the main challenge is finding the most accurate model satisfying time constraints of operational processing. On the other hand, establishing a strategy for developing the best inversion method is a more convoluted task, in that the evaluation of inversion accuracy is an ambiguous question, especially for the case of simultaneous retrieval of many unknowns; - for example, replacing a scalar light-scattering model with a vector one that accounts for polarization results in accuracy improvements in scattered light reproduction. In contrast, identifying a preference between inversion methods is always rather uncertain. A change of inversion scheme in practical multi-parametric retrieval usually is accompanied by rather complex consequences: retrieval accuracy may improve for one parameter but degrade for another parameter and that situation may alter for different observation configurations and circumstances. Hence, identifying a preferable inversion method from comparative tests is not always fruitful and should rely on consideration of rather fundamental principles of inversion optimization. However, existence of a very broad diversity of inversion methodologies leaves the researcher freedom in implementing the actual retrieval. Indeed, there are numerous publications describing different inversion techniques and procedures. On the other hand, the comparisons of different inverse methodologies are rather sparse and often limited to a particular application. Consequently, anyone presently designing a practical retrieval algorithm has to review rather fundamental principles of inversion optimization and make a number of principal decisions and choices in inversion implementation that largely predetermine the successes and limitations of the resulting retrieval. Obviously such "personalization" of inversion implementation raises ambiguity and diversification of retrieval development. For that reason, this study is aimed at analyzing the main principles of inversion optimization with an attempt to outline generalized guidance for inversion development in remote sensing. The considerations and results presented are based on the multi-year efforts of developing inversion algorithms for retrieving comprehensive properties of atmospheric aerosol from light-scattering observations. The proposed concept pursues the idea of establishing a unified formulation combining complementary principles of different inversion approaches.

Detailed reviews of inversion methods can be found in various textbooks [1-6]. However, the details given and descriptions of well established inversion procedures do not provide the reader with sufficient explanations as to which method and why it should be chosen for a particular application. Such a situation is partly a result of the fact that most innovations were proposed under pressure of different specific practical needs and, therefore, derived in scopes of rather different approaches. The inversion strategy described here was proposed and refined in the previous studies [7-9]. The approach is focused on clarifying the connection between different inversion methods established in atmospheric optics and unifying the key ideas of these methods into a single inversion procedure. This strategy is expected to be helpful for building optimized and flexible inversion techniques inheriting benefits from of a variety of methods well established in different applications. For example, considerations of this chapter reveal important connections of retrieval algorithms designed with the inversion methods widely adopted in atmospheric remote sensing and other geophysical applications, such as the methods given by *Kalman* [10], *Phillips* [11], *Tikhonov* [12], *Twomey* [13,14], *Strand and Westwater* [15-16], *Chahine* [17], *Turchin and Nozik* [18], *Rodgers* [19], etc.

Following the elaborations by study [9], the following aspects of inversion optimization will be outlined in the order of their importance and validity, starting from most important and most proven: (i) Optimizing the algorithm to the presence of measurement errors; (ii) Optimizing inclusion of *a priori* and ancillary data; (iii) improving performance of key mathematical operations (linear inversion, non-iteration convergence, etc.); and (iv) adjusting conventional assumption of noise-distribution to account for parameter non-negativity and data redundancy. Each of these aspects is discussed in numerous theoretical and practical studies. However, as a rule, theoretical analyses of inversions overemphasize single aspects of retrieval optimization and therefore the resulting conclusions have limited applicability. This study pursues the idea of formulating an inversion procedure based on harmonized consideration of different aspects of algorithm performance rather than on opposing one principle against another. With this purpose, the present chapter outlines the importance of addressing all above-listed aspects by specifying the role of each optimization principle in the development of successful inversion.

2. Basic inversions of linear systems

Commonly, remote sensing methods are set up to derive N_a unknown parameters a_i from N_f discrete observations f_j and a corresponding retrieval algorithm should solve the following system of equations:

$$\mathbf{f}^* = \mathbf{f(a)} + \geq \mathbf{f}^*, \tag{1}$$

where $\mathbf{f}^*$ is a vector of the measurements f_j, $\geq \mathbf{f}^*$ is a vector of measurement errors $\geq f_j^* = f_j^* - f_j^{real}$, $\mathbf{a}$ is a vector of unknowns a_i, $\mathbf{f(a)}$ denotes a physical forward model that allows adequate simulations of observations f_j from predefined parameters a_i. In remote-sensing applications, $\mathbf{f}^*$ usually includes the atmospheric radiation measurements conducted from ground, satellite or aircraft using detectors with various spectral, angular and polarimetric capabilities. The vector of unknowns $\mathbf{a}$ may include various parameters describing the optical properties of atmospheric or surface compounds, such as concentrations of gases and their vertical distributions, parameters describing composition and size distribution of aerosol, land or ocean reflectance, etc. Correspondingly, $\mathbf{f(a)}$ is usually modeled by solving the radiative-transfer equation accounting for transformations of solar radiation interacting with the atmosphere and the surface. Such physical models $\mathbf{f(a)}$ do not have an analytical inverse transformation and the system of Eq. (1) should be solved numerically. For example, in the simplest physical model $\mathbf{f(a)}$ with characteristics $\mathbf{f}_j$ being linearly dependent on $\mathbf{a}_i$ (i.e., $f_i = \geq_{i=1...Na} K_{ji} a_i$), Eq. (1) is reduced to the system of linear equations:

$$\mathbf{f}^* = \mathbf{f} + \geq \mathbf{f}^* = \mathbf{K\,a}, \tag{2}$$

where $\mathbf{K}$ is the matrix of the coefficients K_{ji}. If the number of measurements is equal to the number of unknowns, the solution of Eq. (2) is straightforward:

$$\hat{\mathbf{a}} = \mathbf{K}^{-1} \mathbf{f}^* \qquad (N_f = N_a), \qquad (3)$$

where $\mathbf{K}^{-1}$ denotes the inverse matrix operator. For the matrix $\mathbf{K}$ with the linearly independent and non-zero rows, Eq. (3) gives a unique solution that always provides the equality of the left and right sides in Eq. (2), i.e. $\mathbf{f}^* = \mathbf{K}\,\hat{\mathbf{a}}$.

Equation (2) also can be solved by other methods without direct implementation of matrix inverse transformation $\mathbf{K}^{-1}$, for example, by means of linear iterations:

$$\mathbf{a}^{p+1} = \mathbf{a}^p - \mathbf{H}_p(\mathbf{K}\,\mathbf{a}^p - \mathbf{f}^*). \qquad (4)$$

There are a number of the methods that use linear iterations, for example, the known *Jacobi* and *Gauss-Seidel* techniques, steepest descent method, etc. differing by the definition of matrix $\mathbf{H}_p$. This matrix should provide convergence of the iterations to a solution $\mathbf{a}^{p+1} \geq \hat{\mathbf{a}}$ attaining equality in Eq. (2):

$$\mathbf{K}\,\mathbf{a}^{p+1} - \mathbf{f}^* = \mathbf{K}\,(\mathbf{a}^p - \mathbf{H}_p\,(\mathbf{K}\,\mathbf{a}^p - \mathbf{f}^*)) - \mathbf{f}^* = (\mathbf{I} - \mathbf{K}\mathbf{H}_p)\,(\mathbf{K}\,\mathbf{a}^{p-1} - \mathbf{f}^*)$$
$$= (\mathbf{I} - \mathbf{K}\mathbf{H}_p)\,(\mathbf{I} - \mathbf{K}\mathbf{H}_{p-1})...(\mathbf{I} - \mathbf{K}\mathbf{H}_0)\,(\mathbf{K}\,\mathbf{a}^0 - \mathbf{f}^*) \geq 0 \quad (\text{for } p \geq \ \ ), \qquad (4a)$$

where $\mathbf{I}$ is the unity matrix. Thus, the iterations converge from any initial guess $\mathbf{a}^0$ to $\hat{\mathbf{a}}$ if the following sequential transformation leads to a zero matrix:

$$(\mathbf{I} - \mathbf{K}\mathbf{H}_p)\,(\mathbf{I} - \mathbf{K}\mathbf{H}_{p-1})...(\mathbf{I} - \mathbf{K}\mathbf{H}_0) \geq 0 \qquad (\text{for } p \geq \ \ ). \qquad (4b)$$

Obviously, if $\mathbf{H}_p = \mathbf{K}^{-1}$, Eq. (4) converges at the first iteration and is fully identical to Eq. (3). An important advantage of iterative techniques is that iterations are stable even if Eq. (2) does not have a unique solution. Indeed, if the matrix $\mathbf{K}$ has linearly dependent rows, applying Eq. (2) is problematic, since matrix $\mathbf{K}^{-1}$ does not exist. Iterating Eq. (4) works even in such a case, with the difference that use of Eq. (4) leads to one of many possible solutions that depend on initial guess.

Equation (2) also can be solved by other methods (see [4]) technically different from Eqs. (3)- (4). All these methods are equivalent in the sense that they lead to the same solution $\hat{\mathbf{a}}$ providing the equality in Eq. (2). Therefore, depending on the developer's preference and the requirements of the particular application, any of these methods can be employed in the retrieval algorithm (see discussion in Section 4.8).

3.   Solution optimization in presence of measurement errors

In many remote-sensing applications the number of measurements N_f exceeds the number of retrieved parameters N_a. This is characteristic of new advanced sensors with multi-spectral, multi-angle [20-23, etc.], and polarimetric capabilities [24-26, etc.]. In addition to the increased volume of physical information, this redundancy ($N_f > N_a$), allows for minimization of retrieval errors in the presence of random noise in the measurement.

The errors $\geq f^*$ in Eq. (2) may have the following two components:

$$\geq f^* = \geq f^*_{sys} + \geq f^*_{ran}, \tag{5}$$

where $\geq f^*_{sys}$ – *systematic errors*, which are repeatable in different measurement realizations and $\geq f^*_{ran}$ – *random errors*, which differ in the different measurements, i. e.:

$$<\geq f^*_{sys}> = b \geq 0 \qquad \text{and} \qquad <\geq f^*_{ran}> = 0, \tag{6}$$

where $<...>$ denotes averaging over measurement realizations, b is the average systematic error or so-called *bias*. The correction of the measured data for the *bias* is straightforward provided that b is identified and evaluated. The correction of the measurements for random errors is not possible, because their values are unpredictable in each individual act of measurement. Nevertheless, the statistical properties of the random errors can be used to improve the statistical properties of the retrievals.

The statistical properties of random errors are characterized by $P(\geq f^*)$ - *Probability Density Function* (PDF) that provides the probabilities of observing various realizations of the errors $\geq f^* = f^* - f^{real}$. The retrieved estimates should be close to the real values of unknowns, i.e. $\hat{a} \Delta a^{real}$. Using an adequate forward model [$f^{real} = f(a^{real})$] the errors $\geq f^*$ can be modeled as

$$\geq f^* = f^* - f(a^{real}) \Delta \geq \hat{f}^* = f^* - f(\hat{a}). \tag{7}$$

The known properties of the PDF can be used to improve the solution $\hat{a}$. Indeed, the modeled measurement errors $\geq \hat{f}^* = f^* - f(\hat{a})$ for $\hat{a} \Delta a^{real}$ should reproduce the known statistical properties of measurement errors as closely as possible. The agreement of modeled $\geq \hat{f}^*$ with known error distribution can be evaluated using the known PDF as a function of modeled errors $P(\geq \hat{f}^*)$: the higher $P(\geq \hat{f}^*)$ the closer the modeled $\geq \hat{f}^*$ to the known statistical properties. Thus, the best solution $\hat{a}^{best}$ should result in modeled errors corresponding to the most probable error realization, i.e. to PDF maximum:

$$P(\geq \hat{f}^*) = P(f(\hat{a}) - f^*) = P(f(\hat{a}) \geq f^*) = max. \tag{8}$$

In essence, this principle is the well-known *Method of Maximum Likelihood* (MML). The PDF written as a function of measurements $P(f(\hat{a}) \geq f^*)$ is called *Likelihood Function*. The MML is one of the strategic principles of statistical estimation that provides statistically the best solution in many senses [27]. For example, the asymptotical error distribution (for infinite number of $\geq f^*$ realizations) of MML estimates $\hat{a}$ have the smallest possible variances of $\geq \hat{a}_i$.

Most statistical properties of the MML solution remain optimum for a limited number of observations [27].

The implementation of MML in the actual retrieval requires an assumption on PDF of errors $\geq \mathbf{f}^*$. The normal (or *Gaussian*) function is most appropriate for describing random noise resulting from numerous additive factors:

$$P\left(\mathbf{f(a)}|\mathbf{f}^-\right) = \exists 2-)^m \det(\mathbf{C})\bar{-}^{-1/2} \exp\bar{-}\frac{1}{-2}\left(\mathbf{f(a)}-\mathbf{f}^-\right)^T \mathbf{C}^{-1}\left(\mathbf{f(a)}-\mathbf{f}^-\right)\bar{-}, \quad (9)$$

where T denotes matrix transposition, $\mathbf{C}$ is the covariance matrix of the vector $\mathbf{f}^*$; $\det(\mathbf{C})$ denotes the determinant of $\mathbf{C}$, and m is the dimension of the vectors $\mathbf{f}$ and $\mathbf{f}^*$. Detailed discussions on the reasoning for using a normal PDF as the best noise assumption can be found in many textbooks [e.g. 3,27].

The maximum of the PDF exponential term in Eq. (9) corresponds to the minimum of the quadratic form in the exponent. Therefore, the MML solution is a vector $\hat{\mathbf{a}}$ corresponding to the minimum of the following quadratic form:

$$* (\mathbf{a}) = \frac{1}{2}(\mathbf{f(a)} * \mathbf{f}^*)^T \mathbf{C}^{*1}(\mathbf{f(a)} * \mathbf{f}^*) = \min. \quad (10)$$

Thus, with the assumption of normal noise $\geq \mathbf{f}^*$, the MML principle requires the search for a minimum in the product of the squared terms of $(\mathbf{f(a)} - \mathbf{f}^*)$ in Eq. (10). This is the basis for the widely known *Least Square Method* (LSM). The minimum of the quadratic form $\Psi(\mathbf{a})$ corresponds to a point with a zero gradient $\Psi\Psi(\mathbf{a})$, i.e. to a point where all partial derivatives of $\Psi(\mathbf{a})$ are equal to zero:

$$\partial\partial \ (\mathbf{a}) = \frac{\partial\partial \ (\mathbf{a})}{\partial a_i} = \mathbf{0}, \quad (i = 1,..,N_a). \quad (11a)$$

The gradient of $\Psi(\mathbf{a})$ can be written as (detailed derivations can be found elsewhere [28-29, etc.]):

$$\geq\geq (\mathbf{a}) = \mathbf{K}_a^T\mathbf{C}^{-1} \mathbf{f(a)} - \mathbf{K}_a^T\mathbf{C}^{-1} \mathbf{f}^* = \mathbf{K}_a^T\mathbf{C}^{-1} (\mathbf{f(a)} - \mathbf{f}^*). \quad (11b)$$

$\mathbf{K}_a$ is a matrix of first partial derivatives in vicinity of $\mathbf{a}$, i.e. $\{\mathbf{K}_a\}_{ji} = \partial f_j/\partial a_i\big|_{\mathbf{a}}$. Correspondingly, for linear forward models, $(\mathbf{f(a)} = \mathbf{K} \ \mathbf{a})$, Eq. (11a) is equivalent to the following system:

$$\mathbf{K}^T\mathbf{C}^{-1} \mathbf{K} \mathbf{a} = \mathbf{K}^T\mathbf{C}^{-1} \mathbf{f}^*. \quad (11c)$$

Using matrix inversion, the LSM solution can be written as

$$\hat{\mathbf{a}} = (\mathbf{K}^T\mathbf{C}^{-1} \mathbf{K})^{-1} \mathbf{K}^T\mathbf{C}^{-1} \mathbf{f}^*. \quad (12)$$

This formula is valid if Eq. (2) has a unique solution, i.e. if $\det(\mathbf{K}^T \mathbf{C}^{-1} \mathbf{K}) \geq 0$. The errors $\geq \hat{\mathbf{a}}$ of the estimate $\hat{\mathbf{a}}$ are normally distributed and have random and systematic components resulting from $\geq \mathbf{f}^*_{sys}$ and $\geq \mathbf{f}^*_{ran}$ in the measurements:

$$\geq \hat{a} = \geq \hat{a}_{ran} + \geq \hat{a}_{sys} = (K^T C^{-1} K)^{-1} K^T C^{-1} (\geq f^*_{ran} + \geq f^*_{sys}). \qquad (13)$$

As follows from Eq. (6) the mean $<\geq \hat{a}>$ is the resultant of measurement *bias*:

$$\hat{a}_{bias} = <\geq \hat{a}> = <\geq \hat{a}_{sys}> = (K^T C^{-1} K)^{-1} K^T C^{-1} b. \qquad (14)$$

The covariance matrices of the estimate errors $\geq \hat{a}$ also have random and systematic components:

$$C_{\hat{a}} = C_{\geq \hat{a}(ran)} + (\hat{a}_{bias})(\hat{a}_{bias})^T = (K^T C^{-1} K)^{-1} + (\hat{a}_{bias})(\hat{a}_{bias})^T. \qquad (15)$$

This equation is derived as follows:

$$C_{\hat{a}} = <\geq \hat{a}(\geq \hat{a})^T> = <(\geq \hat{a}_{ran} + \geq \hat{a}_{sys})(\geq \hat{a}_{ran} + \geq \hat{a}_{sys})^T> =$$

$$= <\geq \hat{a}_{ran}(\geq \hat{a}_{ran})^T> + <\geq \hat{a}_{sys}(\geq \hat{a}_{sys})^T> = C_{\geq \hat{a}(ran)} + (\hat{a}_{bias})(\hat{a}_{bias})^T,$$

where

$$C_{\geq \hat{a}(ran)} = <\geq \hat{a}_{ran}(\geq \hat{a}_{ran})^T> = <(K^T C^{-1} K)^{-1} K^T C^{-1} \geq f^*_{ran}((K^T C^{-1} K)^{-1} K^T C^{-1} \geq f^*_{ran})^T>$$

$$= (K^T C^{-1} K)^{-1} K^T C^{-1} <\geq f^*_{ran}(\geq f^*_{ran})> C^{-1} K(K^T C^{-1} K)^{-1} = (K^T C^{-1} K)^{-1}.$$

The optimality of LSM solution $\hat{a}$ is given by the *Cramer-Rao* inequality [27]:

$$<(\geq g)^2> = < g^T \geq \tilde{a}(g^T \geq \tilde{a})^T> = g^T \geq \tilde{a}(\geq \tilde{a})^T g = g^T C_{\tilde{a}} g \geq g^T C_{LSM} g, \qquad (16)$$

where $\tilde{a}$ denotes any estimate of the vector a with covariance of random errors $C_{\tilde{a}}$, C_{LSM} is the covariance matrix of LSM estimates [Eq. (15)], g is a characteristic linear dependence on a (i.e. $g = g^T a$, g is a vector of coefficients). Thus, according to the *Cramer-Rao* inequality the LSM estimates $\hat{a}$ have the smallest variances of random errors and, moreover, the estimate $g^T \hat{a}$ of any function g obtained using $\hat{a}$ also has the smallest variance determined by Eq. (16), i.e. any product $g^T \hat{a}$ of the LSM estimates $\hat{a}$ is also optimum. These accuracy limits are related to the definition of *Fisher* information [27].

The values of minimized quadratic form of $\geq (\hat{a})$ [Eq. (10)] follows a $\geq^2$ distribution with *m-n* degrees of freedom, i.e. the mean minimum is [27,30]:

$$\left\langle 2\Delta(a) \right\rangle = \left\langle (\hat{f} \Delta f^{\Delta})^T C^{\Delta l} (\hat{f} \Delta f^{\Delta}) \right\rangle = \Delta f^{\Delta T} C^{\Delta l} \Delta f^{\Delta} \Delta \Delta \hat{a}^T C_{\hat{a}}^{\Delta l} \Delta \hat{a} = N_f \Delta N_a, \qquad (17)$$

where $\hat{f} = f(\hat{a})$, $N_f = rank(C)$ and $N_a = rank(C_{\hat{a}})$. For the case when both measurement and estimate vectors have only statistically independent elements then N_f is equal to the number of measurements and N_a is equal to the number of retrieved parameters. The statistical property given by Eq. (17) is used to validate the assumptions on the noise $\geq \hat{f}^*$ in measurement and accuracy of the forward model (see Sections 4,6 and 7).

4. *A priori* constrains

In spite of its optimization properties, the basic LSM given by Eq. (12) is not often used in remote sensing. The modeling of interactions of solar light with the atmosphere and the surface requires a complex theoretical formalism with a large number of internal parameters. The integrative character and confines in viewing geometries limit the sensitivity of remote measurements to unique variations of each internal parameter in the radiative model. Therefore, the remote sensing of natural objects, in general, is inherently underdetermined and belongs to a class of so-called *ill-posed* problems. In fact, the frequent appearance of ill-posed problems in remote sensing and applied optics stipulated the development of methodologies that constrain standard inversion algorithms in order to overcome solution instability.

In terms of considerations given in Section 3, ill-posed problems have a non-unique and/or an unstable solution. For a non-unique solution, the matrix $K^T C^{-1} K$ on the left side of Eq. (11c) has linearly dependent rows (and columns, since it is symmetrical), i.e. $\det(K^T C^{-1} K)=0$ (degenerated matrix) and the inverse operator $(K^T C^{-1} K)^{-1}$ does not exist. For a quasi-degenerated matrix $(\det(K^T C^{-1} K) \gtrsim 0)$ the inverse operator $(K^T C^{-1} K)^{-1}$ exists. However, in this case, the covariances of the retrieval errors [Eq. (15)] become large due to uncertainty of the inverse operator:

$$\{C_{\hat{a}}\}_{ii} \sim \{(K^T C^{-1} K)^{-1}\}_{ii} \gtrsim \ge \quad \text{(for } \det(K^T C^{-1} K) \gtrsim 0). \quad (18)$$

4.1 Basic formulations for constrained inversion

Constraining inversions by *a priori* information is an essential tool for achieving a unique and stable solution of an ill-posed problem. Most remote-sensing inverse techniques are based on the following equations:

$$\hat{a} = (K^T K + \ge \ge)^{-1} K^T f^*, \quad (19)$$

$$\hat{a} = (K^T K + \ge \ge)^{-1} (K^T f^* + \ge a^*). \quad (20)$$

These equations originated the papers by *Phillips* [11], *Tikhonov* [12] and *Twomey* [13]. Equation (19) constrains the solution â by minimizing its k-th differences $\ge^k$:

$$\ge^1 = \hat{a}_{i+1} - \hat{a}_i , \quad \text{(k=1)},$$
$$\ge^2 = \hat{a}_{i+2} - 2\,\hat{a}_{i+1} + \hat{a}_i , \quad \text{(k=2)}, \quad (21a)$$
$$\ge^3 = \hat{a}_{i+3} - 3\,\hat{a}_{i+2} + 3\,\hat{a}_{i+1} - \hat{a}_i , \quad \text{(k=3)}.$$

The minimization of differences in Eq. (19) usually is considered [e.g. 12,13,18] to be an implicit constraint on derivatives. The correspondent smoothness matrix $\ge$ in Eq. (19) can be written as

$$\varkappa = (S_k)^T(S_k),$$ (21b)

where S_k is the matrix of the k-th differences (i.e. $\varkappa^k = S_k \, \hat{a}$). For example, S_2 (k=2) is:

$$S_2 = \begin{vmatrix} 1 & -2 & 1 & 0 & ... & & \\ 0 & 1 & -2 & 1 & 0 & ... & \\ 0 & 0 & 1 & -2 & 1 & 0 & ... \\ \multicolumn{7}{c}{\dotfill} \\ & & & ... & 0 & 1 & -2 & 1 \end{vmatrix}.$$ (21c)

Correspondingly, for minimization of the second differences (introduced by *Phillips* [11]), the required smoothing matrix (written by *Twomey* [13]) is

$$\varkappa = \begin{vmatrix} 1 & |2 & 1 & 0 & 0 & ... & & \\ |2 & 5 & |4 & 1 & 0 & 0 & ... & \\ 1 & |4 & 6 & 4 & 1 & 0 & 0 & ... \\ 0 & 1 & |4 & 6 & 4 & 1 & 0 & 0 & ... \\ & & & & ... & & & \\ & & ... & 0 & 1 & |4 & 5 & |2 \\ & & ... & 0 & 1 & |2 & 1 \end{vmatrix}.$$ (21d)

Twomey [13] also suggested employing the third differences in Eq. (19). *Tikhonov's* formulations consider a generalized definition of the smoothing or "regularization" function that usually is formulated via limitations on magnitudes of unknowns and/or their first differences, i.e. the smoothness matrix $\varkappa$ defined as a sum of the unity matrix (matrix of "zero differences") and matrix of first differences or one of these matrices simply used alone [2,5,12,31]. Equation (20) formulated by *Twomey* [13] constrains the solution $\hat{a}$ to its *a priori* estimates a^* (i.e. formally constrains "zero differences"). The *Lagrange* multiplier $\varkappa$ in Eqs. (20-21) is defined as a nonnegative parameter that controls the strength of *a priori* constraints relative to the contribution of the measurements. The value of $\varkappa$ usually is evaluated by numerical tests and sensitivity studies (Section 4.6).

Unlike LSM, Eqs. (20-21) were derived without direct consideration of noise statistics. Nevertheless, Eqs. (20-21) are based on the minimization of quadratic norms of deviations $(f(a)- f^*)$ which is formally equivalent to assuming normal noise with unit covariance matrix (i.e. $C = I$). Thus, Eqs. (20-21) minimize the quadratic forms that have the additional term

$$2\Omega'(\hat{a}) = (f(\hat{a})\Omega f^\Omega)^T (f(\hat{a})\Omega f^\Omega) + \Omega \hat{a}^T \Omega \, \hat{a} = \min,$$ (22a)

$$2\gamma'(\hat{a}) = (f(\hat{a})\gamma \, f^\gamma)^T (f(\hat{a})\gamma \, f^\gamma) + \gamma \, (\hat{a}\gamma \, a^\gamma)^T (\hat{a}\gamma \, a^\gamma) = \min.$$ (22b)

The inclusion of a second *a priori* term [compare to Eq. (10)] into the minimization process results in the fact that Eqs. (19)-(20) provide stable solutions even for ill-posed problems $(\det(K^T K) \geq 0)$. Formally it can be explained by the fact that an addition of the diagonal (I) or quasi-diagonal

($\geq$) matrices to $(\mathbf{K}^T \mathbf{K})$ results in non-degenerated matrices: $\det(\mathbf{K}^T \mathbf{K} + \geq$ $\geq) > 0$ and $\det(\mathbf{K}^T \mathbf{K} + \geq \geq) > 0$.

In atmospheric remote-sensing applications, the statistical interpretation of constrained inversion often is associated with the studies by *Strand and Westwater* [15-16] and *Rodgers* [19] and the following formulations:

$$\hat{\mathbf{a}} = (\mathbf{K}^T \mathbf{C}^{-1} \mathbf{K} + \mathbf{C}_{a^*}^{-1})^{-1} (\mathbf{K}^T \mathbf{C}^{-1} \mathbf{f}^* + \mathbf{C}_{a^*}^{-1} \mathbf{a}^*) \quad \text{and} \quad (23)$$

$$\hat{\mathbf{a}} = \mathbf{a}^* - \mathbf{C}_{a^*} \mathbf{K}^T (\mathbf{C} + \mathbf{K} \mathbf{C}_{a^*} \mathbf{K}^T)^{-1} (\mathbf{K} \mathbf{a}^* - \mathbf{f}^*), \quad (24)$$

where $\mathbf{a}^*$ is a normally distributed vector of *a priori* estimates called a "virtual solution" [19]. Equation (23) has an obvious similarity to Eqs. (19-20). Indeed, Eq. (23) can be transformed to Eq. (19) if $\mathbf{C}_{a^*}^{-1} = \geq$ and $\mathbf{a}^* = \mathbf{0}$ and to Eq. (20) if $\mathbf{C}_{a^*} = (1/\geq) \geq$ It should be noted that there are other statistical formulas for constrained inversions, for example, a statistical equivalent of Eq. (19) is discussed in studies [18, 32].

Equation (24) is fully equivalent to Eq. (23). This type of constrained inversion is popular (see [19]) in applications of satellite remote sensing for retrieving vertical profiles of atmospheric properties (pressure, temperature, gaseous concentrations, etc.). Equation (24) is also widely used in engineering (e.g. see textbook [30]) and other applications [33], such as assimilation of geophysical parameters [34], where Eq. (24) is known as a "*Kalman filter*" named after the author [10] who originated the technique.

The main difference between Eq. (23) and Eq. (24) is that the matrix $(\mathbf{K}^T \mathbf{C}^{-1} \mathbf{K} + \mathbf{C}_{a^*}^{-1})$ inverted in Eq. (23) has dimension N_f (number of measurements) while $(\mathbf{C} + \mathbf{K} \mathbf{C}_{a^*} \mathbf{K}^T)$ inverted in Eq. (24) has the dimension N_a (number of retrieved parameters). Therefore, Eq. (23-24) are fully equivalent for the situation when $N_f = N_a$ and Eq. (23) generally is preferable for inverting redundant measurements ($N_f > N_a$); whereas, Eq. (24) is preferable when the measurement set is underdetermined ($N_f < N_a$). Indeed, in Eq. (23) [similarly as for Eqs. (19-20)], the estimate $\hat{\mathbf{a}}$ mostly is determined by the measurement term $\mathbf{K}^T \mathbf{C}^{-1} \mathbf{f}^*$ and minor *a priori* terms only are expected to provide uniqueness and stability of the solution. In contrast, in Eq. (24) the solution $\hat{\mathbf{a}}$ is expressed in the form of an *a priori* estimate $\mathbf{a}^*$ corrected or "filtered" by measurements, which is the situation when the small number measurements N_f ($N_f < N_a$) cannot fully determine the set of unknowns $\mathbf{a}$, but can improve *a priori* assumed values $\mathbf{a}^*$.

4.2 Statistically optimized inversion of multi-source data

The similarities of the formulas for constrained inversion with basic non-constrained LSM were mentioned already in the previous section. This section further explores the use of statistical principles for implementing constrained inversion by formulating a statistically optimized inversion of multi-source data that follows the developments [7-9]. Such an approach allows generalizing various inversion formulas into a single formalism.

Formally, both measured and *a priori* data can be written as

$$\mathbf{f}_k^* = \mathbf{f}_k(\mathbf{a}) + \geq \mathbf{f}_k^*, \quad (k=1, 2,...,K), \tag{25}$$

where index k denotes different data sets ("sources"). This assumes that the data from the same source have similar error structure independent of errors in the data from other sources. For example, direct Sun and diffuse sky radiances have different magnitudes and are measured by sensors with a different sensitivity, i.e., errors should be independent (due to different sensors) and likely have different magnitudes. Similarly, *a priori* data are independent of the measurements, i.e. they have errors with a different level of accuracy uncorrelated with remote-sensing errors. Formally, the statistical independence of $\mathbf{f}_k^*$ means that the covariance matrix of joint data $\mathbf{f}^*$ has array structure:

$$\mathbf{f}^{|} = \begin{vmatrix} \mathbf{f}_1^* \\ \mathbf{f}_2^* \\ ... \\ \mathbf{f}_K^* \end{vmatrix} \quad \text{and} \quad \mathbf{C}_{\mathbf{f}^*} = \begin{vmatrix} \mathbf{C}_1 & 0 & 0 & 0 \\ 0 & \mathbf{C}_2 & 0 & 0 \\ ... & ... & ... & ... \\ 0 & 0 & 0 & \mathbf{C}_K \end{vmatrix}, \tag{26}$$

where $\mathbf{f}^*$ is a vector-column with $(\mathbf{f}^*)^T = (\mathbf{f}_1^*, \mathbf{f}_2^*, ..., \mathbf{f}_K^*)^T$ and $\mathbf{C}_k$ is the covariance matrix of the k-th data set $\mathbf{f}_k^*$. Thus, from the formal viewpoint, the only difference of Eq. (25) from Eq. (1) is that Eq. (25) explicitly outlines an expectation of an array structure for the covariance matrix $\mathbf{C}_{\mathbf{f}^*}$. Such explicit differentiation of the input data makes the retrieval more transparent because the statistical optimization of the retrieval is driven by a covariance matrix of random errors. It should, be noted that Eq. (25) does not assume any relations between forward models $\mathbf{f}_k(\mathbf{a})$, i.e. forward models $\mathbf{f}_k(\mathbf{a})$ can be the same or different.

Following Eq. (26), the PDF of joint data $\mathbf{f}^*$ can be obtained by the simple multiplication of the PDFs of data from all K sources:

$$P\left(\mathbf{f}(\mathbf{a})\big|\mathbf{f}^{|1|}\right) = P\left(\mathbf{f}_1(\mathbf{a}),...,\mathbf{f}_K(\mathbf{a})\big|\mathbf{f}_1^{|1|},...,\mathbf{f}_K^{|1|}\right) = \prod_{k=1}^{K} P\left(\mathbf{f}_k(\mathbf{a})\big|\mathbf{f}_k^{|1|}\right). \tag{27}$$

Then, under the assumptions of a normal PDF, one can write

$$P\left(\mathbf{f}(\mathbf{a})\big|\mathbf{f}^{|}\right) = \prod_{k=1}^{K} P\left(\mathbf{f}_k(\mathbf{a})\big|\mathbf{f}_k^{|}\right) \sim \exp\left| \frac{1}{2}\sum_{k=1}^{K} \left(\mathbf{f}_k(\mathbf{a}) \mid \mathbf{f}_k^{|}\right)^T (\mathbf{C}_k)^{|1|}\left(\mathbf{f}_k(\mathbf{a}) \mid \mathbf{f}_k^{|}\right)\right|. \tag{28}$$

Thus, for multi-source data, the LSM condition of Eq. (10) can be written as

$$2\Sigma(\mathbf{a}) = \sum_{k=1}^{K}\left(\mathbf{f}_k(\mathbf{a})\,\Sigma\mathbf{f}_k^{\Sigma}\right)^{\mathrm{T}}(\mathbf{C}_k)^{\Sigma\mathrm{I}}\left(\mathbf{f}_k(\mathbf{a})\,\Sigma\mathbf{f}_k^{\Sigma}\right) = \min. \tag{29}$$

This condition does not prescribe the value of the minimum and, therefore, Eq. (29) can be formulated via weight matrices:

$$2*(\mathbf{a}) = 2*\sum_{k=1}^{K}*_k*_k(\mathbf{a}) = *\sum_{k=1}^{K}*_k\left(\mathbf{f}_k(\mathbf{a})*\mathbf{f}_k^{*}\right)^{\mathrm{T}}(\mathbf{W}_k)^{*\mathrm{l}}\left(\mathbf{f}_k(\mathbf{a})*\mathbf{f}_k^{*}\right) = \min, \tag{30a}$$

where

$$\mathbf{W}_k = \frac{1}{\varepsilon_k^2}\mathbf{C}_k \qquad \text{and} \qquad \varepsilon_k = \frac{\varepsilon_1^2}{\varepsilon_k^2}. \tag{30b}$$

Here ε_k^2 is the first diagonal element of $\mathbf{C}_k$, i.e. $\varepsilon_k^2 = \{\mathbf{C}_k\}_{11}$. Although, Eqs. (29) and (30) are equivalent, sometimes Eq. (30) is more convenient because in Eq. (30) the parameters $\geq_k$ are weighting the contribution of each source relative to the contribution of first data source (obviously, $\geq_l = 1$). Similarly, using weight matrices instead of covariance matrices allows for the analysis of the relative contribution of different measurements within each k-th data set.

The Minimum of the multi-term quadratic form $\geq(\mathbf{a})$ can be found by solving the system of multi-term normal equations, i.e. Eq. (11) can be transformed as

$$\sum_{k=1}^{K}\Sigma_k\,(\mathbf{K}_k)^{\mathrm{T}}(\mathbf{W}_k)^{\Sigma\mathrm{I}}(\mathbf{K}_k)\,\mathbf{a} = \sum_{k=1}^{K}\Sigma_k\,(\mathbf{K}_k)^{\mathrm{T}}(\mathbf{W}_k)^{\Sigma\mathrm{I}}\mathbf{f}_k^{*}. \tag{31}$$

Correspondingly, using matrix inversion the multi-term equivalent of Eq.(12) is

$$\hat{\mathbf{a}} = \left|\sum_{k=1}^{K}I_k\,(\mathbf{K}_k)^{\mathrm{T}}(\mathbf{W}_k)^{\mathrm{l}\,\mathrm{l}}(\mathbf{K}_k)\right|^{-1}\left|\sum_{k=1}^{K}I_k\,(\mathbf{K}_k)^{\mathrm{T}}(\mathbf{W}_k)^{\mathrm{l}\,\mathrm{l}}\mathbf{f}_k^{*}\right|. \tag{32}$$

The generalization of the basic LSM by the multi-term Eqs. (31-32) is useful for utilizing several observational data sets in a single flexible retrieval. In addition, these equations can be a basis for unifying various techniques for constrained inversion techniques. For example, constraining the solution by *a priori* estimates can be considered as a joint inversion of two data sets. For such case, Eq. (25) is

$$\begin{aligned}\overrightarrow{\Xi}_1^{*} &= \mathbf{f}_1^{*}(\mathbf{a}) + \overrightarrow{\Xi}\mathbf{f}_1^{*}\\\overrightarrow{\Xi}_2^{*} &= \mathbf{f}_2^{*}(\mathbf{a}) + \overrightarrow{\Xi}\mathbf{f}_2^{*}\end{aligned} \;\Rightarrow\; \begin{aligned}\overrightarrow{\Xi}^{*} &= \mathbf{f}^{*}(\mathbf{a}) + \overrightarrow{\Xi}\mathbf{f}^{*}\\\overrightarrow{\Xi}\mathbf{a}^{*} &= \mathbf{a} + \overrightarrow{\Xi}\mathbf{a}^{*}\end{aligned}. \tag{33}$$

The matrices $\mathbf{K}_k$ and $\mathbf{W}_k$ required in Eq. (32) are the following:

$$\mathbf{K}_1 = \mathbf{K}, \text{ and } \mathbf{K}_2 = \mathbf{I},$$

$$\mathbf{W}_1 = \mathbf{W} = (1/\gtrsim_{f*})^2 \mathbf{C}_{f*}, \text{ and } \mathbf{W}_2 = \mathbf{W}_{a*} = (1/\gtrsim_{a*})^2 \mathbf{C}_{a*}, \tag{34a}$$

and the two-term Eq. (32) is

$$\hat{\mathbf{a}} = (\mathbf{K}^T \mathbf{W}^{-1} \mathbf{K} + \geq \mathbf{W}_{a*}^{-1})^{-1} (\mathbf{K}^T \mathbf{W}^{-1} \mathbf{f}^* + \geq \mathbf{W}_{a*}^{-1} \mathbf{a}^*). \tag{34b}$$

Equation (34) is an obvious analog of Eq. (23). This is not surprising because both Eq. (23) and Eq. (24) are derived and explained in several previous studies (e.g. [30, 19]) using the approach similar to the above considerations [Eqs. (25-32)]. Also, Eq. (34) can be trivially reduced to the *Twomey* formula [Eq. (20)] by assuming the same accuracy for all measurements f_j^* (i.e. $\mathbf{W} = \mathbf{I}$) and the same accuracy for all *a priori* estimates a_i^* (i.e. $\mathbf{W}_{a*} = \mathbf{I}$). It is interesting that the use of weight matrices gives a clear statistical interpretation to the *Lagrange* multiplier as the ratio of variances:

$$\varepsilon = \varepsilon_{f*}^2 / \varepsilon_{a*}^2 . \tag{35}$$

Such an interpretation of the *Lagrange* multiplier is especially useful for cases when both $\gtrsim_{f*}$ and $\gtrsim_{a*}$ have the same unites or they are unitless. For example, if $\gtrsim_{f*}$ and $\gtrsim_{a*}$ are variances of relative errors: $\gtrsim f_j / f_j$ and $\geq a_i / a_i$. In such a situation, a small value of the *Lagrange* multiplier $\geq$ logically is expected since *a priori* knowledge is always less certain than actual measurements.

4.3 Statistical interpretation of smoothing constraints

Constraining the inversion to a smooth solution as given by *Phillips – Tikhonov – Twomey* Eq. (19) has been proven to be very efficient in numerous applications, e.g. [1, 13, 35-40]. In contrast to Eqs. (20, 23-24) where the solution $\hat{\mathbf{a}}$ is constrained to the actual values of *a priori* estimates $\mathbf{a}^*$, Eq. (19) constrains only differences (derivatives) between elements of retrieved vector $\hat{\mathbf{a}}$ and does not restrict their values. Therefore, smoothing constraints may be preferable in applications where *a priori* magnitudes of unknowns are uncertain. For example, a smooth behavior with no sharp oscillations can naturally be expected for atmospheric characteristic $y(x)$ such as the size distributions of aerosol concentrations. Correspondingly, filtering of the solutions with strong oscillations of $a_i = y(x_i)$ ($i=1, ..., N_a$) appears to be a logical constraint, while finding appropriate *a priori* estimates $a_i^* = y^*(x_i)$ would be problematic because the magnitudes of $a_i = y(x_i)$ (i.e. aerosol loading) may vary by ~ 100 times. Retrieval of vertical profiles of atmospheric gases and aerosol concentrations can be another example where some smoothness in the behavior of unknowns $a_i = y(x_i)$ can be expected.

Studies [11-13] originated Eq. (19) did not imply any statistical meaning to the smoothness constrains. Later studies suggest some statistical interpretation to smoothing constraints. For example, studies [18, 32] considered the smoothness matrix $\geq$ as the inverse matrix to the covariance matrix of *a priori* solutions. *Rodgers* [19] related smoothing constraints with the non-diagonal structure of the covariance matrix $\mathbf{C}_{a*}$ of the *a priori* estimates. The present analysis (as follows from [9]) explicitly considers smoothness constraints as *a priori* estimates of the derivatives of the retrieved characteristic $y(x_i)$.

The values of m-th derivatives g_m of the function $y(x)$ characterize the degree of its non-linearity and, therefore, can be used as a measure of $y(x)$ smoothness. For example, smooth functions $y(x)$, such as a constant, straight line, parabola, etc. can be identified by m-th derivatives as follows:

$$g_1(x) = dy(x)/dx = 0 \quad \Rightarrow \quad y_1(x) = C\,;$$
$$g_2(x) = d^2 y(x)/dx^2 = 0 \Rightarrow y_2(x) = B\,x + C\,; \tag{35}$$
$$g_3(x) = d^3 y(x)/dx^3 = 0 \Rightarrow y_3(x) = A\,x^2 + B\,x + C$$

These derivatives g_m can be approximated by differences between values of the function $a_i = y(x_i)$ in N_a discrete points x_i as:

$$\frac{dy(x_i)}{dx} - \frac{-^1 y(x_i)}{-_1 x_i} = \frac{y(x_i + -x_i) - y(x_i)}{-_1 x_i} = \frac{y(x_{i+1}) - y(x_i)}{-_1 x_i}\,;$$

$$\frac{d^2 y(x_i)}{dx^2} - \frac{-^2 y(x_i)}{-_2(x_i)} = \frac{-^1 y(x_{i+1})/-_1(x_{i+1}) - -^1 y(x_i)/-_1(x_i)}{(-_1 x_i + -_1 x_{i+1})/2} = \dots\,; \tag{36}$$

$$\frac{d^3 y(x_i)}{dz^3} - \frac{-^3 y(x_i)}{-_3(x_i)} = \frac{-^2 y(x_{i+1})/-_2(x_{i+1}) - -^2 y(x_i)/-_2(x_i)}{(-_2(x_i) + -_2(x_{i+1}))/2} = \dots\,;$$

where

$$\geq_1(x_i) = x_{i+1} - x_i;\quad \geq_2(x_i) = (\geq_1(x_i) + \geq_1(x_{i+1}))/2;\quad \geq_3(x_i) = (\geq_2(x_i) + \geq_2(x_{i+1}))/2;$$

$$x_{i'} = x_i + \geq_1(x_i)/2;\quad x_{i''} = x_i + (\geq_1(x_i) + \geq_2(x_i))/2;\quad x_{i'''} = x_i + (\geq_1(x_i) + \geq_2(x_i) + \geq_3(x_i))/2.$$

In retrievals of the function $y(x_i)$ in N_a discrete points x_i, the expectations of limited derivatives of $y(x)$ can be employed explicitly as smoothness constraints. Namely, if the retrieved function is expected to be close to a constant, straight line, parabola, etc., one can use zero m-th derivatives, as follows from Eq. (35), as *a priori* estimates: $\mathbf{g}_m^* = \mathbf{0}$. Using this knowledge as a second source of information about $a_i = y(x_i)$, the multi-source Eq. (25) can be written:

$$\begin{aligned}\vec{\mathbf{f}}_1^* &= \mathbf{f}_1^*(\mathbf{a}) + \vec{\mathbf{f}}_1^* \\ \vec{\mathbf{f}}_2^* &= \mathbf{f}_2^*(\mathbf{a}) + \vec{\mathbf{f}}_2^*\end{aligned} \Rightarrow \begin{aligned}\vec{\mathbf{f}}^* &= \mathbf{f}^*(\mathbf{a}) + \vec{\mathbf{f}}^* \\ \vec{\mathbf{g}}_m^* &= \mathbf{g}_m(\mathbf{a}) + \vec{\mathbf{g}}_m^*\end{aligned} \Rightarrow \begin{aligned}\vec{\mathbf{f}}^* &= \mathbf{f}^*(\mathbf{a}) + \vec{\mathbf{f}}^* \\ \vec{\mathbf{0}}^* &= \mathbf{G}_m \mathbf{a} + \vec{\mathbf{g}}^*\end{aligned}\,, \tag{37}$$

where $\mathbf{g}_m$ is a vector of m-th derivatives $(g_m)_i = g_m(x_i)$ $(i = m+1, \dots, N_a)$, $\mathbf{G}_m$ is the matrix of the coefficients required for matrix form $\mathbf{g}_m = \mathbf{G}_m \mathbf{a}$ of Eq. (32). The errors $\geq_g^*$ reflect the uncertainty in the knowledge of the deviations of $y(x)$

from the assumed constant, straight line, parabola, etc. Correspondingly, assuming that $\geq_g^*$ have a normal distribution with the covariance matrix $\mathbf{C}_{g^*}$, one can use multi-term LSM Eq. (32) with the following matrices $\mathbf{K}_k$ and $\mathbf{W}_k$:

$$\mathbf{K}_1 = \mathbf{K}, \text{ and } \mathbf{K}_2 = \mathbf{G}_m,$$

$$\mathbf{W}_1 = \mathbf{W} = (1/\geq_{f^*})^2 \mathbf{C}_{f^*}, \text{ and } \mathbf{W}_2 = \mathbf{W}_{g^*} = (1/\geq_{g^*})^2 \mathbf{C}_{g^*}, \tag{38a}$$

and the two-term Eq. (32), solving Eq. (37), has the form:

$$\hat{\mathbf{a}} = \left(\mathbf{K}^T \mathbf{W}^{-1} \mathbf{K} + \geq \mathbf{G}_m^T \mathbf{W}_{g^*}^{-1} \mathbf{G}_m\right)^{-1} \mathbf{K}^T \mathbf{W}^{-1} \mathbf{f}^*, \tag{38b}$$

where the multiplier $\geq$ is defined as

$$\varepsilon = \varepsilon_{f^*}^2 / \varepsilon_{g^*}^2, \tag{38c}$$

where $\varepsilon_{g^*}^2$ is the first diagonal element of $\mathbf{C}_{g^*}$, i.e. $\varepsilon_{g^*}^2 = \{\mathbf{C}_{g^*}\}_{11}$.

Thus, Eq. (38b) minimizes the quadratic form [Eq. (30a)] with two terms ($k=1,2$), where the second term $\Psi_2(\mathbf{a})$ represents *a priori* constraints on the m-th derivatives. The inclusion of $\Psi_2(\mathbf{a})$ in the minimization can be considered as applying limitations on the quadratic norm of m-th derivatives of $y(x)$ that are commonly used as a measure of smoothness (e. g. see [32]). Indeed, if one assumes the diagonal covariance matrix $\mathbf{C}_{g^*}$ with diagonal elements

$$\{\mathbf{C}_{g^*}\}_{ii} \sim 1/\Rightarrow_m(x_i) \quad \Rightarrow \quad \{\mathbf{W}_{g^*}\}_{ii} = \{\mathbf{C}_{g^*}\}_{ii}/\{\mathbf{C}_{g^*}\}_{11} = (\Rightarrow_m(x_1))/(\Rightarrow_m(x_i)), \tag{38d}$$

then the quadratic term $\Psi_2(\mathbf{a})$ can be considered as an estimate of the norm of the m-th derivatives obtained using the values of $y(x)$ at N_a discrete points x_i:

$$b_m = \int \left(\frac{d^m y(x)}{d^m x}\right)^2 dx \approx \sum_{i=m+1}^{N_a} \left(\frac{\Psi_m^m y(x_i)}{\Psi_m(x_i)}\right)^2 \Psi_m(x_i) = \mathbf{a}^T \mathbf{G}_m^T \mathbf{C}_{g^*}^{\Psi} \mathbf{G}_m \mathbf{a} \sim \Psi_2(\mathbf{a}). \tag{39}$$

By the means of this equation, one can relate the variance $\varepsilon_{g^*}^2$ and the multiplier $\geq$ to the expected value of the norm b_m. Indeed, the estimates of the derivatives ($\mathbf{g}_m^* = \mathbf{0}^* = \mathbf{0} + \geq_m^*$) employed in Eq. (38) assume the following mean value of the norm b_m:

$$\langle b_m \rangle \varepsilon \varepsilon \sum_{i=1}^{N_a} \left\langle ((\mathbf{g}_m)_i)^2 \right\rangle \varepsilon \, m(x_i) = \sum_{i=m+1}^{N_a} \{\mathbf{C}_{g^*}\}_{ii} \, \varepsilon \, m(x_i) = (N_a \varepsilon \, m) \, \varepsilon \, m(x_1) \, \varepsilon_{g^*}^2. \tag{40}$$

Then, the variance $\varepsilon_{g^*}^2$ and the multiplier $\geq$ can be determined via $\langle b_m \rangle$ as

$$\Delta_{g*}^2 = \frac{1}{\langle b_m \rangle} (N_a \Delta m) \, \Delta_m(x_1) \geq \Delta = \frac{\Delta_{f*}^2}{\langle b_m \rangle} (N_a \Delta m) \, \Delta_m(x_1). \quad (41)$$

Equations (38) explicitly use the discrete approximation of derivatives via ratios of the differences of the function $\geq^m y(x_i)$ and differences of the arguments $\geq_m(x_i)$, while Eq. (19) uses only differences of the function $\geq^m y(x_i)$. Obviously, Eqs. (38) and Eq. (19) are nearly analogous when the differences of arguments $\geq_m(x_i)$ can be trivially accounted, i.e. when $y(x)$ is retrieved in N_a equidistant points $x_{i+1} = x_i + \geq x$ $(i=1,\ldots,N_a-1)$. For this situation, $\geq_m(x_i) = \geq x$ and the derivatives and differences of the function $y(x)$ differ by a constant only:

$$\frac{d^m y(x)}{d^m x} \Delta \frac{\Delta^m y(x_i)}{(\Delta x)^m} \quad \Rightarrow \quad \mathbf{G}_m = (\geq x)^{-m} \, \mathbf{S}_m. \quad (42)$$

Correspondingly, Eq. (37) can be written using the differences $\geq_m y(x_i)$:

$$\left\lfloor \begin{array}{l} \mathbf{f}^* = \mathbf{f}^*(a) + \lfloor \mathbf{f}^* \\ (\lfloor {}^m \mathbf{a})^* = \mathbf{S}_m a + \lfloor (\lfloor {}^m \mathbf{a})^* \end{array} \right. \quad \left\lfloor \begin{array}{l} \mathbf{f}^* = \mathbf{f}^*(a) + \lfloor \mathbf{f}^* \\ \mathbf{0}^* = \mathbf{S}_m a + \lfloor {}^*_m \end{array} \right., \quad (43)$$

where the vectors $\mathbf{0}^*$ and $\geq_m^*$ contain estimates of differences and errors of these estimates, respectively. Also, for equidistant points $x_{i+1} = x_i + \geq$, the covariance matrix $\mathbf{C}_{\Delta*}$ of the differences differs from $\mathbf{C}_{g*}$ by a constant only:

$$(\Rightarrow_m(x_i) = \Rightarrow x) \quad \Rightarrow \quad (-x_i)^{-2m} \mathbf{C}_{-*} = \mathbf{C}_{g*} \text{ and } \mathbf{W}_{\Delta*} = \mathbf{W}_{g*} = 1. \quad (44a)$$

Correspondingly, Eq. (40) relates $\mathbf{C}_{\Delta*}$ with the norm b_m as follows:

$$\langle b_m \rangle \varepsilon \sum_{i=m+1}^{N_a} \varepsilon \{\mathbf{C}_{g*}\}_{ii} \varepsilon_m(x_i) = \varepsilon \frac{1}{\varepsilon} \frac{\varepsilon^{2m+1}}{\varepsilon \times \varepsilon} \sum_{i=m+1}^{N_a} \varepsilon \{\mathbf{C}_{\varepsilon*}\}_{ii} = \varepsilon \frac{1}{\varepsilon} \frac{\varepsilon^{2m+1}}{\varepsilon \times \varepsilon} (N_a \varepsilon m) \varepsilon_\varepsilon^2 * . (44b)$$

Finally, Eqs. (38) can be reduced to the equivalent of Eq. (19) as:

$$\mathbf{K}_1 = \mathbf{K}, \text{ and } \mathbf{K}_2 = \mathbf{S}_m,$$

$$\mathbf{W}_1 = \mathbf{W} = (1/\geq_{f*})^2 \mathbf{C}_{f*}, \text{ and } \mathbf{W}_2 = \mathbf{W}_{\geq*} = (1/\geq_{\geq*})^2 \mathbf{C}_{\geq*} = 1, \quad (45a)$$

$$\hat{a} = (\mathbf{K}^T \mathbf{W}^{-1} \mathbf{K} + \geq \mathbf{S}_m^T \mathbf{S}_m)^{-1} \mathbf{K}^T \mathbf{W}^{-1} \mathbf{f}^*, \quad (45b)$$

$$\Delta = \frac{\Delta_{f*}^2}{\Delta_{g*}^2} = \frac{\Delta_{f*}^2}{\langle b_m \rangle} (N_a \Delta m) (\Delta x)^{\Delta 2m+1}. \quad (45c)$$

where $\Delta_{\Delta*}^2$ and ε_{g*}^2 are the first diagonal elements of the covariance matrices: $\Delta_{\Delta*}^2 = \{\mathbf{C}_{\Delta*}\}_{11}$ and $\varepsilon_{g*}^2 = \{\mathbf{C}_{g*}\}_{11}$.

Thus the multi-term LSM is a useful approach for deriving the *Phillips – Tikhonov – Twomey* constrained inversion [Eq. (19)]. Also, it is shown above that constraining the solution by adding a smoothness term in Eq. (19) can be considered as explicit use of knowledge about the m-th derivatives of retrieved functions $y(x)$. In other words, the inversion of measurements $\mathbf{f}^*$ is replaced by the joint inversion of the measurements $\mathbf{f}^*$ and "measured" derivatives $\mathbf{g}_m^*$. In the scope of such considerations, the *Lagrange* multiplier $\geq$ has a clear quantitative interpretation [Eqs. (41) and (45)]. In addition, Eqs. (38-41) can be used in situations where utilizing the original Eq. (19) is not transparent. For example, Eq. (38) generalizes the use of smoothness constraints on situations where the retrieved function $y(x)$ is defined at points with non-equidistant ordinates x_i. Also Eq. (38) allows differentiating the smoothing strength for different x_i by weight matrix $\mathbf{W}_{g^*}$ non-equal to unity matrix. Although, using $\mathbf{W}_{g^*}$ other than defined by Eq. (38d) requires modifications in definition of $\varepsilon_{g^*}^2$ and $\geq$ Namely, in addition to $\langle b_m \rangle$ the information about smoothness differentiation should be available. Illustrations of applying Eq. (38) can be found in the manuscript [9], where the spectral dependence $n(\geq_i)$ of the aerosol refractive index is retrieved in non-equidistant $\geq_i$ fixed by the measurement specifications.

4.4 Combining multiple *a priori* constraints in the inversion

The consideration of *a priori* constraints as an equal component in the multi-term inversion Eqs. (30-32) is a useful tool for applying multiple constraints in a retrieval algorithm. For example, a simultaneously constraining solution by both *a priori* estimates and smoothness assumptions can be considered as inversion of the data from three independent sources and Eq. (25) is

$$
\begin{array}{ll}
\vec{\mathbf{f}}_1^* = \mathbf{f}_1^*(\mathbf{a}) + \vec{\mathbf{f}}_1^* & \vec{\mathbf{f}}^* = \mathbf{f}^*(\mathbf{a}) + \vec{\mathbf{f}}^* \\
\vec{\mathbf{f}}_2^* = \mathbf{f}_2^*(\mathbf{a}) + \vec{\mathbf{f}}_2^* \Rightarrow & \vec{\boldsymbol{\theta}}^* = \mathbf{S}_m + \vec{(}\vec{\Rightarrow}^m \mathbf{a})^* . \\
\vec{\mathbf{f}}_3^* = \mathbf{f}_3^*(\mathbf{a}) + \vec{\mathbf{f}}_3^* & \vec{\mathbf{a}}^* = \mathbf{a} + \vec{\mathbf{a}}^*
\end{array}
\tag{46}
$$

The matrices $\mathbf{K}_k$ and $\mathbf{W}_k$ required in Eq. (32) are the following:

$$
\mathbf{K}_1 = \mathbf{K}; \quad \mathbf{K}_2 = \mathbf{S}_m; \quad \text{and} \quad \mathbf{K}_3 = \mathbf{I},
$$

$$
\mathbf{W}_1 = \mathbf{W} = (1/\geq_{f^*})^2 \, \mathbf{C}_{f^*}; \; \mathbf{W}_2 = \mathbf{W}_{\geq^*} = 1; \text{ and } \mathbf{W}_3 = \mathbf{W}_{a^*} = (1/\geq_{a^*})^2 \, \mathbf{C}_{a^*},
\tag{47a}
$$

and three-term Eq. (32) is:

$$
\hat{\mathbf{a}} = (\mathbf{K}^T \mathbf{W}^{-1} \mathbf{K} + \geq_2 \mathbf{S}_m^{\ T} \mathbf{S}_m + \geq_3 \mathbf{W}_{a^*}^{\ -1})^{-1} (\mathbf{K}^T \mathbf{W}^{-1} \mathbf{f}^* + \geq_3 \mathbf{W}_{a^*}^{\ -1} \mathbf{a}^*),
\tag{47b}
$$

$$
\Delta_2 = \frac{\Delta_{f^*}^2}{\Delta_{g^*}^2} = \frac{\Delta_{f^*}^2}{\langle b_m \rangle} \, (N_a \Delta m) \, (\Delta x)^{\Delta 2m+1} \quad \text{and} \quad \varepsilon_3 = \frac{\varepsilon_{f^*}^2}{\varepsilon_{g^*}^2} .
\tag{47c}
$$

This equation minimizes three quadratic forms simultaneously:

$$2\Omega(\hat{a}) = 2\sum_{k=1}^{3}\Omega_k\Omega_k(\hat{a}) = \left(f(\hat{a})\Omega f^{\Omega}\right)^{T} W^{\Omega}\left(f(\hat{a})\Omega f^{\Omega}\right) + \Omega_2\hat{a}^{T}\Omega_m\hat{a} + \Omega_3\hat{a}^{T}W_{a*}^{\Omega}\hat{a} = \min. \quad (48)$$

Thus, applying multiple *a priori* constraints is straightforward using multi-term LSM formulations, while multiple *a priori* constraints usually are not considered in the scope of basic formulas [Eqs. (19-20) and (23-24)].

4.5 Error evaluation

Equations (13-15) estimating errors of LSM solutions can be generalized in the case of multi-term solutions by Eqs. (31-32) as

$$C_{\hat{a}} = C_{\geq\hat{a}(ran)} + (\hat{a}_{bias})(\hat{a}_{bias})^{T}, \quad (49a)$$

$$\hat{a}_{bias} = \left|\sum_{k=1}^{K} l_k (K_k)^{T}(W_k)^{|1}(K_k)\right|^{-1} \left|\sum_{k=1}^{K} l_k (K_k)^{T}(W_k)^{|1}b_k^{*}\right|, \quad (49b)$$

$$C_{\varepsilon\hat{a}_{(ran)}} = \left\langle\varepsilon\hat{a}_{(ran)}\left(\varepsilon\hat{a}_{(ran)}\right)^{T}\right\rangle = \sum_{\varepsilon k=1}^{\varepsilon K} \varepsilon_k (K_k)^{T}(W_k)^{\varepsilon 1}(K_k)_{\varepsilon}^{\varepsilon-1} \varepsilon_1^{2}, \quad (49c)$$

where b_k denotes the *bias* vector in the *k*-th data set f_k.

For example, for the three term solution by Eqs. (47), the retrieval bias $\hat{a}_{bias}$ and the covariance matrix of random errors $C_{\geq\hat{a}(ran)}$ can be written as

$$\hat{a}_{bias} = (K^{T}W^{-1}K + \geq_2 \Sigma_m + \geq_3 W_{a*}^{-1})^{-1}(K^{T}W^{-1}b_{f*} + \geq_2 \Sigma_m b_{\geq*} + \geq_3 W_{a*}^{-1}b_{a*}) \quad (50a)$$

$$C_{\geq\hat{a}(ran)} = (K^{T}W^{-1}K + \geq_2 \Sigma_m + \geq_3 W_{a*}^{-1})^{-1}\geq_{*}^{2}, \quad (50b)$$

$$b_f = \langle f* - f(a^{real})\rangle = \langle\geq f*\rangle; \{b_{\geq*}\}_{ii} = (\geq x)^{2m}\left(d^{m}y(x_i)/d^{m}x\right)_{real}; b_{a*} = a^{real} - a^{*}, \quad (50c)$$

where b_f is a *bias* in the measurements $f*$ or in the forward model: $f^{real} - f(a^{real})$; $\Sigma_m = S_m^{T}S_m$ and vector $b_{\geq}$ (with elements $\{b_{\geq*}\}_{ii}$) denotes a *bias* introduced by assuming zeros $0*$ as estimates of *m*-th differences; b_{a*} is a *bias* in *a priori* estimates a^{*}.

Equations (50) are helpful for analyzing the effects of constraints on the solution. It follows from Eq. (50b) that by strengthening *a priori* constraints (by $\geq_2$ and $\geq_3$), one formally can suppress the random errors of the retrieval to any desirable level. However, Eq. (50a) shows that, if *a priori* biases are non-zero, increasing $\geq_2$ and $\geq_3$ leads to increasing systematic errors. Therefore, *a priori* constraints are useful only in the case when the increase of the systematic component $(\hat{a}_{bias})(\hat{a}_{bias})^{T}$ does not exceed the decrease of the random component of the retrieval errors in Eq. (49). Unfortunately, in practice the *a priori* biases are uncertain and, therefore, the selection of optimum *a priori* constraints is a very challenging issue in inversion developments. Generally, *a priori* estimates $\mathbf{a}^*$ always have non-zero bias $\mathbf{b}_{\mathbf{a}*}= \mathbf{a}^{real}-\mathbf{a}^*$. In contrast, the smoothness constraints are likely unbiased (i.e. $\mathbf{b}_{\geq *} \geq 0$), because for smooth functions, m-derivatives are close to zero. This is why smoothness constraints are preferable for the retrieval of smooth functions.

It should be noted that the multi-term estimates [Eqs. (32), (34), (47)] retain the optimality of LSM estimates, i.e. they have smallest errors as determined by the *Cramer-Rao* inequality, Eq. (16). However, the *Cramer-Rao* inequality is valid only if all assumptions about noise in both the measurements and the *a priori* terms are correct. Therefore, the validation of the assumptions is important, while problematic in reality. A useful consistency check can be performed using the achieved value of the minimized quadratic form $\geq (\hat{\mathbf{a}})$ [Eq. (30)]. For example, in case of zero biases, the minimum value of the three-term $\geq (\hat{\mathbf{a}})$ [Eq (48)] has a $\geq^2$ distribution with mean

$$\left\langle \left(2\Delta(\hat{\mathbf{a}})\right)_{min}\right\rangle = \left\langle 2\Delta_{i=1,...,3.}\Delta_i(\hat{\mathbf{a}})\right\rangle = \left\langle \left(\hat{\mathbf{f}}\Delta\mathbf{f}^\Delta\right)^{\mathrm{T}}\mathbf{W}^{\Delta\mathrm{l}}\left(\hat{\mathbf{f}}\Delta\mathbf{f}^\Delta\right) + \Delta_2\hat{\mathbf{a}}^{\mathrm{T}}\Delta_m\hat{\mathbf{a}} + \Delta_3\hat{\mathbf{a}}^{\mathrm{T}}\mathbf{W}^{\Delta\mathrm{l}}_{\mathbf{a}*}\hat{\mathbf{a}}\right\rangle = \quad (51)$$

$$= \left(\Delta_{i=1,...,3}(N_{\mathbf{f}_i})\Delta N_{\mathbf{a}}\right)\Delta_1^2 = \left(N_{\mathbf{f}*}+N_\Delta*+N_{\mathbf{a}*}\Delta N_{\mathbf{a}}\right)\Delta_1^2 = \left(N_{\mathbf{f}*}+N_{\mathbf{a}}\,\Delta m\right)\Delta_1^2$$

where $N_{-*}=N_{\mathbf{a}}-m$ and $N_{\mathbf{a}*}=N_{\mathbf{a}}$. Using this equation one can estimate the value $\geq_1^2$ from a minimum value of $\geq (\hat{\mathbf{a}})$ often called a *residual*:

$$\hat{\geq}_1^2 - \frac{\left(2-(\hat{\mathbf{a}})\right)_{min}}{-_{i=1,...,3}(N_{\mathbf{f}_i})-N_{\mathbf{a}}} = \frac{\left(2-(\hat{\mathbf{a}})\right)_{min}}{N_{\mathbf{f}*}+N_{\mathbf{a}}-m}. \quad (52)$$

If all assumptions are correct, the estimation by Eq. (52) should be close to the assumed $\geq_1^2$. A significant increase of the estimated $\hat{\varepsilon}_1^2$ over expected $\geq_1^2$ can be considered as an indication of unaccounted biases and/or inadequate assumptions about random errors in measurements or *a priori* data sets. The consistency checks relying on estimates $\hat{\varepsilon}_1^2$ from the residual commonly is used in remote sensing and other applications. For example, the effects of unaccounted biases in both measurements and forward modeling on retrievals of aerosol properties from ground-based observations using the residuals in observation fits have been analyzed [8].

4.6 *Lagrange multiplier* selection

Sections 4.3-4.4 provide quantitative definitions of *Lagrange* multipliers. However, in reality the detailed information required for an explicit definition of *a priori* constraints may not be available. In such situations the following recipes and discussions may be useful.

For constraining a retrieval by *a priori* estimates $\mathbf{a}^*$ as in Eqs. (33-35), one can use information about typical magnitudes and variabilities of the parameters $\mathbf{a}$. For example, in atmospheric remote sensing applications, climatological data sets are often used as $\mathbf{a}^*$ [9]. If actual observations are not available one can imply *a priori* estimates $\mathbf{a}^*$ from known ranges of a_i variability:

$$a^* = <a> = (a_{max} - a_{min})/2 \text{ and } \gtrsim_{a^*} = (a_{max} - a_{min})/4. \tag{53}$$

This equation assumes the interval $[a_{max}; a_{min}]$ as 95% confidence interval $[<a> + 2 \gtrsim; <a> - 2 \gtrsim]$. These estimates a^* are biased to the middle of the interval $[a_{max}; a_{min}]$ (or to the climatological values). However, this bias is usually suppressed by a small $\gtrsim$. Indeed, the standard deviations $\gtrsim_a$ determined from Eq. (53) (or from climatology) are usually much larger than the measurement errors. Correspondingly, the *Lagrange* multipliers defined via the ratio of variances defined in Eq. (35) are likely to have small values.

The strength of smoothness constraints in Eqs. (43-45) is linked with known values of the derivatives of the retrieved $y(x)$. If an explicit analysis of derivatives $\gtrsim^m y/\gtrsim x^m$ is not feasible, the strength of smoothing can be implied from known least smooth of all *a priori* known $y(x)$. For example, Eq. (45) can be replaced by the inequality [9]:

$$\Delta\Delta \frac{4^2_{*}}{(b_m)_{max}} (N_a\Delta m) (\Delta x)^{\Delta 2m+1}, \tag{54}$$

where $(b_m)_{max}$ is the norm of the m-th derivatives of "most unsmooth" function $y(x)$. Indeed, the constrained inversion Eq. (45) with $\gtrsim$ given by Eq. (54) limits the retrievals to the functions $y(x)$ with the norm of the m-th derivatives being comparable or smaller than $(b_m)_{max}$, i.e. the retrieval of $y(x)$ much less smooth than the "most unsmooth" $y(x)$ is not allowed.

Thus, even if actual *a priori* data are not available, the values of *Lagrange* multipliers can be determined using Eqs. (53-54) before implementing actual inversion, or in another words *prior* to performing the inversion. This is a difference and a possible advantage of the approach [7-9] described here with respect to a majority of techniques established for determining *Lagrange* multipliers. Conventionally (see discussions [19,36,41]), the *Lagrange* multiplier is chosen from analysis of the sensitivity of the minimized quadratic form $\gtrsim (\mathbf{a})$ [such as given by Eq. (22)] to the weighting balance between contributions of measurements and *a priori* terms. The main idea, employed with some technical differences in many developments [1, 2, 36, 42] is that $\gtrsim$ should be both large enough ($\gtrsim > 0$) for enforcing (via constraints) a stable unique solution and small enough for allowing the algorithm to achieve a reasonably small value of minimized quadratic form $\gtrsim (\mathbf{a})$. Usually a reasonably small value means a value that can be explained by expected presence of normal noise (with no biases) in the measurements. As discussed above [Eqs. (17, 51)], for such noise the residual is $\gtrsim^2$ distributed with m-n degrees of freedom and an expected value of $<2\gtrsim (\mathbf{a})> = (N_{mes}-N_{par})\gtrsim^2_{mes}$ for $\gtrsim (\mathbf{a})$ defined via weight matrices.

The same idea is utilized in the *L-curve method* [42] where a burden between measurements and the *a priori* terms is visualized by plotting the *a priori* norm, i.e. the *a priori* term in the total $\geq$ (**a**), versus the measurements norm, i.e. the measurements term in the total $\geq$ (**a**), with $\geq$ as a function parameter. This plot has an L-shaped corner showing a point with a specific value of $\geq$ of optimum balance between measurement minimization and *a priori* terms.

In spite of the clear rationale and wide use of this optimum balance criterion for determining the *Lagrange* multiplier, there are shortcomings in employing this principle. For example, the technical implementation of this principle is very challenging if more than one *a priori* constraint is needed in the same inversion algorithm; i.e., if the determination of more than one *Lagrange* multiplier is required. Also, the implementation of a conventional determination of the *Lagrange* parameter is rather unclear in inversions of non-linear equations **f**(**a**). Non-linear inversions (see Section 5) use the first derivatives that are functions of the solutions. Therefore the optimum balance between measurements and *a priori* constraints is also a function of the solution. Correspondingly, finding optimum constraints for non-linear inversions requires extensive effort considering the entire space of possible solutions. In contrast, determining optimum constraints, e.g. by Eqs. (53-54), *prior* to the inversion relies only on the knowledge of the variances of errors in measured and *a priori* data. Such definition of a *Lagrange* multiplier is independent of the forward model and can be employed equally in both linear and non-linear algorithms. Also, knowledge about error statistics usually is established independently for each data set. Therefore, once the *Lagrange* multiplier is determined for a single type of *a priori* constraint it can be used with no changes in the inversions employing other types of *a priori* constraints.

Another and more fundamental issue is that the principle of optimum balance is based on an understanding of the limited sensitivity of observations with respect to the retrieved characteristic, but not on considerations of the available *a priori* information. For example, if observations are not sensitive to sharp oscillations of $y(x)$ then a unique inversion of such measurements is possible only if the search for solutions is restricted to smooth functions $y(x)$. Applying smoothness constraints enforces such restrictions and therefore enforces a unique and stable solution. However, the fact of achieving uniqueness and stability of the solution does not guarantee the reality of the solution. In principle, any unsmooth $y(x)$ resulting in the same observations can be a real solution. In other words, the development of a successful retrieval scheme should include two kinds of efforts: (i) identifying type and strength of constraints required for assuring the uniqueness of the solution, (ii) clarifying the realism of *a priori* information assumed by applying identified constraints. In this regard the approach discussed here gives useful insight for relating employed *a priori* constraints to actual properties of retrieved characteristics; for example, Eqs. (53-54) allow researcher to relate the values of *Lagrange* parameters with variability ranges of magnitudes and derivatives of retrieved $y(x)$. In actual applications (e.g. see [9]), the strength of constraints with *Lagrange* multipliers determined by Eqs. (53-54) acquired from knowledge available *prior* to the inversion may be not sufficient for providing satisfactory solutions. In such situations, the estimated ≥can be corrected by sensitivity studies similar to conventional approaches. Obviously, using these corrected (increased) constraints, researchers would face the above raised issue of constraint realism, since the increased constraints would exceed the actual available *a priori* knowledge. In these regards, Eqs. (53-54) can be useful for a quantitative evaluation of possible biases caused by increased *a priori* constraints.

4.7 Limitations of linear constrains.

The difficulty of enforcing non-negative solutions is an essential limitation of linear inversion methods. Indeed, the constrained *Phillips-Tikhonov-Twomey* type linear inversions defined by Eqs. (19-20) do not have a mathematical structure that allows filtering negative solutions even if the retrieved characteristic is physically positively defined. For example, remote sensing is known to suffer from the appearance of unrealistic negative values for retrieved atmospheric aerosol or gas concentrations that are positive by nature. Known techniques of securing non-negative solutions by Eqs. (19-20) force positive retrievals through enhancement of *a priori* smoothness constraints. For example, several studies [36, 42] suggest repeating linear inversion by changing strength of *a priori* constraints until the final solution both satisfies the positivity constraints and provides an admissibly accurate fit of the measurements. In such manner, *King* [36] iteratively adjusted the value of *Lagrange* parameter in Eq. (19). Similarly, *Turchin et al.* [42] iteratively corrected *a priori* terms in the statistical equivalent of Eq. (19), where ≥≥ = $\mathbf{C}_a^{-1}$ and $\mathbf{C}_a$ is considered an *a priori* "correlation" matrix. Such iterative adjustments of ≥ or *a priori* matrix $\mathbf{C}_a$ require more computations than basic constrained inversions by Eqs. (19-20). However, the major concern of implying non-negativity constraints relates to difficulties of conforming these techniques to general methodological basis of constrained linear inversions. Indeed, as was mentioned in Section 4.1, minimization of quadratic forms in the constrained inversions formally is equivalent to assuming errors normality. Correspondingly, Eqs. (19-20) are harmonized with statistical LSM optimization. Study [42] attempted to integrate the non-negativity constraints within a normal-noise framework by introducing a "cutting" normal curve; i.e., forcing zero probability for negative values and retaining a normal distribution for positive values. Such artificial cutting can hardly be accepted because it contradicts the proven symmetry of a Gaussian curve that is a fundamental property of normal noise distribution.

In contrast to Eqs. (19-20), certain types of non-linear iterations invert linear systems and naturally provide non-negative solutions. For example, in atmospheric optics, the relaxation techniques [14, 17] often are considered as alternatives to linear methods (e.g., see discussions [1, 19]).

The solution of the linear system $\mathbf{f}^* = \mathbf{K}\,\mathbf{a}$ by non-linear iterations

$$a_i^{p+1} = a_i^p \left| f_i^1 / f_i^p \right|. \tag{55}$$

is developed by *Chahine* [17]. This method is limited to application where measured and retrieved characteristics are positively defined and the number of measurements and unknowns are equal (i.e. $\mathbf{K}$ is square). Also, for convergence, square matrix $\mathbf{K}$ must be diagonally dominant (i.e., $K_{jj} > K_{jj \neq j}$). One can see that *Chahine's* formula is different from both LSM Eqs. (12) and constrained inversions Eqs. (19-20), (22-23). Namely, instead of addition and subtraction in the linear methods, Eq. (50) is non-linear and includes multiplication and division, thereby eliminating the negative and highly oscillatory solutions occasionally appearing in linear matrix inversions. However, the applicability of Eq. (55) to square and diagonal matrices $\mathbf{K}$ is a serious limitation of *Chahine's* iterations. Many inversion studies adopted *Chahine's* iterations to other situations. The most known generalization of Eq. (55) was proposed by *Twomey* [14] for inverting overdetermined systems $\mathbf{f}^* = \mathbf{K}\,\mathbf{a}$ (where $N_f > N_a$ and $\mathbf{K}$ is rectangular):

$$a_i^{p+1} = a_i^p \prod_{j=1}^{N_f} \left\{ 1 + \left[f_j^* / f_j^p \right] \Pi 1 \right\} \tilde{K}_{ji}. \tag{56}$$

$\tilde{K}_{ji}$ denotes the elements of matrix $\mathbf{K}$ that are scaled to be less than unity. Eq. (56) provides non-negative solutions while it has much broader applicability than original Eq. (55). Nevertheless, Eq. (56) has been derived without formalized analysis of the noise effects in the initial data. Such empirical character of *Chahine*-like iterations makes it difficult for a researcher to use Eqs. (55-56) as a basis for rigorous inversion optimization.

Thus, the non-negativity of solution is not an established constraint in the theoretical foundation of linear methods. On the other hand, the empirically formulated non-linear methods [Eqs. (55-56)] effectively secure positive and stable solutions. Such a "weakness" of the rigorous linear methods indicates a possible inadequacy in criteria employed for formulating the optimum solutions. In Section 6 we discuss possible revisions in assumptions employed for accounting for random noise in inversions. For example, it will be shown that by using log-normal noise assumptions the non-negativity constraints can be imposed into inversion in a fashion consistent with the presented approach inasmuch as one considers the solution as a noise optimization procedure.

4.8 Alternatives to matrix inversion methods

The main formulas for implementing LSM [Eq. (12)] and linear constraint inversions [Eqs. (19, 20, 23, 24, 32)] are written via a matrix inversion operator. This operator is uncertain for ill-posed problems where the matrices to be inverted [$\mathbf{K}$ in Eq(3), $\mathbf{K}^T\mathbf{K}$ in Eqs. (19-21), $\mathbf{K}^T\mathbf{C}^{-1}\mathbf{K}$ in Eqs. (12, 23), etc.] tend to have zero determinant. This is why a number of studies associate solving ill-posed problems with the use of advanced numerical procedures introducing an inverse operator to degenerated matrices. For example, an inverse operator may be formulated by excluding eigensolutions, linear combinations of unknowns, with zero eigenvalues (e.g. see [43]). Singular value decomposition (SVD) is a particularly popular approach for inverting degenerated matrices. SVD is an operation of linear algebra (see details in [4]), that allow one to decompose a square matrix $\mathbf{M}$ as $\mathbf{M} = \mathbf{VI}_w\mathbf{A}$, where matrices $\mathbf{V}$ and $\mathbf{A}$ are orthogonal in the sense that $\mathbf{V}^T\mathbf{V} = \mathbf{I}$ and $\mathbf{A}^T\mathbf{A} = \mathbf{I}$. Matrix $\mathbf{I}_w$ is diagonal with the elements on the diagonal equal to w_i. Inversion of matrix $\mathbf{M}$ trivially follows from this decomposition as $\mathbf{M} = \mathbf{A}^T\mathbf{I}_{1/w}\mathbf{V}^T$. In the case of a singular matrix $\mathbf{M}$, the inverse matrix of $\mathbf{M}^{-1}$ is uncertain, because some values w_i are equal or close to zero. Correspondingly, by means of replacing $w_i = 0$ by a moderately small non-zero w_i, singular matrix $\mathbf{M}$ can be replaced by a reasonably similar, non-singular "truncated" matrix $\mathbf{M}'$ that can be trivially inverted. Therefore, some theoretical developments consider applying SVD as an alternative way of constraining linear-system solutions. For example, the theoretical review by *Hansen* [42] considers a truncated SVD method as an essential equivalent of *Pillips-Tikhonov-Twomey* constrained inversion by Eqs. (19-20). The main concern of using the SVD technique instead of direct *a priori* constraints comes from the fact that replacement of matrix $\mathbf{M}$ with truncated matrix $\mathbf{M}'$ is formal and has no relation to the physics of an application. Therefore using SVD should be accompanied by an analysis clarifying how the solution space was restricted by using truncated matrix $\mathbf{M}'$. Such analysis is challenging, since some linear combinations of unknown parameters excluded by truncation may not have clear physical meaning. Also, SVD analysis becomes even more uncertain in non-linear inversions where matrix $\mathbf{M}$ and it's truncated analog $\mathbf{M}'$ changes during iterations. On the other hand, using SVD instead of direct matrix inversion is undoubtedly a useful tool for improving implementation of constrained inversion. For example, study [9] used SVD to solve a multi-term normal system [Eq. (32)]. In this way the initially ill-posed problem is constrained by *a priori* terms in quadratic form given by Eq. (30), and SVD is applied only for solving Eq. (32) that improves the technical performance of inversion algorithm. Indeed, in case of large matrices $\mathbf{M}$, applying standard methods for matrix inversion can be problematic even for non-singular $\mathbf{M}$, while SVD always gives an inverse operator.

Another alternative to matrix inversion is using linear iterations written by Eq. (4). As discussed in Section 2, linear iteration always provides a solution, even if a linear system [Eq. (2)] has singular matrix $\mathbf{K}$. For example, steepest descent method (e.g. see [44-45]) always converges to a solution (more in Section 5). However, in case of singular matrices $\mathbf{K}$, iterative methods provide only one solution from many possible. Repeating iterations using different initial guesses may provide information about the entire space of possible solutions. However, building a domain of solutions in such a way is not straightforward because it requires establishing a set of initializations providing complete coverage of solution space. Also, in general, iterative methods are more time consuming than matrix inversion. For example, steepest decent may require an enormous number of iterations for convergence (e.g. see [3]).

The semi-iterative method of conjugated gradients is another popular method of solving linear systems of Eq. (1) via N_a iterations [4]. However, applying conjugated-gradients algorithms to solving a quasi-degenerated linear system $(\det(\mathbf{K}) \geq 0)$ results in similar problems as conventional numerical procedures used for inverting matrix (e.g., *Gauss-Jordan* elimination [4]). Although, in the framework of the conjugated-gradients technique, some uncertain components of solution can be suppressed and, as discussed in [41], conjugated gradients may provide a solution close to that of truncated SVD.

There are many other non-matrix inversion methods that are not considered here. Some techniques are based on concepts that are very different standard methods of numerical inversion. For example, using neural networks [46] is a technique that is popular in many applications for observations analysis. The basic idea of neural networking is that prior to interpretation of observations, the researcher establishes unique relations between observations and unknowns via network "training". Network "training" is an analysis that is, in some sense, similar to identifying non-linear regressions between output and input of forward model. Another example is generic-inversions methods that rely entirely on forward simulations [47-48]. Such techniques implement inversion by straightforward computer search for all solutions that admissibly agree with the observations to be inverted. Study [40] proposed a technique that can be considered as a combination of generic inversion with conventional linear inversion. Specifically, the technique implements a large number of inversions using *Phillips-Tikhonov-Twomey* [Eq. (19)] with different values of *Lagrange* multipliers. The average result of such inversions is considered as a suggested solution.

Thus, there are many techniques, only some of which are mentioned here, that can provide an appropriate inverse transformation without implementing direct matrix inversion. Some of these techniques may provide a reasonable solution of ill-posed problems without using explicit *a priori* constraints. However, in comparing those methods against LSM-based constrained inversion [Eqs. (19, 20, 23, 24, 32)], one should realize that each method *a priori* limits a space of possible solutions. For instance, SVD inversion excludes some solutions via matrix truncation; iterative solutions depend on the initial guess; generic inversion considers only solutions included in a search; neural networks are constrained by training process, etc. Therefore, applying all these techniques to an ill-posed problem may result in different solutions from different techniques. These differences reflect differences in employed *a priori* constraints. Hence, the methods adopting the most reliable *a priori* constraints provide the best solution. In this regards, applying *a priori* constraints in a manner consistent with statistical optimization, as shown in Section 4.2, give rigorous and a clear concept for using *a priori* constraints and combining various data in single inversion. Contrary, absence of direct optimization of statistical properties in some inversion approaches can be considered disadvantageous.

5. Optimization of non-linear inversion

Sections 2-4 discussed inversions procedures only for the case of linear forward model $\mathbf{f(a)}$ in Eq. (1). However in practice, and particularly in remote-sensing applications, the majority of physical dependencies $\mathbf{f(a)}$ are non-linear. The purpose of this Section is to discuss inversion of a non-linear Eq. (1) and to outline the differences and similarities between linear and non-linear cases.

5.1. Basic inversions of non-linear equation system

For a case of non-linear functions $f_j(\mathbf{a})$, Eq. (1) usually is solved numerically by iterations relying on linear approximations. Namely, for points $\hat{\mathbf{a}}$ in the close neighborhood of solution $\mathbf{a}'$, $\mathbf{f(a)}$ can be expanded in *Taylor* series:

$$\mathbf{f(a)} = \mathbf{f(\hat{a})} + \mathbf{K}_{\hat{a}}\ (\mathbf{a} - \hat{\mathbf{a}}) + o(\mathbf{a} - \hat{\mathbf{a}})^2 + \dots \ , \tag{57}$$

where $\mathbf{K}_{\hat{a}}$ is the *Jacobi* matrix of the first derivatives $\partial f_j / \partial a_i$ in the near vicinity of $\hat{\mathbf{a}}$; $o(\mathbf{a} - \hat{\mathbf{a}})^2$ denotes the function that approaches zero as $(\mathbf{a} - \hat{\mathbf{a}})^2$ when $(\mathbf{a} \to \hat{\mathbf{a}}) \to 0$. Hence, neglecting all terms of second or higher order in Eq. (57), $f_j(\mathbf{a})$ can be considered as linear functions. Such a linear approximation is insufficient to solve Eq. (1) by Eq. (3) directly through inverse transformation, but can be employed successfully for iterative correction of guessed solution:

$$\mathbf{a}^{p+1} = \mathbf{a}^p - \geq \mathbf{a}^p, \tag{58a}$$

$$\mathbf{K}_p \geq \mathbf{a}^p \geq \mathbf{f(a}^p) - \mathbf{f}^*, \tag{58b}$$

where $\mathbf{a}^p$ is p-th approximation of solution, $\geq \mathbf{a}^p$ is a correction of $\mathbf{a}^p$ that is given by the solution of Eq. (58b), where $\mathbf{K}_p$ is the *Jacobi* matrix of $\partial f_j / \partial a_i$ calculated in the vicinity of $\mathbf{a}^p$. In the situation where $\mathbf{K}_p$ is square ($N_f = N_a$), the successive iterations can be implemented by employing inverse matrices:

$$\mathbf{a}^{p+1} = \mathbf{a}^p - \mathbf{K}_p^{-1}\ (\mathbf{f(a}^p) - \mathbf{f}^*). \tag{59}$$

This is the basic formula of *Newton* iterations to solve non-linear systems.

5.2 Optimization of non-linear solution in presence of random noise

In case of over determined non-linear $f_j(\mathbf{a})$ in Eq. (1), statistical optimization of a solution can be included in the iterations by following MML as described in Section 2. Namely, the solution of Eq. (1) should be performed as minimization of quadratic form $\geq (\mathbf{a})$ given by Eq. (10) and the resulting non-linear Eq. (11a) can be solved by Newton iterations. Replacing $\mathbf{f}(\mathbf{a}^p)$ by gradient $\geq\geq (\mathbf{a}^p)$ and $\mathbf{f}^*$ by $\mathbf{0}$, Eqs. (58) can be written as:

$$\mathbf{a}^{p+1} = \mathbf{a}^p - \geq\mathbf{a}^p, \tag{60a}$$

$$\mathbf{K}_{\geq,p} \geq \mathbf{a}^p \geq \geq\geq (\mathbf{a}^p), \tag{60b}$$

where $\mathbf{K}_\geq$ is a matrix of partial derivatives with elements

$$\left\{ \mathbf{K}_{\Psi,p} \right\}_{ji} = \left. \frac{\Psi(\Psi\Psi)_j}{\Psi_i} \right|_{\mathbf{a}^p}. \tag{61a}$$

The matrix $\mathbf{K}_{\geq,p}$ follows from Eq. (10c) as:

$$\mathbf{K}_{\geq,p} = \mathbf{K}_p^T \mathbf{C}^{-1} \mathbf{K}_p . \tag{61b}$$

Finally, using inverse matrices, MML solution can be written as:

$$\mathbf{a}^{p+1} = \mathbf{a}^p - (\mathbf{K}_p^T \mathbf{C}^{-1} \mathbf{K}_p)^{-1} \geq\geq (\mathbf{a})$$
$$= \mathbf{a}^p - (\mathbf{K}_p^T \mathbf{C}^{-1} \mathbf{K}_p)^{-1} \mathbf{K}_p^T \mathbf{C}^{-1}(\mathbf{f}(\mathbf{a}^p)-\mathbf{f}^*). \tag{62}$$

This equation is known as the *Gauss-Newton* method [49]. For square matrices $\mathbf{K}$, Eq. (62) can be reduced to *Newton* iterations using matrix identity: $(\mathbf{K}^T\mathbf{C}^{-1} \mathbf{K})^{-1} = \mathbf{K}^{-1}\mathbf{C}(\mathbf{K}^T)^{-1}$. This is why solving Eq. (2) with square $\mathbf{K}$, as well as, *Newton* method also can be considered as a minimization of quadratic form [45].

It should be noted that quadratic form $\geq (\mathbf{a})$ [Eq. (11a)] can be minimized by iterations different from Eq. (62). Many such methods also utilize gradient $\geq\geq (\mathbf{a})$ for the solution search. The steepest descent method deserves particular attention among all other techniques. This method correct solution guess $\mathbf{a}^p$ relying only on gradient $\geq\geq (\mathbf{a}^p)$ in point $\mathbf{a}^p$:

$$\mathbf{a}^{p+1} = \mathbf{a}^p - t_p \geq\geq (\mathbf{a})$$
$$= \mathbf{a}^p - t_p \mathbf{K}_p^T \mathbf{C}^{-1}(\mathbf{f}(\mathbf{a}^p)-\mathbf{f}^*), \tag{63}$$

where the coefficient $0 < t_p \geq 1$ is selected empirically to provide convergence. Since gradient $\geq\geq (\mathbf{a}^p)$ shows the direction of the strongest local change of $\geq (\mathbf{a}^p)$, the steepest descent always converges [3, 44-45]. However, implementing this method may take a very long time [3].

5.3 *Levenberg- Marquardt* optimization of iteration convergence

Implementing non-linear inversion by *Newton*-like methods requires assurance of iteration convergence. Iteration by Eqs. (59, 62) may not converge or converge to a wrong solution. The convergence difficulties may be caused by inadequate choice of the initial guess and/or limitations of the linear approximation used for guess correction. Indeed, for strongly non-linear functions $f_j(\mathbf{a})$, the minimized form $\geq (\mathbf{a})$ may have a complex structure with several minima. The analysis of this structure is desirable prior to inversion. However, when three or more unknowns are to be retrieved, such analysis is practically not feasible. Usually, researchers repeat retrieval with a set of initializations and select the best solution. The initializations and the criteria for selecting the best solution are commonly established based on the physical constraints of the application, experience, and intuition of the developers. Also, a convergence of non-linear solutions can be improved by modifying Eqs. (59, 62). The most established modification of *Gauss-Newton* iterations is widely known as the *Levenberg-Marquardt* method [4,49]:

$$\mathbf{a}^{p+1} = \mathbf{a}^p - t_p \, (\mathbf{K}_p{}^T\mathbf{C}^{-1}\mathbf{K}_p + \geq \mathbf{D})^{-1} \, \mathbf{K}_p{}^T\mathbf{C}^{-1}(\mathbf{f}(\mathbf{a}^p)\text{-}\mathbf{f}^*), \qquad (64)$$

where matrix $\mathbf{D}$ and the coefficients $0 < t_p$ and $0 \geq \geq$ are selected empirically to provide convergence. The matrix $\mathbf{D}$ is predominantly diagonal (unity matrix is often chosen as $\mathbf{D}$) and addition of the term $\geq \mathbf{D}$ to $\mathbf{K}_p{}^T\mathbf{C}^{-1}\mathbf{K}_p$ in Eq. (64) is analogous to using *a priori* constraints in linear inversions. Specifically, the matrix $\mathbf{K}_p{}^T\mathbf{C}^{-1}\mathbf{K}_p$ can be singular on some of *p*-th iterations even if it is non-singular in the solution neighborhood. Adding the term $\geq \mathbf{D}$ to $\mathbf{K}_p{}^T\mathbf{C}^{-1}\mathbf{K}_p$ helps to pass the iteration process through areas of $\mathbf{K}_p{}^T\mathbf{C}^{-1}\mathbf{K}_p$ singularities. As pointed out in [4], the *Levenberg-Marquardt* formula generalizes the steepest descent method. Namely, Eq. (64) can be reduced to (63) by defining matrix $\mathbf{D}$ in Eq. (64) as the unit matrix $\mathbf{I}$ and prescribing a large value to the parameter $\geq$. Thus, Eq. (64) always converge with appropriate $\geq$.

The multiplier $0 < t_p \geq 1$ in Eq. (64) is invoked mainly to decrease the length of $\geq \mathbf{a}^p$, because the linear approximation may overestimate the correction $\geq \mathbf{a}^p$. Usually, t_p is decreased by a factor (e.g. by 2) until a condition $\geq (\mathbf{a}^{p+1}) < \geq (\mathbf{a}^p)$ is satisfied. Underestimation of $\geq \mathbf{a}^p$ does not lead to a convergence failure and may only slow down the arrival to a solution. The addition of the term $\geq \mathbf{D}$ also reduces $\geq \mathbf{a}^p$. Correspondingly, using both $\geq \mathbf{D}$ ($\geq > 0$) and $0 < t_p \geq 1$ in the same iteration may seem redundant because both operations reduce $\geq \mathbf{a}^p$. However, using the multiplier t_p is straightforward (compare to adding $\geq \mathbf{D}$) and sufficient in the application with moderately non-linear forward model with non-singular $\mathbf{K}_p{}^T\mathbf{C}^{-1}\mathbf{K}_p$. On the other hand, if the matrix $\mathbf{K}_p{}^T\mathbf{C}^{-1}\mathbf{K}_p$ is singular in some points $\mathbf{a}^p$, using $t_p \geq 1$ does not help and inclusion of constraining term $\geq \mathbf{D}$ is necessary. Thus, use of both t_p and $\geq \mathbf{D}$ modifications in *Levenberg-Marquardt* [Eq. (64)] complement each other in practice.

5.4 Formulation of *Levenberg- Marquardt* iterations using statistical formalism

This Section is aimed to show that statistical considerations analogous to those in Section 4 can be useful for optimizing *Levenberg-Marquardt* iterations.

Gauss-Newton Eq. (62) trivially can be generalized for simultaneous inversion of multi-source data. Specifically, Eq. (25) with non-linear forward models $\mathbf{f}_k(\mathbf{a})$ can be solved by multi-term equivalent of non-linear LSM :

$$\hat{a}^{p+1} = \hat{a}^p \mid \left| \sum_{k=1}^{K} l_k \left(\mathbf{K}_{k,p}\right)^{\mathrm{T}} \left(\mathbf{w}_k\right)^{11} \left(\mathbf{K}_{k,p}\right) \right|^{-1} \left| \sum_{k=1}^{K} l_k \left(\mathbf{K}_{k,p}\right)^{\mathrm{T}} \left(\mathbf{w}_k\right)^{11} \left(\mathbf{f}_k\left(\hat{a}^p\right) \mid \mathbf{f}_k^*\right) \right|, \quad (65)$$

where $\mathbf{K}_{k,p}$ is *Jakobi* matrix of the first derivatives from $\mathbf{f}_k(\mathbf{a})$ in the vicinity of $\mathbf{a}^p$. As discussed above, employing linear approximations for non-linear functions $\mathbf{f}_k(\mathbf{a})$ in Eq. (65) may result in a convergence failure. Therefore, if some $\mathbf{f}_k(\mathbf{a})$ are linear, it seems logical to expect fewer problems with convergence. This idea can be elaborated by considering linear constraints applied to non-linear iterations. Namely, the non-linear equivalent of Eq. (47b) that solves Eq. (46) with non-linear $\mathbf{f}(\mathbf{a})$ can be written as:

$$\hat{a}^p = \hat{a}^{p+1} \Delta \Delta \hat{a}^p, \quad (66a)$$

$$\overset{\Delta}{\underset{\Delta}{\mathbf{K}}}_p^{\mathrm{T}} \mathbf{W}^{\Delta 1} \mathbf{K}_p + \varDelta_2 \Delta_{\mathrm{m}} + \varDelta_3 \mathbf{W}_{\mathbf{a}^*}^{\Delta 1} \overset{\Delta}{\underset{\Delta}{\Delta}} \hat{a}^p =$$

$$= \mathbf{K}_p^{\mathrm{T}} \mathbf{W}^{\Delta 1} \left(\mathbf{f}\left(\hat{a}^p\right) \Delta \mathbf{f}^*\right) + \varDelta_2 \Delta_{\mathrm{m}} \hat{a}^p + \varDelta_3 \mathbf{W}_{\mathbf{a}^*}^{\Delta 1} \left(\hat{a}^p \Delta \hat{a}^*\right) \quad , \quad (66b)$$

where $\mathbf{\geq}_{\mathrm{m}} = \mathbf{S}_{\mathrm{m}}^{\mathrm{T}} \mathbf{S}_{\mathrm{m}}$. Although, Eq. (66b) is constrained by *a priori* terms, solution $\hat{a}^p$ may fail to converge because at initial iterations ($p=1,2,...$) the constrained non-linear Eq. (66) does not differ significantly from non-constrained Eq. (62), similar to basic *Gauss-Newton* iterations by Eq. (62). Indeed, if the initial guess is far from the solution, the values $(f_j(\mathbf{a}^p) - f_j^*)$ are large and the measurement term dominates over *a priori* terms because the values of the *Lagrange* multipliers $\geq_2$ and $\geq_3$ are typically small. *A priori* terms start to matter only when fitting differences $(\mathbf{f}(\mathbf{a}^p) - \mathbf{f}^*)$ reach the level of measurement accuracy $\geq_1$. Therefore, some enhancement of *a priori* terms at initial iterations may improve performance of Eq. (66). This idea can be elaborated in the following considerations.

Each p-th iteration in Eq. (66) assumes the solution of the following overdetermined linear system:

$$\begin{aligned}
&\overset{\Rightarrow}{\underset{\Rightarrow}{\mathbf{K}}}_{1,p} = \mathbf{a}^p = \mathbf{f}_1(\mathbf{a}^p) = \mathbf{f}_1^* + \mathbf{f}_1^* + \mathbf{f}_{1,p}^{\mathrm{lin}} & \overset{\Rightarrow}{\underset{\Rightarrow}{\mathbf{K}}}_p = \mathbf{a}^p = \mathbf{f}(\mathbf{a}^p) = \mathbf{f}^* + \mathbf{f}^* + \mathbf{f}_p^{\mathrm{lin}} \\
&\overset{\Rightarrow}{\mathbf{K}}_{2,p} = \mathbf{a}^p = \mathbf{f}_2(\mathbf{a}^p) = \mathbf{f}_2^* + \mathbf{f}_2^* + \mathbf{f}_{2,p}^{\mathrm{lin}} \Rightarrow & \overset{\Rightarrow}{\mathbf{S}}_{\mathrm{m}} = \mathbf{a}^p = \mathbf{S}_{\mathrm{m}} \mathbf{a}^p = \mathbf{0}^* + \overset{m}{\Rightarrow}{\mathbf{a}})^* \quad , (67) \\
&\overset{\Rightarrow}{\underset{\Rightarrow}{\mathbf{K}}}_{3,p} = \mathbf{a}^p = \mathbf{f}_3(\mathbf{a}^p) = \mathbf{f}_3^* + \mathbf{f}_3^* + \mathbf{f}_{3,p}^{\mathrm{lin}} & = \mathbf{a}^p = \mathbf{a}^p = \mathbf{a}^* + \mathbf{a}
\end{aligned}$$

where $\Delta \mathbf{f}_{k,p}^{\mathrm{lin}}$ denotes the errors of using the linear approximation of ($\mathbf{f}_k(\mathbf{a}^p) - \mathbf{f}_k^*$) in the vicinity of $\mathbf{a}^p$. In contrast to the linear case [Eq. (46)], Eq. (67) is written via differences $\geq \mathbf{a}^p$. Another difference is that the first equation in system (66) includes linearization errors $\Delta \mathbf{f}^{\mathrm{lin}}$. As discussed in Section 4, LSM optimization weights the contributions inversely to variances $\geq_k^2$ of errors $\geq \mathbf{f}_k^*$ [see Eqs. (29-31)]. Such weighting does not account for linearization errors $\Delta \mathbf{f}^{\mathrm{lin}}$ and, therefore, optimizes results only in close vicinity to the actual solution where $\Delta \mathbf{f}^{\mathrm{lin}}$ are small. Accounting for $\Delta \mathbf{f}^{\mathrm{lin}}$ can be introduced into LSM weighting by using $\geq_l^2 + (\geq_{l,\mathrm{lin}})^2$ instead of $\geq_l^2$ in the *Lagrange* multipliers definition [Eq. (33b)]. The value of the $\geq \mathbf{f}_1^{lin}$ variance is not known at each point $\mathbf{a}^p$, but can be estimated from the value of the residual, i.e. analogously to Eq. (52) one can write the following:

$$\hat{\gamma}^2 + (\hat{\gamma}_{1,lin}(\mathbf{a}^P))^2 - \frac{2-(\mathbf{a}^P)}{-\sum_{i=1,\dots,3}(N_{\mathbf{f}_i})-N_{\mathbf{a}}} = \frac{2-(\mathbf{a}^P)}{N_{\mathbf{f}*}+N_{\mathbf{a}}-m}. \tag{68}$$

Using this equation, Eq. (30) can be re-written for non-linear iterations:

$$\gamma_k(\mathbf{a}^P) = \frac{\hat{\gamma}^2 + (\hat{\gamma}_{1,lin}(\mathbf{a}^P))^2}{\gamma_k^2} - \frac{2-(\mathbf{a}^P)}{\gamma_k^2 \left(-\sum_{i=1,\dots,3}(N_{\mathbf{f}_i})-N_{\mathbf{a}}\right)}. \tag{69}$$

This definition of the *Lagrange* multiplier accounts for higher linearization errors at earlier iterations. In close vicinity of the solution $\mathbf{a}'$, where $\geq \mathbf{f}_1^{lin}$ is close to zero, Eq. (68) is reduced to Eq. (52). Hence, utilizing "adjustable" $\gamma_k (k \geq 2)$ in Eq. (66) improves convergence while the final solution retains the same statistical properties.

Derivations similar to those given by Eqs. (67-69) can be used even if actual *a priori* information is not available. For instance, if there is no *a priori* knowledge about magnitudes of unknowns or their correlations (smoothness), one can require such constraints on corrections $\geq \mathbf{a}^P$. Indeed, the restrictions on $\geq \mathbf{a}^P = \mathbf{a}^P - \mathbf{a}^{p+1}$ would not restrict the area of admissible solutions. For example, assuming that linearization may cause an overestimation of $\geq \mathbf{a}^P$, one may force $\geq \mathbf{a}^P$ to small values in order to retain monotonic convergence. Such constraint limits departures of $\mathbf{a}^{p+1}$ from $\mathbf{a}^P$, but it does not limit the values $\mathbf{a}^P$. The possible negative side effect of limiting $\geq \mathbf{a}^P$ is a larger number of iterations if the initial guess $\mathbf{a}^0$ is taken far from the real solution. Similarly, retrieving functions $y(x)$ of $a_i = y(x_i)$, one may require smooth corrections $-y^P(x) = y^{P+1}(x) - y^P(x)$. This does not put constraints on retrieved $y(x)$ and, even if $y^0(x)$ was smooth, a large number of smooth corrections $\geq y^P(x)$ may result in unsmooth $y^{P+1}(x)$. Thus, *Gauss-Newton* iterations can be implemented with use of constraints on $\geq \mathbf{a}^P$ as follows:

$$\mathbf{K}_p \lfloor \mathbf{a}^P \lfloor \mathbf{f}(\mathbf{a}^P) \lfloor \mathbf{f}^* + \lfloor \mathbf{f}^* + \lfloor \mathbf{f}_p^{lin} \quad \begin{cases} \mathbf{K}_p \lfloor \mathbf{a}^P \lfloor \mathbf{f}(\mathbf{a}^P) \lfloor \mathbf{f}^* + \lfloor \mathbf{f}^* + \lfloor \mathbf{f}_p^{lin} \\ \mathbf{S}_m \lfloor \mathbf{a}^P = 0* + \lfloor (\lfloor^m \mathbf{a})^* \\ \lfloor \mathbf{a}^P = 0* + \lfloor \mathbf{a}^* \end{cases} . \tag{70}$$

Here the second and third equations constrain smoothness and magnitudes of the corrections $y^P(x)$, respectively. Hence, *Gauss-Newton* iterations of Eq. (62) can be implemented as a multi-term LSM solving Eq. (70):

$$\lfloor \hat{\mathbf{a}}^{P+1} = \hat{\mathbf{a}}^P - \left| \mathbf{K}_p^T \mathbf{W}^{l\,l} \mathbf{K}_p + l_2 \right|_m + l_3 \mathbf{W}_{\mathbf{a}*}^{l\,l} \left|^{l\,l} \mathbf{K}_p^T \mathbf{W}^{l\,l} \left(\mathbf{f}(\hat{\mathbf{a}}^P) \lfloor \mathbf{f}_k^* \right). \tag{71a}$$

Lagrange multipliers $\gimel_2$ and $\gimel_3$ can be determined analogously to Eq. (69) as:

$$-_k\left(\mathbf{a}^p\right) = \frac{\hat{-}_1^2 + (\hat{-}_{1,\lin}(\mathbf{a}^p))^2}{-_k^2} - \frac{2-(\mathbf{a}^p)}{-_k^2\left(N_{\mathbf{f}*} - N_{\mathbf{a}}\right)}. \tag{71b}$$

Following considerations of Section 4.6, the errors $\geqslant_2$ and $\geqslant_3$ can be established from the general knowledge of physically admissible variability of magnitudes of a_i and/or derivatives of retrieved $y^p(x)$.

Thus, the above derivation that optimizes the solution by constraining $\geq \mathbf{a}^p$ resulting in Eq. (71), which is analogous to Eq. (64), provides additional insight to the formulating term $\geq \mathbf{D}$ in the *Levenberg-Marquardt* iterations. Such an approach is employed in inversion algorithm [9] and shown to be efficient in practice for deriving aerosol properties from remote-sensing observations.

6. Possible adjustments to assumption of Normal noise

Most inversion-algorithm-optimizing solutions are based on the normal noise assumption when in the presence of random noise (see Sections 3-4). This includes even algorithms that are not based on statistical formalism, since minimization of quadratic forms is formally equivalent to assuming of Normal noise. However, in scientific literature, one can find numerous attempts of using alternative noise assumptions. Indeed, MML given by Eq. (8) does not assume this specific type of PDF and gives an optimized solution for any noise distribution, provided the assumed noise distribution is close to reality. For example, assuming that $P(f_j(\mathbf{a})\text{-}f_j^*) \sim \exp(\text{-}\geq(f_j(\mathbf{a})\text{-}f_j^*)\geq)$ leads to the *minimax* methods that differ from the LSM search for the least sum of absolute deviations. The details of implementing *minimax* and other methods based on the noise assumptions alternative to normal noise can be found in various textbooks [3-4]. This Section discusses only a few modifications of the *Gaussian* noise assumption aimed to overcome particular difficulties in performance of the LSM.

6.1. Non-negativity constraints

As was discussed in Section 4.7, the difficulty in securing positive solutions in the retrieval of non-negative characteristics is a limitation of constrained linear inversion. This issue can be addressed by using a lognormal noise assumption in retrieval optimization [7-9]. Such assumption of lognormal noise leads to implementing inversions in logarithmic space, i.e. employing logarithmic transformation of forward model.

Retrieval of logarithms of a physical characteristic, instead of absolute values is an obvious way to avoid negative values for positively defined values. However, the literature devoted to inversion techniques tends to consider this apparently useful tactic as an artificial trick rather than a scientific approach to optimize solutions. Such misconception is probably caused by the fact that the pioneering efforts on inversion optimization by

Phillips [11], *Tikhonov* [12] and *Twomey* [13] and many later theoretical considerations (e.g. *Hansen* [41]) were devoted to solving the *Fredholm* integral equation of the first kind, i.e. a system of linear equations produced by quadrature. The problems addressed by these methods are the retrieval of aerosol size distribution [35] or temperature profile of the atmosphere [19] by inverting spectral dependence of optical thickness. Considering optical thickness as a function of the logarithm of the aerosol concentrations or temperature profile requires replacing the initial linear equation $\mathbf{f} = \mathbf{K}\,\mathbf{a}$ by nonlinear ones $f_j = f_j(\ln a_i)$. On the face of it, such a transformation of linear problems to non-linear ones is not enthusiastically accepted by the scientific community as an optimization. On the other hand, in cases when a forward model is a nonlinear function of parameters to be retrieved (e.g., atmospheric remote sensing in cases when multiple scattering effects are significant), the retrieval of logarithms is accepted as a logical approach. Besides, as discussed in Section 4.7 non-linear *Chahine*-like iterations have proven to be efficient for inverting linear system. Rigorous statistical considerations also reveal some limitations in applying *Gaussian* functions for modeling errors in measurement of positively defined characteristics. It is well known that the curve of the normal distribution is symmetric. In other words, one may affirm that the assumption of a normal PDF is equivalent to the assumption of the principal possibility of obtaining negative results even in the case of physically nonnegative values. For such nonnegative characteristics as intensities, fluxes, etc., the choice of a log-normal distribution for describing the measurement noise seems to be more correct due to the following considerations: (i) log-normally distributed values are positively-defined; (ii) there are a number of theoretical and experimental reasons showing that for positively defined characteristics, the log-normal curve with its multiplicative errors (see [27]) is closer to reality than normal noise with additive errors. Also, as follows from the discussion of statistical experiments [3], the lognormal distribution is best at modeling random deviations in non-negatives values. Besides, using the lognormal PDF for noise optimization does not require any revision of normal concepts and can be implemented by simple transformation of the problem to the space of normally distributed logarithms. This fact is very important from the viewpoint of both theoretical consideration and practical implementation of MML under the lognormal noise assumption. For example, due to the problem of differentiating $P(f_j(\mathbf{a})\text{-}f_j^*) \sim \exp(\text{-}\geq(f_j(\mathbf{a})\text{-}f_j^*)\geq)$, formulation of basic equations for *minimax* solutions is questionable.

Similar to the above considerations of non-negative measurements, there is a clear rational in retrieving logarithms of unknowns instead of their absolute values, e.g., $\ln(y(x_i))$ instead of $y(x_i)$, provided the retrieved characteristics are positively defined. Although, the MML does not implicitly assume a distribution of errors in the final solution, the statistical properties of the MML solution are well studied (see [27]) and, therefore can be projected in algorithm developments. In fact, according to statistical estimation theory, if PDF is normal, the MML estimates are also normally distributed. It is obvious then, that the LSM algorithm retrieving $y(x_i)$ would provide normally distributed estimates $y(x_i)$ and, therefore, it cannot provide zero probability for $y(x_i) < 0$, even if $y(x_i)$ are positively defined by nature. On the other hand, the retrieval of logarithms instead of absolute values eliminates the above contradiction because the LSM estimates of $\ln(y(x_i))$ would have a normal distribution of $\ln(y(x_i))$, i.e. a lognormal distribution of $y(x_i)$ that assures positivity of non-negative $y(x_i)$. Moreover, studies [7, 9] suggest considering the logarithmic transformation as one of the cornerstones of the practical efficiency of *Chahine's* iterative procedures. The derivations of Eqs. (55-56) from LSM formulated in logarithmic space are given in the Appendices of reference [9].

Thus, accounting for non-negativity of solutions and/or non-negativity of measurements can be implemented in the retrieval by using logarithms of unknowns ($a_i \geq \ln a_i$) and/or measurements ($f_j \geq \ln f_j$). In many situations, retrieval of absolute values or their logarithms is practically similar. This is because narrow lognormal or normal noise distributions are almost equivalent. For example, for small variations of non-negative value, the following relationship between $\geq a$ and $\geq a/a$ is valid:

$$\geq \ln a = \ln(a + \geq a) - \ln(a) \geq \geq a/a, \quad (\text{if } \geq \ln a \ll 1). \tag{72a}$$

Then, if only small relative variations of value a are allowed, the normal distribution of $\ln a$ is almost equivalent to the normal distribution of absolute values a. The covariance matrices of these distributions are connected as:

$$\mathbf{C}_{\ln a} \geq (\mathbf{I}_a)^{-1} \mathbf{C}_a (\mathbf{I}_a)^{-1}, \tag{72b}$$

where $\mathbf{I}_a$ is a diagonal matrix with elements $\{\mathbf{I}_a\}_{ii} = a_i$. Hence, for measurements with small relative errors, use of lognormal or normal PDFs with covariance matrices related by Eq. (72) should give similar results. Also, since logarithmic errors can be considered approximately as relative errors, the variances $(\geq_{\ln f})^2$ are unitless and, therefore, Eq. (30b) defining *Lagrange* multipliers as the variance ratio becomes particularly useful. Practical illustrations of using logarithmic transformations in inversion can be found in reference [9].

6.2. Accounting for the data redundancy

A difficulty in accounting for data redundancy is another unresloved issue in implementing optimized inversion. This issue has very high practical importance, although it is not often addressed in the literature on inversion methodologies. For example, infinite enhancement of spectral and/or angular resolutions in remote-sensing observation does not lead to accuracy improvements in retrievals above a certain level. Based on common sense this can be explained by the fact that simple increase of the number of observations N_f may lead to an increasing number of redundant measurements that do not help to improve retrievals. Theoretical considerations (e.g. in Section 3), however, do not assume any "redundant" or "useless" observations. Indeed, performing N_{f_j} straightforward repetitions of the same observation with established unchanged accuracy, from a statistical viewpoint, simply means that the variance of this particular observation f_j should decrease by factor N_{f_j}. Accordingly, the j-th elements of covariance matrix $\mathbf{C_f}$ should decrease and the errors of retrieved parameters [Eqs. (15), (49)] should decrease appropriately. Thus, from a theoretical viewpoint, repeating similar or even the same observation always results in some enhancements of retrieval accuracy. Such contradiction between practical experience and theoretical derivations seriously limits the efforts on estimating retrieval errors, evaluating information content of measurements and planning of optimum experiments. For the multi-term LSM approach presented here, accounting for data redundancy is also of particular importance. Indeed, individual data points from observations of the same type usually are comparable in accuracy. Therefore, it is unlikely, although possible, that inverting single-source data would not lead to a discrimination of some individual observations. In a multi-source inversion, the situation is different because an increase in the number of observations in one of several inverted sets of data would lead to an increase of the weight of this data set, even if the added observations were redundant from a practical point of view. Indeed, in the minimized quadratic form of $\Psi(\mathbf{a})$ in Eq. (29), the higher the value of the k-th term $\geq_k$, the stronger the contribution of the k-th data set on the solution. Using known relationships for the $\geq^2$ distribution, $\geq_k$ can be estimated as $\geq_k \geq N_k$, i.e. the weight of the k-th term in Eq. (29) is proportional to the number of measurements N_k in the k-th data set. In order to eliminate this dependence of $\geq_k$ on N_k, it was suggested [9] that for redundant observations, the accuracy of a single measurement degrades as $1/N_k$ if several measurements are taken simultaneously, i.e.:

$$\varepsilon_k^2(\text{multiple}) = N_k \ \varepsilon_k^2(\text{single}), \tag{73}$$

where the term "multiple" indicates that several analogous measurements are made simultaneously. Correspondingly, Eq. (30b) can be written via accuracy of "single" measurement as follows:

$$\varepsilon_k = \frac{N_1 \ \varepsilon_1^2(\text{single})}{N_k \ \varepsilon_k^2(\text{single})}. \tag{74}$$

This definition of $\geq_k$ makes the relative contributions of the terms $\geq_k \geq_k$ in Eq. (30a) independent of N_k, and therefore equalizes the data sets with different numbers of observation. Relationship (73) assumes that for data set with "redundant" observations $\geq_k$ increases as $\sqrt{N_k}$. Such an increase can be caused by the fact that the number of sources of random errors may increase proportionally to the number of simultaneous measurements. For example, increasing spectral and/or angular resolution in remote-sensing measurements likely results in a decrease of the quality of a single measurement due to increased complexity of the instrumentation and calibration. However, the assumption given by Eq. (73) is of intuitive character since it is not based on actual error analysis. Moreover, the developers of the instrumentation may argue justifiably that accuracy should not degrade if several measurements are taken at the same time. Therefore, it should be noted that Eq. (73) is appropriate only for data sets where actual redundancy has been achieved. Actually, the redundancy may be caused by other factors than instrumentation limits. For example, increasing angular and spectral resolution of satellite observations generally requires larger spatial integration and longer measurement time. Both these factors contribute to an increase of retrieval errors due to natural temporal and spatial variability of the atmosphere and surface.

Thus, identification of measurement redundancy in practice is a difficult effort that strongly relies on the experience of the developer. Nevertheless, it can be advisable to consider data redundancy as a practical factor that may affect retrieval. Namely, if Eq. (52) gives values much higher than the level of expected measurement errors (and retrieval errors are much higher than estimated from Eq.(49)), then it is likely that noise assumptions need to be verified. In such cases the ratios N_1/N_k can be good indications of magnitude and direction of required adjustments in ε_k^2 in order to address domination of the large inverted data sets over smaller ones. For example, assumption (73) was employed successfully in aerosol remote-sensing retrievals [9], where harmonization of the contribution of large sets of angular sky radiance measurements with much fewer observations of spectral optical thickness is beneficial. A similar principle was used in earlier studies [50].

7. Final recommendations

The considerations presented in this chapter were aimed to demonstrate that many important and well established ideas of numerical inversion can be combined and compliment each other in a single inversion methodology. Namely, it is suggested to combine all measured and *a priori* data in a single inversion procedure using the fundamental approach of MML. Under an assumption of normal noise, such an approach results in a multi-term LSM given by Eq. (32), where the contribution of each term is weighted by the values of the errors in the corresponding data set. The discussion in Sections 4-7 concludes that using LSM in the multi-term form allows fruitful connections between such well known and established techniques and methodologies as standard LSM, *Phillips-Tikhonov-Twomey* constrained inversion, and *Kalman-filter* type inversion methods, advocated in remote sensing by studies of *Rodgers* [19]. From a technical viewpoint, the derivation of a multi-term LSM is trivial and the main value of the approach presented is a deliberate consideration of various inversion aspects and approaches with the purpose of consolidating analogies and differences into a single unified concept. As a result, in addition to some generalization of inversion equations, a number of practically important conclusions and recommendations are proposed for implementing numerical inversions. For example, Section 4 suggests considering *Lagrange* multipliers as a ratio of error variances [see Eq. (30)], where variances of *a priori* constraints are related explicitly to knowledge of magnitudes of retrieved parameters. In the case of retrieval of smooth function $y(x)$, *Lagrange* multipliers can be written directly via maximum values of derivatives of $y(x)$. Section 4 also shows how smoothness constraints can be implemented in the retrieval of non-equidistantly binned functions $y(x)$ and how different types of *a priori* constraints can be employed in a single algorithm. Section 6 discusses the use of the multi-term LSM in non-linear Newtonian iterations for optimizing accuracy of the non-linear retrieval in the vicinity of a solution. Moreover, Section 6 shows that multi-term LSM can be used for implementing *Levenberg-Marquardt*-type modifications improving convergence of the non-linear iterations. Section 7 suggests modifications to normal noise assumptions to account for non-negativity of the physical values and addressing data redundancy. The lognormal noise assumption is applied for non-negative values. Under such assumptions, the MML principle results in a multi-term LSM written in logarithmic space. Also, Section 4 emphasizes distinction between two aspects of solution optimization: (i) accounting for distribution of errors in inverted data, and (ii) improving performance of mathematical inverse operations, e.g. replacing matrix inversion by other techniques. It is suggested that uniqueness of the solution should be assured by combining all available measurements and *a priori* information in a multi-term LSM. Then, potentially advantageous mathematical techniques such as SVD, conjugated gradients, iterative search etc. can be used at the stage of solving normal Eqs. (31) for improving the performance of the technical implementation of multi-term LSM. For example, using steepest descent iterations for implementing logarithmic LSM allows (see [9]) the derivation *Chahine*-like iterations [17,14].

Finally, the derivations of the present chapter can be illustrated and summarized by a single formula written for the rather general case when the forward model is non-linear and *a priori* information on both magnitudes and smoothness of retrieved function $y(x)$ is available. If $y(x)$ needs to be retrieved from observations of two different characteristics $z_1(\gtrless) = z_1(\gtrless;y(x))$ and $z_2(\gtrless) = z_2(\gtrless;y(x))$ measured in a range of λ_i ($\gtrless$ can be angle, wavelength, etc.) then the optimized solution can be obtained by iterations:

$$\hat{\mathbf{a}}^p = \hat{\mathbf{a}}^{p+1} \Delta t_p \Delta \hat{\mathbf{a}}^p , \tag{75a}$$

where $\Delta \hat{\mathbf{a}}^p$ is a solution of the normal system:

$$\sum_{k=1}^{2} \Delta_k \mathbf{K}_{1,p}^T \mathbf{W}_k^{\Delta 1} \mathbf{K}_{1,p} + \Delta_3 \Delta_m + \Delta_4 \mathbf{W}_{a*}^{\Delta 1} \Delta \hat{\mathbf{a}}^P =$$

$$= \sum_{k=1}^{2} \Delta_k \mathbf{K}_{k,p}^T \mathbf{W}_k^{\Delta 1} \left(\mathbf{f}_k(\hat{\mathbf{a}}^P) \Delta \mathbf{f}_k^* \right) + \Delta_3 \Delta_m \hat{\mathbf{a}}^P + \Delta_4 \mathbf{W}_{a*}^{\Delta 1} \left(\hat{\mathbf{a}}^P \Delta \hat{\mathbf{a}}^* \right)$$. (75b)

Here, $\mathbf{f}_k$ is a measurement $f_k(\lambda_i)$ and $\mathbf{a}$ is a vector of $a(x_j)$. If the measured functions $z_k(\geq)$ can be both positive and negative then normal noise is assumed and $f_k(\lambda_i) = z_k(\lambda_i)$. If $z_k(\geq)$ are positively defined (e.g. intensities) then the lognormal noise is expected and $f_k(\lambda_i) = \ln(z_k(\lambda_i))$. Similarly, if $y(x)$ can be both positive and negative then normally distributed errors are expected in retrievals and $a_i = y(x_i)$. If $y(x)$ is positively defined (e.g. concentration) then lognormal retrieval errors are expected and $a_i = \ln(y(x_i))$. Symbols $\mathbf{K}_{k,p}$ - matrices of the first derivatives

$$\left\{ \mathbf{K}_{k,p} \right\}_{ji} = \frac{\lambda f_j(\lambda_i)}{\lambda a_i} \bigg|_{\mathbf{a}^P}$$

calculated in the vicinity of $\mathbf{a}^P$; $\mathbf{W}_{...}$ - weighting matrices defined by Eq. (30b). The smoothness matrix Ω_m is determined via matrices of m-th differences $\mathbf{S}_m$ as $\Omega_m = \mathbf{S}_m^T \mathbf{S}_m$ if x_i are equidistant. If x_i are not equidistant, Ω_m is determined via matrices of the m-th derivatives $\mathbf{G}_m$ as $\Omega_m = \mathbf{G}_m^T \mathbf{W}_{g*}^{-1} \mathbf{G}_m$ (see Section 4.3). *Lagrange* multipliers $\geq_k$ are determined by ratios of variances, i.e.:

$$\gamma_1 = 1, \quad \text{and} \quad (\text{for } k \geq 2) \quad \approx_k = \approx_1^2 / \approx_k^2 \approx \hat{\approx}_1^2 (\mathbf{a}^P) / \approx_k^2 \quad , \tag{76}$$

where $\hat{\varepsilon}_1^2(\mathbf{a}^P)$ is an estimate of ε_1^2:

$$\hat{\approx}_1^2(\mathbf{a}^P) - 2 - (\mathbf{a}^P) / (- \sum_{i=1,...,4} N_{f_i} - N_{\mathbf{a}}). \tag{77}$$

Here $\geq(\mathbf{a}^P)$ denotes the value of the residual of the p-th iteration defined as

$$2\Omega(\hat{\mathbf{a}}) = \sum_{k=1}^{2} \Omega_k \left(\mathbf{f}_k(\hat{\mathbf{a}}) \Omega \mathbf{f}_k^* \right)^T \mathbf{W}_k^{\Omega 1} \left(\mathbf{f}_k(\hat{\mathbf{a}}) \Omega \mathbf{f}_k^* \right) + \Omega_3 \hat{\mathbf{a}}^T \Omega_m \mathbf{a} + \Omega_4 \left(\hat{\mathbf{a}} \Omega \hat{\mathbf{a}}^* \right)^T \mathbf{W}_{a*}^{\Omega 1} \left(\hat{\mathbf{a}} \Omega \hat{\mathbf{a}}^* \right) .\tag{78}$$

The variances ε_k^2 (for $k \geq 2$) are determined before implementing iterations. $\geq_2^2$ is the variance of the errors in the second set of measurements $z_2(\geq)$. For the smoothness term ($k = 3$), $\geq_3^2$ can be implied from the knowledge of $y(x)$ m-th derivatives:

$$\approx_3^* - \frac{\langle b_m \rangle (-x)^{2m+1}}{N_a - m} \quad \text{or} \quad \Delta_3^* \Delta \frac{\langle b_m \rangle}{(N_a \Delta m) \Delta_m(x_1)} \quad (\text{if } \geq x_i \geq const) , \tag{79}$$

where the first equation is for equidistant x_i, the second one is for non-equidistant x_i (Section 4.3), $<b_m>$ is the average norm of the of $y(x)$ m-th derivatives [Eq. (40)]. If average derivatives are unknown, $<b_m>$ can be implied using the m-th derivatives of the most unsmooth function $y(x)$ (see Section 4.6). For *a priori* estimates ($k = 4$), $\geq_4^2$ is the variance of *a priori* estimate of a_1^* and $\mathbf{W}_{a*} = \mathbf{C}_{a*}/\geq_4^2$. If actual *a priori* estimates a_i^* are not available, then a_i^*, $\geq_4^2$ and $\mathbf{W}_{a*}$ can be implied from the known variability ranges of a_i (Section 4.6).

Utilization of *Lagrange* multipliers dependent on residual $\geq (\mathbf{a}^p)$ in Eqs. (76-77) helps to provide monotonic convergence of iterations. This operation is functionally analogous to the *Levenberg-Marquardt* method (see Section 5.4). This modification helps to provide a monotonic decrease of $\geq (\mathbf{a}^p)$, i.e. monotonic convergence of iterations. Moreover, if no *a priori* information about the solution is available, the constraints can be applied to the *p-th* correction $\geq\mathbf{a}^p$ instead of $\mathbf{a}^p$. Using such constraints affects only the convergence and does not bias the solution $\hat{\mathbf{a}}$. The correspondent multi-term LSM (same as Eq. (75b) with no *a priori* terms in the right part) is a full equivalent to the *Levenberg-Marquardt* method, with the only difference being that the terms added for improving convergence are clearly related with constraints on smoothness and magnitudes of $\geq\mathbf{a}^p$ (see details in Section 5.4). The coefficient $1 \geq t_p > 0$ is used in Eq. (75a) similar to the *Levenberg-Marquardt* method. If $\geq (\mathbf{a}^{p+1}) > \geq (\mathbf{a}^p)$, t_p should be decreased (e.g. as $t_p \geq t_p/2$) until $\geq (\mathbf{a}^p)$ is decreased.

If forward models $f_k(\lambda_i)$ are linear, no iterations are needed and   Eqs. (75) can be simplified as

$$\left|\sum_{k=1}^{2} /_k \mathbf{K}_k^T \mathbf{W}_k^{|1} \mathbf{K}_k + /_3\right|_m + /_4 \mathbf{W}_{a*}^{|1} \left|\hat{\mathbf{a}} = \sum_{k=1}^{2} /_k \mathbf{K}_k^T \mathbf{W}_k^{|1} \left(\mathbf{f}_k(\hat{\mathbf{a}}) \mid \mathbf{f}_k^*\right) + /_4 \mathbf{W}_{a*}^{|1} \hat{\mathbf{a}}^*. \quad (80)$$

Here $\geq_k$ are given by Eq. (76), with the difference that ε_1^2 is fixed to the error variance in the first data set ($k = 1$). The assumed ε_1^2 should be close to the estimated $\hat{\varepsilon}_1^2(\hat{\mathbf{a}})$ obtained by   Eq. (77) from the residual. A value of $\hat{\varepsilon}_1^2(\hat{\mathbf{a}})$ higher than assumed ε_1^2 indicates inconsistency in the assumptions made. One possibility is that the forward model needs corrections. Otherwise, adjustments are needed in assumptions about errors in measurements or *a priori* data. For example, in case the number of measurements N_k in the first ($k = 1$) and second ($k = 2$) sets of observations are very different, the $\geq_2^2$ can be adjusted by a factor N_1/N_2 in order to account for data redundancy in one of the sets (see Section 6.2).

The covariance matrix of the random errors in the solution $\hat{\mathbf{a}}$ can be estimated in the linear approximation (Section 4.4):

$$\mathbf{C}_{\hat{\mathbf{a}}} = \left(\sum_{k=1}^{2} \gamma_k \mathbf{K}_k^T \mathbf{W}_k^{-1} \mathbf{K}_k + \gamma_3 \Omega_m + \gamma_4 \mathbf{W}_{\mathbf{a}*}^{-1} \right)^{-1} \hat{\varepsilon}_1^2(\hat{\mathbf{a}}).$$ (81)

For the non-linear case, the derivative matrix $\mathbf{K}_k$ is simulated in the vicinity of $\hat{\mathbf{a}}$.

Finally, linear equations (75b) and (79) can be solved by different methods. For example, using inverse matrices reduce Eqs. (75) and (80) to a traditional form of constrained inversion. Alternatively (Section 4.8), other numerical or computer techniques, such as, SVD, conjugated gradients, iterations, generic inversion, etc. can be used for solving Eqs. (75b) and (80).

8. Acknowledgments

I thank Alexander Sinyuk, Valery N. Shcherbakov, Gorden Videen and Ben Veihelmann for reading the chapter and providing useful comments.

References

1. S. Twomey, "Introduction to the Mathematics of Inversion in Remote Sensing and Indirect Measurements", (Elsevier, Amsterdam, 1977).
2. A.N. Tikhonov and V.Y. Arsenin, "Solution of Ill-Posed Problems" (Wiley, New York, 1977).
3. A. Tarantola, "Inverse Problem Theory: Methods for Data Fitting and Model Parameter Estimation", (Elsevier, Amsterdam, 1987).
4. W.H. Press, S.A. Teukolsky, W.T. Vetterling, B.P. Flannery, "Numerical Recipes in FORTRAN. The art of Scientific Computing", (Cambridge University Press, 1992).
5. A. Bakushinsky and A. Goncharsky, "Ill-Posed Problems: Theory and Applications" (Kluwer, Dordrecht, 1994).
6. C.D. Rodgers, "Inverse methods for atmospheric sounding: theory and practice", (World Scientific, Singapore, 2000).
7. O.V. Dubovik, T.V. Lapyonok, and S.L. Oshchepkov, Appl. Opt., 34, 8422 (1995).
8. O. Dubovik, A. Smirnov, B. N. Holben, et al., J. Geophys. Res., 105, 9791(2000).
9. O. Dubovik and M. D. King, J. Geophys. Res., 105, 20673(2000).
10. R.E. Kalman, J. Basic. Engrg., 35, 82 (1960).
11. B.L. Phillips, J. Assoc. Comp. Mach, 9, 84 (1962).
12. A.N. Tikhonov, Dokl. Akad. Nauk SSSR, 151, 501 (1963).
13. S. Twomey, J. Assoc. Comp. Mach., 10, 97 (1963).
14. S. Twomey, J. Comp. Phys, 18, 188 (1975).
15. Strand, O. N., and E. R. Westwater, J. Assoc. Comput. Mach., 15,100 (1968).
16. Strand, O. N., and E. R. Westwater, SIAM J. Numer. Anal., 5, 287 (1968).
17. M.T. Chahine, J. Opt. Soc. Am., 12, 1634 (1968).
18. V.F. Turchin and V.Z. Nozik, Izv. Acad. Nauk SSSR Fiz. Atmos. Okeana, 5, 29 (1969).
19. C.D. Rodgers, Rev. Geophys. Space Phys., 14, 609 (1976).
20. B.N. Holben, T.F. Eck, I. Slutsker et al., Remote Sens. Envir., 66, 1 (1998).
21. M.D. King, M.G. Strange, P. Leone, et al., J. Atmos. Oceanic Technol., 3, 513 (1986).
22. D.J. Diner, J.C. Beckert, T.H. Reilly, et al., IEEE Trans. Geosci. Remote Sens., 36, 1072 (1998).
23. Y. Sasano, M. Suzuki, T. Yokota, et al., Geophys. Res. Lett., 26, 197 (1999).
24. P. Goloub, D. Tanre, J.L. Deuze, et al., IEEE Trans. Geosci. Remote Sens., 37, 1586 (1999).
25. J. Chowdhary, B. Cairns, M. Mishchenko, et al., Geophys. Res. Lett., 28, 243 (2001).
26. J. Chowdhary, B. Cairns, L.D. Travis, J. Atmos. Sci., 59, 383 (2002).
27. W.T. Edie, D. Dryard, F.E. James, M. Roos, B. Sadoulet, "Statistical Methods in Experimental Physics", (North-Holland Publishing Company, Amsterdam, 1971)

28. C. R. Rao, "Linear Statistical Inference and Its Applications" (Wiley, New York, 1965).
29. G. A. Serber, "Linear Regression Analysis", (Wiley, New York, 1977).
30. A. Alpert, "Regression and the Moore-Penrose Pseudoiniverse" (Academic Press, New York, 1972).
31. A. N. Tikhonov, A. S. Leonov and A. G. Yagola, "Nonlinear Ill-Posed Problems", (Chapman & Hall, London 1998).
32. V.F. Turchin,V.P. Kozlov and M.S. Malkevich, Sov. Phys. Usp. Fiz.-USSR, 13, 681 (1971).
33. A. Gelb, "Applied Optimal Estimation", (MIT Press, Cambridge, Mass., 1988).
34. D. Hartely and R. Prinn, J. Geophys. Res., 98, 5183(1993).
35. M.D. King, D. M. Byrne, B. M. Herman et al., J. Atmos. Sci., 21, 2153 (1978).
36. M.D. King, J. Atmos. Sci., 39, 1356 (1982).
37. T. Nakajima, G. Tonna, R. Rao, et al., Appl. Opt., 35, 2672 (1996).
38. D. Muller, U. Wandinger, A. Ansmann, Appl. Opt., 38, 2346 (1999).
39. O.P. Hasekamp and J. Landgraf, J. Geophys. Res., 106, 8077 (2001).
40. I. Veselovskii, A. Kolgotin, V. Griaznov et al., Appl. Opt., 41, 3685 (2002).
41. P.S. Hansen, Inverse Problems, 8, 849 (1992)
42. V.F. Turchin and L.S. Turovtse, Optika I Spectroskopiya, 36, 280 (1974)
43. A.M. Obuhov, Izv. Acad, Sci. SSSR Geophys., 432 (1960)
44. G., Forsythe, and W. Wasow, "Finite Difference Methods for Partial Differential Equations", (Wiley, New York, 1960).
45. J.M. Ortega, "Introduction to Parallel and Vector Solution of Linear System", (Plenum Press, New York, 1988).
46. V. Krasnopolsky, L.C. Breaker, and W.H. Gemmill, J. Geophys. Res, 100, 11,033 (1995).
47. D.E. Goldberg," Genetic algorithms in search, optimization, and machine learning", (Addison-Wesley Pub.,1989)
48. B.R. Lienert, J.N. Porter, S.K. Sharma J. Atmos. Ocean. Tech., 20, 1403, (2003).
49. J.M. Ortega and W.C. Reinboldt, Iterative Solution of Nonlinear Equations in Several Variables, (Academic Press, New York, 1970).
50. S.L. Oshchepkov and O.V. Dubovik, J. Phys. D Appl. Phys., 26, 728 (1993).

Oleg and Margot Dubovik with children Ivan and Alexandrine

Khersonesus Tavricheskiy near Sevastopol

EIGENVALUE SHIFTING - A NEW ANALYTICAL-COMPUTATIONAL METHOD IN RADIATIVE TRANSFER THEORY

HELMUT DOMKE

Ludwig Richter Str. 31, D-14467 Potsdam, Germany

Abstract. Linear integral functional relations holding for the leading regular eigenmodes are employed to transform the radiative transfer equation for polarized light in plane-parallel homogeneous media into an equivalent transfer equation of the same form with a modified scattering integral (pseudo-scattering). In its general form, this transformation incorporates two free function vectors. On specifying them appropriately, the effective single-scattering albedo for the transformed transfer equation can be reduced substantially. This drastically accelerates the convergence of iterative computational approaches. As a particular case, it is demonstrated how to employ the *F*- and *K*-integrals to transform the original transfer equation for conservative isotropic scattering into an equivalent transfer equation for nonconservative pseudo-scattering. A general method of transforming the transfer equation is described for the azimuthally averaged *I,Q*-component of the transfer equation for polarized radiation. Simple formulae are derived that express the original albedo matrix and surface Green's function matrix in terms of the corresponding function matrices related to the transformed transfer equation. It is shown that the transformation described here selectively affects the asymptotic eigenmodes of the transfer equation and shifts the leading pair of discrete eigenvalues.

1. Introduction

The theory of multiple scattering of radiation in optically thick plane homogeneous atmospheres confronts us with a well known paradox: Computations of multiple scattering in optically thick media by iterating the radiative transfer equation encounter difficulties for small true absorption. Then, even very high orders of scattering significantly contribute to the

G. Videen et al. (eds.), Photopolarimetry in Remote Sensing, 107-124.

radiation field. On the other hand, after many acts of scattering, the photons lose any detailed information about their initial angular distribution and polarization. As a result, the distribution of radiative intensity and polarization in deep layers of an optically thick source-free atmosphere illuminated at its top from outside depend only on the local scattering properties of the medium, but not on the characteristics of the incident radiation. For a homogeneous medium consisting of macroscopically isotropic and mirror-symmetric volume elements, the azimuthal dependence of intensity and polarization as well as the elliptical polarization is lost in deep layers, and the radiative depth regime may asymptotically be described by a two-component intensity vector

$$\mathbf{I}(\tau,u) = \begin{pmatrix} I(\tau,u) \\ Q(\tau,u) \end{pmatrix},$$

consisting of the well known Stokes parameters I and Q (cf., e.g., [1]) that describe the intensity and the linear polarization of the radiation. The intensity vector depends on the optical depth τ, and on the cosine u of the polar angle of the propagation direction of the radiation with respect to the outer normal to the top surface of the plane medium. Mathematically, the deep layer regime is described by the leading eigenmode of the I,Q-component of the azimuthally averaged radiative transfer equation (e.g. [2])

$$u \frac{\partial}{\partial \tau} \mathbf{I}(\tau,u) = -\mathbf{I}(\tau,u) + \frac{a}{2} \int_{-1}^{+1} dv \mathbf{W}(u,v) \mathbf{I}(\tau,v). \tag{1}$$

Here, a denotes the single-scattering albedo, while the 2x2-matrix $\mathbf{W}(u,v)$ is the I,Q-component of the azimuthally averaged phase matrix, which obeys the relations of mirror symmetry and reciprocity, respectively:

$$\mathbf{W}(u,v) = \mathbf{W}(-u,-v) = \tilde{\mathbf{W}}(-v,-u). \tag{2}$$

Here, $\tilde{\mathbf{W}}$ denotes the transposed matrix. There hold also the integral relations

$$\frac{1}{2} \int_{-1}^{+1} dv \mathbf{W}(u,v) \mathbf{i}_0 = \mathbf{i}_0, \tag{3}$$

and

$$\frac{1}{2} \int_{-1}^{+1} dv \mathbf{W}(u,v) v \mathbf{i}_0 = \frac{1}{3} \beta_1 u \mathbf{i}_0, \tag{4}$$

with $\tilde{\mathbf{i}}_0 = (1,0)$. The parameter $\beta_1/3$ is the mean cosine of the scattering angle.

Up to a scalar multiplier of normalization, the deep layer regime is described by the leading eigenmode $\mathbf{i}(\tau, u, \eta_1)$ of the transfer equation (1), i.e.,

$$\mathbf{I}(\tau \gg 1, u) \propto \mathbf{i}(\tau, u, \eta_1) = e^{-\tau/\eta_1} \boldsymbol{\varphi}(u, \eta_1), \qquad (5)$$

where the eigenfunction vector $\boldsymbol{\varphi}(u, \eta_1)$ is the solution to the characteristic equation

$$(\eta_1 - u)\boldsymbol{\varphi}(u, \eta_1) = \frac{a\eta_1}{2} \int\limits_{-1}^{+1} dv \mathbf{W}(u, v)\boldsymbol{\varphi}(v, \eta_1) \qquad (6)$$

corresponding to the maximum eigenvalue $\eta_1 = 1/k$. The parameter k is the characteristic exponent of the asymptotic depth regime. The eigenfunction vector is normalized according to

$$\int\limits_{-1}^{+1} dv \, \tilde{\mathbf{i}}_0 \boldsymbol{\varphi}(v, \eta_1) = 1. \qquad (7)$$

Several methods for calculating the leading eigensolution of the transfer equation for polarized light have been described in the literature [3,4,5]. As follows from Eqs. (3) and (6), the leading eigenvalue for conservative scattering, where $a = 1$, tends toward infinity, i.e., $k = 0$, with

$$\mathbf{i}(\tau, u, \infty) = \boldsymbol{\varphi}(u, \infty) = \frac{1}{2} \mathbf{i}_0, \qquad (8)$$

being the corresponding eigenmode. Thus, independent of the single-scattering properties of the medium, the radiation field deep inside of a conservatively scattering atmosphere always is extremely simple. On reaching the deep layers after a large number of scattering events, the photons form there a uniform and isotropic sea of unpolarized radiation.

Obviously, the simplicity of the depth regime in the limit of conservative scattering contrasts with the extremely low rate of convergence observed for iterative methods to solve the transfer equation for small true absorption. Obviously, the contribution of high orders of scattering are mainly to fill up a deep layer regime of simple structure given by the leading eigenmode (5), which can be obtained a priori by solving Eq. (6). This provokes the question: Could knowledge of the leading eigenmode be employed to speed up the convergence of iterations by modifying appropriately the original tranfer equation? The answer is: Yes, it could! And we will describe here how to do that.

2. Isotropic scattering

Before turning to the general transfer equation (1), we want to demonstrate, as an example, the transformation of the transfer equation for conservative isotropic scattering. To this end, we consider the standard Milne problem, which has been studied extensively in the literature (cf. [1], [6]). It amounts to the solution of the homogeneous transfer equation

$$u\frac{\partial}{\partial\tau}I(\tau,u) = -I(\tau,u) + \frac{1}{2}\int_{-1}^{+1}dvI(\tau,v) \tag{9}$$

for the intensity of the radiation field subject to the boundary conditions

$$I(0,\mu)=0, \quad \mu>0, \quad \text{and} \quad \lim_{\tau\to\infty}I(\tau,u)e^{-\varepsilon\tau}=0, \quad \text{for } \varepsilon>0. \tag{10}$$

Physically, the Milne problem describes the radiation field in a semi-infinite medium, where primary sources are assumed to be located only at infinitely deep layers. We find from Eq. (9) that the radiative energy upward net flux

$$F = \frac{1}{2}\int_{-1}^{+1}dvvI(\tau,-v) = -\frac{1}{2}\int_{-1}^{+1}dvvI(\tau,v) \tag{11}$$

is constant throughout the atmosphere, while the so called K-integral,

$$K(\tau) = \frac{1}{2}\int_{-1}^{+1}dvv^2I(\tau,v), \tag{12}$$

depends linearly on the optical depth according to

$$K(\tau) = F\tau + K(0). \tag{13}$$

We note that with Eq. (13), important knowledge is available about the radiation field, which enables us to rewrite the transfer equation (9) in the form

$$u\frac{\partial}{\partial\tau}I(\tau,u) = -I(\tau,u) + \frac{1}{2}\int_{-1}^{+1}dv(1-Cv^2)I(\tau,v)+C[F\tau+K(0)], \tag{14}$$

where C is a free parameter, and the constant $K(0)$ may be determined *a posteriori*.

Thus, by means of the K-integral (13), the original transfer equation (9) for conservative isotropic scattering has been transformed to a new equivalent transfer equation (14) characterized by a modified scattering integral term (pseudo-scattering), and another term describing fictitious internal sources. Contrary to the original transfer equation, the transformed transfer equation

(14), with an appropriate value of the free parameter C, say 3 or 4, can be solved easily by rapidly converging iterations. Obviously, the solution of the Milne problem can be constructed now as a linear combination of two solutions to the transformed transfer equation (14) corresponding, respectively, to constant internal pseudo-sources, and to internal pseudo-sources proportional to τ. The resulting solution for $I(\tau, u)$ can be employed to determine the constant $K(0)$ by means of Eq. (12).

The exact solution to Eq. (14) is well known from the standard radiative transfer theory (cf. [1] or [6]). In particular, the intensity of emergent radiation can be expressed in terms of Chandrasekhar's H-function in the form

$$I(0,-\mu) = C H_c(\mu) H_c(\infty)\{K(0) + F[\mu + \tfrac{1}{2} H_1 H_c(\infty)]\}, \tag{15}$$

where $H_c(\mu)$ satisfies the nonlinear integral equation

$$H_c(\mu) = 1 + \frac{1}{2}\mu H_c(\mu) \int_0^1 dv \frac{(1 - Cv^2) H_c(v)}{\mu + v}, \tag{16}$$

and H_1 denotes the moment

$$H_1 = \int_0^1 dv v (1 - Cv^2) H_c(v). \tag{17}$$

On using Eq. (12) to eliminate the parameter $K(0)$ in Eq. (15), and employing also some relations between the integral moments of the function $H_c(\mu)$ resulting from the nonlinear integral equation (16), we finally obtain

$$I(0,-\mu) = \sqrt{3} F H_c(\mu)(1 + \sqrt{C}\mu). \tag{18}$$

On comparing this with Chandrasekhar's result for the Milne problem, we find that Chandrasekhar's H-function $H(\mu)$ for conservative isotropic scattering can be expressed in terms of the H-function for non-conservative pseudo-scattering $H_c(\mu)$ as

$$H(\mu) = H_c(\mu)(1 + \sqrt{C}\mu). \tag{19}$$

In an earlier paper [7], this formula has been derived directly by means of the nonlinear integral equation Eq. (16).

Thus, instead of calculating Chandrasekhar's H-function for conservative isotropic scattering immediately by iterating the corresponding nonlinear integral equation, i.e., Eq. (16) with $C = 0$, one could compute $H_c(\mu)$ by rapidly converging iteration of Eq. (16), with $C = 3$ or 4, and retrieve subsequently the original H-function $H(\mu)$ by means of Eq. (19). In this way, an essential computational advantage would be achieved.

3. The φ-transformation of the transfer equation

3.1.   The transfer equation for the surface Green's function matrix

Let us now return to the general transport equation (1) for polarized radiation in a plane medium. As a typical transfer problem, we consider the surface Green's function matrix $\mathbf{G}(\tau, u; 0, \mu_0)$, $\mu_0 \in [0,1]$, for a slab with an optical thickness b defined as the solution to the homogeneous transfer equation

$$u\frac{\partial}{\partial \tau}\mathbf{G}(\tau, u; 0, \mu_0) = -\mathbf{G}(\tau, u; 0, \mu_0) + \frac{a}{2}\int_{-1}^{+1}dv\mathbf{W}(u,v)\mathbf{G}(\tau, v; 0, \mu_0), \qquad (20)$$

subject to the boundary conditions

$$\mathbf{G}(+0, \mu; 0, \mu_0) = \frac{1}{\mu}\delta(\mu - \mu_0)\mathbf{E}, \text{ for } \mu \in [0,1] \qquad (21)$$

at the top surface, $\tau = 0$, and

$$\mathbf{G}(b, -\mu; 0, \mu_0) = \mathbf{0}, \text{ for } \mu \in [0,1] \qquad (22)$$

at the bottom surface, $\tau = b$. Here, $\mathbf{E}$ denotes the 2x2 unit matrix. For semi-infinite media, the boundary condition at the bottom has to be replaced by the prescription

$$\lim_{\tau \to \infty} \mathbf{G}(\tau, -\mu; 0, \mu_0) < \infty, \qquad (22a)$$

i.e., the surface Green's function matrix should be bounded at large optical depth.

Note that

$$\mathbf{R}(\mu, \mu_0) = \frac{1}{2}\mathbf{G}(+0, -\mu; 0, \mu_0), \qquad (23)$$

and

$$\mathbf{T}(\mu, \mu_0) = \frac{1}{2}\left(\mathbf{G}(b, \mu; 0, \mu_0) - \frac{1}{\mu}\delta(\mu - \mu_0)e^{-b\tau}\mathbf{E}\right), \qquad (24)$$

respectively, are the matrices of diffuse reflection and transmission according to the definitons commonly used in the literature (cf., e.g., [5]).

In the following, the leading eigenmodes of the transfer equation (20) and some of their functional properties will be employed to transform the original transfer equation into an equivalent transfer equation with a modified scattering term.

3.2.   Functional relations

On taking into account the characteristic equation Eq. (6) together with the symmetry properties of the phase matrix (2), we observe that, with η_1 being the leading positive eigenvalue, $\eta_{-1} = -\eta_1$ also is an eigenvalue, with

$$\varphi(u, \eta_{-1}) = \varphi(-u, \eta_1) \tag{25}$$

being the corresponding eigenfunction vector. From Eq. (6), one gets the orthogonality relations

$$\int_{-1}^{1} dv v \widetilde{\varphi}(v, \eta_i) \varphi(v, \eta_k) = N_i \delta_{ik} , \quad i, k = -1, +1. \tag{26}$$

For our analysis, it is convenient to combine the two leading eigensolutions for η_1 and η_{-1} and use the pair of function vectors

$$\mathbf{j}^+(\tau, u, \eta_1) = \frac{1}{2} \left[\varphi(u, \eta_1) e^{-k\tau} + \varphi(u, \eta_{-1}) e^{k\tau} \right],$$
$$\mathbf{j}^-(\tau, u, \eta_1) = \frac{1}{2N_1} \left[\varphi(u, \eta_1) e^{-k\tau} - \varphi(u, \eta_{-1}) e^{k\tau} \right], \tag{27}$$

instead. We note that

$$\mathbf{j}^+(\tau, u, \eta_1) = \cosh(\tau / \eta_1) \mathbf{j}^+(0, u, \eta_1) - N_1 \sinh(\tau / \eta_1) \mathbf{j}^-(0, u, \eta_1),$$
$$\mathbf{j}^-(\tau, u, \eta_1) = \cosh(\tau / \eta_1) \mathbf{j}^-(0, u, \eta_1) - \frac{1}{N_1} \sinh(\tau / \eta_1) \mathbf{j}^+(0, u, \eta_1). \tag{28}$$

From Eqs. (26) and (27), we find the orthogonality relations

$$\int_{-1}^{1} dv v \widetilde{\mathbf{j}}^+(0, v, \eta_1) \mathbf{j}^+(0, v, \eta_1) = \int_{-1}^{1} dv v \widetilde{\mathbf{j}}^-(0, v, \eta_1) \mathbf{j}^-(0, v, \eta_1) = 0 ,$$
$$\int_{-1}^{1} dv v \widetilde{\mathbf{j}}^+(0, v, \eta_1) \mathbf{j}^-(0, v, \eta_1) = \int_{-1}^{1} dv v \widetilde{\mathbf{j}}^-(0, v, \eta_1) \mathbf{j}^+(0, v, \eta_1) = \frac{1}{2}. \tag{29}$$

As mentioned above, for conservative scattering, where $a = 1$, the leading eigenvalue η_1 turns to infinity, i.e., the characteristic exponent k turns to zero. On expanding the eigenfunction vector for small true absorption in powers of k, we find, by means of Eqs. (2), (3), (4), and (6), the following asymptotic formulae for $(1 - a) \ll 1$:

$$k = 1/\eta_1 = -1/\eta_{-1} = \sqrt{(1-a)(3-\beta_1)} + O[(1-a)^{3/2}],$$

$$\varphi(u, \eta_1) = \varphi(-u, \eta_{-1}) = \frac{1}{2}\left(1 + \frac{3k}{3-\beta_1}u\right)\mathbf{i}_0 + O(1-a), \qquad (30)$$

$$N_1 = -N_{-1} = \frac{k}{3-\beta_1} + O[(1-a)^{3/2}],$$

and

$$\mathbf{j}^+(0, u, \eta_1) = \frac{1}{2}\mathbf{i}_0 + O(1-a),$$

$$\mathbf{j}^-(0, u, \eta_1) = \frac{3}{2}u\mathbf{i}_0 + O(1-a). \qquad (31)$$

Now we proceed to derive some functional relations, which, similar to the K-integral (13) for conservative isotropic scattering considered above, is the cornerstone for the transformation of the transfer equation for the general case. On multiplying Eq. (20) from the left by $\widetilde{\varphi}(u, \eta_{\pm 1})$ and integrating subsequently over the angular variable, we find after elementary analysis the functional relations

$$\int_{-1}^{+1} du u \widetilde{\varphi}(u, \eta_{\pm 1}) \mathbf{G}(\tau, u; 0, \mu_0) = e^{-\tau/\eta_{\pm 1}} \int_{-1}^{+1} du u \widetilde{\varphi}(u, \eta_{\pm 1}) \mathbf{G}(+0, u; 0, \mu_0), \quad (32)$$

or, expressed in terms of the $\mathbf{j}$-function vectors defined by Eq. (27),

$$\int_{-1}^{+1} du u \widetilde{\mathbf{j}}^{\pm}(0, u, \eta_1) \mathbf{G}(\tau, u; 0, \mu_0) = \int_{-1}^{+1} du u \widetilde{\mathbf{j}}^{\pm}(\tau, u, \eta_1) \mathbf{G}(+0, u; 0, \mu_0). \qquad (33)$$

For semi-infinite media, we obtain

$$\int_{-1}^{+1} du u \widetilde{\varphi}(u, \eta_{-1}) \mathbf{G}(\tau, u; 0, \mu_0) = \mathbf{0}. \qquad (33a)$$

Note that for semi-infinite media in case of conservative scattering, $a = 1$, we find from Eqs. (33a), (33) and (30) that the energy flux integral for the surface Green's function matrix will be equal to zero,

$$\widetilde{\mathbf{F}} = -\frac{1}{2}\int_{-1}^{+1} du u \widetilde{\mathbf{i}}_0 \mathbf{G}(\tau, u; 0, \mu_0) \equiv \mathbf{0}, \qquad (34)$$

while the K-integral is constant,

$$\widetilde{\mathbf{K}}(\tau) = \frac{1}{2}\int\limits_{-1}^{+1} du\, u^2\,\widetilde{\mathbf{i}}_0 \mathbf{G}(\tau, u; 0, \mu_0) = \widetilde{\mathbf{K}}(0). \tag{35}$$

The functional relations Eq. (33) can be employed to transform the transfer equation (20) into an equivalent transfer equation for multiple pseudo-scattering with a modified phase matrix.

3.3.   The φ-transformation of the transfer equation

Let us define the modified phase matrix

$$\mathbf{W}_c(u,v) = \mathbf{W}(u,v) - [\mathbf{C}_1(u)\,\widetilde{\mathbf{j}}^-(0,v,\eta_1) + \mathbf{C}_2(u)\,\widetilde{\mathbf{j}}^+(0,v,\eta_1)]v, \tag{36}$$

where the function vectors $\mathbf{C}_1(u)$ and $\mathbf{C}_2(u)$ are free for convenient choice.

On taking into account Eq. (33), and also Eq. (28), we rewrite the transfer equation (20) for the surface Green's function matrix in the form

$$u\frac{\partial}{\partial \tau}\mathbf{G}(\tau, u; 0, \mu_0) = -\mathbf{G}(\tau, u; 0, \mu_0) + \frac{a}{2}\int\limits_{-1}^{+1} dv\, \mathbf{W}_c(u,v)\mathbf{G}(\tau, v; 0, \mu_0)$$
$$+ \mathbf{q}_1(\tau, u)\,\widetilde{\mathbf{g}}^-(0, \mu_0) + \mathbf{q}_2(\tau, u)\,\widetilde{\mathbf{g}}^+(0, \mu_0), \tag{37}$$

with

$$\widetilde{\mathbf{g}}^\pm(0, \mu_0) = 2\int\limits_{-1}^{+1} du\, u\,\widetilde{\mathbf{j}}^\pm(0, u, \eta_1)\mathbf{G}(+0, u; 0, \mu_0), \tag{38}$$

and

$$\mathbf{q}_1(\tau, u) = \frac{a}{4}[\mathbf{C}_1(u)\cosh(\tau/\eta_1) - \mathbf{C}_2(u)N_1\sinh(\tau/\eta_1)],$$
$$\mathbf{q}_2(\tau, u) = \frac{a}{4}\left[\mathbf{C}_2(u)\cosh(\tau/\eta_1) - \mathbf{C}_1(u)\frac{1}{N_1}\sinh(\tau/\eta_1)\right]. \tag{39}$$

This completes the φ-transformation of the transport equation.

As a result, the original homogeneous transfer equation (20) for the surface Green's function matrix has been transformed into the transfer equation (37) with a modified phase matrix (36) and fictitious internal primary sources. When the transfer equation (37) is solved separately for the two internal primary source function vectors $\mathbf{q}_1(\tau, u)$ and $\mathbf{q}_2(\tau, u)$ given by Eq. (39), the vector parameters $\widetilde{\mathbf{g}}^\pm(0, \mu_0)$ needed for compiling the complete solution can be determined by means of Eq. (38).

It is quite remarkable that the φ-transformation incorporates two free function vectors [see Eq. (36)]. In particular, Eq. (37) reproduces the transformed transfer equation (14) considered above as resulting from conservative isotropic scattering, if we recall that $a = 1$, $\eta_1 = \infty$, and take into account Eq. (36) with $\mathbf{W}(u,v) = \mathbf{i}_0\tilde{\mathbf{i}}_0$, $\mathbf{C}_1(u) = (2C/3)\mathbf{i}_0$ and $\mathbf{C}_2(u) = \mathbf{0}$, while $\mathbf{j}^-(0,u,\infty)$ is provided by Eq. (31).

For semi-infinite media, the φ-transformation greatly simplifies. In this case, due to Eq. (33a), we find from Eq. (36) that

$$\tilde{\mathbf{g}}^-(0,\mu_0) = \frac{1}{N_1}\tilde{\mathbf{g}}^+(0,\mu_0) = 2\int_{-1}^{+1} du\, u\, \tilde{\mathbf{j}}^-(0,u,\eta_1)\mathbf{G}(+0,u;0,\mu_0), \qquad (40)$$

and the transformed transfer equation takes the form

$$u\frac{\partial}{\partial\tau}\mathbf{G}(\tau,u;0,\mu_0) = -\mathbf{G}(\tau,u;0,\mu_0) + \frac{a}{2}\int_{-1}^{+1} dv\, \mathbf{W}_c(u,v)\mathbf{G}(\tau,v;0,\mu_0)$$

$$+ \frac{a}{4}e^{-\tau/\eta_1}[\mathbf{C}_1(u) + N_1\mathbf{C}_2(u)]\tilde{\mathbf{g}}^-(0,\mu_0). \qquad (41)$$

Here, only one parameter vector $\tilde{\mathbf{g}}^-(0,\mu_0)$ remains to be determined *a posteriori*.

In the following, it will be shown that the original surface Green's function matrix can be expressed immediately in terms of the surface Green's function matrix for the transformed transfer equation.

4. The φ-transformation of the surface Green's function matrix of a semi-infinte medium

Let the semi-infinite medium surface Green's function $\mathbf{G}_c(\tau,u;0,\mu_0)$ be the solution to the homogeneous transfer equation with the modified phase matrix Eq. (36),

$$u\frac{\partial}{\partial\tau}\mathbf{G}_c(\tau,u;0,\mu_0) = -\mathbf{G}_c(\tau,u;0,\mu_0) + \frac{a}{2}\int_{-1}^{+1} dv\, \mathbf{W}_c(u,v)\mathbf{G}_c(\tau,v;0,\mu_0), \quad (42)$$

subject to the same boundary conditions like those applying to the original surface Green's function matrix, i.e. Eqs. (21), and (22a). On using Eq. (36) and employing the orthogonality relations (26), one finds that

$$\mathbf{I}_{ih}(\tau,u) = \mathbf{i}(\tau,u,\eta_1)\tilde{\mathbf{g}}^-(0,\mu_0) = e^{-\tau/\eta_1}\varphi(u,\eta_1)\tilde{\mathbf{g}}^-(0,\mu_0), \qquad (43)$$

is a particular solution of Eq. (41). The complete solution obeying the boundary conditions, Eqs. (20) and (21a), can be expressed by means of the surface Green's function matrix of the transformed transfer equation in the form

$$\mathbf{G}(\tau,u;0,\mu_0)=\mathbf{G}_c(\tau,u;0,\mu_0)+\mathbf{I}_{ih}(\tau,u)-\int_0^1 d\mu\mu\mathbf{G}_c(\tau,u;0,\mu)\mathbf{I}_{ih}(0,\mu). \quad (44)$$

Now, on inserting the particular solution Eq. (43) and taking into account also Eq. (40), one can find from Eq. (44) the original surface function matrix in terms of the surface Green's function matrix of the transformed transfer equation,

$$\mathbf{G}(\tau,u;0,\mu_0)=\mathbf{G}_c(\tau,u;0,\mu_0)$$

$$+\frac{1}{D}\left(e^{-\tau/\eta_1}\varphi(u,\eta_1)-\int_0^1 d\mu\mu\mathbf{G}_c(\tau,u;0,\mu)\varphi(\mu,\eta_1)\right)\tilde{\mathbf{g}}_c^-(0,\mu_0), \quad (45)$$

where

$$\tilde{\mathbf{g}}_c^-(0,\mu_0)=2\int_{-1}^{+1}du u\,\tilde{\mathbf{j}}^-(0,u,\eta_1)\mathbf{G}_c(+0,u;0,\mu_0), \quad (46)$$

and

$$D=\int_0^1 d\mu\mu\tilde{\mathbf{g}}_c^-(0,\mu)\varphi(\mu,\eta_1). \quad (47)$$

Thus, Eq. (43) is the φ-transformation formula for the surface Green's function matrix of a semi-infinite medium, which enables one to retrieve the original surface Green's function matrix, if the surface Green's function matrix of the transformed transfer equation (41) is known. With an appropriate choice of the free parameters in Eq. (36), it may be much easier to solve the transformed transfer equation (41) rather than to solve the original transport equation Eq. (20).

Let us suppose now, that the effective single-scattering albedo of the modified phase matrix $\mathbf{W}_c(u,v)$ has been reduced substantially by means of an appropriate choice of the free parameter vectors in Eq. (36). Then, with increasing optical depth, the surface Green's function matrix $\mathbf{G}_c(\tau,u;0,\mu_0)$ rapidly fades away, and Eq. (45) asymptotically yields

$$\mathbf{G}(\tau,u;0,\mu_0)\approx\frac{1}{D}e^{-\tau/\eta_1}\varphi(u,\eta_1)\tilde{\mathbf{g}}_c^-(0,\mu_0) \quad \text{if } \tau\gg1, \quad (48)$$

Obviously, this reproduces the well known asymptotic depth regime discussed above [cf. Eq. (5)]. The φ-transformation formula (45) explicitly separates the asymptotic depth regime described by the leading eigenmode.

The φ-transformation formula for the albedo matrix emerges from Eq. (45) as a particular case, for $\tau = +0$,

$$
\mathbf{R}(\mu,\mu_0) = \mathbf{R}_c(\mu,\mu_0)
$$
$$
+ \frac{1}{2D}\left(\varphi(-\mu,\eta_1) - 2\int_0^1 d\eta\eta \mathbf{R}_c(\mu,\eta)\varphi(\eta,\eta_1) \right)\tilde{\mathbf{g}}_c^{\,-}(0,\mu_0), \tag{49}
$$

if one takes into account also Eq. (23). Here, $\mathbf{R}_c(\mu,\mu_0)$ denotes the albedo matrix corresponding to the reduced phase matrix (36). On using the boundary condition (21), the function vector $\tilde{\mathbf{g}}_c^{\,-}(0,\mu_0)$ defined by Eq. (46) can be expressed in terms of the reduced albedo matrix as

$$
\tilde{\mathbf{g}}_c^{\,-}(0,\mu_0) = 2\left(\tilde{\mathbf{j}}^{\,-}(0,\mu_0,\eta_1) - 2\int_0^1 d\eta\eta \tilde{\mathbf{j}}^{\,-}(0,-\eta,\eta_1)\mathbf{R}_c(\eta,\mu_0) \right). \tag{50}
$$

Note that the φ-transformation formula (49) for the albedo matrix $\mathbf{R}(\mu,\mu_0)$ guarantees that the linear constraint

$$
\varphi(-\mu,\eta_1) = 2\int_0^1 d\eta\eta \mathbf{R}(\mu,\eta)\varphi(\eta,\eta_1) \tag{51}
$$

is fulfilled identically.

It should be mentioned that many of the various computational, analytical or semi-analytical standard approaches of the radiative transfer theory can be applied also to the transformed radiative transfer equation Eq. (42). In particular, this is true also for principles of invariance (cf., e.g., [1] or [5]), which can be applied to derive the well-known nonlinear integral equation also for the albedo matrix $\mathbf{R}_c(\mu,\mu_0)$,

$$
(\mu+\mu_0)\mathbf{R}_c(\mu,\mu_0) = \frac{a}{4}[\mathbf{W}_c(-\mu,\mu_0) + 2\mu\int_0^1 d\eta \mathbf{W}_c(-\mu,-\eta)\mathbf{R}_c(\eta,\mu_0)
$$
$$
+ 2\mu_0\int_0^1 d\eta \mathbf{R}_c(\mu,\eta)\mathbf{W}_c(\eta,\mu_0) \tag{52}
$$
$$
+ 4\mu\mu_0\int_0^1 d\eta\int_0^1 dv \mathbf{R}_c(\mu,\eta)\mathbf{W}_c(\eta,-v)\mathbf{R}_c(v,\mu_0)].
$$

With an appropriate choice of the function vectors $\mathbf{C}_1(u)$ and $\mathbf{C}_2(u)$ in $\mathbf{W}_c(u,v)$ defined by Eq. (36), the iteration of Eq. (52) rapidly converges.

It is remarkable that the free function vectors $\mathbf{C}_1(u)$ and $\mathbf{C}_2(u)$ determining the φ-transformation, do not appear explicitly in the φ-transformation formulae for the albedo matrix or for the surface Green's function matrix, Eqs. (49) and (45), respectively.

It should be noticed that the general φ-transformation defined by Eq. (36) does not necessarily retain the symmetry properties of the original physical phase matrix given by Eq. (2). However, these symmetry properties also are preserved for reduced phase matrices of the form

$$
\begin{aligned}
\mathbf{W}_c(u,v) = \mathbf{W}(u,v) \\
-u[C_1 \mathbf{j}^-(0,u,\eta_1)\widetilde{\mathbf{j}}^-(0,v,\eta_1) + C_2 \mathbf{j}^+(0,u,\eta_1)\widetilde{\mathbf{j}}^+(0,v,\eta_1)]v,
\end{aligned}
\tag{53}
$$

with free scalar parameters C_1 and C_2. In this case, the reciprocity relation holds for the reduced albedo matrix in the form

$$
\mathbf{R}_c(\mu,\mu_0) = \widetilde{\mathbf{R}}_c(\mu_0,\mu).
\tag{54}
$$

Note that the transformation of the scalar transfer equation (9) for conservative isotropic scattering considered above, which resulted to Eq. (14), does not fit the class of symmetric φ-transformations defined by Eq. (53). However, the simplicity of that transformation has some other advantages.

5. The surface Green's function matrix for a finite medium

For a medium of finite optical thickness b, the φ-transformation formulae become a bit more complex. On starting with the transfer equation Eq. (37) for the surface Green's function, we find a partial solution

$$
\mathbf{I}_{ih}(\tau,u) = \mathbf{j}^+(\tau,u,\eta_1)\widetilde{\mathbf{g}}^-(0,\mu_0) + \mathbf{j}^-(\tau,u,\eta_1)\widetilde{\mathbf{g}}^+(0,\mu_0).
\tag{55}
$$

In order to get the complete solution, we employ the complete Green's function matrix $\mathbf{G}_c(\tau,u;\tau_0,u_0)$ defined as the solution to the transformed transfer equation

$$
u\frac{\partial}{\partial\tau}\mathbf{G}_c(\tau,u;\tau_0,u_0) = -\mathbf{G}_c(\tau,u;\tau_0,u_0) + \frac{a}{2}\int_{-1}^{+1}dv\mathbf{W}_c(u,v)\mathbf{G}_c(\tau,v;\tau_0,u_0),
\tag{56}
$$

$$
0 < \tau, \tau_0 < b,
$$

subject to the jump condition at $\tau = \tau_0$,

$$
\mathbf{G}_c(\tau_0+0,u;\tau_0,u_0) - \mathbf{G}_c(\tau_0-0,u;\tau_0,u_0) = \frac{1}{u}\delta(u-u_0)\mathbf{E},
\tag{57}
$$

and obeying the boundary conditions

$$\mathbf{G}_c(0,\mu;\tau_0,u_0) = \mathbf{G}_c(b,-\mu;\tau_0,u_0) = \mathbf{0} \quad \text{for} \quad \mu \in [0,1] \quad \text{and} \quad 0 < \tau_0 < b, \quad (58)$$

at the top and the bottom surfaces, respectively.

The solution to Eq. (37) subject to the boundary conditions, Eqs. (21) and (22), can be expressed now as

$$\mathbf{G}(\tau,u;0,\mu_0) = \mathbf{G}_c(\tau,u;0,\mu_0) + \mathbf{I}_{ih}(\tau,u) - \int_0^1 d\mu\mu\mathbf{G}_c(\tau,u;0,\mu)\mathbf{I}_{ih}(0,\mu)$$

$$(59)$$

$$- \int_0^1 d\mu\mu\mathbf{G}_c(\tau,u;b,-\mu)\mathbf{I}_{ih}(b,-\mu), \quad 0 < \tau < b.$$

It remains now to find the parameter vectors $\tilde{\mathbf{g}}^\pm(0,\mu_0)$ incorporated in the partial solution $\mathbf{I}_{ih}(\tau,u)$ given by Eq. (55). On using the definitions Eq. (38), and taking into account the orthogonality relations Eq. (29), we find from Eq. (58) a system of two algebraic equations. Its solution yields

$$\tilde{\mathbf{g}}^+(0,\mu_0) = \frac{1}{d}[g_{-+}\tilde{\mathbf{g}}_c^+(0,\mu_0) - g_{++}\tilde{\mathbf{g}}_c^-(0,\mu_0)],$$

$$(60)$$

$$\tilde{\mathbf{g}}^-(0,\mu_0) = \frac{1}{d}[g_{+-}\tilde{\mathbf{g}}_c^-(0,\mu_0) - g_{--}\tilde{\mathbf{g}}_c^+(0,\mu_0)],$$

with

$$g_{ik} = 2\int_{-1}^{+1} duu \int_0^1 d\mu\mu\tilde{\mathbf{j}}^i(0,u,\eta_1)\mathbf{G}_c(+0,u;0,\mu)\mathbf{j}^k(0,\mu,\eta_1)$$

$$(61)$$

$$+ 2\int_{-1}^{+1} duu \int_0^1 d\mu\mu\tilde{\mathbf{j}}^i(0,u,\eta_1)\mathbf{G}_c(0,u;b,-\mu)\mathbf{j}^k(b,-\mu,\eta_1), \quad i,k=+,-,$$

and

$$d = g_{+-}g_{-+} - g_{++}g_{--}. \quad (62)$$

This completes the φ-transformation for the surface Green's function matrix of a finite medium.

If the reduced phase matrix $\mathbf{W}_c(u,v)$ has the symmetric form (53), the φ-transformation retains the mirror symmetry of the Green's function matrix with respect to the midplane of the medium in the form

$$\mathbf{G}_c(\tau,u;\tau_0,u_0) = \mathbf{G}_c(b-\tau,-u;b-\tau_0,-u_0). \quad (63)$$

Then, obviously, the transformation formulae (59) and (61) can be formulated solely in terms of the surface Green's function matrix $\mathbf{G}_c(\tau,u;0,\mu_0)$.

6. Conclusions

The possibility to change the effective albedo of transfer problems by means of appropriate transformations has been pointed out for particular cases many years ago by Ivanov [8], Domke [7], and Ivanov, Rybicki, and Kasaurov [9]. Subsequently, Nagirner [10], Ivanov [11], and Ivanov and Kasaurov [12] have analyzed the "albedo shifting" in some detail for isotropic and anisotropic scalar scattering, dealing with integral equations for the source functions. Different from that approach, the φ-transformation described above starts with the integro-differential form of the transfer equation, which is transformed into an equivalent transport equation of the same form with a modified scattering term and, as a consequence, with a changed effective single-scattering albedo. This approach not only reveals a larger degree of freedom than recognized earlier, but also quite naturally permits us to deal with polarized radiative transfer and general phase matrices.

The φ-transformation of the transport equation crucially depends on the existence of leading discrete eigensolutions that determine the long-range behavior of the radiation field or, equivalently, the contribution of high orders of scattering to the radiation field. The nature of the transformation easily can be understood in terms of eigenmodes: On modifying the phase matrix as described by Eq. (36), the pair of leading eigenmodes of the original transfer equation is replaced by a new pair of eigenmodes with changed discrete eigenvalues. All other eigenmodes of the original transfer equation (1) and of the transformed transport equation (42) remain the same. In other words, on choosing the function vectors $\mathbf{C}_1(u)$ and $\mathbf{C}_2(u)$ in Eq. (36) appropriately, the φ-transformation enables one to replace the eigenmodes responsible for the long-range characteristics of multiple scattering by new eigenmodes of shorter range. Therefore, the φ-transformation primarily shifts the leading eigenvalues, with an albedo shifting appearing as a consequence.

Thus, for example, it can be shown readily that the transformation of the phase matrix described by Eq. (53) induces a shift of the leading discrete eigenvalues $\eta_{\pm 1} = \pm 1/k$ of the original transfer equation to the new discrete eigenvalues $\eta^c_{\pm 1} = \pm 1/k_c$ of the transformed transfer equation (42), where

$$k_c = \sqrt{\left(kN_1 + \frac{a}{4}C_1\right)\left(\frac{k}{N_1} + \frac{a}{4}C_2\right)}, \tag{64}$$

is the new characteristic exponent, and

$$\varphi(u, \eta^c{}_1) = \mathbf{j}^+(0, u, \eta_1) + \frac{1}{k_c}\left(kN_1 + \frac{a}{4}C_1\right)\mathbf{j}^-(0, u, \eta_1) \tag{65}$$

is the corresponding eigenfunction vector, i.e., the solution of the characteristic equation (6) with the phase matrix given by Eq. (53). For the transformed transfer equation (14), the shifted characteristic exponent is found to be $k_c = \sqrt{C}$, with

$$i(\tau, u, \pm 1/k_c) = \frac{1}{2(1 \mp k_c u)}e^{\mp k_c \tau} \tag{66}$$

being the corresponding pair of discrete eigensolutions of Eq. (14).

Obviously, the φ-transformation may be applied also to the leading eigenmodes of the U,V-component of the azimuthally averaged transfer equation as well as to higher azimuthal Fourier components. This may be justified if leading eigenmodes with sufficiently large eigenvalues occur. In practice, this may be relevant to the azimuthally averaged U,V-component as well as to some of the first azimuthal Fourier components.

From the computational point of view, the φ-transformation is most advantageous for multiple scattering in optically thick media with small true absorption. In particular, for semi-infinite media, the simple transformation formulae, Eqs. (44) and (48), offer the possibilities for highly efficient computational schemes. The φ-transformation also may be useful for avoiding problems of convergence, which may occur for adding and doubling computational schemes.

Numerical experiments of Ivanov and Kasaurov [12], who studied the behavior of iterations for the transformed integral equation for the source function in the case of scalar anisotropic scattering, revealed that with an appropriate choice of parameters, a dramatic acceleration of convergence can always be achieved. One should expect that the same also is true for iterative computational methods based on the φ-transformed transfer equation for polarized radiation derived above.

References

1. S. Chandrasekhar, "Radiative Transfer" (Dover, New York, 1960).

2. J. W. Hovenier, and C. V. M. van der Mee, Astron. Astrophys. **128**, 1 (1983).

3. H. Domke, Astrophysics **10**, 125 (1975).

4. M. Benassi, R. D. M. Garcia, and C. E. Siewert, Z. Angew. Math. Phys. **35**, 308 (1984).

5. W. A. De Rooij, "Reflection and Transmission of Polarized Light by Planetary Atmospheres" (Ph.D. Thesis, Free University, Amsterdam, 1985).

6. V. V. Ivanov, "Transfer of Radiation in Spectral Lines" (National Bureau Standards Spec.

Publ. 385, V.S. 9001, U.S. Govt. Printing Office, Washington, D.C., 1973). (Original Russian edition: Nauka, Moscow, 1969.)

7.   H. Domke, J. Quant. Spectrosc. Radiat. Transfer **39**, 283 (1988).

8.   V. V. Ivanov, Astrophysics **13**, 284 (1977).

9.   V. V. Ivanov, G. B. Rybicki, and A. M. Kasaurov, "Albedo shifting: a new facet of classical radative transfer" (Harvard–Smithsonian Center for Astrophysics Preprint Series No. 3478, 1992).

10.   D. I. Nagirner, Dokl. Akad. Nauk SSSR **343**, 191 (1995).

11.   V. V. Ivanov, Astron. Zh. **75**, 102 (1998).

12.   V. V. Ivanov, and A. M. Kasaurov, Astrophysics **42**, 485 (1999).

On the walk to the ancient sites at Bakhchissarai
T. Viik (l) and H. Domke (r)

Wine-tasting at Livadia Palace with Alla Rostopchina- Shakhovskaya, (top) Nikolai Voshchinnikov, Marina Prokopjeva, Daria Dubkova, Dmitry Shakhovskoy, (bottom) Nikolai Kiselev, Vera Rosenbush, Jim Hough, and Leonid Shulman.

Michael Mishchenko, happy again to be at CrAO.

Ludmila Chaikovskaya, Alina Ponyavina, Valery Loiko, Liudmila Astafyeva, and Anatoliy Gavrilovich pose with Churchill's favorite lion at Alupka Palace.

QUADRATIC INTEGRALS IN INVERSE PROBLEMS WITH MULTIPLE SCATTERING

T. VIIK[1] AND N. J. MCCORMICK[2]
[1] *Tartu Observatory,*
Tõravere, Tartumaa, 61602, Estonia
[2] *University of Washington,*
Mechanical Engineering, Seattle, WA 98195-2600, USA

Abstract.
    Procedures are described and tested for determining the single-scattering albedo using the measurement of specific intensities within or at the surfaces of homogeneous, finite or infinite atmospheres. Unpolarized radiation in an isotropically scattering semi-infinite atmosphere and polarized radiation in an atmosphere that scatters according to the Rayleigh-Cabannes law are considered. According to our numerical experiments the albedo of single scattering can be derived with great accuracy even when the measurements are not so accurate while the determination of other characteristics of the medium is much more complicated.

1. Introduction

It is well known that the equation of radiative transfer for a conservative medium has two exact integrals: the flux integral and the so-called K-integral which is closely connected with the radiative pressure. These integrals are linear with respect to the specific intensity. The transfer equation also admits quadratic integrals, first considered in detail by Rybicki [13] for a homogeneous semi-infinite atmosphere that scatters isotropically. He has shown that some of his integrals are related to the $\sqrt{\epsilon}$ - law (where ϵ is the probability of photon destruction in a single scattering event), a law much discussed in the literature (e.g., by Frisch and Frisch [5]).

    Rybicki's quadratic integrals were generalized by Ivanov [6] in such a way that they relate the radiation fields at two different optical depths in an atmosphere. Ivanov showed that by using these integrals for determining

125

G. Videen et al. (eds.), *Photopolarimetry in Remote Sensing,* 125-136.

the radiation field in an optically semi-infinite atmosphere the problem can be reduced to finding only the specific intensity for the downward radiation. He also noticed that one may obtain the quadratic integrals proceeding from completely different considerations [1].

Derivation and use of another type of quadratic integrals was initiated independently by Siewert in 1978 [15]. He showed how to determine the single-scattering albedo for unpolarized radiation in a semi-infinite atmosphere with isotropic scattering, and later extended the work to find the albedo and the coefficients for two-term [17] and three-term [18] phase functions for anisotropic scattering in optically finite or infinite atmospheres. For the two-term phase function no numbers were given, but according to some (unpublished) calculations by one of the authors (TV) the single-scattering albedo easily can be found, although the determination of the parameter of the phase function is extremely sensitive to errors in measuring the specific intensities. The errors in determining the numerical characteristics in the case of three-term phase function have been analyzed by Dunn and Maiorino [4].

McCormick developed two complementary sets of equations involving quadratic integrals for determining the albedo and an arbitrary number of scattering expansion coefficients, provided unpolarized intensity measurements are dependent on both the polar and azimuthal angles [7]. Numerical tests were performed by Oelund and McCormick [11] that demonstrated the sensitivity of the estimated parameters to simulated errors in the measurements, and provided insight into which of the two independent sets of equations was better.

For polarized radiation with a 2-vector intensity, Siewert [16] showed how to determine the single-scattering albedo from the measurements of the specific intensity of the emergent radiation from a Rayleigh-scattering half-space. Similarly, Siewert and Maiorino [19] exploited quadratic integrals to determine the single-scattering albedo λ and the albedo of a Lambertian bottom, λ_0, in a finite Rayleigh-scattering atmosphere, but the authors noted that accurate values of λ and λ_0 can be obtained only when exceptionally accurate experimental data become available.

Applying an approach similar to that of Siewert and Maiorino, McCormick [8] developed a procedure to determine the single-scattering albedo from polarization measurements of the specific intensity at two levels in a finite homogeneous atmosphere that scatters radiation according to the Rayleigh-Cabannes law with true absorption. Viik and McCormick [24] tested McCormick's procedure by solving the forward radiative transfer problem for the 2-vector specific intensity at two levels in the atmosphere or on its boundaries, [21]– [23] and then inducing random errors and trying to extract the single-scattering albedo and the depolarization factor.

In this report we demonstrate the similarity of the quadratic integrals for unpolarized radiation, as employed by Rybicki, to the quadratic integrals used by Siewert and McCormick for the 2-vector polarized radiation problem. We have assumed that both I- and Q-components may be measured to the same accuracy. This may not be the case, as pointed out by Mishchenko, [10] because the Q-component of the unpolarized incident flux is much smaller than I-component. How the difference in "measurement" errors influences the results remains to be studied in a forthcoming paper.

2. Rybicki's quadratic integrals for unpolarized radiation

To make the formulas more transparent we omit (where possible) the arguments of functions. We consider an atmosphere that is optically semi-infinite, homogeneous, and plane-parallel in which the radiation is scattered isotropically and monochromatically. Radiative transfer in such an atmosphere is described by

$$\mu \frac{\partial I}{\partial \tau} = I - S, \tag{1}$$

where $I(\tau, \mu)$ is the specific intensity at optical depth τ at an angle $\cos^{-1} \mu$ to the *negative* τ-axis. $S(\tau)$ is the source function

$$S = \lambda J + (1 - \lambda) B, \tag{2}$$

where λ is the albedo of single scattering, and

$$J = \frac{1}{2} \int_{-1}^{1} I(\tau, \mu) d\mu. \tag{3}$$

Next we define

$$I_+(\tau, \mu) \equiv I(\tau, \mu), \quad I_-(\tau, \mu) \equiv I(\tau, -\mu). \tag{4}$$

For these quantities

$$\mu \frac{\partial I_+}{\partial \tau} = I_+ - S, \quad -\mu \frac{\partial I_-}{\partial \tau} = I_- - S. \tag{5}$$

While

$$\frac{\partial}{\partial \tau}(I_+ I_-) = (S + \mu \frac{\partial I_+}{\partial \tau}) \frac{\partial I_-}{\partial \tau} + (S - \mu \frac{\partial I_-}{\partial \tau}) \frac{\partial I_+}{\partial \tau}, \tag{6}$$

we find that

$$\frac{\partial}{\partial \tau}(I_+ I_-) = S \frac{\partial}{\partial \tau}(I_+ + I_-). \tag{7}$$

From this equation Rybicki's quadratic integral $Q(\tau)$, defined by [13]

$$Q(\tau) = \frac{1}{2} \int_{-1}^{1} I(\tau, -\mu) I(\tau, \mu) d\mu, \tag{8}$$

is found to satisfy the equation

$$\frac{dQ}{d\tau} = 2S \frac{dJ}{d\tau}. \tag{9}$$

Rybicki also considered the quadratic integral

$$R(\tau) = \int_{-1}^{1} \mu^2 I(\tau, -\mu) I(\tau, \mu) d\mu. \tag{10}$$

Here this integral is considered only in the planetary problem. (In passing we note that these same two quadratic integrals were utilized by Siewert and McCormick for their applications.)

In the following we study only three kinds of internal sources: $B = 0$, corresponding to the Milne problem where the sources of radiation are deep in the atmosphere, the case with a constant source B, and the case with exponential sources in the planetary problem.

2.1. THE MILNE PROBLEM

For the Milne problem

$$\frac{dQ}{d\tau} = 2\lambda J \frac{dJ}{d\tau} = \lambda \frac{dJ^2}{d\tau}, \tag{11}$$

which can be integrated from τ_1 to τ_2 to give

$$Q(\tau_2) - Q(\tau_1) = \lambda \left[J^2(\tau_2) - J^2(\tau_1) \right]. \tag{12}$$

Because $Q(0) = 0$ it immediately follows that

$$\lambda = Q(\tau) \left[J^2(\tau) - J^2(0) \right]^{-1}. \tag{13}$$

This means that if we measure the specific intensities at optical depths $\tau > 0$ and $\tau = 0$ (preferably doing this at Legendre-Gauss nodes for better integrations) we find the single-scattering albedo.

2.2. THE CONSTANT SOURCE PROBLEM

Rybicki [13] also has shown that the Q-integral can be used for the case with a spatially homogeneous distribution of internal sources of radiation of magnitude B. This case represents a model of an isothermal atmosphere with B as the Planck function and, according to the Kirchhoff law, the internal source of radiative energy is proportional to ϵB, where ϵ is the probability of a photon's destruction in a single scattering event given by

$$\epsilon = 1 - \lambda. \tag{14}$$

With this definition we can rewrite Eq. (2) for the source function

$$S = (1 - \epsilon)J + \epsilon B. \tag{15}$$

After incorporating this equation into Eq. (9) it follows that

$$\frac{dQ}{d\tau} = (1 - \epsilon)\frac{dJ^2}{d\tau} + 2\epsilon B\frac{dJ}{d\tau}, \tag{16}$$

which immediately can be integrated to give

$$Q = (1 - \epsilon)J^2 + 2\epsilon BJ + \text{const.} \tag{17}$$

Rybicki went on to show, using the asymptotics at infinite optical depth, where $S = Q = J = B$, that

$$S^2(\tau) = (1 - \epsilon)Q(\tau) + \epsilon B^2. \tag{18}$$

At the outer boundary of an atmosphere, where $Q(0) = 0$ because there is no incoming radiation, the source function becomes

$$S(0) = \epsilon^{1/2}B, \tag{19}$$

which is called the $\sqrt{\epsilon}$-law for monochromatic scattering. Equation (18) is the generalization of this law to all optical depths in the atmosphere.

One may use either Eq. (18) or Eq. (19) for determining the destruction probability. Because it is easier to measure the intensities at the surface of an atmosphere, we use Eq. (19) for that purpose. After using Eq. (15) in Eq. (19) and remembering that all values are taken at $\tau = 0$ we get

$$(J - B)^2\epsilon^2 + (2JB - 2J^2 - B^2)\epsilon + J^2 = 0. \tag{20}$$

This quadratic equation can easily be solved for ϵ and its physically meaningful root is

$$\epsilon = \left(\frac{J}{J - B}\right)^2. \tag{21}$$

2.3. THE PLANETARY PROBLEM

Next we consider the planetary problem for a source function of the type

$$S = \lambda J + S^* \exp(-\tau/\mu_0), \tag{22}$$

where

$$S^* = \frac{1}{4}\lambda F \tag{23}$$

with πF the net flux of a parallel beam of radiation incident on a plane-parallel atmosphere per unit area normal to itself, and μ_0 is the direction cosine of the angle of incidence referred to the *outward* normal.

When generalizing the results by Rybicki [13] Ivanov [6] found an equation involving Q and R that can be evaluated for $\tau_1 = \tau_2 = \tau$ to give

$$S^2 = \lambda Q - \lambda R/\mu_0^2 + (1 - \lambda)^{-1}\left[S^* + \lambda H/\mu_0\right]^2, \tag{24}$$

where

$$H = \frac{1}{2}\int_{-1}^{1} I(\tau, \mu)\mu d\mu. \tag{25}$$

For simplicity we consider here only the case $\tau = 0$ so that $Q = R = 0$. From Eq. (24) we have

$$\lambda = 1 - \left[\frac{4H + \mu_0 F}{\mu_0(4J + F)}\right]^2. \tag{26}$$

Thus, to determine the albedo of single scattering for the planetary problem one has to measure the mean intensity and the flux at the surface of the atmosphere.

3. Quadratic integrals and the inverse problem for polarized radiation

Next we consider the application of quadratic integrals for polarized radiation. We analyze an optically semi-infinite or finite homogeneous atmosphere that scatters radiation according to the azimuthally-averaged Rayleigh-Cabannes law for molecular scattering.

We consider two problems: the planetary problem with boundary conditions

$$\begin{aligned}
\mathbf{I}(0, \mu) &= \mathbf{F}\delta(\mu - \mu_0), \quad 0 \le \mu \le 1, \\
\mathbf{I}(\tau_0, \mu) &= 0, \quad -1 \le \mu \le 0,
\end{aligned} \tag{27}$$

for incident flux $\mathbf{F}$, and the Milne problem with a source deep in the interior, for which

$$\mathbf{I}(0, \mu) = 0, \quad 0 \le \mu \le 1, \tag{28}$$

and the intensity in deep layers does not increase more rapidly than $\exp(\tau)$. When dealing with the planetary problem we have to remember that in the following all the specific intensities of the downward radiation are sums of direct and diffuse components.

We proceed from the equation of transfer

$$\mu \frac{\partial}{\partial \tau} \mathbf{I}(\tau, \mu) + \mathbf{I}(\tau, \mu) = \frac{\lambda}{2} \mathbf{Q}(\mu) \int_{-1}^{1} \mathbf{Q}^T(\mu') \mathbf{I}(\tau, \mu') d\mu', \quad 0 \le \tau \le \tau_0, \tag{29}$$

where μ is measured from the *positive* axis of the optical depth τ and the intensity $\mathbf{I}$ is a 2-vector with components $I_\ell(\tau, \mu)$ and $I_r(\tau, \mu)$. The matrix $\mathbf{Q}(\mu)$ is defined as [2, 14]

$$\mathbf{Q}(\mu) = \frac{3}{2(c+2)^{1/2}} \left[\begin{array}{cc} c\mu^2 + \frac{2}{3}(1-c) & (2c)^{1/2}(1-\mu^2) \\ \frac{1}{3}(c+2) & 0 \end{array} \right]. \tag{30}$$

We first operate on Eq. (29) with the transpose vector $2\mathbf{I}^T(\tau, -\mu)$ and integrate over μ to obtain the scalar equation

$$2 \int_{-1}^{1} \mathbf{I}^T(\tau, -\mu) \mu \frac{\partial}{\partial \tau} \mathbf{I}(\tau, \mu) d\mu = \lambda \mathbf{P}_0^T(\tau) \mathbf{P}_0(\tau) - S_0(\tau), \tag{31}$$

where

$$\mathbf{P}_n(\tau) = \int_{-1}^{1} \mu^n \mathbf{Q}^T(\mu) \mathbf{I}(\tau, \mu) d\mu \tag{32}$$

and

$$S_n(\tau) = 4 \int_0^1 \mu^{2n} \mathbf{I}^T(\tau, -\mu) \mathbf{I}(\tau, \mu) d\mu. \tag{33}$$

After differentiating Eq. (31) and substituting Eq. (29) into Eq. (31) we get

$$-2 \int_{-1}^{1} \mathbf{I}^T(\tau, -\mu) \frac{\partial}{\partial \tau} \mathbf{I}(\tau, \mu) d\mu + \lambda \mathbf{P}_0^T(\tau) \frac{d}{d\tau} \mathbf{P}_0(\tau) =$$
$$\frac{d}{d\tau} \left[\lambda \mathbf{P}_0^T(\tau) \mathbf{P}_0(\tau) - S_0(\tau) \right]. \tag{34}$$

Next we add Eq. (34) and its transpose (with a reversal in the sign of μ) to obtain

$$\frac{d}{d\tau} \left[\lambda \mathbf{P}_0^T(\tau) \mathbf{P}_0(\tau) - S_0(\tau) \right] = 0. \tag{35}$$

This result immediately can be integrated over $\tau_1 \le \tau \le \tau_2$ to give

$$\lambda = [\mathbf{P}_0^T(\tau_1)\mathbf{P}_0(\tau_1) - \mathbf{P}_0^T(\tau_2)\mathbf{P}_0(\tau_2)]^{-1}[S_0(\tau_1) - S_0(\tau_2)]. \qquad (36)$$

This is the equation involving the quadratic integral $S_0(\tau)$.

We can repeat the described procedure by operating on Eq. (29) with $2\mu\mathbf{I}^T(\tau, -\mu)$ and integrating over μ. Eventually we get [8]

$$\lambda[\mathbf{P}_1^T(\tau_1)(\mathbf{U} - \lambda\mathbf{R})^{-1}\mathbf{P}_1(\tau_1) - \mathbf{P}_1^T(\tau_2)(\mathbf{U} - \lambda\mathbf{R})^{-1}\mathbf{P}_1(\tau_2)] = \\ S_1(\tau_1) - S_1(\tau_2), \qquad (37)$$

where

$$(\mathbf{U} - \lambda\mathbf{R})^{-1} = \Delta^{-1} \begin{bmatrix} c + 2 - 12\lambda c/5 & \lambda(2c)^{1/2}(1 - 7c/10) \\ \lambda(2c)^{1/2}(1 - 7c/10) & c + 2 - \lambda(2 + 7c^2/10) \end{bmatrix} \qquad (38)$$

for $\Delta = (1 - \lambda)(c + 2)(1 - 7\lambda c/10)$. Eq. (37) is an equation involving the second quadratic integral, $S_1(\tau)$.

4. Numerical tests

We performed numerical tests on the presented schemes by solving the respective forward radiative transfer problem and getting the specific intensities at the Gauss nodes. These results next were used as the mean values to which the normally distributed random errors

$$\varepsilon(x) = \delta \exp(-x^2/2\rho^2) \qquad (39)$$

were added by sampling 10,000 "measurements." Here ρ is the dispersion of random errors of an individual measurement and δ is the amplitude of the error (i.e., $\delta = 0.1$ corresponds to an error magnitude of 10%).

To obtain the normally distributed pseudorandom errors we first used the RAN1 program [12] for getting uniformly distributed pseudorandom errors on (0,1). Then the Box-Muller method [12] was applied to transform these errors to normally distributed ones.

The direct problems for isotropic scattering were solved by applying the method of Viik et al. [20] and for polarized transfer by applying the method described in references 21-23.

In the following figures each line of mean values is bounded on either side by the values of the mean plus or minus two standard deviations. If there were no errors in specific intensities the curves would have been diagonal straight lines.

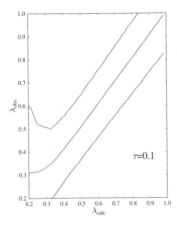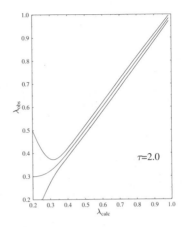

Figure 1.   The Milne problem. λ_{obs} obtained from Eq. (13) versus λ_{calc} of forward problem for $\tau = 0.1$ and $\tau = 2.0$ with error amplitude $\delta = 0.05$ and individual dispersion $\rho = 1.0$.

4.1. THE MILNE PROBLEM

Determination of single-scattering albedos for this problem is very sensitive to "measurement" errors. Even for $\delta = 0$, with no errors at all, (consequently, the "measurements" were made to machine accuracy) we could not obtain good results even for small τ. However, in Fig. (1) we see that the accuracy was much better when $\lambda \to 1$ and the measurements were made deep inside the atmosphere. For example, if $\delta = 0.05$ we could obtain λ to an accuracy of 0.001 for $\lambda = 1$ and $\tau = 5.0$.

4.2. THE CONSTANT SOURCE PROBLEM

The calculations showed that the destruction probability (or the single-scattering albedo) can be obtained in a stable manner even for large error amplitudes of a single measurement ($\delta \leq 0.2$) if we only measure the specific intensity at the boundary of an atmosphere. Figure (2) shows how quickly this stability deteriorates if we try to transfer the measurements into the atmosphere according to Eq. (18). More-or-less trustworthy results can be obtained only for $\tau \leq 0.1$, which is understandable because for large optical

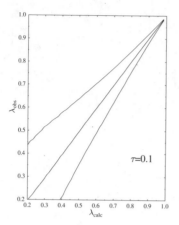
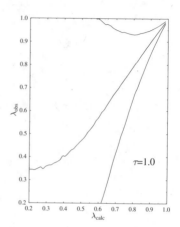

Figure 2. The constant source problem. $\lambda_{obs} = 1 - \epsilon_{obs}$ obtained from Eq. (21) versus λ_{calc} of forward problem for $\tau = 0.1$ and $\tau = 1.0$ with error amplitude $\delta = 0.10$ and individual dispersion $\rho = 1.0$.

depths the intensity I in the isothermal medium turns to B, independently of the destruction probability. If we want to determine ϵ from measurements in deeper layers we have to measure the specific intensity very accurately.

4.3. PLANETARY PROBLEM FOR POLARIZED RADIATION

Our calculations have shown that the single-scattering albedo can be obtained from Eq. (36) even if the error amplitude is $\delta \approx 0.25 - 0.3$. The results for λ in Fig. (3) with $\mu_0 = 0.5$ are very nearly the same as for $\mu_0 = 0.1$ reported earlier [24]. After having found λ from Eq. (36) we use it in Eq. (37) first to determine the number of possible values of the depolarization factor c. If there are two or three roots we have to use each c-value in our model to determine which minimizes the quadratic form

$$q = \int_{-1}^{1} [\mathbf{\Delta I}^T(\tau_2, \mu)\mathbf{\Delta I}(\tau_2, \mu) - \mathbf{\Delta I}^T(\tau_1, \mu)\mathbf{\Delta I}(\tau_1, \mu)]d\mu, \qquad (40)$$

where $\mathbf{\Delta I}(\tau, \mu)$ is the difference between measured values and those computed with the selected value of c.

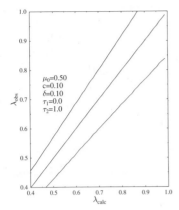
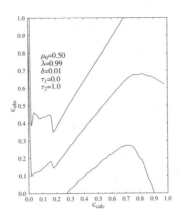

Figure 3. Molecular scattering. Left panel: λ_{obs} obtained from Eq. (36) versus λ_{calc} of forward problem for $c = 0.1$ with error amplitude $\delta = 0.1$. Right panel: c_{obs} obtained from Eq. (37) versus c_{calc} of forward problem for $\lambda = 0.99$ with error amplitude $\delta = 0.01$. For both figures the individual dispersion is $\rho = 1.0$ and planetary illumination is at $\mu_0 = 0.5$. The levels of "measurement" are $\tau_1 = 0.0$ and $\tau = 1.0$, respectively.

Here we demonstrate the influence of the two sets of parameters on the results of calculations. It is clear that even errors of only 1% in measured specific intensities may make the procedure unstable. The instabilities in determining c are very similar to those encountered by Siewert and Maiorino [19] when they considered a finite atmosphere with Lambertian reflection at the ground. They too found that λ could be determined more accurately than the Lambertian reflection coefficient.

5. Conclusions

As shown, the quadratic integrals Q and R of Rybicki for unpolarized radiation and the quadratic integrals S_0 and S_1 of Siewert and McCormick for polarized radiation are closely related. These integrals of radiative transfer provide us with a convenient tool for solving some elementary inverse problems. Numerical experiments have shown that for these problems the single-scattering albedo can be derived with great accuracy even when the measurements are not so accurate. The determination of other characteristics of the medium is much more complicated.

6. Acknowledgments

The authors thank an anonymous referee for the constructive review. One of the authors (TV) has benefited from discussions with Michael Mishchenko, Helmut Domke and Vsevolod Ivanov. This work was supported by the Estonian Ministry of Education within the project No. TO 0060059S98 and by the Estonian Science Foundation under grant No. 4701.

References

1. D. Anderson, J. Inst. Maths. Applics. **12**, 55 (1973).
2. G.R. Bond, C.E. Siewert, Astrophys. J. **164**, 97 (1971).
3. S. Chandrasekhar: "Radiative Transfer" (Oxford University Press, 1950, [Dover, 1960]).
4. W.L. Dunn, J.R. Maiorino, J. Quant. Spectrosc. Rad. Transfer **24**, 203 (1980).
5. U. Frisch, H. Frisch, Mon. Not. Roy. Astron. Soc. **173**, 167 (1975).
6. V.V. Ivanov, Astronomicheskij J. **55**, 1072 (1978).
7. N.J. McCormick, J. Math. Phys. **20**, 1504 (1979).
8. N.J. McCormick, Astrophys. Space Sci. **71**, 235 (1980).
9. N.J. McCormick, R. Sanchez, J. Quant. Spectrosc. Rad. Transfer **30**, 527 (1983).
10. M.I. Mishchenko, private communication (2003).
11. J.C. Oelund, N.J. McCormick, J. Opt. Soc. Am. A **2**, 1972 (1985).
12. H.W. Press, B.P. Flannery, S.A. Teukolsky, T.W. Vetterling: "Numerical Recipes" (Cambridge University Press, 1986).
13. G.B. Rybicki, Astrophys. J. **213**, 165 (1976).
14. C.E. Siewert, E.E. Burniston, Astrophys. J. **174**, 629 (1972).
15. C.E. Siewert, J. Math. Phys. **19**, 1587 (1978).
16. C.E. Siewert, Astrophys. Space Sci. **60**, 237 (1979).
17. C.E. Siewert, J. Appl. Math. Phys. (ZAMP) **30**, 522 (1979).
18. C.E. Siewert, J. Quant. Spectrosc. Rad. Transfer **22**, 441 (1979).
19. C.E. Siewert, J.R. Maiorino, J. Appl. Math. Phys. (ZAMP) **31**, 767 (1980).
20. T. Viik, R. Rõõm, A. Heinlo, Tartu Teated **76**, 3 (1976).
21. T. Viik, Earth, Moon, and Planets **46**, 261 (1989).
22. T. Viik, Earth, Moon, and Planets **49**, 163 (1990).
23. T. Viik, J. Quant. Spectrosc. Rad. Transfer **66**, 581 (2000).
24. T. Viik, N.J. McCormick, J. Quant. Spectrosc. Rad. Transfer **78**, 235 (2002).

V. Ivanov and T. Viik (left) before the conquest of the Swallow's nest.

POLARIZATION FLUCTUATION SPECTROSCOPY

K.I. HOPCRAFT[1], P.C.Y. CHANG[1], E. JAKEMAN[1,2],
J.G. WALKER[2]

[1]*School of Mathematical Sciences, University of Nottingham, Nottingham, NG7 2RD, UK.*
[2]*School of Electrical and Electronic Engineering, University of Nottingham, Nottingham, NG7 2RD, UK.*

Abstract. Polarization fluctuation spectroscopy is a dynamic light scattering technique that extends photon correlation spectroscopy to account for particles changing the polarization state of an incident beam of coherent light. This extension enables the shape in addition to the size of the particles to be sensed. The technique requires the particles to be sufficiently numerous for the scattered light to be in the Gaussian scattering regime, but sufficiently sparse for multiple-scattering effects to be neglected. The temporal cross-correlation function of intensities scattered into separate polarization states depends on the polarization state of the input beam, the scattering angle, and the relative refractive index, size and shape of the particles. Measurement of the temporal cross-correlation function enables independent determination of the size and aspect ratio of mono-disperse samples. The theory underpinning the method is developed from first principles and the processing and experimental techniques are discussed, together with the accuracy of measurement that can be achieved. The viability of the technique is demonstrated experimentally.

1. Introduction

Photon correlation spectroscopy (PCS) is a well established technique that exploits the fluctuations coherent radiation scattered by small particles in order to determine their size. Its formulation, implementation and exploitation have been treated thoroughly in two previous NATO Advanced Study Institutes [1,2]. The concepts upon which the technique is founded are not unique to metrological applications however, for they also underpin interferometric methods used in radio astronomy. The original formulation of PCS was for

G. Videen et al. (eds.), Photopolarimetry in Remote Sensing, 137-174.

scalar light, but the inclusion of polarization into the development furnishes an additional function for the technique, namely the extraction of particle shape information in addition to the particles' size. This additional functionality can be achieved with modest adaptation to existing PCS measurement systems, but leads to a new instrumental technique, termed Polarization Fluctuation Spectroscopy (PFS).

The purpose of this chapter is to provide a self-contained introduction to the technique of PFS that ranges from its theoretical foundations, via the experimental and data processing tricks of the trade that make the method work, to the experimental results and the problems raised by inverting the data for particle characteristics. The tone is meant to be pedagogic and illustrative of the physical principals rather than exhaustive and general. Neither is the chapter meant as a replacement for the two previous NATO ASI volumes on PCS, nor for others that have been written since on allied topics [3,4]. Taken collectively this body of work provides an indispensable compendium to which the interested reader should consult for those details that are omitted here.

When an ensemble of small particles, usually held in suspension, is illuminated by a coherent source of radiation, the light scattered by the particles forms a speckle pattern in the far field. If the particles are sufficiently sparse in number, the light detected at a location outside the measurement volume will not have interacted with any other particles comprising the ensemble: the single-scattering regime. If the particles are randomly distributed throughout the measurement volume, the relative positions of the particles and the detector will also be random. Hence a speckle pattern results from the random interference between the individually scattered field contributions from each of the particles at the detector. If, on the other hand, the particles are sufficiently numerous, the field at the detector can be described by summing a 'large' number of individual and independent random contributions. In such circumstances the central limit theorem of classical statistics can be invoked, with result that the detected field is described by a Gaussian random *variable*. If the particles are in relative motion between themselves and the detector, the speckle pattern will fluctuate in time, in which case a Gaussian random *process* describes the evolution of the field and the speckle pattern. The correlation function of the process has a temporal behavior that can be related to the random movement of the particles through the suspending medium, and this in turn is related to the particles' size. Hence PCS requires a measurement of the rate of decay of the intensity correlation function to determine particle size.

The essential ingredients of achieving such a measurement in a PCS system is a source of coherent radiation, usually provided by a laser; a means of detecting the light scattered by an ensemble of particles, usually a photomultiplier tube or avalanche detector and a means for forming the intensity autocorrelation function. Nowadays this can be achieved in real time on the raw signal without the preconditioning and special processing of the signal that the first systems required [5,6,7].

It is well known that non-spherical particles scatter polarized light differently from spherical particles. If the illuminating light has a specific polarization state, the amount by which the particles corrupt this state on scattering is then related to the particles' departure from a spherical shape. This perturbation from the initial polarization state can be gauged by measuring the correlation function between two scattered intensities detected in different (and preferably orthogonal) polarization states. The difference between this measurement and that obtained on assuming the particles to be spherical is a function of the aspect ratio of the particles. This is the simple principle that underpins PFS.

The additional components for achieving a PFS measurement are polarizing optics for manipulating both the illumination and scattered polarization states: a polarizing beam splitter for directing orthogonal states into *two* detectors, and a means of forming three polarization-sensitive intensity correlation functions.

Both PCS and PFS are predicated on the measurement of fluctuations of scattered light that are described by Gaussian random processes. It is therefore important to appreciate the properties of these processes, their impact on the results obtained and their implications for obtaining data. Section 2 contains a review of the basic statistical notions required for treating measurements made on random processes. It continues by concentrating on the key factorization properties of real and complex Gaussian processes that inform the development of PCS and PFS. Section 3 distills the ideas of PCS and Section 4 then shows how polarization can be incorporated to obtain PFS. Section 5 provides an analytical illustration of the technique through a consideration of scattering by spheroidal Rayleigh particles, modeled by dipoles of unequal strength. Although something of an idealization, the Rayleigh model has the virtue of being analytical whilst encapsulating some of the problems that remain when treating particles of finite size and independent refractive index that are encountered in practice. The PFS instrumental configuration is discussed in Section 6 together with the techniques employed to ameliorate instrument and data-processing imperfections. The accuracy of the measurements is treated in Section 7. Section 8 shows experimental results and discusses the means by which particle shape and size information can be independently assessed. The concluding section discusses the merits and limitations of the technique.

2. Statistical Preliminaries

The purpose of this section is to provide definitions for the statistical measures that will be used throughout, and for establishing the principal properties of real and complex Gaussian processes.

Let $V(t)$ denote a real-valued signal that varies randomly with some parameter t, which usually denotes time. The result of a series of measurements of V can be expressed statistically through the single-fold interval probability density function (pdf) $P(V(t))$, for which $P(V(t))dV$ gives the likelihood of obtaining a measurement in the range $V \rightarrow V + dV$ at time t. The likelihood of obtaining measurements V_1 at t_1 and V_2 at t_2 is given by the joint pdf $P(V(t_1), V(t_2))$, and knowledge of the n-fold multiple joint distributions $P(\{V(t_i)\})$, where $1 \leq i \leq n$, provides a complete statistical description for the process $V(t)$. Single-fold distributions are rarely measured however, joint-distributions even less so. Instead, a less complete description is obtained through measuring a subset of the more easily obtained moments and correlations of the single fold and joint distributions. This information is equivalent to that encapsulated in the distributions if all the moments:

$$\langle V^n(t) \rangle \equiv \int_{-\infty}^{\infty} V^n P(V) dV \qquad (2.1)$$

and correlation functions:

$$\langle \prod_i V^{n_i}(t_i) \rangle \equiv \int_{-\infty}^{\infty} P(\{V_i\}) \prod_i V_i^{n_i} dV_i . \qquad (2.2)$$

can be determined.

If the probability distributions do not depend on time, so that

$$P(\{V(t_i)\}) = P(\{V(t_i + \tau)\})$$

the process is *stationary* in the *strict sense*, which implies that the origin of time is immaterial when determining correlation functions of arbitrary order. A less stringent, and more commonly encountered criterion for a process is that $\langle V(t) \rangle$ is independent of time and that:

$$\langle V(t)V(t+\tau) \rangle = \langle V(0)V(\tau) \rangle ,$$

in other words the correlations depend only on a time increment and not the origin of time. In this case the process is said to be stationary in the *wide-sense*. Any process that is stationary in the strict sense is also stationary in the wide sense, but not vice-versa.

If ensemble averages (2.1) are equal to corresponding time averages, i.e if

$$\langle f(V) \rangle \equiv \int_{-\infty}^{\infty} f(V) P(V) dV = \lim_{T \to \infty} \frac{1}{T} \int_{-T/2}^{T/2} f(V(t)) dt$$

the process is *ergodic*. Loosely speaking, the concept of ergodicity is obtained when each state of the random process is realized during an observation period that is of sufficiently long duration. If the observation time is too short, not all states of the process can occur and a time average will not provide an accurate estimate of the corresponding ensemble average. This situation is equivalent to a determination of how many data points are required to obtain accurate and unbiased estimates of the moments and correlations. This issue will be considered further in Section 6, together with other uncertainties in the measurement process that affect accuracy. The subtle differences between strict and wide sense stationarity and ergodicity are discussed with illuminating examples in [8].

The autocorrelation function of a stationary ergodic signal is

$$G(\tau) \equiv \langle V(0)V(\tau) \rangle = \lim_{T \to \infty} \frac{1}{T} \int_{-\infty}^{\infty} V(T)V(T+\tau)\, dT$$

and so the correlation function is symmetric, $G(\tau) = G(-\tau)$. For an entirely random process, all memory is erased at large delay times so that the measurements become independent of each other. Hence it follows that:

$$\lim_{T \to \infty} G(\tau) = \langle V \rangle^2,$$

whereas for vanishingly small τ, the measurements are not independent and so the autocorrelation tends to the second moment $\langle V^2 \rangle$. It is convenient to define a normalized correlation function, the form of which is dictated by whether V adopts positive and negative values, like a field variable:

$$g^{(1)}(\tau) = \frac{\langle V(0)V(\tau) \rangle}{\langle V^2 \rangle} \qquad (2.3)$$

or whether V is strictly positive, such as an intensity, in which case:

$$g^{(2)}(\tau) = \frac{\langle V(0)V(\tau) \rangle}{\langle V \rangle^2} \qquad (2.4).$$

We shall see that for the special case of a complex Gaussian random process, $g^{(2)}(\tau)$ is expressible in terms of $g^{(1)}(\tau)$.

The Gaussian random process is ubiquitous because of the central limit theorem of classical statistics [9]. This states that a random variable formed from a sum of random variables whose distributions are arbitrary (save for the proviso that the variance for each of the variables exists) has a Gaussian distribution when the number comprising the sum is large. Hence if an observable is the result of a series of independent random effects, then a Gaussian random variable, and indeed a Gaussian random process, often describes its statistical behavior. In addition to its wide-ranging applicability, the Gaussian process has particularly attractive analytical properties. Suppose that the values of a real random variable V, sampled at n times $t_1, t_2, \ldots, t_n$, are denoted by the column vector $\mathbf{V} = \{V_1, V_2, \ldots, V_n\}^T$, then if $\mathbf{V}$ is described by a Gaussian random process, its pdf is

$$P(\mathbf{V}) = \frac{1}{(2\pi)^{n/2} |\underline{\underline{C}}|^{1/2}} \exp\left(-\frac{1}{2}(\mathbf{V} - \langle \mathbf{V} \rangle)^T \underline{\underline{C}}^{-1} (\mathbf{V} - \langle \mathbf{V} \rangle)\right)$$

where the correlation matrix has elements

$$C_{ij} = \langle \left(V_i - \langle V \rangle_i\right)\left(V_j - \langle V \rangle_j\right)\rangle$$

and $|\underline{\underline{C}}|$ and $\underline{\underline{C}}^{-1}$ are the determinant and inverse of the correlation matrix $\underline{\underline{C}}$ respectively. Note that pdf for the process depends only upon the second order correlations, hence all higher order correlations are expressible in terms of these second order quantities. In particular, if all the n Gaussian variables have zero mean, i.e. $\langle V_i \rangle = 0$, the correlation between an odd number of variables all vanish

$$\langle V_1 V_2 \ldots V_{2k+1} \rangle = 0$$

and the correlation between an even number of variables factorize:

$$\langle V_1 V_2 \cdots V_{2k} \rangle = \sum_Q \left(\langle V_j V_m \rangle \langle V_k V_n \rangle \ldots \langle V_l V_p \rangle\right)\big|_{j \neq m, k \neq n, l \neq p}$$

where the summation is taken over Q which denotes all possible distinct groupings of the $2k$ variables when taken in pairs. For example, the case when $k = 2$ gives:

$$\langle V_1 V_2 V_3 V_4 \rangle = \langle V_1 V_2 \rangle \langle V_3 V_4 \rangle + \langle V_1 V_3 \rangle \langle V_2 V_4 \rangle + \langle V_1 V_4 \rangle \langle V_2 V_3 \rangle \qquad (2.5)$$

illustrating that the fourth order correlation is formed from a sum of second order correlation functions.

An important extension to this real Gaussian random process is its complex generalization, which has particular relevance to applications in optics. Here the column vector $\mathbf{V}$ is complex

$$\mathbf{V} = \{\mathrm{Re}(V_1), \mathrm{Re}(V_2), \ldots, \mathrm{Re}(V_n), \mathrm{Im}(V_1), \mathrm{Im}(V_2), \ldots, \mathrm{Im}(V_n)\}^T$$

which has $2n$ elements. The pdf for a complex Gaussian random process is:

$$P(\mathbf{V}) = \frac{1}{(2\pi)^n |\underline{\underline{C}}|^{1/2}} \exp\left(-\frac{1}{2}(\mathbf{V} - \langle \mathbf{V} \rangle)^T \underline{\underline{C}}^{-1} (\mathbf{V} - \langle \mathbf{V} \rangle)\right)$$

where the correlation matrix has the same structure as that for the real process, but there are now correlations between the real and imaginary parts of $\mathbf{V}$. An analogous factorization theorem to (2.5) can be deduced if the $2n$ variables have zero mean, so that $\langle \mathrm{Re}(V_i) \rangle = \langle \mathrm{Im}(V_i) \rangle = 0$, and in addition, the variables are *circular*, by which is meant the correlations have the structure:

$$\langle \mathrm{Re}(V_i)\mathrm{Re}(V_j) \rangle = \langle \mathrm{Im}(V_i)\mathrm{Im}(V_j) \rangle,$$
$$\langle \mathrm{Re}(V_i)\mathrm{Im}(V_j) \rangle = -\langle \mathrm{Im}(V_i)\mathrm{Re}(V_j) \rangle$$

and the correlation matrix is symmetric. Care has to be taken to ensure that the correlations are real valued, in which case the factorization property has a particularly simple structure:

$$\left\langle V_1^* \ldots V_k^* V_{k+1} \cdots V_{2k} \right\rangle = \sum_Q \left(\left\langle V_1^* V_m \right\rangle \left\langle V_2^* V_n \right\rangle \ldots \left\langle V_k^* V_p \right\rangle \right)$$

where the summation is taken over Q which denotes the $k!$ permutations of the labels $(m, n, \ldots, p)$ with $(1, 2, \ldots, k)$. If the $\{V_i\}$ correspond to field components, then the case $k = 2$ assumes a particular importance, since it corresponds to an intensity correlation. In this case the factorization formula gives

$$\left\langle V_1^* V_2^* V_3 V_4 \right\rangle = \left\langle V_1^* V_3 \right\rangle \left\langle V_2^* V_4 \right\rangle + \left\langle V_1^* V_4 \right\rangle \left\langle V_2^* V_3 \right\rangle .$$

Selecting the $\{V_i\}$ to be the following field components $V_1 = V_3 = E(t), V_2 = V_4 = E(t + \tau)$ yields

$$\left\langle |E(t)|^2 |E(t + \tau)|^2 \right\rangle = \left\langle |E(t)|^2 \right\rangle \left\langle |E(t + \tau)|^2 \right\rangle + \left| \left\langle E(t)^* E(t + \tau) \right\rangle \right|^2 . \quad (2.6)$$

For stationary processes $\left\langle |E(t)|^2 \right\rangle = \left\langle |E(t + \tau)|^2 \right\rangle = \langle I \rangle$, so that (2.6) becomes, on normalizing by $\langle I \rangle^2$:

$$\frac{\langle I^2 \rangle}{\langle I \rangle^2} = 1 + \frac{\left| \left\langle E(0)^* E(\tau) \right\rangle \right|^2}{\langle I \rangle^2}$$

and from (2.3) and (2.4) this can be written succinctly as

$$g^{(2)}(\tau) = 1 + \left| g^{(1)}(\tau) \right|^2 \quad (2.7)$$

which is the factorization property of a complex Gaussian random process, or Siegert relationship [10]. Note that for a zero time delay, $\left| g^{(1)}(0) \right| = 1$, in which case the intensity correlation function is the normalized second moment, or speckle contrast, and has value 2. Since for classical fields, $0 \leq \left| g^{(1)}(\tau) \right| \leq 1$, it follows that $1 \le g^{(2)}(\tau) \le 2$, the smaller value being attained at large times when the fields decorrelate and become statistically independent of each other. We shall see that this correlation time can be related to the characteristic diffusion time of scatterers when they are in suspension, and this enables the size of the particles to be evaluated.

One last statistical notion is a valuable concept in what is to follow, that of the real and complex random walk. A random walker moves to and fro along the x-axis, the length of each step being a random variable with finite variance, and the decision to move either to the left or right being equally weighted and independent of the previous step. Many quantities of interest can be couched in terms of knowing the probability for the walker being a prescribed distance from a point of choice, and a useful tool for facilitating this is either the characteristic function or moment generating function. The characteristic function is the Fourier transform of the pdf, viz:

$$C(u) = \int_{-\infty}^{\infty} dx \, \exp(iux) p(x)$$

whereas the moment generating function is the Laplace transform of the pdf:

$$Q(s) = \int_{0}^{\infty} dx \, \exp(-ux) p(x)$$

the epithet coming from the fact that successive differentiations of the moment generating function obtains the moments of the distribution, as can be verified expanding the exponentials in either of the above functions and integrating term by term.

The distribution for the distance from the origin that results from a random walk comprising N steps is the N-fold convolution of the probability density with itself. This is easily obtained in the space of the Fourier variable u or Laplace variable s, as being the characteristic or moment generating function for a single step raised to the power N:

$$C_N(u) = C(u)^N$$

from which the probability density for the resultant can be obtained by Fourier inversion.

For a Gaussian variable of zero mean, the probability density

$$p(x) = \frac{1}{(2\pi)^{1/2} \sigma} \exp\left(-\frac{x^2}{2\sigma^2}\right)$$

where σ^2 is the variance, and the characteristic function has a similar form:

$$C(u) = \exp\left(-\frac{1}{2}\sigma^2 u^2\right).$$

The resultant of an N-step random walk, where the length of each step is an independent Gaussian random variable is therefore

$$C_N(u) = \exp\left(-\frac{1}{2}\sigma^2 u^2\right)^N = \exp\left(-\frac{1}{2}N\sigma^2 u^2\right)$$

which is the characteristic function of another Gaussian random variable with variance $N\sigma^2$. Hence scaling the variance by N leaves the resultant invariant.

The characteristic function is also of value for determining the properties of random walks for when the number of steps in the walk is not deterministic, but rather is a fluctuating quantity with probability distribution $P(N)$. The resultant averaged characteristic function for a random walk having statistically identical step lengths *and* a fluctuating number of steps is

$$C_{\overline{N}}(u) = \sum_{N=0}^{\infty} P(N) C(u)^N$$

which expresses the fact that a random walk can comprise N steps with frequency given by the probability of obtaining N. For example, if N is purely random, it can be described by a Poisson distribution. The characteristic function for the resultant is:

$$C_{\overline{N}}(u) = \sum_{N=0}^{\infty} \frac{(\overline{N})^N \exp(-\overline{N})}{N!} \exp\left(-\frac{1}{2}\sigma^2 u^2\right)^N$$

$$= \exp\left(-\overline{N}\left(1 - \exp\left(-\frac{1}{2}\sigma^2 u^2\right)\right)\right)$$

and for small values of u, which describe the tails of the resultant distribution, the exponential can be expanded, $\exp\left(-u^2\sigma^2/2\right) \approx 1 - u^2\sigma^2/2$, whereupon the averaged characteristic function is again found to be asymptotically that of a Gaussian random variable with variance $N\sigma^2$.

3. Photon Correlation Spectroscopy: obtaining particle size

Figure 1 illustrates the essential properties of PCS. A volume contains an ensemble of particles, usually held in suspension, which are illuminated by a coherent radiation source, usually a laser of frequency ω. The particles, labeled by a superscript j, have instantaneous position vectors $\mathbf{r}^j$ relative to the detector that collects the light scattered by each of them. Each particle scatters with amplitude a^j which is a function of the particle morphology and composition. The optical distance between each particle and the detector is $\mathbf{k}^i \cdot \mathbf{r}^i - \omega t$. If the particles are sufficiently sparse, the radiation leaves the scattering volume having been scattered only once. The detected field then comprises a coherent sum of the N (fluctuating or otherwise) illuminated particles:

$$E(t) = \sum_{j=0}^{N} a^{j}(t) \exp\left[i\left(\mathbf{k}^{j} \cdot \mathbf{r}^{j}(t) - \omega t\right)\right]. \tag{3.1}$$

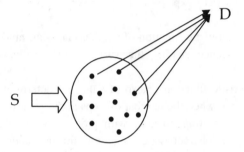

Figure 1. Schematic of a PCS instrument. Particles are illuminated by a coherent source S and the light scattered by them is detected at D. The field at D is a sum of elemental scattered fields and so admits the interpretation of being the resultant of a random walk.

Because the location of the particles relative to the detector are random, so too are the phases and consequently each of the terms appearing in the summation is effectively a (complex) random number. Hence the detected field is the resultant of a complex random walk. If, on the other hand, the number of particles is sufficiently numerous, the statistics of E will be described by a complex Gaussian process, by virtue of the central limit theorem. The random interference that occurs produces a characteristic speckle pattern, the fluctuations at any instant being described by a complex random *variable*. As the particles diffuse due to their Brownian motion through the suspending medium, the optical paths continuously change, leading to an evolving or 'boiling' speckle pattern whose fluctuations are described by a complex Gaussian *process*. Thus the rate at which the speckles evolve is related to the diffusion of the particles through the suspending medium, and this in turn is related to the size of the particles, as we shall now demonstrate.

Assuming that we have a simple square-law detection scheme, the intensity correlation function can be formed according to (2.4), and this serves to quantify the evolution of the speckle. However, if N is sufficiently large, the statistics of the field are described by a Gaussian random process and so this intensity correlation necessarily is given by the Siegert relation (2.7). Thus even though we measure the *intensity* correlation function, we are required to determine the structure of the *field* correlation function, for this is the fluctuating quantity.

$$\left|G^{(1)}(\tau)\right| \equiv \left|\left\langle E(0)E(\tau)^{*}\right\rangle\right|$$

$$\propto \left|\sum_{j} \sum_{k} \left\langle a^{j}(0)a^{k}(\tau) \exp\left[i\mathbf{k} \cdot \left(\mathbf{r}^{j}(0) - \mathbf{r}^{k}(\tau)\right)\right]\right\rangle\right|$$

If the particles are statistically identical then

$$\left|G^{(1)}(\tau)\right| \propto N\left|\left\langle a(0)a(\tau)\exp[i\mathbf{k}\cdot(\mathbf{r}(0)-\mathbf{r}(\tau))]\right\rangle\right|$$

and if the particles' center of mass motion is also described by a statistically identical random variable, the amplitudes and phases are independent, and the above factorises:

$$\left|G^{(1)}(\tau)\right| \propto N\left\langle a(0)a(\tau)\right\rangle\left|\left\langle \exp[i\mathbf{k}\cdot(\mathbf{r}(0)-\mathbf{r}(\tau))]\right\rangle\right|$$

For the sake of simplicity, if we assume the particles to be spherical, then the amplitude correlation is constant, and the temporal variation is attributable solely to the phase fluctuations. Note that the phase is proportional to the particles' displacement and this is related to their velocity:

$$\mathbf{r}(\tau)-\mathbf{r}(0)=\int_0^\tau \mathbf{v}(t)dt$$

For simplicity, if it is assumed that k is in the x-direction then

$$\left|G^{(1)}(\tau)\right| \propto N\left\langle a^2\right\rangle\left\langle \exp\left(-ik\left(\int_0^\tau v_x(t)dt\right)\right)\right\rangle$$

so that

$$\left|g^{(1)}(\tau)\right| \equiv \frac{\left|G^{(1)}(\tau)\right|}{\langle I\rangle}=\left\langle \exp\left(-ik\left(\int_0^\tau v_x(t)dt\right)\right)\right\rangle$$

The exponential function can be expanded in powers of k and integrated term by term, viz:

$$\left|g^{(1)}(\tau)\right|=1-ik\int_0^\tau\left\langle v_x(t)\right\rangle dt-\frac{k^2}{2!}\int_0^\tau dt_1\int_0^\tau\left\langle v_x(t_1)v_x(t_2)\right\rangle dt_2$$

$$-\frac{ik^3}{3!}\int_0^\tau dt_1\int_0^\tau dt_2\int_0^\tau\left\langle v_x(t_1)v_x(t_2)v_x(t_3)\right\rangle dt_3+$$

$$+\frac{k^4}{4!}\int_0^\tau dt_1\int_0^\tau dt_2\int_0^\tau dt_3\int_0^\tau\left\langle v_x(t_1)v_x(t_2)v_x(t_3)v_x(t_4)\right\rangle dt_4+\ldots$$

Assuming the particles undergo Brownian motion, the velocity is by definition an isotropic random variable described by a real Gaussian process of zero mean. Hence all the odd-ordered velocity correlation functions are zero, and the factorization property for a real Gaussian process can be used to determine the even ordered velocity correlations in terms of the *second* order velocity correlation function. Noting in particular that

$$\int_0^\tau dt_1\int_0^\tau\left\langle v_x(t_1)\ v_x(t_2)\right\rangle dt_2 =2\tau\int_0^\infty\left\langle v_x(0)\ v_x(t)\right\rangle dt$$

and using (2.5) obtains, after a little algebra, the result

$$\left|g^{(1)}(\tau)\right| = 1 + \sum_{m=1}^{\infty} \frac{\left(-k^2\right)^m}{(2m)!}(2m-1)\ (2m-3)...5\cdot3\cdot1 \times \left(2\tau \int_0^{\infty} \langle v_x(0)\ v_x(t)\rangle dt\right)^m$$

$$= \exp\left(-k^2\tau \int_0^{\infty} \langle v_x(0)\ v_x(t)\rangle dt\right) \equiv \exp\left(-k^2\ D\tau\right) \qquad (3.2)$$

hence the field correlation function, and thereby the measured intensity correlation function, decays exponentially with delay time, the constant of proportionality depending on the square of the wave number of the scattered radiation, and the translational diffusion coefficient D. This latter quantity is related to the hydrodynamics of the particles in the suspending medium, and therefore intrinsically to their size. Crucially, $g^{(1)}(\tau)$ is independent of the optical properties of the particles. The characteristic diffusion time $t_D \sim 1/Dk^2$ is much greater than the characteristic decay time τ_v of the velocity correlation function by virtue of the particles undergoing Brownian motion. Using a simple Langevin treatment, $\tau_v = m/f$ where m is the mass of the particle and f the coefficient of friction. The diffusion coefficient can now be related to f through Einstein's relation

$$D = \frac{\kappa T}{f} \qquad (3.3)$$

where κ is Boltzmann's constant and T the temperature of the suspending medium. If the particles are spherical, f is given by Stokes' expression for the force acting on a sphere of radius R moving through a medium of viscosity η:

$$f = 6\pi\eta R \qquad (3.4)$$

Thus by measuring the rate of decay of the autocorrelation function from the speckle fluctuations, and knowing the temperature of the solvent, the particle size can be gauged.

To summarize, PCS exploits the fact that the speckle fluctuations evolve as a Gaussian random process with a characteristic timescale that is related to the diffusion of particles through a solvent, and this in turn is related to the size of the particles. The method is realized by measuring the intensity autocorrelation function which decays exponentially from value two to unity. By determining the decay constant, the particle size can be inferred from equations (3.2) to (3.4). The result is independent of the optical parameters of the particles and polarization state of the illuminating and detected radiation. Note that (3.3) assumes the particles are spheres. Any departure from sphericity would cause the frictional drag on the particle to be modified, and would be manifested by the particles' motion through the solvent being changed by the hydrodynamics. This effect has been discussed at length in [11,12,13], where expressions equivalent to (3.4) have been derived for a variety of shapes, together with the further corrections that are required to account for rotational diffusion. Thus the *shape* of the particles is implicitly interwoven into the reconstruction procedure for their size. To determine the shape of the particle requires an independent measurement, and the means of achieving this is afforded by polarization.

4. Polarization Fluctuation Spectroscopy: obtaining particle shape

In this section we shall see that the incorporation of polarization into the formulation for PCS enables the shape of the particles to be inferred. This is achieved by a modest augmentation of PCS equipment that facilitates the measurement of the cross-correlation of the scattered intensity that has been resolved into different polarization states. The temporal decay of the polarized intensity cross-correlation function is identical to that of the intensity auto-correlation function used in PCS, however it decays from value *less* than 2, and the difference from 2 is related to the particles' departure from a sphere. Before demonstrating precisely how this occurs, it is first necessary to see how polarization generalizes the Siegert relation.

We commence from the random-walk model (3.1) for the detected field that has been scattered from an ensemble of particles. Assuming the electric field is polarized gives, at any instant in time:

$$\begin{pmatrix} E_{\alpha\beta} \\ E_{\alpha\gamma} \end{pmatrix} = \sum_{j=1}^{N} \begin{pmatrix} c^j \\ d^j \end{pmatrix} \exp(i\eta^j). \tag{4.1}$$

The N particles comprising the ensemble are illuminated in polarization state α. $E_{\alpha\beta}$ denotes the scattered electric field at the detector that is in polarization state β. These detected fields are formed from a coherent sum of polarized fields, where c^j and d^j denote the scattering amplitudes from the j-th particle when detected in states β and γ respectively. The η^j denotes the optical path-length from the j-th particle to the detectors, which are independent of polarization.

We may now construct the scattered intensity in each polarization state and the correlations between these intensities. If the particles are statistically identical, the intensities are simply

$$I_{\alpha\beta} = \overline{N}\langle cc^* \rangle = \overline{N}\langle |c|^2 \rangle$$

$$I_{\alpha\gamma} = \overline{N}\langle dd^* \rangle = \overline{N}\langle |d|^2 \rangle$$

where the angled brackets are ensemble averages taken over the orientation of the particles. We have also averaged over number fluctuations, which are assumed to be independent of the particle characteristics. The correlation between the intensities is formed from

$$
I_{\alpha\beta}I_{\alpha\gamma} = \sum_{j=1}^{N} c^j \left(c^j\right)^* d^j \left(d^j\right)^* + \sum_{j=1}^{N} \sum_{m \neq j} c^j \left(d^m\right)^* \left(c^m\right)^* d^j +
$$

$$
\sum_{j=1}^{N} \sum_{m \neq j} c^j \left(c^j\right)^* d^m \left(d^m\right)^* + h(\exp(i\eta))
$$

and averaging over the particle orientations gives

$$
\langle I_{\alpha\beta}I_{\alpha\gamma} \rangle = \sum_{j=1}^{N} P(j) \left\{ j \langle |c|^2 |d|^2 \rangle + j(j-1) \left| \langle cd^* \rangle \right|^2 + \right.
$$

$$
\left. j(j-1) \langle |c|^2 \rangle \langle |d|^2 \rangle \right\}
$$

where $P(n)$ is the probability density for the particle number fluctuations.
    Performing the average over N obtains

$$
\langle I_\beta I_\gamma \rangle = \overline{N} \langle |c|^2 |d|^2 \rangle + \langle N(N-1) \rangle \left| \langle cd^* \rangle \right|^2 + \langle N(N-1) \rangle \langle |c|^2 \rangle \langle |d|^2 \rangle \qquad (4.2)
$$

where the pre-suffix α, denoting the illumination polarization state, has been suppressed since it is the same for both detected states β and γ. However, it should be borne in mind that the above expression depends implicitly on the illumination state by virtue of the forms adopted by the scattered amplitudes c and d. Equation (4.2) can be normalized by the product of the intensities in the two detected polarization states to form the dimensionless intensity cross-correlation function:

$$
\frac{\langle I_\beta I_\gamma \rangle}{\langle I_\beta \rangle \langle I_\gamma \rangle} = \frac{1}{\overline{N}} \frac{\langle |c|^2 |d|^2 \rangle}{\langle |c|^2 \rangle \langle |d|^2 \rangle} + \frac{\langle N(N-1) \rangle}{\overline{N}^2} \left(1 + \frac{\left| \langle cd^* \rangle \right|^2}{\langle |c|^2 \rangle \langle |d|^2 \rangle} \right).
$$

The first of these terms is proportional to the reciprocal of the mean number of particles comprising the ensemble. Since the mean number of particles is typically large, this first 'non-Gaussian term' is usually very much less than unity. If the mean number of particles is small however (it is possible for $\overline{N} < 1$) this term may become dominant. If the number of particles passing into and out of the illumination volume is random, as is the case if their motion is Brownian, the distribution for N is Poisson, in which case $\langle N^2 \rangle = \overline{N}^2 + \overline{N}$

and so if $\overline{N} \gg 1$, the light is once again described by a Gaussian complex process and the intensity cross-correlation becomes

$$\frac{\langle I_\beta I_\gamma \rangle}{\langle I_\beta \rangle \langle I_\gamma \rangle} = 1 + \frac{\left|\langle cd^* \rangle\right|^2}{\langle |c|^2 \rangle \langle |d|^2 \rangle}$$

or equivalently:

$$g_{\beta\gamma}^{(2)}(0) = 1 + \left| g_{\beta\gamma}^{(1)}(0) \right|^2 .$$

An arbitrary delay time can be inserted without loss of generality to give the Siegert relationship generalized to include polarization:

$$g_{\beta\gamma}^{(2)}(\tau) = 1 + \left| g_{\beta\gamma}^{(1)}(\tau) \right|^2 \tag{4.3}$$

This has a similar structure to the form given by equation (2.7). The left-hand side is the measured correlation function between two intensities in polarization states β and γ: if $\beta = \gamma$ then (2.7) is recovered. The right-hand side is formed from the field correlation function. Note that the normalized correlations for the ensemble depend on the correlations obtained from a *single* particle, this being a consequence of the single-scattering assumption. To discover how (4.3), when evaluated at $\tau = 0$, can be used to determine particle shape requires a model for the scattering. For the purposes of illustrating the principles of the method, we shall assume the radiation is scattered from a spheroidal particle, modeled by orthogonally arranged dipoles of different strength. This model [14,15] has the advantage of being simple and analytical in nature, whilst retaining some of the subtleties that occur for more realistic particle models.

5.   Dipole Scattering Model

The generalized Siegert relation (4.3) contains quantities that depend on the field correlation function of a single particle. Figure 2 shows a particle that is illuminated by radiation, incident along the z-axis, which is in a specified polarization state with components $\mathbf{E}_\alpha = (E_1, E_2)$ being denoted by the subscript α. The particle scatters the radiation that is observed at angle θ. The scattered field is polarized with amplitude components $(a_{\alpha\beta}, a_{\alpha\gamma})$, the first subscript denoting the incident state, the second the detection state. The initial polarization state has been corrupted through interaction with the non-spherical particle, and the depolarization therefore is a measure of the particles' departure from a sphere, which would not cause depolarization to occur.

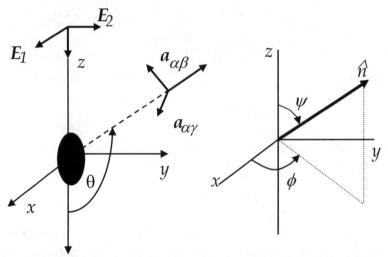

Figure 2. A particle is illuminated with coherent light directed along the z-axis that is in polarization state (E_1, E_2,0) and measured at scattering angle θ. The instantaneous orientation of the particle is described by Euler angles ψ and ϕ.

If the dipole is a spheroid, then it can be modeled by a polarizability tensor, which in the rest-frame of the particle has the form

$$\underline{\underline{\alpha}} = \begin{pmatrix} \alpha_1 & 0 & 0 \\ 0 & \alpha_2 & 0 \\ 0 & 0 & \alpha_2 \end{pmatrix}$$

in which case the scattered field can be found from $a_{\alpha\beta} \propto \hat{\mathbf{e}}_\beta \cdot (\underline{\underline{\alpha}} \cdot \mathbf{E}_\alpha)$, where $\hat{\mathbf{e}}_\beta$ is a unit vector that projects out the desired detected polarization state. For example the unit vectors for detection in the s and p states are

$$\hat{\mathbf{e}}_s = (1,0,0),$$
$$\hat{\mathbf{e}}_p = (0,-\cos\theta,\sin\theta)$$

In the laboratory frame, the particle will have orientation described by the Euler angles ϕ and ψ, as also shown in Figure (2). The instantaneous normal of the particle is

$$\hat{\mathbf{n}} = (\sin\psi \cos\phi, \sin\psi \sin\phi, \cos\psi)$$

hence any combination of incident and detected polarization state can be determined for any instantaneous particle orientation. For example, if the incident radiation is s-polarized and it is detected in either the parallel s or orthogonal p states, the scattering amplitudes are found to be

$$a_{ss} = E_s\left[(\alpha_1 - \alpha_2)\sin^2\psi \cos^2\phi + \alpha_2\right]$$
$$a_{sp} = -E_s(\alpha_1 - \alpha_2)\sin\psi \cos\phi(\cos\psi \sin\theta + \sin\psi \sin\phi \cos\theta)$$

whereas if the incident light is p-polarized, the amplitudes are

$$a_{ps} = -E_p(\alpha_1 - \alpha_2)\sin^2\psi\sin\phi\cos\phi$$

$$a_{pp} = -E_p[(\alpha_1 - \alpha_2)\sin\psi\sin\phi(\sin\psi\sin\phi\cos\theta + \cos\psi\sin\theta) + \alpha_2\cos\theta]$$

which depend on the polarizabilities, the instantaneous orientation of the particle and the observation angle. If we assume that the particles are freely tumbling in space, any orientation of the particle is as likely as another, so the Euler angles are then random variables with probability densities

$$p(\phi)d\phi = \frac{d\phi}{2\pi}, \quad 0 \le \phi < 2\pi,$$

$$p(\psi)d\psi = \frac{1}{2}\sin\psi\, d\psi, \quad 0 \le \psi < \pi$$

We can now calculate the average intensities and field correlations that appear in the Siegert relation by forming appropriate products of the above scattered amplitudes and averaging over all orientations of the particles. These averaged intensity correlations can, in principle, be functions of the polarizabilities and scattering angle. Many of the correlations are identically zero, indicating that there are no polarization fluctuations, however, particular combinations yield non-trivial results. For example,

$$\frac{\left|\langle a_{ss}a_{pp}\rangle\right|^2}{\langle |a_{ss}|^2\rangle\langle |a_{pp}|^2\rangle} = \frac{\left[(r-1)^2 + 10(r-1) + 15\right]^2\cos^2\theta}{(3r^2 + 4r + 8)\left[(r-1)^2(3\cos^2\theta + \sin^2\theta) + 5(2r+1)\cos^2\theta\right]} \quad (5.1)$$

which can be achieved by illuminating with an equal mixture of s and p states and detecting separate s and p states concurrently. Here the polarizability ratio $r = \alpha_1/\alpha_2$ gives the aspect ratio of the particles, being prolate, spherical and oblate as r ranges through values greater than unity, equal to unity and less than unity respectively. Note in (5.1), if $r = 1$, the expression is unity, so that the intensity cross-correlation function adopts the Gaussian value of two – this being consistent with spheres not depolarizing light. Figure 3 shows the form of $\left|g^{(1)}\right|^2$ as a function of scattering angle for a variety of values for r ranging

from thin needle-like particles, through spheres to flat discs. Note that $\left|g^{(1)}\right|^2$

is always less than the value of unity adopted by spherical particles, and this discrepancy is the key property that enables the particle shape to be gauged. However the form of the curve is not unique for a given value of r, there being matching pairs for $r > 1$ and $r < 1$. For example, the curve for $r = 0.4$ is identical to that for $r = 2$ shown in Figure 3. In many circumstances there exists prior knowledge for whether the particles are prolate or oblate, but if this information is unavailable, a single measurement of the cross-correlation function is unable to resolve the ambiguity. The procedure for obtaining a unique reconstruction without prior knowledge about the particles' shape is discussed in Section 8.

The dipole model adequately illustrates the principle of PFS, however real particles of interest are rarely sufficiently small to be described by the Rayleigh scattering regime. Particles of interest often have sizes similar to or greater than a wavelength, and values of refractive index that violate the assumptions of dipole scattering. Fortunately, computational methods for calculating the scattered amplitudes from particles of large size (compared with the wavelength) and complex shape can now be performed with comparative ease [16]. Hence the equivalent field correlation functions can be calculated, albeit computationally. Because particles are not generally in the Rayleigh scattering regime, the form adopted by the scattered amplitudes and those functions that depend on them, are modified substantially. Nevertheless, the decay of the intensity correlation function still gauges the particle size, and the departure of $g^{(1)}(0)$ from unity still measures the departure of the particles' shape from a sphere. The next section considers the configuration of a PFS instrument, and how the data it measures is processed to gauge size and shape information.

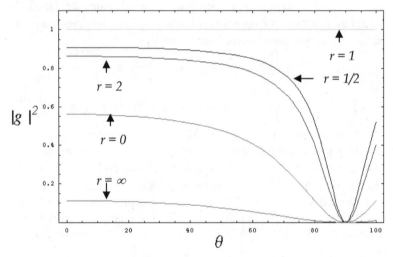

Figure 3. Equation (5.1) plotted as a function of scattering angle θ for a variety of particle aspect ratios r. Note that spheres do not produce any depolarization.

6.   Instrument configuration

The practical implementation of dynamic light-scattering techniques is not a difficult task, but there are a number of choices to be made and a few pitfalls that should be avoided. The following describes the simplest configuration of a PFS instrument and provides a guide to these choices and problems.

The apparatus is shown in Figure 4 and comprises a coherent radiation source, usually a laser. In selecting a laser to illuminate the particles, the main parameters to consider are the following:
- power level

- power stability
- coherence
- wavelength

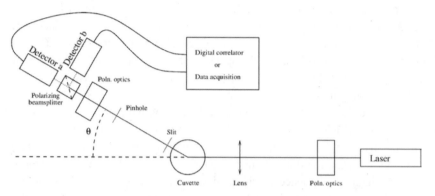

Figure 4. Schematic diagram of a PFS measurement system.

Most dynamic scattering experiments are dominated by photon noise, so the power level generally should be as high as is practicable. Most systems use a power level in the 15 – 50 mWatt range. Ideally the laser output should be completely stable since any variations are in addition to those arising from the particle dynamics. Variations less than ~1% r.m.s. usually are considered acceptable because the timescales of the laser fluctuations are often very much longer than those arising from the particles. The temporal coherence length only needs to be greater than the typical path-length differences in the scattering cell, i.e. typically less than 1 mm, so almost any laser will fulfill this criterion. Wavelength is also not a critical parameter unless the particles of interest are highly absorbing. However the choice of wavelength may be influenced by the choice of detector (see below). The lasers that have been used most commonly are stabilized HeNe and Argon Ion. The new range of diode pumped solid state lasers now offer powers of up to 50 mWatt in very compact packages and seem ideally suited to this application, although they are still expensive compared with HeNe lasers. Because only scattered light is detected, good beam quality is not a requirement.

The laser can be oriented to provide an arbitrary linear polarization state with respect to the vertical. Should a state different from linear be required for illumination purposes, polarizing optics, comprising a quarter-wave plate can be placed between the source and the sample.

The lens is used to reduce the scattering volume since an important requirement is to sample an area comparable with the size of a single speckle in the far field of the scattered light. This requires that the effective scattering volume is sufficiently small to place the far field at a suitable working distance and to make the speckles of a suitable size, so that a detector pinhole may isolate a single speckle. A scattering volume with a linear size ~ 200 μm at 90° is suitable to fulfill both these requirements. This may be achieved by working at the waist of the beam formed by a lens and by placing a vertical slit, of width = 200 μm, close to the scattering cell. The required lens focal length depends on the beam diameter emitted by the laser, but for most lasers this is ~200 mm. For this size of scattering volume, the far-field occurs at a distance of about 100 mm, and the speckles at this distance are of the order of 200 μm and may be sampled by a pinhole of this size. Precision slits and pinholes having these sizes are readily available.

As discussed in the development of the theory, the particle concentration should be sufficiently low to exclude significant multiple scattering but sufficiently high to have many particles illuminated in the scattering volume, the latter ensuring the fluctuations are Gaussian and that there is a sufficiently strong signal and hence reliable data. This imposes two conflicting requirements on the cell size. To meet the requirement of single scattering, the particle density should be limited to a value such that the mean free path of the light is many times greater than the cell size. This may be calculated, given an estimate of the particles' size and optical properties, but may also be checked visually by the appearance of the beam that should appear as a clean pencil-like ray passing through the cell with no evidence of light being emitted from elsewhere in the cell. If a sample is examined in normal room lighting conditions it should appear completely clear. This condition is easier to achieve for a cell of small dimensions. A small cell also has the obvious but sometimes important advantage of requiring less of the material to be sampled. However, if too small a cell is used, then laser light scattered at the cell walls (particularly the outer wall) may contribute to the signal. Index matching may be used to reduce this effect. A commonly used arrangement is to use a small sample cell surrounded by a larger cell filled with an index matching fluid. Such an arrangement also may be used to maintain a near constant temperature so that thermal fluctuations do not cause fluctuations in the translational diffusion coefficient. If scattering angles close to 90° are to be used then square or rectangular cells may be employed. Otherwise cylindrical cells are more appropriate, especially if a variety of scattering angles are to be used. Whatever the geometry of the cell, it must be made with high quality glass that does not introduce spurious changes in polarization.

Perhaps the most obvious, but sometimes neglected, precaution to ensure that the light arriving at the detector is dominated by light singly-scattered by the particles of interest is to ensure that they are the only particles in the scattering cell. One essential step is to pay close attention to cleanliness in the sample preparation and to thoroughly clean the scattering cell. In addition, care must be taken to ensure that the particles to be characterized are not agglomerated. This is usually achieved by breaking up the particle agglomerates prior to measurement using an ultrasonic bath, typically for a few minutes. This process can cause damage to delicate cells, so expensive optical quality cells should not be subjected to this process. It is often difficult to know if the insonification process has been totally effective. But one check that can be made is a visible inspection of the laser beam passing through the cell. If any significant fluctuations in the light level within the beam i.e. glinting or twinkling can be seen then there is most likely a problem either with agglomerates or contamination.

The detection optics work in combination with the polarizing beam-splitter, which selects the two detection polarization states. A combination of half-wave and quarter-wave plates can effect the detection of most orthogonal state pairs. The detection optics is mounted on a computer-controlled movable arm that rotates about the center of the scattering cell through angles $0 < \theta < 130°$. The polarizing beam-splitter divides the light into the selected orthogonal polarization states and directs each to its own detector.

Because of the low light levels typically achieved, photon counting detectors are an essential part of a dynamic scattering system. The choice is between photo-multiplier tubes (PMTs) and photon-counting avalanche photo-diodes (APDs). Both types of detector are now available in convenient modules that are of similar sizes and require only low voltage supplies. There is not a great difference in price. The APDs are solid state devices and therefore inherently more rugged than the glass envelope PMTs. Hence if the instrument is likely to be subjected to any violence, PMTs are not a sensible choice. If this is unlikely, the pros and cons are that APDs have somewhat higher quantum efficiency than PMTs, especially at the red end of the visible spectrum but it should be borne in mind they usually have a significantly higher dark count. The optimum choice ultimately depends on the scattering strength of the particles to be characterized.

Figure 5 shows a photograph of the experimental apparatus. The photo-counts are recorded using a data acquisition system for off-line processing. This enables the data to be checked for outliers that can corrupt the cross- and two auto-correlations required for the measurement process. The photon counts from the two PMTs are initially stored separately in two files in a raw binary format having 32-bit integer counts per sample interval. Although these files can be up to hundreds of megabytes in size, nowadays hard disk drives are sufficiently capacious and inexpensive for this not to be a limiting factor. Moreover, storing the raw data permits a range of different processing operations to be post-performed: for example, higher order spectra can be calculated which can serve to check the assumptions that the process is Gaussian.

The sample interval depends on the particle size. Typically a sample interval of 10 or 50 microseconds is used and the sampling period is 600 seconds. The pair of files is processed together by a correlation program that splits the streams into user-specified batches that are normally 10 seconds in length. Each batch is used to calculate two intensity auto-correlation functions and the cross-correlation function using a fast Fourier transform method. The number of correlation channels is usually 1024. These are examined to find outlier batches, which may contain spurious 'glitches' from noise sources such as dust particles contaminating the sample cell, power supply spikes and electromagnetic interference. Such outliers can produce spurious correlations.

Figure 5. A photograph of PFS experimental apparatus, from [17].

The method for finding outliers takes averages around two positions in each auto-correlation and calculates the medians from the distribution of values over the batches. The first position is at zero time-lag where a single, spurious high count distorts the mean squared value. The second position is at two-thirds of the maximum lag where the correlation curve has decayed to its value in the tail and a train of non-zero counts tends to extend the tail. The four median values then are used to form threshold criteria to remove batches that contain auto-correlations that exceed the medians by a factor greater than the ratios of median values to the lowest ranked members.

Estimates of the statistical errors in the correlation data are determined by subdividing batches into five sets. Each set contains a triplet of correlation curves. These curves are each fitted to a single exponential model according to (3.2) using the Marquardt-Levenberg nonlinear least-squares algorithm, which is a standard feature in many proprietary data processing packages. The extracted values from each set are taken together to form sample means and errors for the amplitudes and decay rates of the exponentials. The three amplitudes then are used to calculate the cross-correlation coefficient.

Direct measurement of $g_{\alpha\beta}^{(2)}(0)$ is especially prone to artifacts introduced by the detectors such as after pulsing and dead-time. Hence this value is deduced from the known exponential dependence of $g_{\alpha\beta}^{(2)}(\tau)$, the value at $\tau = 0$ being inferred by extrapolation to the origin.

Another effect that affects the measurement of $g_{\alpha\beta}^{(2)}(0)$ is spatial averaging caused by the pin-hole, whereby the fluctuations of more than one speckle are recorded and averaged, leading to a reduction in the measured value of the correlation coefficients. The distortions caused by spatial averaging can be compensated through the way the data is processed [18]. In particular the quantity:

$$h_{\alpha\beta}(\tau) \equiv \frac{g_{\alpha\beta}^{(2)}(\tau)-1}{\left[\left(g_{\alpha\alpha}^{(2)}(\tau)-1\right)\left(g_{\beta\beta}^{(2)}(\tau)-1\right)\right]^{1/2}} \tag{6.1}$$

is identically equal to $\left|g_{\alpha\beta}^{(1)}(0)\right|^2$ when measurements are taken in the Gaussian scattering regime, for then $g_{\alpha\alpha}^{(2)} = g_{\beta\beta}^{(2)} = 2$ when evaluated at $\tau = 0$, and this normalization compensates for spatial averaging. Moreover it also reduces the effects of dark counts or incoherent additive stray light. Suppose that each detector is subject to additional contributions, denoted as B_α and B_β which are Gaussian processes, independent of each other and the intensity scattered by the particles. The measured correlation coefficient is therefore,

$$g_{\alpha\beta}^{(2)}(\tau) \equiv \frac{\left\langle \left(I_\alpha(\tau) + B_\alpha\right)\left(I_\beta(\tau) + B_\beta\right)\right\rangle}{\left\langle I_\alpha(\tau) + B_\alpha\right\rangle \left\langle I_\beta(\tau) + B_\beta\right\rangle}$$

and the measured speckle contrast in each detector is

$$g_{\alpha\alpha}^{(2)}(\tau) \equiv \frac{\left\langle \left(I_\alpha(\tau) + B_\alpha\right)^2\right\rangle}{\left\langle I_\alpha(\tau) + B_\alpha\right\rangle^2}.$$

It is simple to show that $h_{\alpha\beta}$ is then independent of B_α and B_β, depending on I_α and I_β alone.

The PFS instrument, for all its comparative simplicity, has an infinite number of configurations provided by the combinations between input and detection polarization states. However many of these are redundant or unsuitable because of their insensitivity for detecting particle shape. The sensitivity problem is of primary concern and will be treated in the next section.

7.   Accuracy analysis

The most effective configuration for a PFS instrument is determined through a complex mixture of the illumination and detection optics and the physical nature of the particles themselves.

The most easily realized configurations are, for the illumination state:
- linear polarized – vertical
- linear polarized – horizontal
- linear polarized – 45°

which can be achieved by orienting the laser appropriately, or
- circular – right
- circular – left

which can be achieved with the aid of quarter-wave a plate.

The two detection states are
- linear – vertical and horizontal: using a polarizing beam splitter
- linear – at +45° and −45° from the vertical: using a half-wave plate at 22.5°
- circular – right and left: using a quarter-wave plate at 45°.

Hence there are fifteen separate configurations that must be tested for their sensitivity to particle shape detection and accuracy of reconstruction.

The factors that dictate the best choice for the configuration of a PFS instrument are that

1. $g^{(1)}_{\alpha\beta}(0)$ should be sufficiently sensitive to particle aspect ratio, and

2. $g^{(2)}_{\alpha\beta}(\tau)$ should be accurately measurable.

The first condition is principally dependent on the particles' optical and physical parameters, and these can be calculated with the aid of computer codes that calculate scattering properties of particles of complex composition and shape. The second condition requires there to be sufficient light in *both* of the detection channels for a reliable measurement of the cross correlation function to be made, and this involves the stochastic nature of the detection process itself.

The detection of photoelectrons is a doubly-stochastic Poisson process wherein the light intensity fluctuates and causes a random modulation in discrete number of measured photoelectrons, which are themselves subject to intrinsic fluctuations in the shot-noise limit. Suppose there are n photoelectrons counted in an integration time t_s, the mean photo-count is proportional to the mean intensity. An estimate for the mean photo-count can be derived from N samples:

$$\hat{n} = \frac{1}{N} \sum_{m=1}^{N} n(mt_s)$$

which is evidently unbiased since $\langle \hat{n} \rangle = \bar{n} = q\langle I \rangle$, where q is the quantum efficiency of the detector tubes. The variance of the estimated mean photo-count is easily determined [19] to be $Var(\hat{n}) = \bar{n}/N$, so that the error in estimating the mean scales with the reciprocal of the number of samples. The normalized cross-correlation function formed from the photo-counts

$$\varepsilon_{\alpha\beta}(\tau) = \frac{\langle n_\alpha(0) n_\beta(\tau) \rangle}{\bar{n}_\alpha \bar{n}_\beta}$$

is equal to the intensity cross-correlation function $g_{\alpha\beta}^{(2)}(\tau)$. A biased estimator can be constructed from the data according to:

$$\hat{\varepsilon}_{\alpha\beta}(\tau) = \frac{1}{\langle \hat{n}_\alpha \rangle \langle \hat{n}_\beta \rangle} \frac{1}{N} \sum_{m=1}^{N} n_\alpha(mt_s) n_\beta(mt_s + \tau).$$

If $N \gg 1$, the mean of this is

$$\langle \hat{\varepsilon}_{\alpha\beta}(\tau) \rangle = \varepsilon_{\alpha\beta}(\tau) + O(1/N)$$

and in the photon-limited case:

$$Var(\hat{\varepsilon}_{\alpha\beta}(\tau)) = \frac{\varepsilon_{\alpha\beta}(\tau)}{N\bar{n}_\alpha \bar{n}_\beta}$$

the variation described by which being attributable to those fluctuations inherent in the detection process. However, other sources of noise also can influence the estimator for the correlation coefficients, fluctuations in the laser source being one example. Assuming that all these additional independent errors are described by a Gaussian process enables them to be incorporated into the variance for the estimator of the correlation coefficient thus:

$$Var(\hat{\varepsilon}_{\alpha\beta}(\tau)) = \frac{\varepsilon_{\alpha\beta}(\tau)}{N\bar{n}_\alpha \bar{n}_\beta} + B_{\alpha\beta}.$$

As explained earlier, the inversion is not performed on $g_{\alpha\beta}^{(2)}(\tau)$ but rather on the quantity

$$h_{\alpha\beta}(\tau) \equiv \frac{g_{\alpha\beta}^{(2)}(\tau) - 1}{\left[\left(g_{\alpha\alpha}^{(2)}(\tau) - 1\right) \left(g_{\beta\beta}^{(2)}(\tau) - 1\right) \right]^{1/2}}$$

$$= \frac{\varepsilon_{\alpha\beta}(\tau) - 1}{\left[\left(\varepsilon_{\alpha\alpha}(\tau) - 1\right) \left(\varepsilon_{\beta\beta}(\tau) - 1\right) \right]^{1/2}}$$

which compensates for spatial averaging of the speckles. Hence given knowledge of the volume and scattering parameters of the particles, we may determine the error $\Delta h_{\alpha\beta}$ in measuring $h_{\alpha\beta}$ due to an error Δr in the aspect ratio of the particle. At a given scattering angle θ, the error in measuring $h_{\alpha\beta}$ can be deduced from:

$$\Delta h_{\alpha\beta} = \frac{\partial h_{\alpha\beta}}{\partial r}\Delta r$$

$$= \frac{\partial h_{\alpha\beta}}{\partial \varepsilon_{\alpha\beta}}\frac{\partial \varepsilon_{\alpha\beta}}{\partial r}\Delta r + \frac{\partial h_{\alpha\alpha}}{\partial \varepsilon_{\alpha\alpha}}\frac{\partial \varepsilon_{\alpha\alpha}}{\partial r}\Delta r + \frac{\partial h_{\beta\beta}}{\partial \varepsilon_{\beta\beta}}\frac{\partial \varepsilon_{\beta\beta}}{\partial r}\Delta r$$

$$= \frac{\partial h_{\alpha\beta}}{\partial \varepsilon_{\alpha\beta}}\Delta \varepsilon_{\alpha\beta} + \frac{\partial h_{\alpha\alpha}}{\partial \varepsilon_{\alpha\alpha}}\Delta \varepsilon_{\alpha\alpha} + \frac{\partial h_{\beta\beta}}{\partial \varepsilon_{\beta\beta}}\Delta \varepsilon_{\beta\beta}$$

and so writing the left-hand side in terms of the mean-square error in r gives:

$$\left(\frac{\partial h_{\alpha\beta}}{\partial r}\right)^2 \langle \Delta r^2 \rangle = \left(\frac{\partial h_{\alpha\beta}}{\partial \varepsilon_{\alpha\beta}}\right)^2 \left\langle \left(\Delta \varepsilon_{\alpha\beta}\right)^2 \right\rangle + \left(\frac{\partial h_{\alpha\alpha}}{\partial \varepsilon_{\alpha\alpha}}\right)^2 \left\langle \left(\Delta \varepsilon_{\alpha\alpha}\right)^2 \right\rangle$$

$$+ \left(\frac{\partial h_{\beta\beta}}{\partial \varepsilon_{\beta\beta}}\right)^2 \left\langle \left(\Delta \varepsilon_{\beta\beta}\right)^2 \right\rangle.$$

Exploiting the fact that the measurements are taken in the Gaussian scattering regime, we obtain

$$\langle \Delta r^2 \rangle = \left(\frac{\partial \varepsilon_{\alpha\beta}}{\partial r}\right)^{-2} \left\{ \frac{\varepsilon_{\alpha\beta}}{N\,\bar{n}_\alpha \bar{n}_\beta} + \frac{\left(\varepsilon_{\alpha\beta}-1\right)^2}{4N}\left(\frac{1}{\bar{n}_\alpha^2} + \frac{1}{\bar{n}_\beta^2}\right) + B\left(1 + \frac{\left(\varepsilon_{\alpha\beta}-1\right)^2}{2}\right) \right\}.$$

$$(7.1)$$

The terms within the braces correspond to the errors made in measuring the cross-correlation coefficient, auto-correlation coefficients and the other errors respectively. Taking sufficiently many samples can minimize the first two sources of error. Note that if the mean count rate in one of the channels is low, the error is caused principally by the measurement of the autocorrelation function in that channel. The 'other' errors cannot be suppressed by obtaining more data. The multiplying prefactor encapsulates the sensitivity of the measurement to the shape of the particles. It indicates that good measurements will be obtained when $\varepsilon_{\alpha\beta}$ is a rapidly changing function of aspect ratio. Thus the error is minimized when there is sufficient light in either polarization channel, *and* the cross-correlation function is a rapidly changing function of particle aspect ratio. When considered as a function of scattering angle, these optimum conditions do not occur coincidentally and so there is a trade-off between the two effects. If the physical attributes of the particles are assumed, the dependence of $\varepsilon_{\alpha\beta}$ on r can be calculated and the mean-square error determined as a function of θ. Thus the optimum scattering angle at which to collect data can be evaluated.

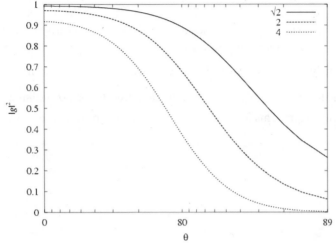

Figure 6. The field correlation function for the linear polarization configuration of the instrument, shown for a range of particle aspect ratios.

Figure (6) shows results for when the input state is linearly polarized at $45°$, so there is an equal mixture of s and p states, and the detection states are vertical and horizontal linear states. The particles are assumed to be hematite spheroids of aspect ratio $r \sim 1.8$, a volume equivalent radius $0.044\ \mu m$ and a complex refractive index $n = 1.698-0.0149i$. The variation of $\left|g_{\alpha\beta}^{(1)}(0)\right|^2$ with θ is shown for three values of aspect ratio. Note the unusual logarithmic angular scale for the abscissa, which highlights the fact that for $0 < \theta \leq 70°$, $\left|g_{\alpha\beta}^{(1)}(0)\right|^2$ does not change significantly. Indeed, nor is $\left|g_{\alpha\beta}^{(1)}(0)\right|^2$ particularly sensitive to changes in r in this range. It is only for $\theta > 80°$ that $\left|g_{\alpha\beta}^{(1)}(0)\right|^2$ changes substantially with r and θ.

Figure (7) shows the mean photo-counts calculated for the two tubes. Although the photo-counts differ in magnitude, the overall shapes of the mean intensity in either polarization channel are rather similar when displayed as a function of θ, illustrating the comparative insensitive of the mean intensity to the shape of the particles, here taken to have an aspect ratio $r = 2$. The mean photo-counts are large and similar in magnitude at small θ, and referring to equation (7.1), the terms in braces are small in this region. However, the multiplying prefactor is large, since $\left|g_{\alpha\beta}^{(1)}(0)\right|^2$ is essentially flat. In the region $\theta > 80°$ there is a disparity of nearly two orders of magnitude between the photo-counts measured by the two channels, hence the relative size of the two terms in equation (7.1) is essentially reversed.

Figure (8) shows the size of the mean-square error for reconstructing r from the data. In general, the error increases nonlinearly with r. Clearly making measurements in the forward scattering direction should be avoided, principally because $\left|g_{\alpha\beta}^{(1)}(0)\right|^2$ is insensitive to r in this region. The error curves all display a relatively narrow minimum (recall the logarithmic scale) in scattering angle. This indicates the detector should be located at $\sim 84°$.

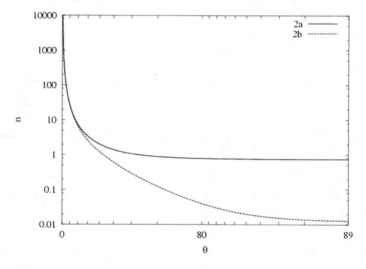

Figure 7. The mean intensity in the two detectors for the linear polarization configuration, shown for a particle aspect ratio $r = 2$.

Figures (9-11) show graphs similar to the above, but for the instrument configured with a right circularly polarized input state and detection with right and left circularly polarized states. Note again the 'logarithmic' angular scale which, this time, is expanded at smaller angles. Most light is scattered near the forward-scattering direction, whereas $\left|g_{\alpha\beta}^{(1)}(0)\right|^2$ changes most rapidly for $30° < \theta < 60°$. The errors are minimized for a scattering angle $\sim 40°$. The absolute value of the error is about the same as for the linear case however, so there is little advantage in using one configuration in preference to another in this instance.

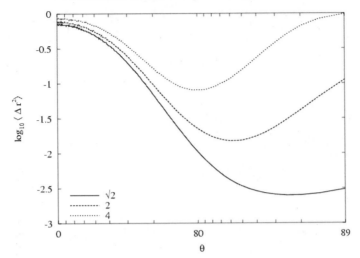

Figure 8. The mean square error for a recovered aspect ratio as a function of scattering angle for a section of particle aspect ratios.

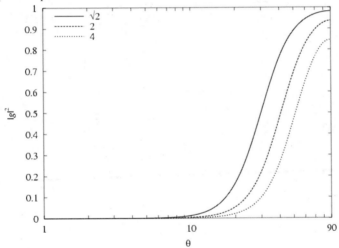

Figure 9. The field correlation function for the circular polarization configuration of the instrument, shown for a range of particle aspect ratios.

The next section demonstrates the robustness of this theory by performing experiments on well characterized particles.

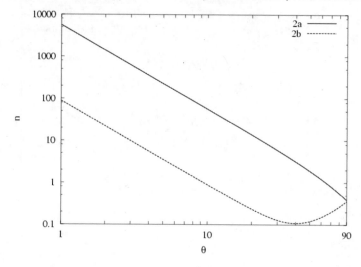

Figure 10. Mean intensity in the two detectors for the circular polarization configuration for $r = 2$.

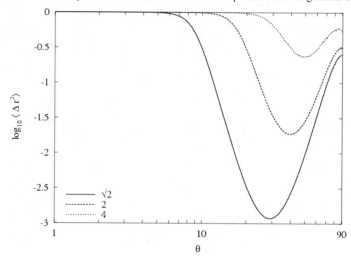

Figure 11. Mean square error for a recovered aspect ratio using a the circular polarization configuration for a selection of particle aspect ratios.

8.   Experimental Results

This section presents results performed on specially characterized hematite spheroidal particles of major/minor axes 0.13 ± 0.02 μm and 0.07 ± 0.01 μm respectively. The particles are suspended in water. The illumination source is a 10mW stabilized 632.8nm HeNe laser. The relative refractive index of the particles at this wavelength is $n = 1.698 - 0.0149$. The detectors employed are photon-counting photomultipliers with red-enhanced photocathodes.

    For an experiment designed to measure aspect ratio, the error analysis of the previous section shows that for the optimum linear configuration, the detector should be located ideally at $\sim 84°$. We shall illustrate the sensitivity by performing reconstructions at a full range of angles $0° < \theta < 89°$. Figure 12 shows the experimentally determined values of $\left| g_{\alpha\beta}^{(1)}(0) \right|$ displayed as a function of θ together with the calculated form assuming a particle aspect ratio $r = 1.8$. Figure 13 shows the measured scattering efficiency in each channel, displayed as crosses, together with the theoretical predictions. Figure 14 shows the aspect ratios deduced from the measured values of $g_{\alpha\beta}^{(2)}(0)$ as a function of θ.

The full line is $r = 1.8$ with the dotted lines denoting a scatter of ± 0.1 about this nominal value. There are a number of noteworthy features that this figure reveals. First, the aspect ratio is underestimated at small scattering angles and this is most probably a result of the particles agglomerating to form larger and approximately more spherically shaped structures. The phase function of larger particles is more sensitive to size rather than shape and tend to scatter preferentially in the forward direction, thereby distorting the data in this region. Second, poor estimates of the aspect ratio are recovered when the detector is located between $\sim 50° < \theta < 75°$. This is predicted by the results shown in Figure 14 and occurs principally because of the weak dependence of $\left| g_{\alpha\beta}^{(1)}(0) \right|$ on aspect ratio throughout this range of angles. Third, the location for obtaining the most consistently accurate reconstructions of the aspect ratio are for $\sim 83° < \theta < 87°$ which is in accord with the error analysis.

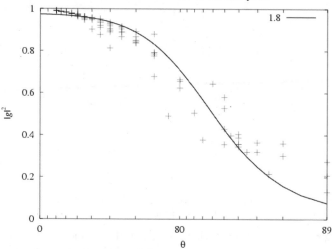

Figure 12. Crosses show measured values of $|g|$ as a function of scattering angle together with the predicted form of $|g|$ for hematite spheroids with aspect ratio of 1.8.

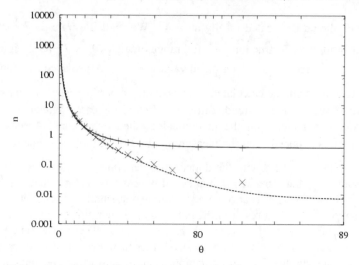

Figure 13. Crosses show measured average photon count in either detector as a function of scattering angle together with the predicted values.

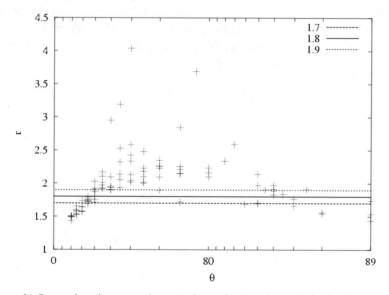

Figure 14. Crosses show the recovered aspect ratio as a function of scattering angle. The horizontal lines are the nominal values $r = 1.8 \pm 0.1$.

Figure 14 illustrates the viability of the PFS method, but here it is worth taking stock of the assumptions that underpin PCS and PFS. A principal virtue of size estimates derived from PCS are that they are independent of the optical properties of the particle. They are however implicitly dependent on the particle shape, for the shape is required to determine the translational diffusion coefficient in the suspending medium. The particles are usually assumed to be spherical. The aspect ratios displayed in Figure 13 were reconstructed assuming the size of the particles and their refractive index was known, and moreover that the particles were prolate. Recall from Section 5 that another solution exists for oblate particles, which is consistent with the measured value of $g_{\alpha\beta}^{(2)}(0)$. Thus particle size, which is deduced from the temporal decay of $g_{\alpha\beta}^{(2)}(\tau)$, depends on particle shape. Particle shape is deduced from the behavior of $g_{\alpha\beta}^{(2)}(0)$ with scattering angle, but requires the particle size. Hence size and shape are not independent measures and should be obtained self-consistently. We now show how this can be achieved, together with a procedure for ascertaining whether a particle is prolate or oblate, should that prior knowledge be unavailable.

A convenient way to envisage how a self-consistent inversion for size and shape is effected is shown in Figure 15. The full lines denote contours of constant $\left|g_{\alpha\beta}^{(1)}(0)\right|$, whose values are labeled. The dotted lines are contours of constant decay rate of $g_{\alpha\beta}^{(2)}(\tau)$, this being the reciprocal of the correlation time τ_D, hence the labeled values are in units of s^{-1}. These contours are displayed as functions of particle aspect ratio r along the ordinate and the 'effective' or spherical volume radius a along the abscissa. The contours have been evaluated for the configuration with illumination and detection in linearly polarized states, hence the detection is assumed to take place at the optimum angle of $84°$. It has been assumed that $r > 1$. From Figure 15, observe that a measured value of $\left|g_{\alpha\beta}^{(1)}(0)\right| = 0.5$ could correspond to a particle of aspect ratio r ~ 1.8 and size a ~ 50 nm, as denoted by the '+' but could equally correspond to any combination of r and a along that contour. Measuring the correlation decay time τ_D provides another contour that can intersect with that for $\left|g_{\alpha\beta}^{(1)}(0)\right|$ and thereby provide a precise reconstruction for a and $r > 1$. For example, a decay rate of $560\ s^{-1}$ intersects the $\left|g_{\alpha\beta}^{(1)}(0)\right| = 0.5$ just once where $r = 3$ and $a = 110\text{nm}$, and is depicted by a '×'. Note however that if $a > 150\text{nm}$, the contours of $\left|g_{\alpha\beta}^{(1)}(0)\right|$ and $1/\tau_D$ no longer intersect at a single point in the $r\text{-}a$ plane, frustrating the determination of a unique reconstruction. This effectively prescribes the upper bound of particle size for which the PFS technique can be used.

If it is not known whether the particles are prolate or oblate, another solution for $r < 1$ exists and this is shown in Figure 16. The intersection of the contours for $\left|g_{\alpha\beta}^{(1)}(0)\right| = 0.5$ and $1/\tau_D = 560\ s^{-1}$ give the same effective radius (as indeed it must) but an aspect ratio of $r = 0.38$. How might the $r = 1.8$ and 0.38 solutions be distinguished from each other? The answer comes by performing another experiment at a different angle.

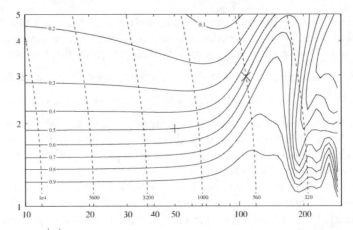

Figure 15. Maps of $|g|$ and the inverse correlation time displayed as functions of aspect ratio $r > 1$ along the ordinate, and effective radius in nm along the abscissa.

Detection at $88°$ produces a value of $g_{\alpha\beta}^{(2)}(\tau)$ that is sufficiently different from those obtained at $84°$, but is still sufficiently accurate according to the error analysis. The box and square in Figure 17 show the reconstructed values for $\theta = 84°$ and $88°$ respectively. Note that reconstructions for the $r > 1$ portion of the graph are similar, whereas there is a discernable discrepancy between the two in the $r < 1$ region. Hence $r = 1.8$ is the correct solution to select.

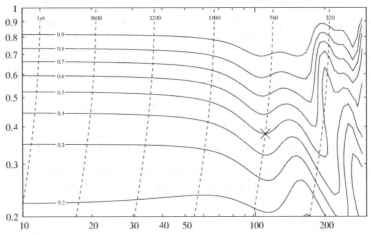

Figure 16. Maps of $|g|$ and the inverse correlation time displayed as functions of aspect ratio $r < 1$ along the ordinate, and effective radius in nm along the abscissa.

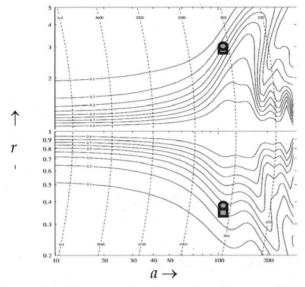

Figure 17. Particle reconstructions shown on the parameter maps illustrating how ambiguity in the value of *r* can be resolved through repeating the experiment at a different angle. The boxes are the reconstructions at 84°, the circles at 88°. The reconstructions in the region *r* < 1 are displaced from each other, indicating that this is the incorrect solution.

The contours in the above figures were calculated assuming the refractive index of the particles $n = 1.7 + 0.15\ i$. Figure 18 shows how the contours for $\left|g_{\alpha\beta}^{(1)}(0)\right|$ broaden into bands should there be uncertainty in the value assumed for *n*. The contours for $1/\tau_D$ are independent of refractive index. The values of the refractive index in water are taken to be $n = 1.7 \pm 0.05 + 0.15\ i$ It is apparent that the influence of refractive index becomes a problem for uncertainty in *n* greater than about 0.05.

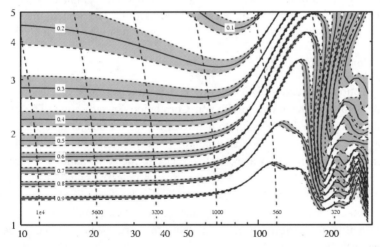

Figure 18. The effect of variation in the real part of the refractive index on the reconstruction maps. The center of the band is for n = 1.7, the lower limit for n=1.65 and the upper limit n=1.75.

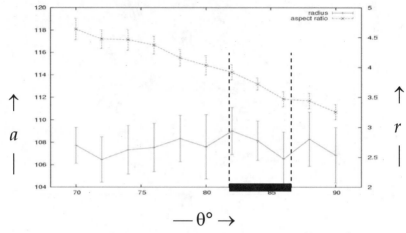

$$— \theta° \to$$

Figure 19: Self-consistent reconstructions for the effective radius in nm and aspect ratio for hematite prolate spheroids. The optimum region for performing measurements is shown.

To test the effectiveness of applying this procedure in practice, experiments were again performed with hematite prolate particles with major/minor axes 220nm and 60nm respectively, giving an aspect ratio of $r \sim$ 3.7. Experiments were performed at scattering angles $70° < \theta < 90°$. For each measurement, the value of $\left| g_{\alpha\beta}^{(1)}(0) \right|$ and the τ_D were estimated by least-squares fitting to the measured correlation functions together with an estimate of the standard error in each measurable. Figure 19 shows a reconstruction for the effective radius to be essentially independent of scattering angle, with value $a \sim 108$nm. The value of r is a decreasing function of θ with $3.5 < r < 3.9$ in the most accurate region, which is indicated on the figure. These agree well with the known values of particle parameters that were independently obtained using electron microscopy. The dotted lines show where most accurate results are to be found according to the error analysis.

9.   Conclusions

Polarization Fluctuation Spectroscopy is a relatively simple extension to the well established technique of Photon Correlation Spectroscopy. The optical equipment required in addition to that used for PCS is standard, as also is the data processing hardware for determining the correlation functions. The means to analyze the data is now achievable because of the computational techniques for calculating the scattering properties for particles of complex shape and composition that have been developed in the last decade. The additional functionality afforded by PFS is achieved at modest additional cost, and importantly, by performing measurements that are essentially the same as those for PCS.

The technique has distinct advantages insofar as it performs reconstructions on data that are formed as dimensionless quantities. This means that the technique is largely independent of experimental artifacts and therefore does not require calibration with some standard particle type. The principal disadvantages are that it is essentially a laboratory based technique, the samples require careful preparation to ensure they are free from contamination, and that the measurements are performed in the single-scattering regime. The latter constraint implies that dense mixtures, which are often of interest, cannot be assessed.

Frequently it is required to assay mixtures of particles. This creates potential problems for the PCS element for the characterization of particle size, but these can nevertheless be overcome. In such a situation, each species of particle has an associated diffusion constant giving rise to exponentially decaying correlation functions. The measured correlation functions are convolutions over the particle species which embody the different timescales. Nevertheless, careful processing of the data using singular valued decomposition techniques [e.g. 20,21] can extract the sizes of the particles comprising the ensemble. If there is a distribution of aspect ratios, the polarization element of the measurement process extracts the average value of the distribution [22].

Additional functionality that could be incorporated into an instrument is to perform measurements at different wavelengths. The optical size of the particle would thereby change, but the imaginary refractive index would change too, providing access to the imaginary part of the refractive index.

Acknowledgments

This work is supported by the UK Engineering and Physical Science Research Council.

References

1. "Photon Correlation and Light Beating Spectroscopy", H.Z. Cummins and E.R. Pike Eds. (Plenum, New York, 1973).
2. "Photon Correlation Spectroscopy and Velocimetry", H.Z. Cummins and E.R. Pike Eds. (Plenum, New York, 1976).
3. C. F. Bohren and D. R. Huffman, "Absorption and Scattering of Light by Small Particles" (Wiley, New York, 1983).
4. "Measurement of Suspended Particles by Quasi-Elastic Light Scattering," B.D. Dahneke Ed. (Wiley, New York, 1983).
5. E. Jakeman, and E.R. Pike, J. Phys. A: Gen Phys, **2**, 411 (1969)
6. E. Jakeman, J. Phys. A: Math. Gen. **3**, 201 (1969).
7. E. Jakeman, C.J. Oliver, E.R. Pike, **5**, L93 (1972).
8. J.W. Goodman, "Statistical Optics," (Wiley, New York, 1985).

9. W. Feller, "An Introduction to Probability Theory and Its Applications," Vol II, 2nd Edition, (Wiley, New York, 1971).

10. A.J.F. Siegert, MIT Rad. Lab. Rep. 465 (1943).

11. B.J. Berne, R. Pecora, "Dynamic Light Scattering with Applications to Chemistry, Biology and Physics," (Wiley, New York, 1976).

12. F. Perrin, Le Journal de Physique et Le Radium (Paris), Serie VII, Tome V, No. 10, 497 (1934), Le Journal de Physique et Le Radium (Paris), Serie VII, Tome VII, No. 1, 1 (1936) .

13. T.G.M van de Ven, "Colloidal Hydrodynamics," (Academic Press, London 1989).

14. E. Jakeman, Waves Random Media, **5**, 427 (1995)

15. A.P. Bates, K.I. Hopcraft, E. Jakeman, **8**, 235 (1998)

16. M. I. Mishchenko, L.D. Travis and D. W. Mackowski, J. Quant. Spectrosc. Radiat. Transfer **55**, 535 (1996).

17. P.C.Y. Chang, K.I. Hopcraft, E. Jakeman, J.G. Walker, Meas. Sci. Technol. **13**, 535 (2002).

18. M. Pitter, K.I. Hopcraft, E. Jakeman, J.G. Walker, J. Quant. Spectrosc. Radiat. Transfer **63**, 433 (1999).

19. E. Jakeman, E.R. Pike, S. Swain, J.Phys. A: Math. Gen. **4**, 517 (1971).

20. S. Provencher. Makromol. Chem, **180**,201,(1979).

21. E.R. Pike, B.M. McNally, in "Scientific Computing", F.T. Luk and R.J. Plemmons, Eds. (Springer, Berlin 1997).

22. M.C. Pitter, K.I. Hopcraft, E. Jakeman, J.G. Walker, Proc. SPIE, **375**, 55 (1999).

Keith Hopcraft, Rosario Vilaplana, José María Saiz, and Francisco González at the reception.

INTENSITY AND POLARIZATION FLUCTUATION STATISTICS OF LIGHT SCATTERED BY SYSTEMS OF PARTICLES

F. GONZÁLEZ, F. MORENO, J.M. SAIZ AND
J.L DE LA PEÑA
Universidad de Cantabria. Dep. de Física Aplicada
Av. Los Castros s/n 39005 Santander, Spain

Abstract. The analysis of intensity and polarization fluctuations of light scattered by systems with particles is of interest in the resolution of the so-called inverse problem. In this chapter we study the possibility of obtaining information, by means of light scattering techniques, about the geometrical and optical properties of systems composed of two Rayleigh particles, either in volume or located on a substrate. Intensity fluctuations are studied using the normalized second order moment. Polarization fluctuations are studied using the probability of getting null values when measuring the cross-polarized component $P(I_{cross}=0)$. It is observed that the behavior of these parameters depends on both the polarizability of the particles and the distance between them. Some experimental results are shown with measurements of the parameter $P(I_{cross}=0)$ for the case of metallic particles on a flat conducting substrate.

1. Introduction

The non-invasive analysis of systems of particles using light-scattering measurements is of great interest in various fields. Two general morphologies of interest include 1) light scattering characterization of particles in a volume, with applications, for instance, in astronomy, atmospheric contamination and biology; and 2) light scattering by particles on substrates, with applications, for instance, in the semiconductor industry, surface control and inspection, or photodetection enhancement.

G. Videen et al. (eds.), Photopolarimetry in Remote Sensing, 175-190.

The characterization of a system from the light it scatters is an *inverse problem*, i.e., a procedure to obtain information of the scattering system from the electromagnetic properties of the scattered radiation: intensity, angular distribution, polarization, etc.

In the case of a simple system, like a single Rayleigh particle or a regular particle of which we know an exact solution, it is possible to obtain the scattering patterns and, inversely, from the features found in these patterns, obtain, for instance, morphological properties of the particles [1]. If the system consists of either irregular particles, anisotropic particles, inhomogeneous particles, or particles larger than the incident wavelength, the calculation of the scattering patterns may require numerical methods like the discrete dipole approximation (DDA) or finite-difference time-domain (FDTD) algorithms. In such cases, the inverse problem is a difficult task to perform because computational time becomes a significant factor. In additional, many solutions are not unique [2].

Other systems often found in nature are those usually referred to as *random media*, where scattering elements are randomly located in a medium. Though the inverse problem drastically increases its complexity, the use of numerical and simulation methods may allow for a statistical description of some observable parameters like, for instance, the scattered intensity. By these means, it is possible sometimes to obtain information of the optical or morphological properties of individual scatterers within the system [3]. In the case of a high number of coherently illuminated independent scatterers, the resulting field is a circular complex Gaussian random variable. The probability density function of the scattered intensity is a negative exponential and the normalized mth order moment follows a simple factorial law, given by [4]:

$$\frac{\langle I^2 \rangle}{\langle I \rangle^2} = m! \tag{1}$$

In this case it is very difficult to obtain any information about the individual scatterers from the statistical analysis of the scattered intensity. However, this behavior may be different if the number of independent scatterers is small so that the central limit theorem does not apply. This happens, for instance, in the case of a dilute sample or for a small illuminated surface area. In this situation of a small number of independent scatterers there are still two main cases: first, when the number of scatterers is deterministic (i.e. N is fixed); and second, when the number of scatterers fluctuate between each realization. For fixed N and if the scatterers are independent (uncorrelated amplitude and phase of the individual scattered fields), and if all contributions to the total field are statistically identical, then the second moment of the intensity is given by [4]:

$$\frac{\langle I^2 \rangle}{\langle I \rangle^2} = 2\left(1 - \frac{1}{N}\right) + \frac{1}{N} \frac{\langle a^4 \rangle}{\langle a^2 \rangle^2} \tag{2}$$

where N is the number of scatterers and a is the electric field amplitude scattered by one particle. From Eq.2 we may conclude that fluctuations of the scattered intensity not only depend on the number of illuminated particles, but also on the statistics of the amplitude a that can produce an enhancement of the fluctuations.

For particles with spherical symmetry: $\langle a^2 \rangle^2 = \langle a^4 \rangle$ and therefore:

$$\frac{\langle I^2 \rangle}{\langle I \rangle^2} = 2 - \frac{1}{N} \tag{3}$$

In the case of a fluctuating number of scattering particles and keeping the above made assumptions of independence, statistical identity and spherical symmetry, a new expression for the second moment of the intensity can be deduced, provided that the variable number of particles follows Poisson statistics:

$$\frac{\langle I^2 \rangle}{\langle I \rangle^2} = 2 + \frac{1}{\overline{N}} \tag{4}$$

where $\overline{N}$ is the mean number of scattering particles. Eqs. 3 and 4 show how the measurement of the intensity fluctuations may be used to determine the number of scattering particles in a non-Gaussian regime.

A more realistic situation is that the scattering of each particle is influenced by the presence of neighboring particles; this can be caused by the proximity or by particular optical properties, resonances etc. Now, the amplitude and phase of the electric field scattered by particle i is affected by the field scattered by the rest of the particles interacting with particle i. Multiple scattering in the near-field region may change the statistics of the scattered intensity fluctuation and/or the scattered light polarization.

The objective of this chapter is to investigate the intensity and polarization fluctuations that appear in the light scattering by a coherently illuminated two-particle system. We want to obtain information about the geometry and/or optical properties of the system from the statistical knowledge of the scattered light. We study a two-particle because it is the simplest non-Gaussian system where multiple scattering may appear. The analysis of the intensity fluctuation is made by means of the second normalized moment of the scattered intensity ($<I^2>/<I>^2$) and the analysis of the polarization fluctuations is made through the calculation of the probability of detecting a null cross-polarized component, $P(I_{cross}=0$). These two parameters have been used in previous works where they have proven to be efficient in showing changes in the statistics [5].

The chapter is structured as follows: Section 2 shows the different two-particle configurations studied and the numerical and calculation methods used to find the scattered electric far field. In Section 3 the main results obtained for the same systems are shown, and Section 4 is devoted to the experiment corresponding to one of the parameters studied, $P(I_{cross}=0)$, for the case of a sample constituted by metallic particles on a conducting substrate.

2.   Theory

2.1 Scattering geometries

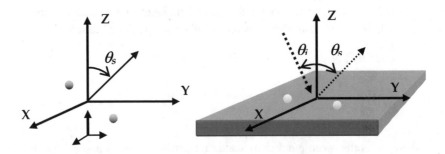

Figure 1: Scattering geometries. (a) Particles distributed inside a volume, (b) Particles distributed on a substrate.

The two scattering models used in the numerical calculations are shown in Fig. 1. They consist of two particles distributed in free space (Fig. 1.a) or on a substrate of dielectric constant ε (Fig. 1.b). For simplicity, we call them volume and surface cases respectively. In both systems the particles are assumed to be spheres of radius a and dielectric constant ε_p. Their sizes are much smaller than the incident wavelength λ. Thus the polarizability of the particles α, is given by [6]:

$$\alpha = a^3 \frac{\varepsilon_p - \varepsilon_0}{\varepsilon_p + 2\varepsilon_0} \qquad (5)$$

where ε_0 is the relative dielectric constant of the surrounding medium: in our case $\varepsilon_0 = 1$.

In the volume case, the particle system is illuminated by a monochromatic plane wave whose propagation direction is along the Z-direction. In the surface case, the incident wave direction forms an angle θ_i with respect to the surface normal and it is contained in the Z-Y plane. The scattered intensity is analyzed in the plane Z-Y (scattering plane) as shown in Fig. 1. We consider two linear polarization states of the incident field. The incident wave is S-polarized when the electric field vibrates parallel to the X axis, and it is P-polarized when the electric field vibrates within the YZ plane. Similarly, we consider two polarization states for the scattered field: The co-polarized case when both incident and scattered field have the same polarization states (SS or PP) and the cross-polarized case when both fields are perpendicular to each other (SP or PS).

In both systems, the particles are randomly distributed. In the surface case, the particles are homogeneously distributed inside a circle of radius R_{max} (the radius of the illuminated area at normal incidence). Similarly, in the volume case, the particles are homogeneously distributed inside a sphere of radius R_{max} (the radius of the scattering volume). Both distributions are calculated using the *inverse transform method* based on Monte Carlo techniques [7,8].

2.2 Calculation procedure and approximations

In general, the study of the scattered intensity by systems where multiple scattering is non-negligible is a difficult task. In order to resolve these difficulties we have used several approximations: the particle size is assumed to be much smaller than the incident wavelength, so the *Rayleigh dipole approximation* is used to characterize the particle scattering. For the surface case, the *perfect conductor approximation* (PCA) ($\varepsilon = -\infty$) for the substrate is applied. In both systems, the scattered field is calculated in the far-field approximation. For the purpose of this research, these approximations do not constitute serious restrictions and facilitate the numerical calculations. For real substrates (metallic or dielectric) and/or finite size particles, the main conclusions are unaffected [9]. The scattered intensity is calculated by means of the *coupled dipole method* (CDM) [10] for both the volume case and the surface case. In the latter, the PCA allows us to apply the *image theory* [11].

In the scattering system constituting N dipole-like particles of the CDM, the total field at the position of particle i, E_i, is the sum of the incident field E_{oi} plus the fields that are scattered by the rest of the (N-1) particles:

$$\vec{E}_i = \vec{E}_i^0 + \sum_{j \neq i} \left[A_{ij} \tilde{\alpha} \vec{E}_j + B_{ij} (\tilde{\alpha} \vec{E}_j \vec{n}_{ji}) \vec{n}_{ji} \right] \tag{6}$$

In Eq.6, $\vec{n}_{ji}$ is the unit vector from the ith to the jth particle and $\tilde{\alpha}$ is the complex polarizability tensor (which for the case of isotropic particles is a complex constant). A_{ij} and B_{ij} are the interaction coefficients whose expression is in [10].

This set of equations for the local field can be written in matrix form as:

$$\vec{E} = \vec{E}^0 + \tilde{C} \cdot \vec{E} \tag{7}$$

where $\vec{E}$ and $\vec{E}^0$ are vectors of $3N$ components that represent the total local and incident fields, respectively. $\tilde{C}$ is a $3N \times 3N$ matrix containing the interaction tensors $\tilde{C}_{ij}$. Thus, the total local field is obtained from

$$\vec{E} = (\tilde{I} - \tilde{C})^{-1} \cdot \vec{E}^0 \tag{8}$$

where $\tilde{I}$ is the $3N \times 3N$ identity matrix. The total scattered field in the far-field approximation can be written as:

$$\vec{E}_s = \frac{k^2 e^{ikR}}{R} (\tilde{I} - \vec{n}_s \vec{n}_s) \sum_{i=1}^{N} e^{-ik\vec{n}_s \vec{r}_i} \alpha_i \vec{E}_i \tag{9}$$

where $\vec{n}_s$ is the unit vector in the scattering direction, $\vec{r}_i$ is the position vector of the ith particle with respect to the reference system and R is the distance between the scattering system and the detector ($kR \gg 1$).

3.   Numerical results

Numerical calculations of the scattered intensity have been performed for both scattering systems assuming spherical scatterers with radius $a = 0.05\lambda$. The incident field propagates along the Z-direction ($\theta_i = 0°$ for the surface case) and the scattered intensity is detected in the YZ plane at a scattering angle $\theta_s = 30°$. We have chosen the particle polarizability α as the parameter for evaluating the strength of the multiple scattering between particles. For a dipole-like particle, the scattered intensity is proportional to the square of its polarizability, so multiple scattering is enhanced as the polarizability increases. Results are divided in two sub-sections: the first corresponds to S incident polarization and the second to P incident polarization. We include comparisons between both cases.

The choice of $P(I_{cross})$ as a statistical parameter for obtaining information is supported by the shape of its *probability density function* (PDF). This PDF has a maximum at $I_{cross}=0$ and decreases very quickly as the intensity increases. It also may be observed that $P(I_{cross})$ remains finite even for values of I_{cross} in excess of the mean of the cross-polarized scattered light $(<I_{cross}>)$ but the fluctuation of the values of $P(I_{cross})$ increases. This means that some peaks of high cross-polarized intensity appear corresponding to cases in which the particles are close together. So, the values of $P(I_{cross})$ for $I_{cross} = 0$ can be determined with the smallest relative error [5].

In a real experiment, there is always a threshold intensity I_0 below which it is impossible to discriminate between detector noise and the signal. In order to establish this threshold level for calculation purposes, we have considered that $I_{cross}=0$ if the calculated value of I_{cross} fulfills the following inequality:

$$I_{cross} \leq 10^{-6}(I_s+I_p) \tag{10}$$

where I_s and I_p are the scattered intensities by one isolated particle when it is illuminated by field with an S or P polarization, respectively.

3.1 S incident polarization

In Fig. 2 we show the evolution of $P(I_{cross} = 0)$ with α for the volume case $(R_{max} = 1\lambda, 5\lambda)$. We have simulated 10 series of 10^4 positions of the particles. As particle interaction increases, either by increasing α or by reducing the scattering volume, the probability of having no cross-polarization decreases monotonically to zero. However, when particles are on a substrate (same number of samples and $R_{max} = 1\lambda, 5\lambda$), the shape of $P(I_{cross} = 0)$ changes, see Fig. 3. $P(I_{cross} = 0)$ shows a minimum around $\alpha = 5\times10^{-4}$ and then saturates to a constant value. Such a minimum can be explained as being due to the particle–substrate interaction. As α increases, the intensity scattered by a single isolated particle on a substrate increases until a maximum value and remains constant thereafter. The position of this maximum coincides with the zone of the minimum in $P(I_{cross} = 0)$. It also has been pointed out that the substrate severely weakens the particle–particle interaction [12]. For instance, at $\alpha = 10^{-2}$ the change of $P(I_{cross} = 0)$ in the volume case (from $R_{max}=1\lambda$ to 5α) is smaller than in the substrate case.

For comparison, we have calculated the statistical parameter $n^{(2)}$ (second order factorial moment) for the PDF shown in Fig. 2 and Fig. 3. The scattering geometries are the same but the number of samples is higher (10 series of 10^5 positions). The results are shown in Fig. 4 and Fig. 5. In general, $n^{(2)}$ is noisier and its relative errors are much higher than those of $P(I_{cross} = 0)$. For both scattering systems, the second moment of the co-polarized scattered intensity,

$n_{co}^{(2)}$ at low values of α has the same behavior than that corresponding to independent scatterers. They follow the relation [13]:

$$n_{co}^{(2)} = 2 - \frac{1}{N} \qquad (11)$$

In our case, $N = 2$ so that $n_{co}^{(2)} = 1.5$.

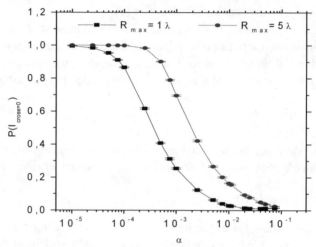

Figure 2. Evolution of $P(I_{cross}=0)$ with α for different values of R_{max} when the particles are illuminated by a S-polarized field. Volume case.

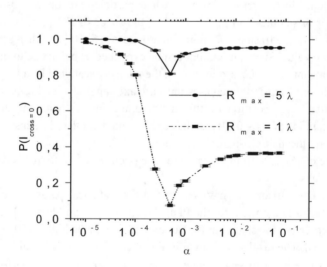

Figure 3. Evolution of $P(I_{cross} = 0)$ with α for two values of R_{max} when the particles are illuminated by a S-polarized field. Surface case.

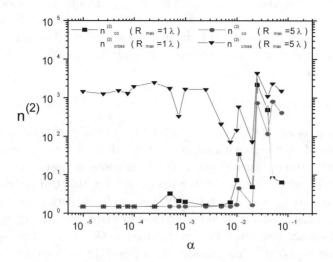

Figure 4. Evolution of $n^{(2)}$ with α for different values of R_{max} when the particles are illuminated by a S-polarized field. Volume case.

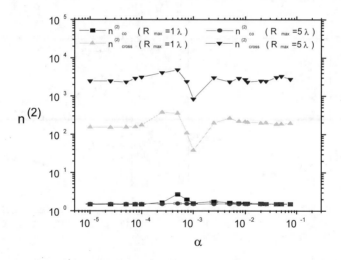

Figure 5. Evolution of $n^{(2)}$ with α for different values of R_{max} when the particles are illuminated by a S-polarized field. Surface case.

When the particles are on a substrate the fluctuations decrease and the relative errors are lower than when they are inside a volume. This is evidence of the weakening of multiple scattering due to the presence of the substrate. In the surface case, the values of $n_{co}^{(2)}$ and $n^{(2)}_{cross}$ are different and they remain constant, except for the region around the minimum of $P(I_{cross} = 0)$. Also if the scattering volume is increased, the multiple scattering effect is reduced and with it, the amount of cross-polarized intensity. This produces higher values, and relative errors of $n^{(2)}_{cross}$.

3.2 P incident polarization

In Fig. 6(*left*) $P(I_{cross} = 0)$ is plotted for the case of particles illuminated by *P*-polarized field and contained in a volume of $R_{max} = 1\lambda$. Ten series of 10^4 different positions are simulated. For comparison purposes, we have also plotted the values of $P(I_{cross} = 0)$ when the incident field is *S* polarized. Both evolutions are equivalent as expected. However for the surface case, the *S* and *P* cases behave differently. In Fig. 6(*right*) we show the values of $P(I_{cross} = 0)$ for the surface case and *P* incident polarization ($R_{max} = 1\lambda$). The *S*-polarized case also is plotted for comparison. Ten series of 10^4 positions are simulated. From these results it can be inferred that the presence of the substrate produces less particle interaction with illumination in *P* polarization than for *S* polarization [12]. Similar conclusions are obtained for $R_{max} = 5\lambda$.

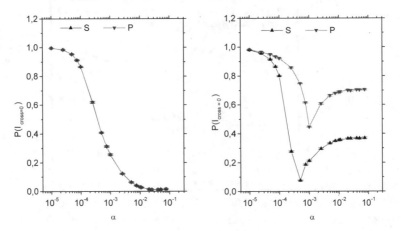

Figure 6. Values of $P(I_{cross} = 0)$ when particles are illuminated by *S* or *P* polarized field. Left: Volume case. Right: Surface case.

The simulated numerical values for $n^{(2)}$ are shown in Fig. 7 for the volume and surface cases (10 series of 10^5 cases each). In both the volume and the surface cases, the values of $n^{(2)}$ corresponding to *P* polarization are almost equal to those corresponding to *S* polarization. The small differences are within the statistical errors. So $n^{(2)}$ cannot discriminate between *S* and *P* either for the volume case or for the surface case. As in the analysis of $P(I_{cross} = 0)$, similar conclusions can be obtained for $R_{max} = 5\lambda$.

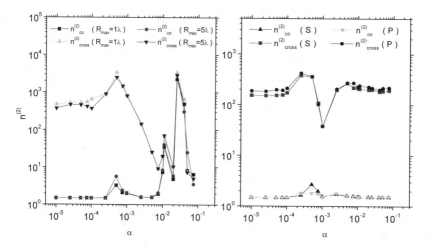

Figure 7. Values of $n^{(2)}$ when particles are illuminated by S or P incident wave. $R_{max}=1\lambda$. Left: Volume case. Right: Surface case

3.3 Other examples

For completeness, we also have analyzed the fluctuations of the intensity scattered by a fluctuating number of particles when the mean number of particles in the scattering zone is $\overline{N} = 2$. The number of illuminated particles is simulated according to a Poisson distribution whose mean is $\overline{N}$. The simulated values of $P(I_{cross} = 0)$ are higher than those obtained when a fixed number of particles is illuminated. This is because there is a non-zero probability of getting 0 or 1 particles in the scattering zone (such cases do not produce cross-polarization). However, the general behavior of the statistical parameters $P(I_{cross} = 0)$ and $n^{(2)}$ are similar to those obtained when we illuminate a fix number of particles. For instance, the shape of $P(I_{cross} = 0)$ is similar and the position of the minimum in the surface case does not change.

Finally in this section, we present for comparison some results for the intensity fluctuations statistics of the scattered light from two dipole-like particles separated by a fixed distance d, moving freely in space. This would constitute the simplest model of an aggregate particle suspended in the atmosphere. In Fig.8 we show the evolution of different parameters as a function of α/d, α being the polarizability of the particles.

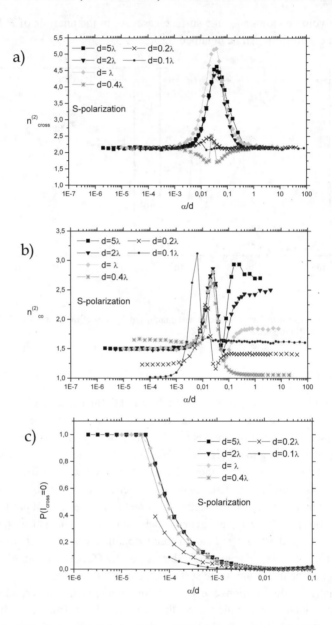

Figure 8. Evolution of several statistical variables as a function of parameter α/d: a) $n^{(2)}_{co}$; b) $n^{(2)}_{cross}$; c) $P(I_{cross}=0)$

The polarization of the incident radiation is S, although similar conclusions can be drawn for P polarization. As can be seen, when plotted against α/d, those parameters have a universal behaviour whenever the interparticle distance is larger than λ. For shorter separations, multiple scattering is important and the statistics of the intensity fluctuations change; for instance, the position of the enhancement in $n^{(2)}_{co}$ moves to smaller values of α/d. This enhancement is due to the presence of a resonance and it shifts to smaller values of α/d as d decreases, i.e. as multiple scattering increases. For low values of α/d, $n^{(2)}_{co}$ reaches the value 1.5 (corresponding to two scatterers c.f. Eqn (3)) except when $d < \lambda$, since the random phase difference between the two scattered fields is not uniformly distributed in the interval $[0,2\pi]$. $n^{(2)}_{cross}$ shows similar behaviour. As expected, the fluctuations of the scattered field when d is fixed for both co- and cross-polarized fields are smaller than when d fluctuates, and for the same number of samples they are much less noisy.

4. Experiment

Backscattering is of interest because of remote-sensing applications. Backscattering measurement can be applied in combination with statistical techniques to extract information about samples. In the case of surfaces and particles, the particle surface density can be the main source of multiple scattering for flat or slightly rough surfaces. For incident polarization, either parallel or perpendicular to the scattering plane, multiple scattering processes may produce changes in the polarization plane of light detected within the plane of incidence. Consequently a nonzero cross-polarized component is observed when the average distance between particles d is sufficiently small; i.e. when the particle surface density d^{-2} is sufficiently large. If no other depolarization mechanism is present and assuming Gaussian regime (high number of scatterers), the probability of detecting zero when observing the cross-polarized component can be expressed as [5]:

$$P(I_{cross} = 0) = \exp\left[-\frac{L\pi ab}{d^3}\right] = \exp(-\gamma ab) \qquad (12)$$

where a and b are the semiaxes of the illuminated area that is dependent on the width of the incident beam and the angle of incidence, and L is the distance at which multiple scattering between two particles can be neglected. The parameter γ includes only intrinsic dependencies on the sample; whereas, the presence of a and b shows the importance of the number of illuminated particles. Experiments carried out for the purpose of fitting statistical expressions like Eq. 12 require a large number of measurements from the sample. It is possible to obtain so many measurements by rotating the sample and detecting the intensity fluctuations [5]. A probability density function (PDF) is found for either the co- and cross-polarized intensities.

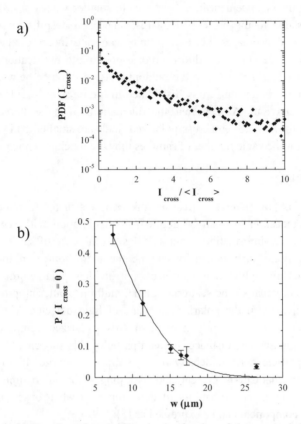

Figure 9. (a) Experimental probability density function of the cross-polarized backscattered intensity produced by interacting spherical scatterers located on a flat substrate. (b) Experimental values of $P(I_{cross} = 0)$ (dots) and their corresponding error bars as a function of the illuminated area. Continuous line corresponds to the fitted function (Eq.12).

In Fig. 9.a, the case of backscattered cross-polarized detection is shown. The sample consisted of gold particles sized approximately 1μm, seeded on a flat gold substrate. If we focus on the probability value at $I_{cross} = 0$ (the highest value in the curve, i.e., the smallest relative error) and measure that probability under different spot sizes, we can use Eq. 12 to obtain γ. Fig. 9.b shows the evolution of $P(I_{cross} = 0)$ for six different illuminating conditions. The fit to Eq. 12 is plotted as a continuous line. From this fitting, $\gamma = 8.6 \times 10^{-3}$ mm^{-2} is found; hence, d can be obtained if L is known and vice versa. For $d = 8$ μm, as seen from the electron microscope, a value of $L = 1.25$μm is obtained, slightly larger than the particle diameter, which is consistent with other experimental works [14]. The parameter γ is now seen as a measurement of the interacting capacity of the scatterers. It is dependent upon the particle density and the efficiency of the combination of scatterers to depolarize the light.

5. Acknowledgments

The authors wish to thank the MCYT (Ministerio de Ciencia y Tecnología) for its financial support: project #BFM2001-1289.

References

1. L.L. de la Peña, F. González, J.M. Saiz, F. Moreno, and P.J. Valle, J. Appl.Phys. **85** 432 (1999).
2. MI Mishchenko, JW Hovenier, LD Travis, editors. Light scattering by nonspherical particles. London: Academic Press, 2000.
3. F. González, J.M. Saiz, F. Moreno, and P.J. Valle, J.Phys.D:Appl.Phys. **25,** 357 (1992).
4. P.N. Pusey in: H.Z. Cummings, E.R. Pike, editors, *Photon correlation spectroscopy and velocimetry*, NATO ASI series, Plenum Press, New York (1977).
5. E.M. Ortiz, F. González, J.M. Saiz and F. Moreno, *Opt Express* **10**, 190 (2002).
6. M. Born and E. Wolf. *Principles of optics*, Pergamon Press, Oxford (1959).
7. W.H. Press. *Numerical recipes in FORTRAN*, Cambridge University Press, Cambridge (1992).
8. R.Y. Rubinstein. *Simulation and the Monte Carlo method*, Wiley, New York (1981).
9. E.M. Ortiz, F. González and F. Moreno in: F. González and F. Moreno, Editors, *Light scattering from microstructures*, Springer, Berlin (1998).
10. S.B. Singham and C.F. Bohren, *J Opt Soc Am A* **5**, 1867 (1988).
11. Lord Rayleigh. On the line dispersed from fine lines rule upon reflecting surfaces or transmitted by very narrow slits. Cambridge University, Cambridge: Scientific Papers, 1912, 5.
12. E.M. Ortiz, P.J. Valle, J.M. Saiz, F. González and F. Moreno, *Waves Random Media* **7**, 319 (1997).
13. J.C. Dainty, editor. *Laser speckle and related phenomena*. Berlin: Springer, 1984.
14. F. González, J.M. Saiz, P.J. Valle and F. Moreno, *Opt. Comm.* **137**, 359 (1997).

Michael Mishchenko, Gorden Videen, Chema Saiz, Francisco González and Keith Hopcraft performing fluctuation research experiments at CrAO

Participants (left to right) Matt Easley, Hal Maring, Michael Mischenko, Oleg Dubovik, Valery Loiko, Olga Kalashnikova, Alina Ponyavina, Vladimir Dick, Anatoliy Gavrilovich, Ludmila Chaikovskaya, Liudmila Astafyeva, Halina Ledneva, Vera Rosenbush, and Nikolai Voshchinnikov prepare for Valery Loiko's welcoming toast (right)

Irina Kulyk, Tamara Bulba, Gorden Videen and Vera Rosenbush at reception.

SCATTERING PROPERTIES OF PLANETARY REGOLITHS NEAR OPPOSITION

Y. SHKURATOV,[1,2] G. VIDEEN,[3] M. KRESLAVSKY[1,4]
I. BELSKAYA,[1] V. KAYDASH,[1] A. OVCHARENKO[1]
V. OMELCHENKO,[1] N. OPANASENKO,[1] E. ZUBKO[1]

[1]*Astronomical Institute of Kharkov National University,
35 Sumskaya Street, Kharkov, 61022, Ukraine*
[2]*Radioastronomical Institute, NASU, 4
Chervonopraporova St., Kharkov, 61002, Ukraine*
[3]*Army Research Laboratory AMSRL-CI-EM, 2800
Powder Mill Road, Adelphi Maryland 20783, USA*
[4]*Dept. Geological Sciences, Brown University,
Providence, RI 02912-1846, USA*

Abstract. We provide a brief overview of photometric and polarimetric data of planetary regoliths obtained through telescopic and spacecraft observations. We apply imaging photometry to study brightness opposition spikes of the Moon and Jupiter satellite Europa. We also show that the opposition phase curves of lunar sites are almost independent of wavelength in the visual-NIR spectral range. The lunar highland surface demonstrates slightly steeper opposition phase curves than the mare surface at phase angles $\leq 1°$. We present an example of an image showing the depth of negative polarization $|P_{min}|$ for the lunar nearside. The image demonstrates a complicated correlation of the $|P_{min}|$ with albedo showing variations of the polarization degree in the range $0.5\% - 1.5\%$. We discuss results of photopolarimetry of Mars addressing in particular to HST observations. Polarimetric images demonstrate significant polarimetric variability attributed to the Martian surface and clouds. The phase ratio $I(0.1°)/I(3°)$ of the Europa surface shows a rapid decrease with growth of the surface albedo. Data for minor planets show that dark asteroids and trans-Neptunian objects demonstrate different phase angle behaviors of brightness and polarization near opposition. This can reflect differences of the surface structure indicating that trans-Neptunian objects may have fluffier surfaces than those of dark asteroids. Two hypotheses for the double-minima shape of the negative polarization of Europa are discussed: In the first the effects of both coherent backscattering and polarization due to single particle scattering are incorporated; in the second, the effect is the result of macroscopical optical inhomogeneities of the surface.

G. Videen et al. (eds.), Photopolarimetry in Remote Sensing, 191-208.
© 2004 *Kluwer Academic Publishers. Printed in the Netherlands.*

1. Introduction

Rapid progress currently is being made in interpreting photometry and polarimetry of particulate media. This is driven by the needs in many fields of science: from laboratory experiments with biological objects to terrestrial and planetary remote sensing. One classic example of particulate media (powder-like surfaces) is a planetary regolith, i.e. soil.

Regoliths form the very upper layers of atmosphereless bodies that include planets, asteroids, and most satellites of major planets. The lunar regolith is well studied, since samples have been obtained and can be examined in laboratories. The lunar regolith, and likely those of asteroids and satellites, are composed of irregular particles with characteristic sizes on the order of 100 μm. Lunar regolith particles are fragments of glasses, rocks and minerals disintegrated in so-called space weathering processes, in particular, meteorite impacts. A significant portion of these particles appear as complicated aggregates [1]. Molten ejecta impacting the lunar surface cement minute regolith particles into aggregates, forming so-called agglutinates that are widely spread on the lunar surface. An example of such agglutinate particles is shown in Figure 1a. Regoliths have a dust-size grain component contaminating the surface of larger particles. Figure 1b demonstrates that the lunar regolith contains fragments and roughness of very small scales, including sizes less than 1 μm. On the other hand, the mass distribution of lunar particle sizes clearly demonstrates a maximum near 60 μm revealing a relative deficiency of small particles [2].

The Martian sand is also an example of extraterrestrial soils. Often, the term "Martian regolith" is used, however soils of Mars did not originate from impacts as was the case for lunar and asteroid regoliths. Recently the Mars rover Spirit, taking microscopic images of the pristine Martian surface, has demonstrated that the size of the soil particles is of about several tens of microns (http://marsrovers.jpl.nasa.gov/gallery/press/spirit/20040116a.html).

Figure 1. Microphotographs of lunar regolith particles at different scales [1].

The planetary regoliths and simulant regolith-like surfaces demonstrate brightness and polarimetric effects at small phase angles α (i.e., near opposition); the former is sometimes referred to as the opposition brightness surge and the latter is often called the negative polarization branch or polarization opposition effect. The opposition brightness surge was first described for the Moon as a phenomenon by Barabashev [3]. The negative polarization branch at small phase angles was discovered by Lyot [4] for the Moon, Mars, asteroids, and laboratory samples. Both these effects can be seen in Figure 2, using lunar data obtained in [4-6]. At the range of phase angles 0-10°, the amplitude of the lunar opposition effect, which we characterize here as the ratio $I(0°)/I(10°)$, is about 1.6. The negative polarization is almost symmetric with the depth $|P_{min}|$ near 1% and the inversion angle α_{min} of approximately 23°.

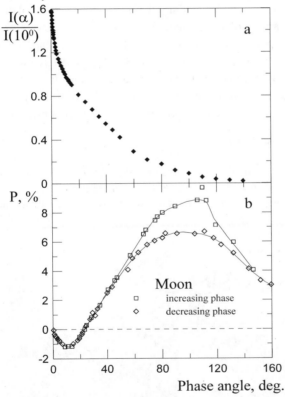

Figure 2. Phase curves of brightness (a) and degree of linear polarization (b) of the Moon. The data were adopted from [4-6].

Although Mars has a thin atmosphere we traditionally consider this celestial body as atmosphereless, implying that optical characteristics of the Martian surface can be retrieved by removing the atmosphere influence.

Photometric and polarimetric data obtained from telescopic observations and laboratory measurements at small phase angles have been accumulating for many years; however, the interpretation of these effects, especially of the polarization branch, was not satisfactory until recently [7-20] when the coherent backscattering (interference) mechanism was developed to explain both effects. Since that time, additional experimental [21-23] and theoretical [18, 24-28] data were obtained, showing that the effects are more complex than initially considered. For example, negative polarization branches with two minima were discovered for some Jupiter satellites and bright asteroids [29,30], and unusual behavior of the opposition spike amplitude with albedo for the Moon, asteroids, and Jupiter's satellite Europa were found [31-33].

In this chapter we overview (1) the progress in telescopic and spacecraft observations of the opposition effects of some atmosphereless celestial bodies and (2) the problems in understanding the photometric and polarimetric properties of planetary regoliths at small phase angles. Since several reviews of photometric and polarimetric observations of the Moon, Mars, asteroids, and planetary satellites have been published [34-36], we focus on findings related predominantly to imaging photometry and polarimetry of planetary surfaces with spacecraft and telescopic techniques. We consider also recent results concerning photometric and polarimetric observations of minor planets.

2.   The Moon, Earth, and Mars

Among a great number of the lunar images taken by the UVVIS and NIR Clementine cameras, a few images contain the point of zero phase angle. This gives a unique opportunity to study the opposition effect of the lunar surface. This effect is seen in the images as a diffuse brightened spot around the zero-phase-angle point (opposition surge).

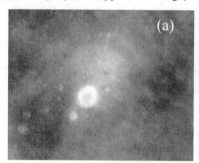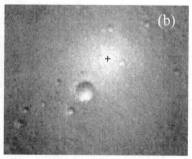

Figure 3. (a) An albedo image of a portion of Sinus Medii acquired at near zero phase angle. The brightness opposition surge cannot be distinguished from the albedo variation. (b) A phase-ratio image for the same site makes the surge clearly visible. The center of the surge (zero-phase angle point) is denoted with the black cross.

As an example, Figure 3a shows the UVVIS frame LUA1983J.168 (415 nm filter). The frame is centered at 1.40°N 1.30°E and includes a portion of

Sinus Medii with crater Blagg (the largest crater near the frame center). The phase angle spans to 2° at the edge of the frame. The contrast of the opposition surge is comparatively low for the dark lunar surface in blue light and may be swamped by albedo variations.

Opposition surges also may be present over terrestrial landscapes. Figure 4 shows a photograph taken by one of the authors (YGS) while traveling via airplane over Arizona.

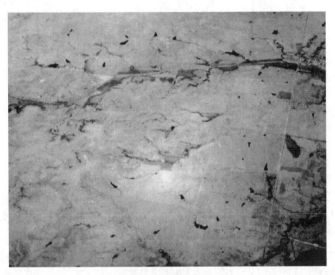

Figure 4. An image of the opposition surge for a region of Arizona.

To retrieve the phase dependence of brightness from lunar images, we make phase ratio images using data obtained at small and comparatively large phase angles for the same sites [6]. This partially compensates for albedo variations and allows the determination of phase curves. Figure 3b shows the phase ratio image $I(\sim 0°)/I(26°)$ for the area shown in Figure 3a. The point of zero phase angle (or the point of spacecraft shadow) is shown with the small cross. The phase angle α in the image is proportional to the distance from the cross. We neglect small ($< 2°$) variations of the phase angle over the larger-phase-angle image, as the estimated error introduced is on the order of the noise level in the ratio image. Using this approach, the average phase curves of several lunar sites has been calculated with the five filters of the UVVIS camera, $\lambda = 0.42$, 0.75, 0.90, 0.95, and 1.00 μm [6]. The results of this examination are unexpected: the phase curves corresponding to different wavelengths are very similar in the phase-angle range $0 - 2°$ [6]. Similar results were obtained for data from the NIR Clementine camera. We consider images taken with the filters 1.10, 1.25, 1.50, and 2.00 μm for several mare sites close to the lunar nearside center (orbits 168 and 300). The phase curves presented below were obtained for three adjacent lunar sites, two mare sites and one highland site. The calculated phase curves are normalized at 1°.

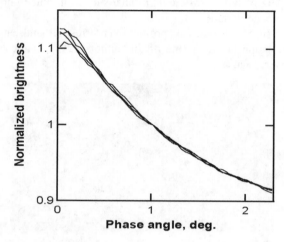

Figure 5. Phase curves of brightness of a lunar mare site obtained from NIR camera in four spectral bands 1.10, 1.25, 1.50, and 2.00 μm [31].

In Figure 5 we show phase curves for one mare site at the four NIR spectral bands. No differences between these curves are observed. This is in agreement with our results obtained from the UVVIS camera. Figure 6 shows that the phase curves at angles 0 – 1.5° are slightly steeper for the highland than for the mare sites (the data for mare sites were averaged). The differences are small, but reproducible.

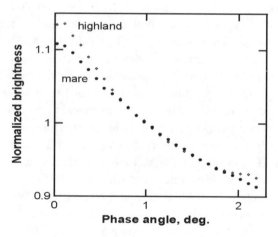

Figure 6. Phase curves of brightness of the lunar mare and highlands at 1 μm [31].

At phase angles < 0.2°, the surge becomes rounded as the phase angle approaches zero. This is due in part to the angular extent of the Sun, as the opposition spikes also depend on light-source parameters.

The shadowing effect, single-particle scattering, incoherent multiple scattering between particles, and coherent enhancement of backscattering are important factors in determining the light scattering from particulate surfaces near opposition. Theories of shadowing and incoherent multiple scattering predict a decrease of the phase-angle slope with albedo growth. In contrast, coherent backscattering models predict an increase of the slope with increasing albedo. The mean mare and highland albedos are 8 % and 16 %, respectively, at 1 μm. Applying these considerations to the observations, we might deduce from the differences in the plots of Figure 6 that the effect is the result of coherent backscattering. Its amplitude is small, because the lunar regolith is rather dark, and there is additional smoothing of the spike by the solar disk.

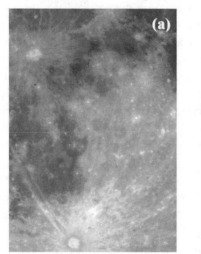 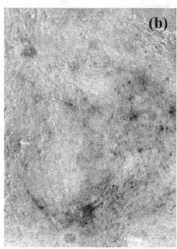

Figure 7. Albedo (a) and $|P_{min}|$ (b) images of the lunar nearside [39]. Darker shades in the right panel correspond to lower values of $|P_{min}|$.

The lunar surface albedo also depends on wavelength, and in the visual and NIR spectral range, the larger the wavelength, the higher the albedo. The lunar albedo in the visible is approximately half that in the NIR, and mare albedos are approximately half those of the highlands. In addition, we expect to observe an inherent wavelength dependence of the opposition spike parameters (e.g., the spike width) characterized by the ratio l/λ, where λ and l are the wavelength and the characteristic interference pathlength in the media, respectively. However, we found in Figure 5, for example, the unexpected absence of wavelength dependence of the phase curves. This contradicts the coherent backscattering mechanism, and the coherent backscattering interpretation of the difference between the mare and highland opposition spikes may have been premature. Surge differences also can be the result of morphological differences in the regoliths; e.g., a more complicated structure (higher porosity) of the highland regolith caused by the difference between particle sizes of mare and highland regoliths. Owing to higher fragility of plagioclases [2] that is the main constituent of the highland regolith, the lunar highlands include a comparatively higher amount of dust component. The morphological differences result in differences in the phase curves. This example illustrates some of the complications of interpreting photometric data.

There are many works devoted to studies of polarimetric properties of the Moon at small phase angles, e.g., [37-40]. However, almost all these studies were carried out for small selected sites. The first attempt to map the parameter $|P_{min}|$ of the Moon was undertaken in [39]. An example of such a mapping is given in Figure 7. The panel (a) shows an albedo image of the central portion of the lunar nearside that shows the craters Tycho (bottom) and Copernicus (upper left). The panel (b) displays a map of $|P_{min}|$ for the same region. Many interesting details are seen in this map. All bright (young) craters show up, having lower values of $|P_{min}|$ than surrounding areas. Mare regions, e.g., Sinus Medii, also have smaller $|P_{min}|$ than regions with intermediate albedo. Thus there is a horseshoe relationship where $|P_{min}|$ has relatively small values at large and small albedo values, and a maximum at intermediate values. This also was found using discrete polarimetry [38,40]. Qualitatively, this dependence was confirmed with laboratory measurements of different particulate surfaces [38]. The total variation of the parameter $|P_{min}|$ for the Moon is about 0.5% – 1.5% [38,40].

Like the Moon, Mars reveals the brightness opposition effect [41,42] and the negative polarization branch [43]. Using data obtained with the spacecraft Phobos-2 we estimate [44] the amplitude of the spike with the brightness ratio $I(0°)/I(16°)$ to be approximately 1.6 at $\lambda = 0.45$ μm for the Pavonis Mons region. Note that this ratio is typical for the lunar surface. Integral polarimetry shows the Martian surface to have comparatively large values of α_{inv} and α_{min}, 24° and 11.5°, respectively; the parameter $|P_{min}| \approx 1.1\%$. All the polarimetric values are spectrally and seasonally dependent. They also depend on Martian atmosphere conditions, however, during the periods of high transparency the atmosphere effects are usually negligible at $\lambda > 0.5$ μm.

During the close approach to the Earth in 2003 observations of Mars were carried out by NASA/ESA Hubble Space Telescope (HST) [45,46]. This includes polarimetric observations. Five series of images of Mars at phase angles about 6°, 8°, 10°, 13° and 16° were taken with polarization filters. The moments for observations were chosen so that Mars was turned to the observer by the same side (disk center at 19°S 20-35°E) containing Valles Marineris and contrasting albedo details of Terra Meridiani and surroundings. Images were taken with the ACS high-resolution channel. Each series consists of four sets of images taken with wide-band spectral filters ($\lambda \approx 250$ nm, 330, nm, 435 nm, and 814 nm). Each set contains three images taken with three polarization filters that differ from the others by 60° orientations of the polarization axis.

We have found that typical values of $|P_{min}|$ are near 1%. Minimal values of polarization are observed for the south-polar cap and some clouds (< 0.1%). There are portions of clouds with polarization as high as 2.5%. The polarization degree is also high at the Martian limb.

3.   Asteroids

Since its discovery for asteroid 20 Massalia by Gehrels [47], the opposition surge in the asteroid magnitude phase curves has been observed many times. At first the value of the opposition effect was considered to be very similar among different types of asteroids. Later a very sharp and narrow opposition surge was discovered for high-albedo asteroids [48] and an almost linear phase curve was seen in some low-albedo asteroids [49]. The increasing number of observations has yielded different phase-curve behavior for asteroids of different composition. The amplitude of the opposition effect depends on asteroid albedo in a non-monotonic way with the maximum for moderate-albedo asteroids, and a decrease both for low- and high-albedo asteroids [32] (see Figure 8). In this case we define the amplitude M as the difference between the magnitude measured at phase angle of 0.3° and the magnitude obtained as a linear extrapolation of the linear part of the asteroid phase curves. We note that the analogous diagram of backscatter intensity enhancement (shown in Figure 8a of [36]), which additionally includes data for planetary satellites, demonstrates similar, but inverted horseshoe behavior, where the horseshoe opening is on the top.

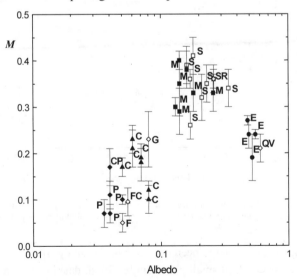

Figure 8. The amplitude of the opposition effect as a function of albedo [32]. Letters indicate the asteroid compositional type.

Low-albedo objects are diverse in opposition behavior. Actually, asteroids of the primitive P and F types, located mainly in the outer asteroid belt, show linear phase curves down to phase angles of 2° [49,50]. The first tentative phase curves for other dark and distant Solar system objects, like Centaurs and trans-Neptunian (Kuiper Belt) objects, also show linear phase dependence at small phase angles. However their phase curve slope is at least twice as large as that of low-albedo asteroids [51]. For both types of dark objects there is an indication of a very narrow opposition surge at sub-degree phase angles.

Laboratory photometric and polarimetric measurements of carbon soot smoked on a flat substrate (albedo about 2%) reveal a small brightness spike and negative polarization at phase angles < 2° that definitely demonstrates the importance of multiple scattering in such dark fluffy surfaces [14,21]. Multiple scattering can exist in low-albedo surfaces if the scattering indicatrix of the particles is strongly oblong in the forward direction. In the regoliths of Kuiper-Belt objects, it can be related to the very fluffy ("fairy castles") structure of the objects as well as to relatively low refractive indices of the regolith material [51]. The latter is characteristic of organic matter with high porosity that can be the result of space weathering processes.

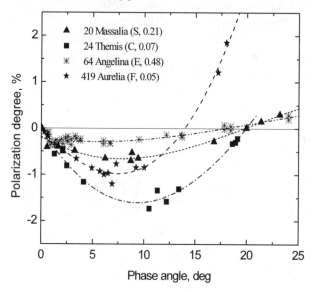

Figure 9. Phase curves of the degree of linear polarization for asteroids (types and albedos are given in the parentheses). Data are taken from [36,50,52] and Asteroid Polarimetry Database (http://pdssbn.astro.umd.edu/sbnhtml).

Recently, great efforts have been invested in the study of the polarization phase curves of asteroids at small phase angles. First, data have been obtained for a few asteroids of different surface composition. The detection of the negative polarization peak of about 0.4% in the V-band (green light) for high-albedo E-type asteroid 64 Angelina has been reported [30]. Polarimetric observations of 20 Massalia have shown an absence of sharp features near opposition for this moderate-albedo S-type asteroid [52].

An interesting result was obtained for the dark F-type asteroid 419 Aurelia [50]. A decrease of both the depth and width of the negative polarization branch was found for this asteroid compared with those of other dark asteroids (see Figure 9). This is the first evidence of negative polarization "saturation" previously observed for very dark laboratory samples [38]. The effect implies that $|P_{min}|$ and α_{inv} increase with decreasing albedo only up to some albedo value and then are constant or even decrease with further albedo decrease. The first polarimetry data of the trans-Neptunian object Ixion (2001 KX$_{76}$) revealed significant negative polarization at phase angles $< 1.3°$ [53].

4.   Europa

Thanks to the Galileo spacecraft, many photometrically valuable images were acquired for the Jupiter satellite Europa [54]. The surface of Europa consists mainly of water ice and can contain noticeable amount of salts, silicates, and organic matter. The range of albedo variation for Europa is very large, exceeding a factor 5. Europa's opposition spike generally was found to be very intense at extremely small phase angles ($< 0.1°$), but varies significantly with albedo, wavelength, and terrain [33,54].

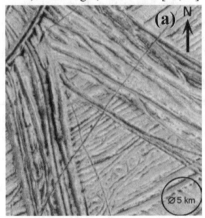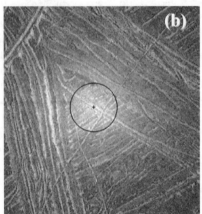

Figure 10. Albedo (a) and phase-ratio $I(\sim0°)/I(3°)$ (b) images of Europa's surface obtained with the Galileo camera using a 0.56 µm filter [33].

During one of its encounters with Europa, the Galileo spacecraft obtained four successive high-resolution images of an area located at 24°N 170°W [54] (frame numbers 440949111, 440949114, 440949121, 440949132). The first image 440949111 in this series contains the spacecraft shadow point; that is, the phase angle changes over the image from 0° to ~0.3°. In other images the phase angle progressively increases: 0.4°, 1.4°, and 2.9°. The first (shown in Figure 10a) and last images in the series were used to construct phase curves.

The resulting phase-ratio image $I(\sim0)/I(3°)$ is presented in Figure 10b. The bright diffuse spot near the center is the "opposition surge". Comparison of the albedo and the phase-ratio images (cf. Figures 10a and 10b) clearly shows that the ratio is systematically higher for increasingly darker surfaces. The phase-ratio image was used to calculate the phase function for the small region of Europa and is shown in Figure 11. The data show an opposition spike with an amplitude of nearly 20% in the phase angle range $0 - 0.2°$.

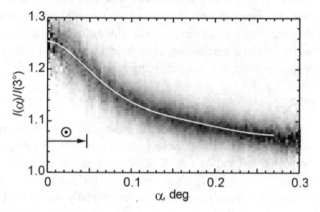

Figure 11. Phase curve of the Europa surface calculated with the image shown in Figure 10b [33].

The angular size of the solar disk for the Jupiter-Sun distance is shown with a circle in Figure 10b and with an arrow in Figure 11. The observed phase function shown in Figure 11 continues to increase well within the solar disk boundaries. This provides evidence for an extremely sharp true phase function (i.e. the phase function after de-convolution of the finite angular size of the solar disk) at $\alpha < 0.05°$. Figure 10 shows the phase ratio $I(0.1°)/I(3°)$ as a function of $I(3°)$, i.e., surface albedo. These data reveal a decrease of the ratio with increasing albedo. Note that the same effect is observed for bright asteroids (see Figure 10). The dependence shown in Figure 12 is steep for dark terrains and almost absent for bright terrains.

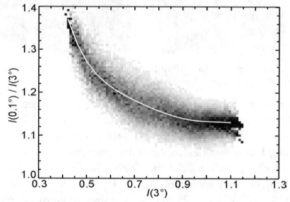

Figure 12. The phase ratio $I(0.1°)/I(3°)$ as a function of albedo for Europa [33].

In general, a decrease of the opposition spike amplitude with increasing albedo contradicts the coherent backscattering mechanism that predicts an increase of the spike amplitude with increasing albedo. One possible explanation of this contradiction is that with the albedo increase, the width of the spike becomes significantly smaller than 0.1° and the phase ratio $I(0.1°)/I(3°)$ is not sensitive to this spike.

Another interesting feature of Europa's light scattering near opposition is the double-minima shape of the negative polarization branch (see Figure 13). This effect recently was seen in telescopic observations of Galilean satellites and some bright asteroids [29,30,36].

This double-minima feature has been measured in laboratory samples that were investigated using the laboratory small-angle photometer/polarimeter located at Kharkov National University. The setup allows measurements of phase curves of reflectance and degree of linear polarization of scattered light, providing studies of surfaces with complicated structure including powders in the phase angle range 0.2° – 17° [14,21,55,56]. A spectral band centered at λ_{red} = 0.65 μm is selected. Measurements are carried out at the standard illumination/observation geometry, when samples are viewed with the detector along the sample normal. The rotated light source arm changes the incidence angle that is the phase angle α.

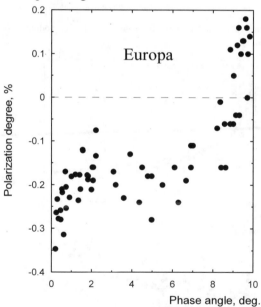

Figure 13. Phase curve of the degree of linear polarization for Europa in V band [36].

Among the homogeneous materials we examined, only one sample had a polarization response with two negative minima. Figure 14 shows photometric and polarimetric phase curves for lycopodium spores. The powder consists of approximately uniform particles with average diameter of about 20 μm. The particles are fairly bright, not transparent, and have roundish shapes.

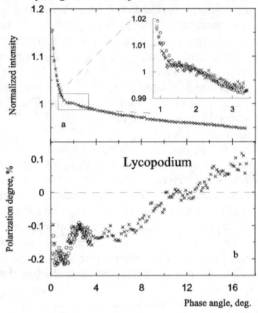

Figure 14. Photometric (a) and polarimetric (b) phase curves for lycopodium powder [14].

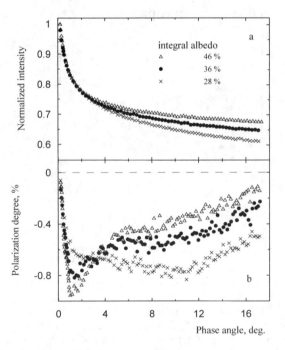

Figure 15. Photometric (a) and polarimetric (b) phase curves for composite samples [14].

Although there is a strong resemblance of the polarimetric branches in Figures 13 and 14, the likelihood of Europa's surface being covered with such spores is extremely small. This resemblance may reflect only similarities in the surface structure and mechanisms that form the minima. For instance, we could suggest that the small-phase-angle minimum corresponds to the coherent backscattering mechanism and the large-phase-angle minimum is the negative polarization from single particle scattering [21, 23]. Like the Europa surface, the lycopodium sample demonstrates a very conspicuous brightness spike. We note also that the sample has a slight feature of the photometric curve at the phase angle near 2° shown in the insertion in Figure 14.

Another point in modeling the double-minima negative polarization branches is the effect of heterogeneity, or composite surfaces [14]; i.e., one sample composed of two sub-samples. We can consider a host sample composed of a mixture of soot and chalk, for instance, with an albedo of approximately 20%. This sub-sample has a wide and deep negative polarization branch. The other sub-sample consists of a thick MgO layer on a small piece of cardboard that has a narrow branch of negative polarization. Figure 15 shows phase angle measurements of brightness and polarization for composite samples consisting of three different-size sub-samples placed at the central location of the host. As can be seen, a superposition of the narrow and wide branches is produced, in which two local minima in polarization phase dependence are observed for the "mixture" with albedo 36%. The corresponding photometric curves of Figure 15 are very similar and cannot be used to distinguish the surfaces. The measurements suggest that the surfaces of the Galilean satellite Europa may contain regions having different surface characteristics; e.g., relatively dark materials having wide branches of negative polarization and bright materials with a MgO-like behavior of polarization. The first type would be, e.g., salt and/or silicate components or dirty ices; the second one could be fresh hoar-frost-like deposits [14].

5. Conclusion and future work

In summary we emphasize the following:
1. It was found that the opposition phase curves of lunar sites are almost the same for wavelengths in the range $1 - 2$ μm, although the wavelength variation is accompanied by a significant albedo change. Meanwhile, the lunar highland surfaces demonstrate slightly steeper polarization phase curves than those of mare surfaces.
2. The phase ratio $I(0.1°)/I(3°)$ of the Europa surface shows a rapid decrease with increasing albedo. Qualitatively, the same effect is observed for bright asteroids. This appears to be at odds with the coherent-backscattering mechanism. It is possible, however, that with the albedo increase, the spike width becomes significantly smaller than $0.1°$ (in the case of Europa) and the phase ratio $I(0.1°)/I(3°)$ does not sense this spike.
3. Imaging photometry and polarimetry of the Moon, Mars, and other atmosphereless celestial bodies with spacecraft are very prospective and

informative tools to study planetary regoliths; however, great care must be taken when interpreting the results.

4. Simultaneous photometry and polarimetry of extremely bright and very dark asteroids seems to be fairly opportune. Bright asteroids as well as Galilean satellites can produce double-minima negative polarization branches and narrow opposition spikes. On the other hand the dark asteroids and trans-Neptunian objects demonstrate different opposition-phase-angle behavior of brightness and polarization that can reflect differences in the surface structure of these bodies [51].

6. Acknowledgments

The authors thank Vera Rosenbush for useful discussions. This work was partially supported by INTAS grant # 2000-0792, the TechBase Program on Chemical and Biological Defense, and by the Battlefield Environment Directorate under the auspices of the U.S. Army Research Office Scientific Services Program administered by Battelle (Delivery Order 291, Contract. No. DAAD19-02-D-0001).

References

1. O. Rode, A. Ivanov, M. Nazarov, A. Cimbalnikova, K. Jurek, and V. Hejl, "Atlas of photomicrographs of the surface structures of lunar regolith particles" (Academia, Prague, 1979).

2. D. McKay, G. Heiken, A. Basu, G. Blanford, S. Simon, R. Reedy, B. French, and J. Papike, "Lunar source-book" p. 285 (Cambridge Univ. Press, NY. 1991).

3. N. P. Barabashev. Astron. Nachr. **217**, 445 (1922).

4. B. Lyot, Ann. Obs. Meudon. **8**, 1 (1929).

5. G. Rougier, Ann. Obs. Strasburg **2**, (1933).

6. Yu. Shkuratov, M. Kreslavsky, A. Ovcharenko, D. Stankevich, E. Zubko, C. Pieters, and G. Arnold, Icarus **141**, 132 (1999).

7. Yu. Shkuratov, Astronomicheskii Tsircular **1400**, 3 (Sternberg Astron. Institute Moscow 1985).

8. Yu. G. Shkuratov, Kinematika i Fisika Nebesnykh Tel. **4**, 33 (1988).

9. Yu. G. Shkuratov, Astronomicheskii Vestnik **23**, 176 (1989).

10. Yu. G. Shkuratov, Solar System Res. **25**, 134 (1991).

11. K. Muinonen, "Proc. 1989 URSI Electromagnetic Theory Sympos". p.428 (Stockholm, Sweden 1989).

12. B. Hapke, Icarus **88**, 407 (1990).

13. M. I. Mishchenko, Astrophys. Space Sci. **194**, 327 (1992).

14. Yu. Shkuratov, K. Muinonen, E. Bowell, K. Lumme, J. Peltoniemi, M. Kreslavsky, D. Stankevich, V. Tishkovetz, N. Opanasenko, and L. Melkumova, The Earth, Moon, and Planets **65**, 201 (1994).

15. M.I. Mishchenko, Phys. Rev. **B44**, 12,579 (1991).

16. M.I. Mishchenko, J. Opt. Soc. Am. **A9**, 978 (1992).

17. M.I. Mishchenko, J.-M. Luck, and T. M. Nieuwenhuizen, J. Opt. Soc. Am. **A17**, 888 (2000).

18. M. Mishchenko, V. Tishkovets, and P. Litvinov, "Optics of Cosmic Dust", ed. by G. Videen and M. Kocifaj. p. 239 (Kluwer Academic Publishers, Dordrecht, 2002).

19. K. Muinonen, A. Sihvola, I. Lindell, and K. Lumme, J. Opt. Soc. Am. A8, 477 (1991).

20. K. Muinonen, "Asteroids, Comets and Meteors", ed. by A. Milani, M. Di Martino, and A. Cellino, p. 271. (Kluwer, Dordrecht, 1994).

21. Yu. Shkuratov, A. Ovcharenko, E. Zubko, O. Miloslavskaya, K. Muinonen, J. Piironen, R. Nelson, W. Smythe, V. Rosenbush, and P. Helfenstein, The opposition effect and negative polarization of structural analogs of planetary regoliths. Icarus 159, 396 (2002).

22. Yu. Shkuratov, A. Ovcharenko, In: "Optics of cosmic dust" / Eds. G. Videen and M. Kocifaj. p 225. NATO Science Series. (Kluwer Academic Publishers, London. 2002).

23. Yu. Shkuratov, A. Ovcharenko, E. Zubko, H. Volten, O. Muñoz, G. Videen, J. Quant. Spectrosc. Radiative Transfer (2004) Submitted.

24. G. Videen, Appl. Opt. 41, 5115 (2002).

25. G. Videen, J. Quant. Spectrosc. Radiative Transfer. 79-80, 1103 (2003).

26. P.V. Litvinov, V.P. Tishkovets, K. Muinonen, and G. Videen, in: "Wave scattering in complex media: from theory to Applications," Eds. B. van Tiggelen and S. Skepetrov. p.567. (Kluwer Academic Publishers, Dordrecht, 2003).

27. V. P. Tishkovets, P. V. Litvinov, and M. V. Lyubchenko, J. Quant. Spectrosc. Radiat. Transfer 72, 803 (2002).

28. V. P. Tishkovets, P. V. Litvinov, and S. V. Tishkovets, Opt. Spectrosc. 93, 899 (2002).

29. V. Rosenbush, V. Avramchuk, and A. Rosenbush, Astrophys. J. 487, 402 (1997).

30. V. Rosenbush, N. Kiselev, K. Jockers, V. Korokhin, N. Shakhovskoy, and Yu. Efimov, Kinematics and physics of celestial bodies. Suppl. 3, 227 (2000).

31. V. Omelchenko, V. Kaydash, and Yu. Shkuratov, NATO Advanced Study Institute on "Photopolarimetry in Remote Sensing" Yalta, Ukraine, 20 September - 3 October 2003, p. 69.

32. I.N. Belskaya and V.G. Shevchenko, Icarus 147, 94 (2000).

33. M. Kreslavsky, P. Helfenstein, and Yu. Shkuratov, Lunar and Planet. Sci. 31. Abstract #1142. LPI Houston. (2000).

34. Yu.G. Shkuratov, Solar System Res. 28, 3 (1994).

35. Yu.G. Shkuratov, Solar System Res. 28, 23 (1994).

36. V.K. Rosenbush, N.N. Kiselev, V.V. Avramchuk, and M.I. Mishchenko, "Optics of cosmic dust," eds. G. Videen and M. Kocifaj, p. 191. NATO Science Series. (Kluwer Academic Publishers, London. 2002).

37. A. Dollfus and E. Bowell. Astron. Astrophys. 10, 29 (1971).

38. Yu. G. Shkuratov, N. V. Opanasenko, and M. A. Kreslavsky, Icarus 95, 283 (1992).

39. N.V. Opanasenko, A.A. Dolukhanyan, Yu.G. Shkuratov, D.G. Stankevich, M.A. Kreslavskii, and V.G. Parusimov, Solar System Research. 28, 98 (1994).

40. N.V. Opanasenko and Yu.G. Shkuratov, Solar System Research 28, 233 (1994).

41. B. O'Leary and D. Rea, Icarus 9, 405 (1968).

42. T. Thorpe, Icarus 49, 398 (1982).

43. A. Dollfus and J. Focas, Ann. Astrophys. 2, 63 (1969).

44. Yu. Shkuratov, D. Stankevich, A. Ovcharenko, L. Ksanfomaliti, E. Petrova, and G. Arnold, Solar Syst. Res. 32, 90 (1998).

45. Yu. Shkuratov, M. Kreslavsky, V. Kaydash, N. Opanasenko, G. Videen, J. Bell, M. Wolff, M. Hubbard, K. Noll, and A. Lubenow, Lunar Planet. Sci. Conf. 35th. Abstract # 1435 LPI Houston. (2004).

46. J. Bell, M. Wolff, K. Noll, A. Lubenow, E. Noe-Dobrea, M. Hubbard, R. Morris, G. Videen, and Yu. Shkuratov, Eos Trans. AGU, 84(46), Fall Meet. Suppl., Abstract P12C-01 (2003).

47. T. Gehrels, Astrophys. J. **123**, 331 (1956).

48. A.W. Harris, J.W. Young, L. Contreiras, T. Dockweiler, L. Belkora, H. Salo, W.D. Harris, E. Bowell, M. Poutanen, R.P. Binzel, D.J. Tholen, and S. Wang, Icarus **81**, 365 (1989).

49. V.G. Shevchenko, V.G. Chiorny, A.V. Kalashnikov, Yu.N. Krugly, R.A. Mohamed, and F.P. Velichko, Astron. Astrophys. Suppl. Ser. **115**, 475 (1996).

50. I.N. Belskaya, V.G. Shevchenko, Yu.S. Efimov, N.M. Shakhovskoy, Yu.G. Shkuratov, N.M. Gaftonyuk, R. Gil-Hutton, Yu.N. Krugly, and V.G. Chiorny, "Proceed. of Asteroids, Comets, Meteors 2002", p. 489 (Berlin, ESA Publishing Division, 2002).

51. I.N. Belskaya, M.A. Barucci, and Yu.G. Shkuratov, Earth, Moon and Planets **92**, N1-4, (2003).

52. I.N. Belskaya, V.G. Shevchenko, N.N. Kiselev, Yu.N. Krugly, N.M. Shakhovskoy, Yu.S. Efimov, N.M. Gaftonyuk, A. Cellino, and R. Gil-Hutton, Icarus, **160**, (2004) In press.

53. H. Boehnhardt, S. Bagnulo, K. Muinonen, A. Barucci, L. Kolokolova, E. Dotto, and G.P. Tozzi, Astron. Astrophys. Letters, (2003). submitted

54. P. Helfenstein and 21 colleagues, Icarus. **135**, 41 (1998).

55. Yu.G. Shkuratov, A. A. Ovcharenko, D.G. Stankevich, and V.V. Korokhin, Solar System Res. **31**, 56 (1997).

56. A. Ovcharenko, Yu. Shkuratov, and R. Nelson, Solar System Res. **35**, 291 (2001).

Yuriy Shkuratov and colleagues at NATO ASI reception.

BACKSCATTERING FROM PARTICLES NEAR PLANAR SURFACES

GORDEN VIDEEN[1] AND KARRI MUINONEN[2]
[1]*Army Research Laboratory AMSRL-CI-EE,*
2800 Powder Mill Rd., Adelphi, MD 20783-1197
[2]*University of Helsinki, Observatory, P.O. Box 14*
FIN-00014, Helsinki, Finland

Abstract. The light scattered by a small particle near a planar interface may be expressed to high accuracy as the superposition of the contributions of four weighted rays. We provide experimental evidence that demonstrates the model may be used to predict polarization of scattered light from larger irregular particles near planar surfaces. This system is of interest because the two reciprocal-ray components, whose interference contributes to the narrow, asymmetric polarization opposition effect, do not decrease significantly relative to the other two ray components when the separation distance between the particle and the planar interface increases. As a result, we demonstrate that for this particular scattering system the wide, parabolically shaped negative polarization branch also is the result of the interference of these reciprocal rays.

1. Introduction

A significant percentage of naturally occurring particle systems can be considered to be agglomerates; i.e., particles located in close proximity to other particles. When one or more of these particles is significantly smaller than a larger host particle, it may be considered as a microstructure or contaminant of the host particle [1]. Naturally occurring dust particles (shown in Figure 1) may constitute such systems. If the local radius of curvature of the host particle tends toward infinity at the position of the contaminant, we may consider the scattering from a contaminant located on a planar interface. Such

G. Videen et al. (eds.), Photopolarimetry in Remote Sensing, 209-220.

systems comprise dew or dust on surfaces like windows or computer wafers, ice crystals on various surfaces, etc.

The polarization state of light scattered in the near backward direction from complex systems may contain a narrow, asymmetric, negatively polarized spike within a few degrees of the backscatter, sometimes referred to as the polarization opposition effect. It may also contain a wide, parabolically shaped negative polarization feature that may stretch out to a few tens of degrees, sometimes referred to as the negative polarization branch. These two features are realizations of what generally may be called the negative polarization surge. Polarization curves containing some or all of these features may be measured from asteroids, satellites, planets, comets, and Saturn's rings [2-5], as well as terrestrial and man-made simulants [6,7].

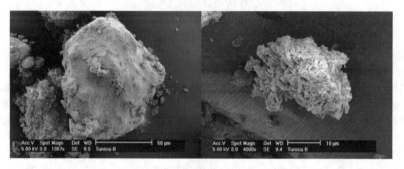

Figure 1. Saharan dust particles showing agglomeration of small particles on the host (photos courtesy of Timo Nousiainen).

A modeling effort has demonstrated that the polarization opposition effect can be reproduced using the enhanced backscatter mechanism, namely the constructive interference of reciprocal rays. Various models have been developed that reproduce this feature. For instance, a vector theory of coherent backscattering for a semi-infinite medium of Rayleigh scatterers is able to reproduce the structure of the polarization feature when the cyclical (reciprocal) terms are included [8-12]. In addition, various ray-tracing methodologies from different types of scattering systems also are able to produce the features of the polarization opposition effect when the constructive interference mechanism is included [13-21]. Several review articles address this topic [2, 7, 12, 18].

2.   Small particles near planar surfaces

In this chapter we present the backscatter polarization signals resulting from small particles located in the proximity of a planar interface. This system was treated by Lindell *et al.* [22] and Muinonen *et al.* [23] using Exact Image Theory (EIT). Simultaneously, and in the same journal issue, Videen [24, 25] provided a general scattering solution for a sphere on a planar interface and later showed [26] that it converged to the EIT results when the sphere becomes Rayleigh. This system is especially interesting because a ray-treatment converges to the exact solution to within a few percent [23, 26, 27].

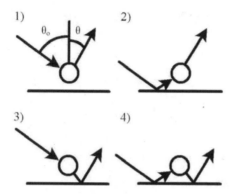

Figure 2. Illustration of the four paths a ray may take to pass from the source to the detector having interacted with the sphere one time.

The light scattered by a small Rayleigh sphere near a planar interface may be solved as the superposition of four components undergoing the four possible ray paths from the source to the detector. These four paths are shown in Figure 2: The incident ray may 1) be scattered by the sphere and then continue to the detector, 2) reflect off the planar interface, be scattered by the sphere, and then continue to the detector, 3) be scattered by the sphere, reflect off the planar interface, and then continue to the detector, and 4) reflect off the planar interface, be scattered by the sphere, reflect off the planar interface, and then continue to the detector. Mathematically, the electric field amplitudes can be written as

$$\begin{pmatrix} E_\theta^{sca} \\ E_\phi^{sca} \end{pmatrix} = \begin{bmatrix} S_2 & S_3 \\ S_4 & S_1 \end{bmatrix} \begin{pmatrix} E_{TM}^{inc} \\ E_{TE}^{inc} \end{pmatrix} \tag{1}$$

where E_{TE}^{inc} and E_{TM}^{inc} are the magnitudes of the TE and TM incident electric fields, and E_θ^{sca} and E_ϕ^{sca} are the $\hat{\theta}$ and $\hat{\phi}$ components of the scattered electric field. The incident angle θ_o and scattering angle θ are measured from the z-axis that is perpendicular to the planar interface. The elements of the scattering amplitude matrix $\mathbf{S}$, which include the sum of all four paths, can be written as [22-27]

$$S_1 = A\mathrm{i} \left[1 + R_{TE}(\theta_o)\exp(2ikd\cos\theta_o)\right]\left[1 + R_{TE}(\theta)\exp(2ikd\cos\theta)\right]\cos\varphi \tag{2}$$

$S_2 = A\mathrm{i} \{[1 - R_{TM}(\theta_o)\exp(2ikd\cos\theta_o)][1 - R_{TM}(\theta)\exp(2ikd\cos\theta)]\sin\theta_o \sin\varphi$

$\quad + [1 + R_{TM}(\theta_o)\exp(2ikd\cos\theta_o)][1 + R_{TM}(\theta)\exp(2ikd\cos\theta)]\cos\theta_o \cos\theta \cos\varphi\}$

$S_3 = A\mathrm{i} [1 + R_{TE}(\theta_o)\exp(2ikd\cos\theta_o)][1 + R_{TM}(\theta)\exp(2ikd\cos\theta)]\cos\theta \sin\varphi$

$S_4 = -A\mathrm{i} [1 + R_{TM}(\theta_o)\exp(2ikd\cos\theta_o)][1 + R_{TE}(\theta)\exp(2ikd\cos\theta)]\sin\varphi \cos\theta_o$

where the Fresnel coefficients are given as

$$R_{TE}(\alpha) = \frac{\cos\alpha - m_{sub}^2[m_{sub}^2 - \sin^2\alpha]^{1/2}}{\cos\alpha + m_{sub}^2[m_{sub}^2 - \sin^2\alpha]^{1/2}} \tag{3}$$

$$R_{TM}(\alpha) = -\frac{m_{sub}^2\cos\alpha - [m_{sub}^2 - \sin^2\alpha]^{1/2}}{m_{sub}^2\cos\alpha + [m_{sub}^2 - \sin^2\alpha]^{1/2}}$$

and

$$A = \frac{\exp(ikr)}{ikr}\frac{m_{sph}^2 - 1}{m_{sph}^2 + 2}(ka)^3 \tag{4}$$

and k is the wavevector, m_{sub} and m_{sph} are the refractive indices of the substrate below the planar interface and the radius a sphere located a distance d above the planar interface, and r is the distance from the sphere to the detector.

3. Some Comparisons

Many naturally occurring particles have polarization responses that resemble those of small, Rayleigh particles, even though the particles may be much larger than the wavelength of the illuminating radiation and the total intensity of their scattered light (Mueller matrix element S_{11}) resembles that of a polydispersion of large particles [28]. Recent modeling suggests that as the particle irregularity increases, the Rayleigh scattering of microstructures plays a dominant role in determining the scattering signal [29]. Because of the size of the microstructures, the interactions between them are relatively small [30]. Bringing such a particle system close to an interface should approximate the scatter from a cluster of Rayleigh particles placed near an interface. Experimental measurements from surface contaminants [31] on optically smooth substrates reveal a structure that closely resembles that of a Rayleigh sphere placed near the planar interface [32]. These results are reproduced in Figure 3.

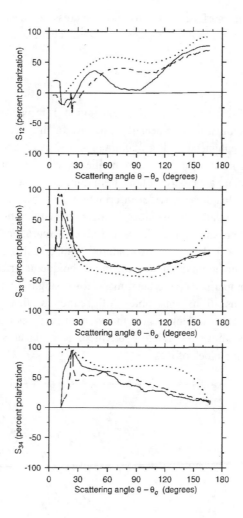

Figure 3. Normalized polarization Mueller matrix elements calculated for a small Rayleigh particle placed a distance $d = \lambda/4.7$ from a substrate ($m_{sub} = 0.5 + 3.5i$) illuminated at $\theta_o = 79°$ (dotted). Superimposed are the experimental matrix elements measured by Iafelice and Bickel [31] of a perfect (solid) and decayed (dashed) mirror surface, $\lambda = 0.4416\mu m$. Iafelice and Bickel noted the presence of defects ranging in size from 0.2 to 1.0 μm. The size range did not vary per sample, but the number density for the "perfect" surface was approximately five times less than for the "decayed" surface.

The polarization state of the scattered light from the surface contaminants shown in Figure 3 strongly resemble those of a single dipole located at the approximate center of the contaminant, even though the contaminant sizes are well outside the Rayleigh limit. It should be noted that the structures in the polarization responses can be modeled more accurately by changing the dipole height above the surface. For instance, increasing distance d slightly reproduces the deeper minimum seen in matrix element S_{12} of the perfect mirror surface.

4. Backscattering

Light scattered back toward the detector region is especially relevant for remote-sensing applications, especially LIDAR. In this detector region, the intensity may be enhanced with a maximum in the exact backscatter region. The polarization state, which is zero in the exact backscatter for random media, may contain a negative (TM) minimum near the exact-backscatter direction. These features provide information on the scattering system, specifically those related to pathlengths of rays within the medium. It is important that approximate techniques are able to reproduce these features. Figure 4 shows the polarization of scattered light calculated using EIT and calculated using the approximate ray-tracing technique described in section 2. While similar comparisons have been made for a single orientation [26], we show comparisons from ensembles of scatterers that have practical applications in remote sensing, where many small particles may be located at various distances above a randomly oriented, near-planar surface, like ice crystals formed on a block of ice.

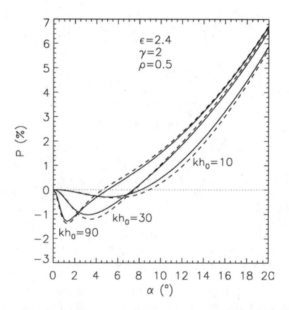

Figure 4. Exact (solid) and ray-tracing (dashed) calculations from small spheres near a planar interface: $\varepsilon = m_{sub}^2 = 2.4$; $\gamma = 2$; $\rho = 0.5$; $kh_o = 10, 30, 90$ (from right to left).

In Figure 4, the probability density for height of the small particle above the surface is assumed to be a gamma distribution with integer γ [23]:

$$p_h(h)dh = \frac{1}{(\gamma-1)!}\left(\frac{\gamma}{h_o}\right)h^{\gamma-1}\exp\left(-\frac{\gamma h}{h_o}\right)dh \tag{5}$$

where h_o is the mean height. Orientation averaging is also carried out, assuming an isotropic Gaussian probability density for slope ($t = \tan\theta$) [23]:

$$p_t(t)dtd\phi = \frac{1}{2\pi\rho^2}\exp\left(-\frac{\tan^2\theta}{2\rho^2}\right)\frac{\sin\theta}{\cos^3\theta}d\theta d\phi \tag{6}$$

The agreement between the EIT and ray-tracing methods shown in Figure 4 is excellent. It is important to stress that this agreement is not the result of the averaging, as similar agreement can be seen for single orientations [27].

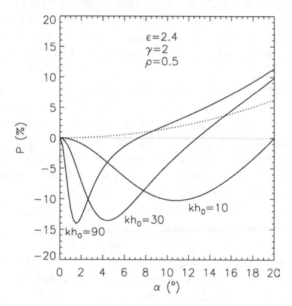

Figure 5. Polarization state of scattered light resulting from reciprocal ray components and from the other components: $\varepsilon = m_{sub}^2 = 2.4$; $\gamma = 2$; $\rho = 0.5$; $kh_o = 10, 30, 90$ (from right to left). Ray paths 1 and 4(dotted) have minimal dependence on kh_o.

The system of a small particle near a planar surface is of interest not only because the ray-tracing solution converges to the exact solution to very high accuracy, but also because the components that contribute to the coherent backscattering can be separated from the other components. Figure 5 shows the contribution of the polarization state of scattered light by the second-order reciprocal components of ray paths 2 and 3 (see Figure 2). The polarization contribution of ray paths 1 and 4 has minimal dependence on separation parameter kh_o and do not contribute to the negative polarization. For all separation parameters, the negative-polarization minimum is determined by the reciprocal rays. Comparing the minima of Figures 4 and 5, demonstrate that the effect of non-reciprocal rays 1 and 4 are to dampen the depth of the minima and to move the minima to smaller phase angles.

5. Discussion

The morphology of a particle near a planar interface is especially relevant in the consideration of constructive interference of reciprocal rays, i.e., enhanced-backscattering phenomena. One reason for this is that the amplitudes of the multiply scattered rays that undergo two reflections and contribute to this effect (ray paths 2 and 3 of Figure 2) are approximately the same as the first-order rays that are only scattered by the dipole, even when the separation parameter d is large; in contrast, for instance, to the scattering by two spherical particles whose coherent-backscattering contribution to the intensity decays as $1/d^2$.

In addition, the system of a dipole near a planar surface may also provide some insight into the nature of the wide, parabolically shaped negative polarization branch whose minimum occurs at large phase angles. There is currently some controversy concerning the mechanism for this particular scattering feature. While constructive interference of reciprocal rays explains the narrow polarization opposition effect quite adequately, it may not be appropriate to explain the wider negative polarization branch because the pathlength differences between reciprocal rays must be on the order of the wavelength of illumination. There are legitimate questions concerning the legitimacy of ray-tracing techniques in this regime. However, for the present scattering system, the ray-tracing solution converges to the exact solution to within a few percent, even for pathlengths d smaller than the wavelength. Figure 4 demonstrates that the physical mechanism for both wide and narrow polarization branches is the same constructive interference of reciprocal rays. We conclude that at least for this particular scattering system, both polarization features can be attributed to constructive interference of reciprocal rays.

6. Conclusion

We present an approximate scattering solution for a dipole located near a planar surface that converges to the exact solution to within a few percent. The solution may also be used to calculate the polarization response from an irregular particle system located near a planar surface. Such a particle system may approximate many systems of interest like contaminants on substrates like computer wafers, or ice crystals located on larger blocks of ice. The latter case has astrophysical implications [5]. The solution may also be appropriate in considering the effects of microstructures on larger bodies. We demonstrate that for the present system, both negative polarization features seen in the near backscatter region, negative polarization branch and polarization opposition effect, are the result of the constructive interference of reciprocal rays.

7. Acknowledgments

GV acknowledges support from the TechBase Program of Chemical and Biological Defense. KM is grateful for the Academy of Finland for partial support of his research.

References

1. F. Moreno and F. González, "Light scattering from microstructures," (Springer-Verlag, Heidelerg, 2000).
2. V.K. Rosenbush, N. Kiselev, V. Avramchuk and M. Mishchenko, "Photometric and polarimetric opposition phenomena exhibited by solar system bodies," in Optics of Cosmic Dust, ed. by G. Videen and M. Kocifaj (Kluwer Academic Publishers, Dordrecht, 2002) 191-224.
3. V.K. Rosenbush, V.V. Avramchuk, A.E. Rosenbush, and M.I. Mishchenko, "Polarization properties of the Galilean satellites of Jupiter: observations and preliminary analysis," Astrophys. J. **487**, 402-414 (1997).
4. A.C. Levasseur-Regourd, E. Hadamcik, and J.B. Renard, "Evidence for two classes of comets from their polarimetric properties at large phase angles," Astron. Astrophys. **313**, 327-333 (1996).
5. M.I. Mishchenko, "On the nature of the polarization opposition effect exhibited by Saturn's rings," Astrophys. J. **411**, 351-361 (1993).
6. Yu. Shkuratov, A. Ovcharenko, E. Zubko, V. Kaydash, D. Stankevich, V. Omelchenko, O. Miloslavskaya, K. Muinonen, J. Piironen, S. Kaasalainen, R. Nelson, W. Smythe, V. Rosenbush, and P. Helfenstein, "The opposition effect and negative polarization of structural analogs of planetary regoliths," Icarus **159**, 396-416 (2002).
7. Yu. G. Shkuratov and A. Ovcharenko, "Experimental modeling of opposition effect and negative polarization of regolith-like samples," in Optics of Cosmic Dust, ed. by G. Videen and M. Kocifaj (Kluwer Academic Publishers, Dordrecht, 2002).
8. V.D. Ozrin, "Exact solution for the coherent backscattering of polarized light from a random medium of Rayleigh scatterers," Waves Random Media **2**, 141-164 (1992).
9. M.I. Mishchenko, "Polarization effects in weak localization of light: Calculation of the copolarized and depolarized backscattering enhancement factors," Phys. Rev. B **44**, 12,579-12,600 (1991).

10. M.I. Mishchenko, "Enhanced backscattering of polarized light from discrete random media," J. Opt. Soc. Am. A **9**, 978-982 (1992).

11. M.I. Mishchenko, J.-M. Luck, and T. M. Nieuwenhuizen, "Full angular profile of the coherent polarization opposition effect," J. Opt. Soc. Am. A **17**, 888-891 (2000).

12. M. Mishchenko, V. Tishkovets, and P. Litvinov, "Exact results of the vector theory of coherent backscattering from discrete random media: an overview," in Optics of Cosmic Dust, ed. by G. Videen and M. Kocifaj (Kluwer Academic Publishers, Dordrecht, 2002) 239-260.

13. K. Muinonen, "Coherent backscattering by solar system dust particles," in Asteroids, Comets and Meteors, ed. by A. Milani, M. Di Martino, and A. Cellino (Kluwer, Dordrecht, 1974) 271-296.

14. Yu. Shkuratov, M. Kreslavsky, A. Ovcharenko, D. Stankevich, E. Zubko, C. Pieters, and G. Arnold, "Opposition effect from Clementine data and mechanisms of backscatter," Icarus **141**, 132-155 (1999).

15. K. Muinonen, "Coherent backscattering by absorbing and scattering media," in Light scattering by nonspherical particles, ed. by B. Gustafson, L. Kolokolova and G. Videen (Army Research Laboratory, Adelphi, 2002) 223-226.

16. G. Videen, "The negative polarization branch and second-order ray-tracing," Appl. Opt. **41**, 5115-5121 (2002).

17. G. Videen, "Polarization opposition effect and second-order ray-tracing: cloud of dipoles," J. Quant. Spectrosc. Radiative Transfer. **79-80**, 1103-1109.

18. K. Muinonen, G. Videen, E. Zubko, and Yu. Shkuratov, "Numerical techniques for backscattering by random media," in Optics of Cosmic Dust, ed. by G. Videen and M. Kocifaj (Kluwer Academic Publishers, Dordrecht, 2002) 261-282.

19. P.V. Litvinov, V.P. Tishkovets, K. Muinonen, and G. Videen, "Coherent opposition effect for discrete random media," in Wave scattering in complex media: from theory to Applications, ed. by B. van Tiggelen and S. Skepetrov (Kluwer Academic Publishers, Dordrecht, 2003) 567-581.

20. V.P. Tishkovets, P.V. Litvinov, and M.V. Lyubchenko, "Coherent opposition effects for semi-infinite discrete random medium in the double-scattering approximation, J. Quant. Spectrosc. Radiat. Transfer **72**, 803-811 (2002).

21. G. Videen, K. Muinonen, and K. Lumme, "Coherence, power laws, and the negative polarization surge," Appl. Opt. **42**, 3647-3652 (2003).

22. I.V. Lindell, A.H. Sihvola, K.O. Muinonen, and P.W. Barber, "Scattering by a small object close to an interface. I. Exact-image theory formulation," J. Opt. Soc. Am. A **8**, 472-476 (1991).

23. K.O. Muinonen, A.H. Sihvola, I.V. Lindell, and K.A. Lumme, "Scattering by a small object close to an interface. II. Study of backscattering," J. Opt. Soc. Am. A **8**, 477-482 (1991).

24. G. Videen, "Light scattering from a sphere on or near a surface," J. Opt. Soc. Am. A **8**, 483-489 (1991); errata. J. Opt. Soc. Am. A **9**, 844-845 (1992).

25. G. Videen, "Light scattering from a particle on or near a perfectly conducting surface," Opt. Comm. **115**, 1-7 (1995).

26. G. Videen, M.G. Turner, V.J. Iafelice, W.S. Bickel, and W.L. Wolfe, "Scattering from a small sphere near a surface," J. Opt. Soc. Am. A **10**, 118-126 (1993).

27. G. Videen, W.L. Wolfe, and W.S. Bickel, "Light scattering Mueller matrix for a furface contaminated by a single particle in the Rayleith limit," Opt. Eng. **31**, 341-349 (1992).

28. O. Muñoz, H. Volten, and J.W. Hovenier, "Experimental light scattering matrices relevant to cosmic dust," in Optics of Cosmic Dust, ed. by G. Videen and M. Kocifaj (Kluwer Academic Publishers, Dordrecht, 2002) 57-70.

29. Sun, W., T. Nousiainen, K. Muinonen, Q. Fu, N. G. Loeb and G. Videen, "Light scattering by Gaussian particles: A solution with finite-difference time domain technique," J. Quant. Spectrosc. Radiative Transfer. **79-80**, 1083-1090 (2003).

30. R.H. Zerull, B.S. Gustafson, K. Schulz, and E. Thiele-Corbach, "Scattering by aggregates with and without absorbing mantle: microwave analog experiments," Appl. Opt. **32**, 4088-4100 (1993).

31. V.J. Iafelice and W.S. Bickel, "Polarized light scattering matrix elements for select perfect and perturbed optical surfaces," Appl. Opt. **26**, 2410-2415 (1987).

32. G. Videen, "Polarized light scattering from surface contaminants," Opt. Comm. **143**, 173-178 (1997).

Gorden Videen with wife Erin Code and daughters Paisley and Scotia (standing) at Crimean Astrophysical Observatory.

Karri Muinonen with wife Marjo, daughter Ida and son Ohto on the beach.

Elena Petrova, Klaus Jockers, Oksana Goryunova, and Victor Tishkovets

Above Left: Pavel Litvinov and Ludmila Litvinova.

Above: Pavel Litvinov, Victor Tishkovets, Ludmilla Kolokolova, and Klaus Ziegler in heated discussions at poster session.

Left:Victor Tishkovets, Ludmila Tishkovets and Nikolai Voshchinnikov

BACKSCATTERING EFFECTS FOR DISCRETE RANDOM MEDIA: THEORETICAL RESULTS

VICTOR TISHKOVETS,[1] PAVEL LITVINOV,[1]
ELENA PETROVA,[2] KLAUS JOCKERS,[3] AND
MICHAEL MISHCHENKO[4]

[1]*Institute of Radioastronomy of NASU, 4 Chervonopraporna St., Kharkiv, Ukraine*
[2]*Space Research Institute of RAS, Profsoyuznaya 84/32, Moscow 117979, Russia*
[3]*Max-Planck-Institute for Aeronomy, Katlenburg-Lindau, D-37191, Germany*
[4]*NASA Goddard Institute for Space Studies, 2880 Broadway, New York, NY 10025, USA*

Abstract. The effect of enhanced backscattering of light from discrete random media, often referred to as the coherent photometric opposition effect (or weak photon localization), is a remarkable optical phenomenon that is being studied actively. When the incident light is unpolarized, the opposition intensity peak may be accompanied by the so-called opposition polarization effect, which manifests itself as a sharp asymmetric negative-polarization feature at small phase angles. The optical phenomenon that causes these effects is the constructive interference of multiply scattered waves propagating along the same light-scattering paths in a medium but in opposite directions. The theoretical description of multiple scattering becomes more complicated for closely packed media because of potentially significant near-field effects that can depress the photometric opposition peak significantly and increase the depth of the negative-polarization feature. In this chapter, we discuss the opposition effects for semi-infinite sparse scattering media and study their dependence on concentration and microphysical properties of the constituent scatterers. Manifestations of the near-field interactions are illustrated by theoretical calculations for randomly oriented clusters of spherical particles.

1. Introduction

The phenomenon of electromagnetic scattering and absorption is exploited widely in remote-sensing and laboratory characterization of various objects

G. Videen et al. (eds.), Photopolarimetry in Remote Sensing, 221-242.

[1–14]. Calculations of different characteristics of radiation scattered by discrete random media are important in atmospheric optics, astrophysics, biophysics, and many other areas of science and engineering. More often than not, multiple-scattering effects on the characteristics of the measured detector response must be taken into account. In many cases measurements of angular dependencies of the emerging radiation demonstrate a backscattering enhancement in the form of a sharp peak of intensity centered at exactly the backscattering direction. This effect is also known as weak photon localization or the coherent opposition effect. In the case of unpolarized incident light, it can be accompanied by the so-called opposition polarization effect in the form of a branch of negative linear polarization limited to a narrow range of backscattering angles [8,9,12,14]. Presently, both effects are explained as the result of constructive interference of multiply scattered waves propagating inside the medium along direct and reverse trajectories [1–13].

The coherent backscattering effect was predicted first theoretically in studies of propagation of electromagnetic waves in turbulent plasmas [15]. Then it has been analyzed in numerous experimental and theoretical studies (see, e.g., [16–18] and references therein). The strong dependence of the angular width of the interference peak on the particle number density has been demonstrated both experimentally and theoretically [1–3]. However, only recently rigorous formulas describing the opposition effects have been derived in the particular case of normal incidence of light on a plane-parallel layer of a sparse discrete random medium. Specifically, a complete analytical solution for nonabsorbing, randomly positioned Rayleigh scatterers has been obtained in [19–21]. The rigorous approach was later extended to randomly positioned and randomly oriented particles with arbitrary sizes, shapes, and refractive indices [22]. A more general case of oblique illumination was described analytically in [23]. Numerical results obtained in the double-scattering approximation revealed a significant dependence of the characteristics of the opposition effects on particle microphysical properties [18,24,25].

Theoretical consideration of multiple light scattering by a closely packed medium is complicated by potentially significant near-field effects [26–28]. Indeed, the scattered electromagnetic wave in the close vicinity of the scatterer is strongly inhomogeneous. The analysis of scattering of this wave by an adjacent particle requires more sophisticated techniques than those used to address the problem of scattering of a plane electromagnetic wave.

In this chapter, we discuss the opposition effects for sparse media and closely packed systems of particles. Section 2 summarizes the basic equations describing the scattering of light by systems of spherical particles. Specific differences in the description of light scattering by closely packed and by sparse systems of particles are discussed. In section 3, equations for the reflection matrix of a layer of sparse medium are given. Numerical examples illustrate considerable dependence of the opposition effects on microscopic characteristics of the constituent scatterers. In section 4, the near-field effects and their manifestations in light scattering by randomly oriented clusters of spherical particles are considered qualitatively.

2. Basic definitions and equations

Over the past few decades, the approach based on the theory of electromagnetic scattering by particle ensembles (clusters) has become quite popular in analyzing the problem of light scattering by discrete random media. We use the theory of electromagnetic scattering by ensembles of spherical particles [29] in order to derive equations describing the process of multiple scattering in discrete random media. An assumption of particle sphericity is not crucial. However, it allows us to avoid more complex and cumbersome calculations. The generalization of the results thus obtained to randomly oriented nonspherical particles forming a low-density medium can be achieved rather easily [22,23].

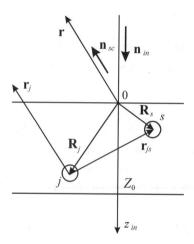

Figure 1. Scattering geometry (see text).

Consider a discrete random medium in the form of a homogeneous and isotropic layer consisting of randomly positioned spherical particles and denote by Z_0 its geometrical thickness. The scattering geometry is specified using the coordinate system shown in Fig. 1. An incident plane wave propagates along the z_{in}-axis of a coordinate system $\hat{\mathbf{n}}_{in}$. Throughout the paper, bold letters with carets $\hat{\mathbf{n}}_i$ denote right-handed coordinate systems (x_i, y_i, z_i) with the z_i-axes along the unit vectors $\mathbf{n}_i$. Coordinates of scatterers are determined in the coordinate system $\hat{\mathbf{n}}_{in}$ whose xy plane coincides with the upper boundary of the medium. The scattered wave propagates along the z_{sc}-axis of the coordinate system $\hat{\mathbf{n}}_{sc}$. The rotation from $\hat{\mathbf{n}}_{in}$ to $\hat{\mathbf{n}}_{sc}$ is determined by the Euler angles $\varphi, \vartheta, \gamma$.

It is convenient to describe wave scattering by using the circular-polarization basis, in which the incident wave can be written as

$$\mathbf{E}_n^0 = \mathbf{e}_n(\hat{\mathbf{n}}_{in})\exp(ik\mathbf{n}_{in}\mathbf{r}), \tag{1}$$

where $n = \pm 1$, $k = 2\pi/\lambda$, λ is the wavelength, and $\mathbf{e}_n(\hat{\mathbf{n}}_{in})$ is a covariant spherical basis vector [30] in the coordinate system $\hat{\mathbf{n}}_{in}$. When $n = 1$, the direction of rotation of the electric vector of the wave (1) corresponds to the clockwise direction when looking in the direction of the vector $\mathbf{n}_{in}$. The linearity of the Maxwell equations and boundary conditions allows one to define the amplitude scattering matrix of the entire layer T_{pn} as

$$\begin{pmatrix} E_1 \\ E_{-1} \end{pmatrix} = \frac{\exp(ikr)}{-ikr} \begin{pmatrix} T_{11} & T_{1-1} \\ T_{-11} & T_{-1-1} \end{pmatrix} \begin{pmatrix} E_1^0 \\ E_{-1}^0 \end{pmatrix}, \tag{2}$$

where r is the distance from the origin of the coordinate system $\hat{\mathbf{n}}_{in}$ to the observation point, and then to express it in the form

$$T_{pn} = \sum_j t_{pn}^{(j)}. \tag{3}$$

Here $t_{pn}^{(j)}$ is the 2×2 amplitude scattering matrix [29] of the j th scatterer. Equation (2) assumes that all linear dimensions of the medium are small relative to r. The 4×4 scattering matrix **S**, which transforms the Stokes parameters of the incident wave into those of the scattered light [29], is defined by the following expression:

$$S_{pn\mu\nu} = \langle \sum_j t_{pn}^{(j)} t_{\mu\nu}^{*(j)} \rangle + \langle \sum_{j,\sigma \neq j} t_{pn}^{(j)} t_{\mu\nu}^{*(\sigma)} \rangle, \tag{4}$$

where the angular brackets denote ensemble averaging, the indices take on the values $p, n, \mu, \nu = \pm 1$, and the asterisk denotes complex conjugation.

We use the standard theory of light scattering by a system of spherical particles to derive the requisite equations. In this case the field scattered by the j th particle can be expressed in the form [22,23,29]

$$\mathbf{E}^{(j)} = \frac{\exp(ikr_j)}{-ikr_j} \sum_{pLM} \frac{2L+1}{2} A_{LM}^{(jpn)} D_{Mp}^{*L}(\hat{\mathbf{n}}_{in}, \hat{\mathbf{n}}_{sc})\mathbf{e}_p(\hat{\mathbf{n}}_{sc}). \tag{5}$$

Here the $D_{Mp}^L(\hat{\mathbf{n}}_{in}, \hat{\mathbf{n}}_{sc}) = \exp(-iM\varphi) d_{Mp}^L(\vartheta)\exp(-ip\gamma)$ are Wigner D functions [30], r_j is the distance from particle j to the observation point, $\mathbf{n}_{sc}$

is the unit vector in the scattering direction, $e_p(\hat{\mathbf{n}}_{sc})$ is a covariant spherical basis vector in the coordinate system $\hat{\mathbf{n}}_{sc}$. It is assumed here that the scattering direction is the same for all particles of the medium.

The coefficients $A_{LM}^{(jpn)}$ are determined by the system of equations [22,29]

$$A_{LM}^{(jpn)} = a_L^{(jpn)} \exp(ik\mathbf{n}_{in}\mathbf{R}_j)\delta_{Mn} + \sum_{q,s\neq j} a_L^{(jpq)} \sum_{lm} A_{lm}^{(sqn)} H_{LMlm}^{(q)}(\hat{\mathbf{n}}_{in},\hat{\mathbf{n}}_{sj}), \qquad (6)$$

where $a_L^{(jpn)} = a_L^{(j)} + pnb_L^{(j)}$, $a_L^{(j)}$ and $b_L^{(j)}$ are the Lorenz-Mie coefficients [29], $q = \pm 1$, $\mathbf{R}_j$ is the radius-vector of particle j (Fig.1), $\hat{\mathbf{n}}_{sj}$ is the coordinate system with the z_{sj} axis along the vector $\mathbf{r}_{sj}$, and the $H_{LMlm}^{(q)}(\hat{\mathbf{n}}_{in},\hat{\mathbf{n}}_{sj})$ are the coefficients of the addition theorem for the vector Helmholtz harmonics [22,26,31]

$$H_{LMlm}^{(q)}(\hat{\mathbf{n}}_{in},\hat{\mathbf{n}}_{sj}) = \frac{2l+1}{2}(-1)^m \sum_{l_1 m_1} i^{-l_1} h_{l_1}(kr_{js}) D_{m_1 0}^{l_1}(\hat{\mathbf{n}}_{in},\hat{\mathbf{n}}_{js}) C_{LMl-m}^{l_1 m_1} C_{Lql-q}^{l_1 0}. \qquad (7)$$

Here $h_l(x)$ is the Hankel spherical function of the first kind, and the Cs are the Clebsch-Gordan coefficients [30].

To determine the matrix $t_{pn}^{(j)}$, let us introduce the basis vectors $e_\perp$ and $e_\parallel$ with respect to the plane through the vectors $\mathbf{n}_{in}$ and $\mathbf{n}_{sc}$. The vector $e_\perp$ is perpendicular to the reference planes, whereas the vector $e_\parallel$ is parallel to them. Transforming these vectors into spherical basis vectors yields the contravariant spherical basis vectors [30] $e^n(\hat{\mathbf{n}}_{in})$ and $e^p(\hat{\mathbf{n}}_{sc})$, which are rotated with respect to the vectors $e_n(\hat{\mathbf{n}}_{in})$ and $e_p(\hat{\mathbf{n}}_{sc})$ through the angles φ and $-\gamma$, respectively. We, therefore, get from Eqs. (2) and (5)

$$t_{pn}^{(j)} = \exp(-ik\mathbf{n}_{sc}\mathbf{R}_j - in\varphi - ip\gamma)\sum_{LM} \frac{2L+1}{2} A_{LM}^{(jpn)} D_{Mp}^{*L}(\hat{\mathbf{n}}_{in},\hat{\mathbf{n}}_{sc}). \qquad (8)$$

The solution of the system of equations (6) can be obtained by iteration. This representation of the solution corresponds to the expansion of the coefficients $A_{LM}^{(jpn)}$ in a multiple-scattering series. The first two terms of this series are

$$A_{LM}^{(jpn)} = a_L^{(jpn)} \exp(ik\mathbf{n}_{in}\mathbf{R}_j)\delta_{Mn} +$$
$$\sum_q a_L^{(jpq)} \sum_{slm} a_l^{(sqn)} H_{LMlm}^{(q)}(\hat{\mathbf{n}}_{in},\hat{\mathbf{n}}_{sj}) \exp(ik\mathbf{n}_{in}\mathbf{R}_s)\delta_{mn} + \dots . \qquad (9)$$

By inserting the series (9) into Eq. (8), writing the same series for the incident wave with the initial polarization ν and scattered polarization μ, and calculating the matrix (4) we obtain the series corresponding to various scenarios of wave scattering. Let us combine the terms corresponding to the same scattering scenario. Then the matrix (4) can be represented as

$$S_{pn\mu\nu} = S^{1}_{pn\mu\nu} + S^{L}_{pn\mu\nu} + S^{C}_{pn\mu\nu} + S^{O}_{pn\mu\nu}. \tag{10}$$

Here the matrix $\mathbf{S}^{1}$ corresponds to the first-order scattering, including the interference of singly scattered waves. The matrix $\mathbf{S}^{L}$ is the sum of all scattering orders corresponding to the case when both waves propagate along the same path in the same direction, thereby describing incoherent scattering. The matrix $\mathbf{S}^{C}$ is the sum of all scattering orders corresponding to the case when both waves propagate along the same path, but in opposite directions. In the backscattering region, the interference of such waves is constructive and results in the opposition effects. The matrix $\mathbf{S}^{O}$ corresponds to the rest of scattering contributions, including the interference of waves scattered once and twice, twice and three times, etc.

Thus, the calculation of the matrix (4) reduces to the calculation of the matrices $\mathbf{S}^{1}$, $\mathbf{S}^{L}$, $\mathbf{S}^{C}$, and $\mathbf{S}^{O}$. It is very difficult to calculate all these matrices for closely packed media comprising scatterers comparable to the wavelength. In this case, the matrix $\mathbf{S}^{O}$ can contribute significantly to the matrix (4), and all the matrices must be calculated with the coefficients of the addition theorem in the form (7). These coefficients describe all peculiarities of the waves in the vicinity of the scatterers including the near-field effects. For low-density media, when the distances between the particles $r_{js} \gg \tilde{a}_{j}$, $\tilde{a}_{s}$ (where the $\tilde{a}_{j}$ and $\tilde{a}_{s}$ are the radii of particles j and s, respectively), these coefficients are

$$H^{(q)}_{LMlm}(\hat{\mathbf{n}}_{in}, \hat{\mathbf{n}}_{sj}) = \frac{2l+1}{2} \frac{\exp(ikr_{js})}{-ikr_{js}} D^{L}_{Mq}(\hat{\mathbf{n}}_{in}, \hat{\mathbf{n}}_{sj}) D^{*l}_{mq}(\hat{\mathbf{n}}_{in}, \hat{\mathbf{n}}_{sj}). \tag{11}$$

If the scatterers are randomly positioned and $r_{js} \gg \lambda$, then the matrix $\mathbf{S}^{O}$ vanishes, and Eq. (10) takes the form

$$S_{pn\mu\nu} = S^{1}_{pn\mu\nu} + S^{L}_{pn\mu\nu} + S^{C}_{pn\mu\nu}. \tag{12}$$

In the next section, we provide equations to calculate the matrix (12) corresponding to the reflection of radiation from a plane-parallel layer. In solving these equations numerically, our main interest is in the dependence of the opposition effects on microphysical properties of the medium such as particle size, refractive index, and concentration.

3. Backscattering by a sparse plane-parallel layer

In this section, we consider multiple scattering by a plane-parallel layer consisting of discrete, randomly positioned scatterers of arbitrary shape and in random orientation. The incident wave is assumed to propagate normally to the boundary of the medium. A derivation of the equations describing the matrix (12) in this case is given in [22]. The more general case of oblique illumination is considered in [23]. In the circular polarization basis, the equations for the matrices $\mathbf{S}^l$ and $\mathbf{S}^L$, defined per unit area of the boundary, take the following form:

$$S_{pn\mu v}^{l} + S_{pn\mu v}^{L} = \frac{\eta}{k}\sum_{l}d_{M_0 N_0}^{l}(\vartheta)\int_{0}^{kZ_0}\exp\left(2z\frac{\mathrm{Im}(\varepsilon)}{\cos\vartheta}\right)\alpha_{l}^{(z)(pn)(\mu v)}dz, \qquad (13)$$

where η is the particle number density, $p, n, \mu, v = \pm 1$, $M_0 = v - n$, $N_0 = \mu - p$, ϑ is the scattering angle (the angle between the incidence and scattering directions), and ε is the complex effective refractive index of the medium. The expansion coefficients $\alpha_{L}^{(z)(pn)(\mu v)}$ are determined from the system of equations

$$\alpha_{L}^{(z)(pn)(\mu v)} = \chi_{L}^{(pn)(\mu v)}\exp(-\tau_z) + \frac{2\pi\eta}{k^3}\sum_{qq_1}\chi_{L}^{(pq)(\mu q_1)}$$

$$\times\int\sum_{l}\alpha_{L}^{(y)(qn)(q_1v)}\exp(-\tau_\rho)d_{M_0 N}^{l}(\omega)d_{M_0 N}^{l}(\omega)d\rho\sin\omega d\omega, \quad (14)$$

where ρ, ω are polar coordinates of the integration point with respect to the point z, the angle ω is measured relative to the backscattering direction, $\tau_x = 2\,\mathrm{Im}(\varepsilon)x$, $N = q_1 - q$ $(q, q_1 = \pm 1)$, and $y = z - \rho\cos\omega$. The upper integration limit over ρ is equal to $z/\cos\omega$ for $\omega < \pi/2$ and to $(z - Z_0 k)/\cos\omega$ for $\omega > \pi/2$. The $\chi_{L}^{(pn)(\mu v)}$ are the coefficients in the expansion of the individual-particle scattering matrix [29] in the Wigner d functions [30]. Equations (13) and (14) are equivalent to the vector radiative transfer equation in the circular-polarization basis.

The corresponding equations for matrix $\mathbf{S}^C$ are as follows:

$$S_{pn v\mu}^{C} = \frac{2\pi\eta^2}{k^4}\sum_{qq_1 LM}(-1)^{L}\zeta_{LM}^{*(q_1\mu)(qp)}\int_{0}^{kZ_0}\beta_{LM}^{(z)(qn)(q_1v)}\exp(-\varepsilon_1 z)dz, \qquad (15)$$

where the matrix $\beta_{LM}^{(z)(pn)(\mu v)}$ is defined by system of equations

$$\beta_{LM}^{(z)(pn)(\mu v)} = B_{LM}^{(pn)(\mu v)}\exp(-\varepsilon_1^* z) + \frac{2\pi\eta}{k^3}\sum_{qq_1 lm}\chi_{l}^{(pq)(\mu q_1)}i^{M-m}\int\beta_{lm}^{(y)(qn)(q_1v)}$$

$$\times d_{MN_0}^{L}(\omega)d_{mN_0}^{l}(\omega)\exp(-\tau_\rho)J_{m-M}(\rho\sin\vartheta\sin\omega)d\rho\sin\omega d\omega. \qquad (16)$$

Here $\zeta_{LM}^{(pn)(\mu v)}$ are the coefficients in the expansion of the phase matrix elements [22–25] in Wigner d functions, $J_m(x)$ are Bessel functions,

$$\varepsilon_1 = \mathrm{Im}(\varepsilon)\left(1 - \frac{1}{\cos\vartheta}\right) + i\left(\cos\vartheta + 1\right)\left(\frac{\mathrm{Re}(\varepsilon)-1}{\cos\vartheta} + 1\right), \tag{17}$$

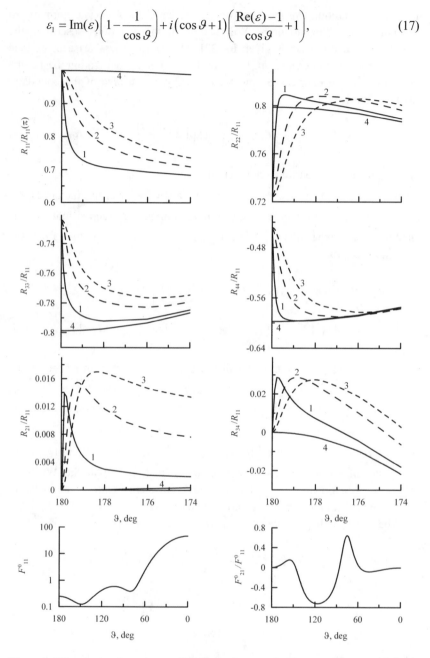

Figure 2. The angular dependence of the reflection matrix elements for a semi-infinite medium composed of spherical particles with $\tilde{x} = 3$ and $\tilde{m} = 1.35 + 0i$ for various values of the filling factor: 1) $\xi = 0.001$, 2) $\xi = 0.005$, 3) $\xi = 0.01$, 4) the incoherent component. The F^0_{11} and F^0_{21} are scattering matrix elements for an individual scatterer.

$$B_{LM}^{(pn)(\mu v)} = \sum_{lm} \varsigma_{lm}^{(pq)(\mu q_1)} i^{M-m} \int d_{MN_0}^L (\omega) d_{mN_0}^l (\omega) J_{m-M} (\rho \sin \vartheta \sin \omega)$$

$$\times \exp(-\tau_\rho + \rho \varepsilon_1^* \cos \omega) d\rho \sin \omega d\omega. \qquad (18)$$

The matrix S^C describes the interference of the conjugate pairs of waves propagating along the same trajectories in the medium but in opposite directions, including looped trajectories.

Equations (13)–(18) are rigorous provided that any scatterer is located in the far-field zone of all other scatterers. They are valid for arbitrary particles in random orientation. It should be noted that in the case of the exact backscattering direction ($\vartheta = \pi$), the matrix S^C can be obtained from the matrix S^L using the equation $S_{pnv\mu}^C = S_{pn\mu v}^L$ [32].

The above equations for a semi-infinite medium can be solved numerically in the approximation of several scattering orders. All the calculations below are given for the reflection matrix $R = -S/2k^2 \cos \vartheta$ in the linear-polarization basis. The transformation of the matrix (12) from the circular-polarization basis to the linear-polarization one can be found in [29]. The effective refractive index of the medium is given by $\varepsilon = 1 + i\eta C_{ext}/2k$, where C_{ext} is the extinction cross section per particle [29].

Figure 2 depicts all nonzero elements of the reflection matrix in the double-scattering approximation for a semi-infinite layer composed of identical spherical particles. The particle size parameter is $\tilde{x} = 2\pi\tilde{a}/\lambda = 3$, where $\tilde{a}$ is the particle radius, and the particle refractive index is $\tilde{m} = 1.35 + 0i$. The results are shown for different values of the filling factor (packing density) $\xi = 4\pi\tilde{a}^3\eta/3$.

As seen from Fig. 2, the interference affects all elements of the reflection matrix. The linear dependence of the width of the intensity peak on the packing density (the R_{11} curves), discussed in theoretical and experimental studies (see [1–3] and references in [17]), is very noticeable. The examples given in Fig. 2 as well as other calculations not shown here reveal the same dependence on ξ in the other components of the coherent reflection matrix. For instance, the angular position of the R_{21} and R_{34} extrema depends linearly on ξ if the corresponding matrix elements for the incoherent component depend weakly on ϑ . Otherwise, the superposition of the coherent and incoherent components disrupts this linear dependence [24].

Figure 3 demonstrates the dependence of the intensity and the degree of linear polarization of the reflected light on the particle properties and on the order of scattering. Since the effective refractive index of the medium is kept the same, the differences in the features of the opposition effects are caused by the differences in the microscopic characteristics of the scatterers. To explain the influence of the particle properties on the R_{11} element, let us consider the interference of doubly scattered waves for two cases of scattering geometry. Let two particles be placed along the direction of the $\mathbf{n}_{in}$ vector in the first case and along the perpendicular direction in the second case. Then the phase difference for the waves propagating along the direct and reverse paths is proportional to $1 + \cos\vartheta$ in the first case and to $\sin\vartheta$ in the second case. For $\vartheta \to \pi$, the difference for the first pair of particles is negligible, whereas for the second one it is proportional to $\pi - \vartheta$. In other words, the first pair produces a broader interference peak than the second one. Particles with $\tilde{x} = 4.5$ scatter radiation more effectively in the forward and backward directions than particles with $\tilde{x} = 3$ (compare curves 2, 3 and 1 of Fig. 3c). As a result, the width of the intensity peak for a medium composed of these particles is larger (compare curves 2, 3 and 1 of Fig. 3a). Particles with $\tilde{x} = 4.5$ and $\tilde{m} = 1.59 + 0i$ display a more symmetric phase function than those with $\tilde{x} = 4.5$ and $\tilde{m} = 1.35 + 0i$, and the peak for these particles is sharper.

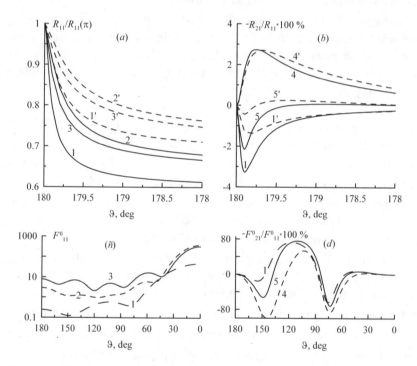

Figure 3. The dependence of the relative intensity (a) and the degree of linear polarization (b) on the particle properties and on the scattering order. The curve numbers correspond to the following parameters: 1) $\tilde{x} = 3$, $\tilde{m} = 1.35 + 0i$, $\xi = 0.0012$; 2) $\tilde{x} = 4.5$, $\tilde{m} = 1.35 + 0i$, $\xi = 0.001$; 3) $\tilde{x} = 4.5$, $\tilde{m} = 1.59 + 0i$, $\xi = 0.001$; 4) $\tilde{x} = 3$, $\tilde{m} = 1.50 + 0i$, $\xi = 0.0007$; 5) $\tilde{x} = 3.1$, $\tilde{m} = 1.35 + 0i$, $\xi = 0.0011$. In all cases, $\text{Im}(\varepsilon) = 0.000286$. The dashed curves are for the double-scattering approximation, whereas the solid ones are computed with the account of the third-order scattering. The F_{11}^0 and F_{21}^0 are scattering matrix elements for an individual scatterer.

The results of Fig. 3(b) demonstrate that the state of polarization of the backscattered radiation is controlled strongly by the intensity and state of polarization of singly scattered radiation (i.e., by the angular profiles of F_{11}^0 and F_{21}^0). It is well known that positive polarization of light scattered by isolated particles leads to a negative-polarization feature in the backscattering direction [8,9]. The angular dependence of the polarization state for wavelength-sized spherical particles is much more complex and oscillatory [29]. The resulting interference of doubly scattered waves can result in positive polarization (curves 4 and 4') as well as in negative polarization (curves 1 and 1' and curves 5). Such behavior of polarization is analyzed in detail in [23,25]. The interference may lead to a more complex dependence of polarization on the scattering angle, with positive and negative polarization regions appearing simultaneously (curve 5' in Fig. 3b).

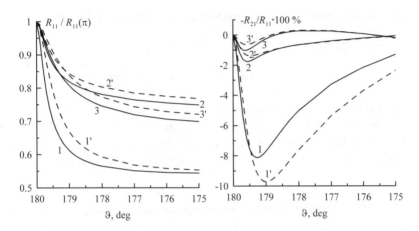

Figure 4. The same as in Fig.3 (a) and (b) but for the third-order-scattering approximation (dashed curves) and an approximate solution including all orders of scattering (solid curves). Curves 1 and 1': $\tilde{x}=3$, $\tilde{m}=1.35+0.1i$, $\xi=0.0076$; curves 2 and 2': $\tilde{x}=2$, $\tilde{m}=1.4+0.5i$, $\xi=0.0048$; curves 3 and 3': $\tilde{x}=3$, $\tilde{m}=1.5+0.5i$, $\xi=0.0061$. In all cases, $\mathrm{Im}(\varepsilon)=0.002$.

As can be seen in Fig. 3, the angular range where the interference contributes noticeably to the scattered radiation becomes narrower with increasing order of scattering (see also [3]). In Fig. 4 the results of the third-order scattering approximation are compared with the results of an approximate calculation including all orders of scattering, as follows. The radiation coming to the particle from above is calculated exactly, and a part of the radiation coming to the particle from below is calculated approximately, as if for an infinite medium. The errors of this approximation are estimated for the incoherent scattering term for particles with $\tilde{x}=3$ and $\tilde{m}=1.5+0.5i$ or

$\tilde{m} = 1.5 + 0.1i$ and do not exceed 10% and 15%, respectively. The explanation of the angular behavior of the intensity and polarization of the multiply scattered radiation is similar to that for the second-order and third-order scattering approximations.

All the results presented in this section are obtained in the approximation of quasi-homogeneous waves [16], which implies that any particle of the medium is in the wave zones of all other particles. This assumption allows the radiation reflected by the medium to be represented as a sum of the matrices (12) with the coefficients (11). Moreover, the characteristics of individual scattering particles, such as the scattering matrix [29], can be used in the calculations of multiply scattered waves. In closely packed media, a wave propagating from a scatterer to another one can be strongly inhomogeneous. This effect, which can significantly influence the opposition phenomena, is the focus of the next section.

4. Opposition effects for closely packed systems of spherical particles

4.1. Near-field effects

The reflection matrix for a closely packed medium composed of wavelength-sized scatterers can be represented as a sum of matrices (10) with the coefficients of the addition theorem (7). These coefficients describe all the details of the field in the vicinity of any scatterer, including the near-field effects [26]. We consider the manifestations of these effects qualitatively using the field configuration near a spherical scatterer as the simplest example.

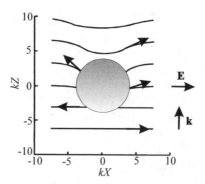

Figure 5. Surfaces of constant phase and directions of vectors **E** (sum of the incident and scattered waves) in the close vicinity of a spherical particle. The incident wave propagates along the Z-axis (indicated by the wave vector **k**) and is polarized along the X-axis. The particle size parameter is $\tilde{x} = 4$ and the refractive index is $\tilde{m} = 1.32 + 0.05i$.

In the immediate vicinity of an individual particle or a system of particles, the scattered wave is strongly inhomogeneous. For such a wave, surfaces of constant phase and amplitude do not coincide, and the amplitude, polarization, and propagation directions depend on the location with respect to the scatterer. When the inhomogeneous wave excites another particle, the resulting scattered light can differ substantially from that predicted by the theory based on the consideration of only plane waves. In what follows, the effects caused by wave inhomogeneity are referred to as near-field effects.

Direct calculations using the Lorentz–Mie theory for spherical particles show that the surfaces of constant phase of the total field are funnel-shaped in the vicinity of the particle (Fig. 5). Such near-field properties are typical for spherical particle with other size and refractive index [26,28].

Let us consider test Rayleigh particles surrounding a constituent particle (CP) of a particle aggregate and adopt a coordinate system in which the Z-axis is along the propagation direction of the incident wave, whereas the XZ-plane defines the scattering plane (Fig. 6). The incident field is assumed to be polarized in the scattering plane. If the test particles are far from the CP and far from each other, they would experience a homogeneous electromagnetic field (Fig. 6a). The dipole moment induced in the test particles would be along the X-axis. However, if the test particles are close to each other and to the CP, a dipole moment induced in test particles 1 and 3 has a non-zero component in the direction of wave propagation Z (Fig. 6b). At the same time, the X-component of the dipole moment is smaller than in the case of a large distance between the CP and the test particles.

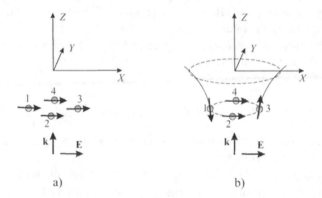

a)   b)

Figure 6. Light scattering by a system of dipoles excited by a homogeneous and an inhomogeneous wave. The incident wave propagates along the Z-axis and is polarized in the XZ-plane. a): The wave is homogeneous, and all dipole moments point in the X-direction. b): The wave is inhomogeneous because of the distortion of the wave front (see Fig. 5), and some dipole moments have a non-zero Z component.

The changes of the dipole moment induced by the high packing density do not depend on the polarization of the incident wave. If the latter are polarized perpendicularly to the scattering plane, the roles of test particles 1 and 3 and 2 and 4 would simply be interchanged, i. e. the gross changes of the induced dipole moment would be the same and would only be caused by other test particles. Consideration of light scattering by an ensemble of test dipoles (without interference of waves) shows the intensity at $\vartheta = 0$ and $\vartheta = \pi$ for Fig. 6b less then for Fig. 6a and vice versa at $\vartheta = \pi / 2$. The angular dependence of the degree of linear polarization is bell-shaped and may be negative for Fig. 6b, whereas for Fig. 6a it is positive with maximum value 100% at $\vartheta = \pi / 2$. In other words, the inhomogeneity of the wave in the particle vicinity and its interaction with neighboring scatterers reduces the scattered intensity in the directions $\vartheta \approx 0$ and $\vartheta \approx \pi$ and leads to the appearance (or enhancement of already existing) negative polarization in the backscattering region.

An analogous analysis of scattering in a densely packed ensemble of particles using the other wavelength-sized CPs instead of the test Rayleigh scatterers requires the consideration of the gradient of the wave inhomogeneity. Unfortunately, this problem is far from being well-studied. We can only note the following. The zone of wave inhomogeneity extends to distances of the order of λ from the particle surface. Consequently, the near-field effects are essential only for CPs with sizes comparable to or less than λ and are negligible for larger particles. A closer examination of the near-field effects including simulation examples for various types of particles is performed in [26–28,34]. These examples display a complex dependence of the near-field effects on the particle properties and the scattering angle. In the first approximation, the effect of the rotation of the field vector, which is mainly caused by the radial component of the scattered field, is described by the interference of waves scattered once and twice. Its contribution strongly depends on the size and refractive index of the CPs, the distance between them, and the scattering angle.

4.2. Opposition effects for randomly oriented clusters of spherical particles

We illustrate the scattering phenomena described above using a bisphere as the simplest example. We assume that the bisphere consists of monomers with the size parameter $\tilde{x} = 1.49$. The complex refractive index of the monomers is $\tilde{m} = 1.80 + 0.01i$. A sphere with these optical parameters has neutral (near zero) polarization at scattering angles close to the backscattering direction and positive polarization elsewhere. In view of the preceding discussion of wave interference and near-field effects, this allows us to demonstrate how the different optical phenomena contribute to intensity and polarization. To explain qualitatively all these phenomena, we restrict the analysis to the double-scattering approximation and use the formulas derived in [26]. The results are given in Fig. 7 computed for the case of touching spheres. Curves 1 correspond to a single particle. Curves 2 are computed by including the contributions of single scattering, incoherent double scattering, and the interference of singly scattered light. Model 3 additionally takes into account the contribution of the interference of waves scattered twice, which leads to the backscattering enhancement of intensity and a weak negative branch in polarization. Model 4 includes all the components of model 3 plus the interference of singly and doubly scattered waves, i. e. all first- and second-order scattering terms. As compared to model 3, the intensity is reduced and the negative branch develops further. Curves 5 shows the exact solution.

Curves 2–5 are obtained by taking the near field effects into account [see Eqs. (7) and (10)]. In order to demonstrate what happens if these effects are ignored, we also show the curves derived for homogeneous waves [see Eqs. (11) and (12)], i.e., when the waves propagating from one particle to another are assumed to be spherical and have only tangential components of the field vector (curves 6). This approach is considered in [8,9]. It is evident that the homogeneous-wave approximation is invalid for touching monomers, i.e. the near-field effects must be taken into account in this case. As seen from Fig. 7, the interference contribution is essential for $\vartheta > 130°$ (compare the behavior of curves 2, 3, and 6), whereas the near-field effects influence a wider angular range (compare curves 4 and 3).

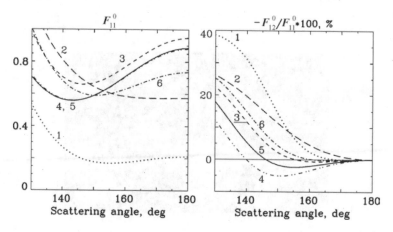

Figure 7. The contributions of interference and near-field effects in the backscattering region as demonstrated by the example of a bisphere. The curves correspond to the following models. Curves 1 (dots): calculations for single sphere; curves 2–6: calculations for bispheres. Curves 2 (long dash): single scattering + incoherent double scattering + interference of light scattered once. Curves 3 (short dash): model 2 + interference of light scattered twice. Curve 4 (dash triple dot): model 3 + interference between the light scattered once and twice. Curves 5 (solid): exact calculations. Curves 6 (dot dash): model 3 but without near-field effects (note that the negative branch of polarization almost vanishes).

Let us now consider the scattering properties of clusters composed of a large number of monomers and discuss how these properties depend on the number of monomers, their refractive index, and packing density. Aggregates producing a negative branch of polarization in the backscattering region are of special interest, because this part of the angular dependence of polarization is important for the interpretation of observational data for various objects. Figure 8 shows the aggregate structures used for the scattering simulations. Clusters 1 have a tetrahedron lattice and the CPs are placed adjacent to each other in the grid points. The overall shape of the cluster is close to spherical. This is the most compact aggregate considered in this work. Clusters 2 also have a tetrahedron lattice, but no condition for equidimensional shape is imposed. Several more CPs are put on their otherwise compact surfaces. They preserve the tetrahedron lattice but otherwise are added randomly. Note that cluster 2 turns out to be more compact in its 50-CP version than in the 100-CP one. This causes an additional peculiarity of the polarization (see below). Clusters 3, 5, and 6 are generated by the diffusion-limited aggregation method [35]. Rather compact random clusters 4 are generated by the ballistic particle-cluster aggregation method [36]. The fractal dimensions and the pre-factors also are indicated in Fig. 8 together with the gyration radius R_g for each cluster. The gyration radius of clusters 4 is somewhat larger than those of clusters 2 and 3 and somewhat smaller than those of the sparse aggregates 5 and 6.

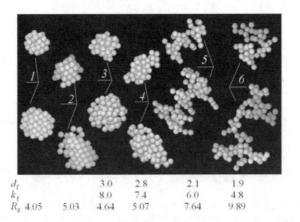

d_f			3.0	2.8	2.1	1.9
k_f			8.0	7.4	6.0	4.8
R_g	4.05	5.03	4.64	5.07	7.64	9.89

Figure 8. The aggregates composed of 50 (top) and 100 (bottom) CPs used in the model calculations. The first two are regular and have a tetrahedron lattice, the others are random fractals with different packing parameters (see text for details). The fractal dimensions, prefactors, and gyration radii (for 100 CPs) are shown in the footnote.

Our numerous calculations for various clusters with a moderate number of CPs and with the real part of the refractive index ranging from 1.4 to 1.9 show that the negative polarization branch at $\vartheta > 150°$ can be produced by compact aggregate structures with CP size parameters $0.7 < \tilde{x} < 2$ depending on $\text{Re}(\tilde{m})$. However, the cluster structure is of significant importance. For example, for the clusters composed of CPs with $\tilde{x} = 1.5$ the value of the polarization inversion angle can vary within a 15° -wide range depending on specific cluster structure. This probably is caused by the interference of singly and doubly scattered waves. In what follows, we present only the results for $\tilde{x} = 1.5$ and $\tilde{m} = 1.65 + 0.05i$. (Note that the angular dependence of the degree of linear polarization for individual CPs with such properties is close to the Rayleigh one, i.e., the polarization is positive for all scattering angles and has a maximum of about 100% near $\vartheta = \pi / 2$.) A detailed analysis of the optical properties of randomly oriented clusters composed of spherical particles, as well as the results for the other values of $\tilde{x}$ and $\tilde{m}$ can be found in [28].

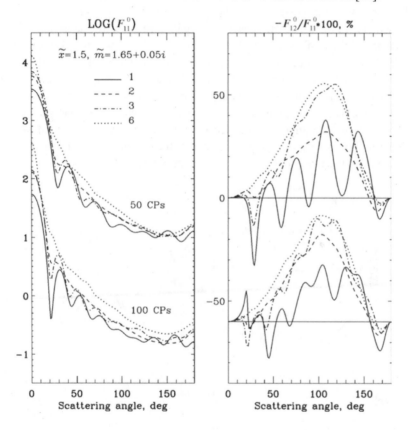

Figure 9. Intensity and polarization for aggregates of various types (1, 2, 3, and 6) consisting of 50 and 100 CPs with $\tilde{x} = 1.5$ and $\tilde{m} = 1.65 + 0.05i$. The sets of curves for 100 CPs are shifted down by 2.0 for intensity and by 60% for polarization.

A common feature of the aggregates considered is a significant decrease of polarization at side-scattering angles compared to the polarization of the individual CPs. As shown in [28], the increase of the real part of the refractive index strengthens this difference. This can be explained by both the multiple-scattering contribution and the near-field effects.

The intensity and polarization of light scattered by four types of clusters (from those shown in Fig. 8) are displayed in Fig. 9 versus scattering angle. Although clusters 2 and 3 are also compact (like cluster 1), their scattering characteristics differ substantially from those of cluster 1, except at scattering angles 15°–40°, where similar behavior is observed for all three clusters. The difference is particularly evident in polarization. Some wavy behavior still appears in the curves for very compact random structures (cluster 3), especially for $\tilde{N} = 100$. However, the regular aggregate 2 with randomly added monomers on its outside shows a rather smooth angular dependence of the intensity and polarization. Similarly, if we add random monomers to the surface of a compact regular cluster (cluster 1) in such a way that each of them contacts only one of the CPs of the existing cluster, the interference structure in the curves of the intensity and polarization are damped. The influence of the structure of the surface layer of the aggregate on the scattering properties is considered in detail below. The curves for the fluffy aggregate 6 show no oscillations. The polarization curve is bell-shaped like that of the individual CPs. However, the angular dependence for the aggregate has a much lower maximum (as compared to the polarization of the CP) and a negative branch at scattering angles close to the backscattering direction.

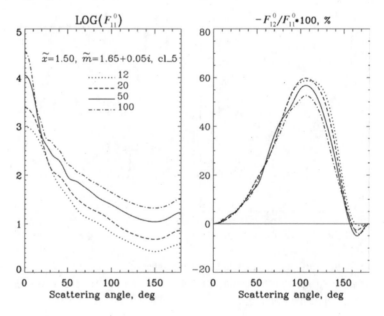

Figure 10. Dependence of intensity and polarization on the number of monomers (or cluster size) for cluster type 5 (see Fig. 8). The intensity increases, the polarization decreases, and the negative branch deepens with increasing the number of CPs.

It is worth noting that the sparse structures 5 and 6 are not very fluffy.

They contain rather compact blocks of CPs connected by chains. While the cluster as a whole determines the level of the intensity of the scattered light, which strongly depends on the cluster size, the blocks are responsible for generating polarization. This is well seen in Fig. 10, where the characteristics of type 5 clusters are displayed for an increasing number $\tilde{N}$ of CPs. The increase in $\tilde{N}$ depresses slightly the polarization maximum and strengthens the negative branch, although for larger $\tilde{N}$ the latter effect becomes less noticeable and the inversion angle remains almost unchanged.

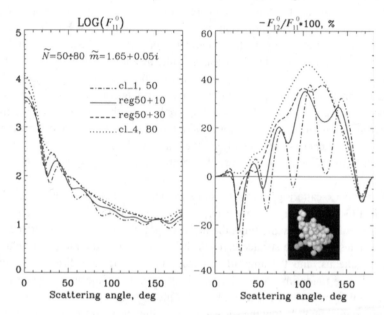

Figure 11. The intensity and polarization of light scattered by clusters with different structure of the surface layer. Additional CPs are placed in a random way on the outside of the completely regular compact cluster 1. The model for the random cluster 4 is also shown for comparison. The number of CPs is shown for each model. A picture of the aggregate after 30 CPs have been added to the surface also is shown.

Let us now consider the manifestations of the near-field effects in the scattering characteristics in more detail using clusters with different surface-layer structure. These clusters have been generated in the following way. The compact regular aggregate of type 1 composed of 50 CPs served as a core. First, 10 CPs are added to its surface in a random way. Then 20 more CPs are added to the surface in the same way. The intensity and polarization of the original cluster and the two modified ones are displayed as functions of scattering angle in Fig. 11 and compared with those of a type 4 cluster consisting of 80 monomers. Even a few CPs added to the surface of the regular type 1 cluster significantly suppress the interference oscillations in the curves of the original type 1 cluster and make the curves to look similar to those of the type 4 cluster with a more random structure. This can be explained as follows.

A layer of random particles added to the regular cluster works as an amplitude-phase inhomogeneity for the incident wave (see Fig. 5). After having passed through this layer, the wave becomes strongly inhomogeneous; the variations of its amplitude, phase, and propagation direction become randomized. If the number of particles placed in the outer layer is large enough, then there is almost no correlation between the phases of the radiation produced in the scattering of such a wave by the individual underlying particles. Consequently, the location of CPs deeper inside the cluster with respect to each other is unimportant for the scattering process. For the angular dependence of the intensity, the inner structure of the cluster is even less important. The intensity level increases with the number of CPs, albeit this growth also depends on the imaginary part of the refractive index [28]. The polarization produced in the scattering of inhomogeneous waves already has been considered in this section. Here we only note that since the scattering properties of the inner CPs are not sensitive to their location, the polarization of the cluster depends only weakly on the number of CPs in the cluster and on the regularity of its inner structure, but more strongly on its packing density (see also Fig. 9).

In the backscattering region, the enhancement of intensity is formed by the CPs in the outer layer of the cluster, where the radiation field is practically homogeneous. At the same time, the negative polarization also is generated by the particles below the surface layer of the cluster where the radiation field is inhomogeneous, and the amplitude, phase, and propagation direction of the wave change randomly. Note that the situation is the same in a powder-like layer, which makes conclusions also relevant to regolith surfaces. Thus, the field inhomogeneity below the surface layer of the cluster (or regolith) reduces the dependence of the negative polarization on the location of CPs in the deeper layers, but not on the compactness. Depending on the structure of the aggregate, the interference of multiply scattered waves or the near-field interactions are more efficient for a given cluster. This means that the opposition effects in intensity and polarization do not always go in parallel. That is, for the models in Fig. 10 the backscattering enhancement is almost the same for the clusters composed of 12 and 100 CPs, while the negative branch is weak for the small aggregate. For the compact regular clusters (Fig. 9), there is practically no backscattering enhancement, whereas the negative branch is more pronounced than for the sparse cluster.

5. Conclusions

We have presented the basic relations for electromagnetic scattering by ensembles of spherical particles and have demonstrated the differences in the description of the light scattering processes by sparse and closely packed systems. In a low-density medium, the wave propagating from a scatterer to any neighbor is quasi-homogeneous. This allows the single-scattering characteristics of the individual particles to be used in describing the process of multiple scattering. We provide equations quantifying the reflection from a plane-parallel layer of a sparse discrete random medium. One of them is a well-known vector radiative transfer equation; another describes the interference of multiply scattered waves resulting in the opposition effects. Numerical solutions of these equations show considerable dependence of the opposition effects on the properties of the medium and, specifically, on the microscopic characteristics of the scatterers.

In closely packed media, the waves in the vicinity of particles are inhomogeneous. Our qualitative analysis demonstrates that the wave inhomogeneity can lead to a suppression of the opposition effect in intensity and to causing (or strengthening) the negative polarization effect in the backscattering angular region. Computations for randomly oriented clusters show that the opposition effect in intensity and in the negative branch of polarization do not always appear simultaneously. Depending on the cluster structure, either the interference of multiply scattered waves or the near-field effects can play the decisive role.

6. Acknowledgments

V. Tishkovets, E. Petrova, and K. Jockers acknowledge the support of the Deusche Forschungsgemeinschaft in the framework of the priority program "Mars and the terrestrial planets" (Grant 436 RUS 113/684/0-1). The work by E. Petrova was supported in part by the RFBR Grant No. 01-02-17072. M. Mishchenko acknowledges support from the NASA Radiation Sciences Program managed by Donald Anderson.

References

1. M. P. van Albada and A. Lagendijk, Phys. Rev. Lett. **55**, 2692 (1985).
2. P. E. Wolf and G. Maret, Phys. Rev. Lett. **55**, 2696 (1985).
3. E. Akkermans, P. E. Wolf, R. Maynard R., and G. J. Maret, J. Phys. (Paris) **49**, 77 (1988).
4. M. P. van Albada, M. B. van der Mark, and A. Lagendijk, Phys. Rev. Lett. **58**, 361 (1987).
5. M. B. van der Mark, M. P. van Albada, and A. Lagendijk, Phys. Rev. B **37**, 3575 (1988).
6. B. Hapke, Icarus **88**, 407 (1990).
7. P. Sheng (Ed.), "Scattering and Localization of Classical Waves in Random Media" (World Scientific, Singapore, 1990).
8. K. Muinonen, Ph.D. thesis, Univ. Helsinki (1990).
9. Yu. G. Shkuratov, Astron. Vestnik **25**, 152 (1991).
10. M. I. Mishchenko and J. M. Dlugach, Mon. Not. R. Astron. Soc. **254**, 15P (1992).

11. M. I. Mishchenko and J. M. Dlugach, Planet. Space Sci. **41**, 173 (1993).

12. M. I. Mishchenko, Astrophys. J. **411**, 351 (1993).

13. A. Lagendijk and B. A. van Tiggelen, Phys. Rep. **270**, 143 (1996).

14. V. K. Rosenbush, V. V. Avramchuk, A. E. Rosenbush, and M. I. Mishchenko, Astrophys. J. **487**, 402 (1997).

15. R. S. Ruffine and D. A. De Wolf, J. Geophys. Res. **70**, 4313 (1965).

16. Yu. N. Barabanenkov, Yu. A. Kravtsov, V. D. Ozrin, and A. I. Saichev, Progr. Opt. **29**, 65 (1991).

17. V. L. Kuz'min and V. P. Romanov, Phys. Usp. **39**, 231 (1996).

18. M. I. Mishchenko, V. P. Tishkovets, and P. V. Litvinov, in G. Videen and M. Kocifaj (Eds.), "Optics of Cosmic Dust," p. 239 (Kluwer, Dordrecht, 2002).

19. V. D. Ozrin, Waves Random Media **2**, 141 (1992).

20. E. Amic, J. M. Luck, and Th. M. Nieuwenhuizen, J. Phys. I **7**, 445 (1997).

21. M. I. Mishchenko, J.-M. Luck, and T. M. Nieuwenhuizen, J. Opt. Soc. Am. A **17**, 888 (2000).

22. V. P. Tishkovets, J. Quant. Spectrosc. Radiat. Transfer **72**, 123 (2002).

23. V. P. Tishkovets, P. V. Litvinov, and M. V. Lyubchenko, J. Quant. Spectrosc. Radiat. Transfer **72**, 803 (2002).

24. V. P. Tishkovets, P. V. Litvinov, and S. V. Tishkovets, Opt. Spectrosc. **93**, 899 (2002).

25. V. P. Tishkovets and M. I. Mishchenko, J. Quant. Spectrosc. Radiat. Transfer (2004) (in press).

26. V. P. Tishkovets, Opt. Spectrosc. **85**, 212 (1998).

27. V. P. Tishkovets, Yu. G. Shkuratov, and P. V. Litvinov, J. Quant. Spectrosc. Radiat. Transfer **61**, 767 (1999).

28. V. P. Tishkovets, E. V. Petrova and K. Jockers, J. Quant. Spectrosc. Radiat. Transfer (2004) (in press).

29. M. I. Mishchenko, L. D. Travis, and A. A. Lacis, "Scattering, Absorption, and Emission of Light by Small Particles" (Cambridge University Press, Cambridge, 2002).

30. D. A. Varshalovich, A. N. Moskalev, and V. K. Khersonskii, "Quantum Theory of Angular Momentum" (World Scientific, Singapore, 1988).

31. B. U. Felderhof and R. B. Jones, J. Math. Phys. **28**, 836 (1987).

32. M. I. Mishchenko, J. Opt. Soc. Am. A **9**, 978 (1992).

33. P. V. Litvinov, V. P. Tishkovets, K. Muinonen, and G. Videen, in B. van Tiggelen and S. Skipetrov (Eds.), "Wave Scattering in Complex Media: From Theory to Applications," p. 567 (Kluwer, Dordrecht, 2003).

34. V. P. Tishkovets and P. V. Litvinov, Solar Syst. Res. **33**, 186 (1999).

35. R. Jullien and R. Botet, "Aggregation and Fractal Aggregates" (Singapore: World Scientific, 1987).

36. P. Meakin, J Colloid Interface Sci. **96**, 415 (1983).

INVERSE POLARIMETRY, AND LIGHT SCATTERING FROM LEAVES

SERGEY N. SAVENKOV[1] AND RANJAN S. MUTTIAH[2]

[1]*Taras Shevchenko University, Department of Radiophysics*
Kiev, Ukraine
[2]*Texas Agricultural Experiment Station,808 East Blackland Road,*
Temple, Texas, USA

Abstract. We describe how Mueller matrices from polarization experiments are obtained in practice. A generalized method to determine Mueller matrices and anisotropy parameters from polarimetry apparatus is given. We then devote a section to leaf light scatter.

1. Introduction

There are two classes of problems involving linear interaction of light with media: direct and inverse problems. The direct problem is to calculate or measure light scatter with known optical parameters for the object. The so-called inverse problem is to characterize an object of interest based on laboratory polarization experiments or remote sensing observations. Our attention in this chapter is on design of polarization experiments that determine Mueller matrices of leaves. From the Mueller matrix, depolarization and anisotropy properties can be determined. Past polarimetry studies [1-3] for a leaf were limited by the nature of incident light, but these experiments provided evidence for the usefulness of scattering matrices to characterize particle features. As shown in [4], measurement of the Mueller matrix secures the completeness of polarimetric studies. In particular, the number of non-zero Mueller matrix elements is found to be four and six, respectively, for transmitted and scattered light. Mueller matrices also offer the possibility of performing imaging polarimetry within a reasonable time, and to estimate measurement error, since leaf imaging can involve upwards of

G. Videen et al. (eds.), Photopolarimetry in Remote Sensing, 243-264.
© 2004 *Kluwer Academic Publishers. Printed in the Netherlands.*

10^6 pixels at large-scale measurement. Our contribution is as follows: section 2 describes light polarization, the Jones and Mueller formulations are presented in section 3, theory of measurement, Stokes vector, and optimization procedures for minimizing errors are given in section 4, and Mueller matrices for leaves are discussed in section 5, and scale-up through multiple scattering is briefly visited in section 6.

2. Light polarization

Light can be completely described by grouping four Stokes parameters into a column vector, the so-called Stokes vector [5,6], $S = \begin{bmatrix} s_1, & s_2, & s_3, & s_4 \end{bmatrix}^T$ (T means transpose). The vector components are given by:

$$s_1 = \langle E_x^2 \rangle + \langle E_y^2 \rangle$$
$$s_2 = \langle E_x^2 \rangle - \langle E_y^2 \rangle \qquad (2.1)$$
$$s_3 = 2 \langle E_x E_y \cos(\Delta) \rangle$$
$$s_4 = 2 \langle E_x E_y \sin(\Delta) \rangle$$

where, E_x, E_y are orthogonal components of electrical vector E of beam of light propagating along the z -axis; Δ - is the phase shift between E_x and E_y; $\langle \rangle$ defines the statistical averaging in time (or the ensemble average, since we assume light to be stationary and ergodic). The first Stokes parameters, s_1, is the total intensity of light. The other three parameters describe the polarization state of light. If light is completely polarized then:

$$s_1^2 = s_2^2 + s_3^2 + s_4^2 \qquad (2.2)$$

The physical meanings of the Stokes vector components (2.1), gives a principle way to determine the state of polarization of an arbitrary beam of light [6,7]. The most convenient set of measurements are those that yield the following information:

(i)   the intensity of light,
(ii)   the degree of linear polarization with respect to vertical and horizontal axes,
(iii)   the degree of linear polarization with respect to the axes oriented at ±45°,
(iv)   the degrees of left and right circular polarizations.

The second and third of these measurements can be made with a precisely oriented linear polarizer, while the fourth assumes the additional use of a wave plate. Stokes vector components can be expressed by means of the parameters of the polarization ellipse:

$$
\begin{bmatrix} s_1 \\ s_2 \\ s_3 \\ s_4 \end{bmatrix} = \begin{bmatrix} s_1 \\ s_1 \cos(2\varepsilon)\cos(2\theta) \\ s_1 \cos(2\varepsilon)\sin(2\theta) \\ s_1 \sin(2\varepsilon) \end{bmatrix}.
\tag{2.3}
$$

where θ, ε are, respectively, the azimuth and ellipticity of the polarization ellipse in a plane on which vector E oscillates. The degree of polarization of partially polarized light is defined by the ratio of the intensity of polarized light to the total intensity of light:

$$
p = \frac{I_{pol}}{I_{tot}} = \frac{\sqrt{s_2^2 + s_3^2 + s_4^2}}{s_1}
\tag{2.4}
$$

Thus, for completely polarized light (2.4) gives $p = 1$, for unpolarized light $p = 0$, and partially polarized $0 \le p \le 1$. It is often convenient to use the degree of linear polarization:

$$
p_l = \frac{\sqrt{s_2^2 + s_3^2}}{s_1},
\tag{2.5}
$$

and/or the degree of circular polarization:

$$
p_c = \frac{\sqrt{s_4^2}}{s_1}.
\tag{2.6}
$$

3. Polarization properties of studied object

When light with arbitrary polarization propagates through or interacts with an object, its polarization changes. Thus, the effect of an object on input light is to transform the Stokes vectors describing the light. Therefore, the object can be represented by a transformation matrix – the Mueller matrix:

$$
\begin{bmatrix} s_1^{out} \\ s_2^{out} \\ s_3^{out} \\ s_4^{out} \end{bmatrix} = \begin{bmatrix} m_{11} & m_{12} & m_{13} & m_{14} \\ m_{21} & m_{22} & m_{23} & m_{24} \\ m_{31} & m_{32} & m_{33} & m_{34} \\ m_{41} & m_{42} & m_{43} & m_{44} \end{bmatrix} \begin{bmatrix} s_1^{inp} \\ s_2^{inp} \\ s_3^{inp} \\ s_4^{inp} \end{bmatrix}
\tag{3.1}
$$

The Mueller matrix is a 4×4 matrix with real elements. Given the directions of input and output light and wavelength, the Mueller matrix contains all polarization properties of an object: parameters of depolarization, and amplitude and phase anisotropy. Additional information about the Mueller matrices can be found elsewhere [7-11]. The main problem, which will be addressed in this section, is the following: suppose we have a real 4×4 matrix. What conditions must be satisfied in order for it to be a Mueller matrix? A large variety of tests can be found in the literature [10-16]. All tests differ in terms of simplicity, convenience and completeness. Some of these tests are:

(i) $m_{11} \geq 0$

(ii) $m_{11} \geq |m_{ij}|$

(iii) $Tr(M) \geq 0$ (3.2)

(iv) $4m_{11}^2 \geq \sum_{i=1}^{4}\sum_{j=1}^{4} m_{ij}^2$

(v) $m_{11} \pm m_{22} \geq m_{33} \pm m_{44}$

Neither of (i)-(v), nor their combinations constitute a complete test. These tests provide only necessary conditions.

(vi)   Another test is based on the coherency matrix T which is derived from the Mueller matrix [14,16].

The elements of matrix T linearly depend on elements of M. Matrix T has four real eigenvalues since T is Hermitian, i.e., $t_{ij} = t_{ji}^*$. If three of the eigenvalues of T vanish, M is a deterministic Mueller matrix. If all four eigenvalues of T are not equal to zero, and if the entropy H of the studied object (that is its measure of "polarimetric disorder") defined by:

$$H = -\sum_{i=1}^{N} P_i \log_N P_i \; ; \; P_i = \frac{\lambda_i}{\sum_j \lambda_j}$$ (3.3)

is low (<0.5), then the object is weakly depolarizing. The Mueller matrix, which corresponds to the maximal eigenvalue, is a dominant type of deterministic polarization transformation. In [16], N (the number of non-zero eigen values of the coherency matrix) is chosen so that H is in range: $0 \leq H \leq 1$. Given eigenvalues λ_r of coherency matrix T, the initial Mueller matrix has the form:

$$M = \sum_{r=1}^{4} \lambda_r M_{J_r},$$ (3.4)

where, M_{Jr} are the deterministic Mueller matrices obtained from the Jones matrices J_r [9,10]:

$$J_r = \begin{bmatrix} d_1^i + d_2^i & d_3^i - id_4^i \\ d_3^i + id_4^i & d_1^i - d_2^i \end{bmatrix},$$ (3.5)

where, vectors $[d_1, d_2, d_3, d_4]_i^T$ are the eigenvectors of the coherency matrix T.

If entropy H is high (> 0.5), the object is strongly depolarizing and it can no longer be assumed that the object has a single dominant type of deterministic polarization transformation. In such cases, it is suggested that all four eigenvectors be used.

The coherency matrix method permits the so-called dominant type of deterministic polarization transformation i.e., the corresponding deterministic part, Mueller-Jones matrix, of the initial Mueller matrix. A Jones matrix J is a 2×2 complex valued matrix containing generally eight independent parameters from the real and imaginary parts for each the four matrix elements, or seven parameters if the absolute (isotropic) phase which is not of interest for polarizations is excluded. Every Jones matrix can be transformed into an equivalent Mueller matrix but the converse assertion is not necessarily true. Between Jones J and Mueller M_J matrices that describe deterministic objects there exist a one-to-one correspondence:

$$M_J = A\left(J \otimes J^*\right)A^{-1} \tag{3.6}$$

where, A is the following matrix:

$$A = \begin{bmatrix} 1 & 0 & 0 & 1 \\ 1 & 0 & 0 & -1 \\ 0 & 1 & 1 & 0 \\ 0 & i & -i & 0 \end{bmatrix}, \tag{3.7}$$

and $\otimes$ - denotes the Kronecker product, * - complex conjugate.

The Mueller-Jones matrix provides a complete description of the anisotropy properties of an object [9,10]. However, the information in the matrix is in implicit form. The history of the problem of analysis of the Jones and Mueller-Jones matrix goes back to the derivation of three equivalence theorems by Hurwitz and Jones [17]. According to the first theorem, an optical system (object) composed of any number of retardation plates (that is an object with linear phase anisotropy) and rotators (circular phase anisotropy) is optically equivalent to a system containing only two elements – a retardation plate, and a rotator. The second theorem is analogous to the first and but is concerned with partial polarizers (linear amplitude anisotropy) and rotators. The third theorem claims that an optical system composed of any number of partial polarizers, retardation plates, and rotators is optically equivalent to a system containing only four elements: two retardation plates, a partial polarizer, and rotator.

Another approach to the analysis of Jones and Mueller-Jones matrix exploits the polar decomposition theorem [18]. This approach was first suggested in [19] and was explored in [20,21]. The polar decomposition of a Jones matrix J can be represented as:

$$J = J_P J_R \text{ or } J = J_R J'_P \tag{3.8}$$

where, J_P, J_P' are Hermitian matrices, and J_R is a unitary matrix. The Hermitian matrix is analogous to a partial polarizer, and a unitary matrix is analogous to a retardation plate.

Proceeding from the demand of physical clearness of decomposition [22] to determine the exact values of anisotropy parameters inherent in the object of interest from its Jones or Mueller-Jones matrices, the method presented in [23] can be used. The substance of the method is the following: generally, there are four "anisotropy" transformation matrices for polarized light interaction with an object. These are phase of linear and circular anisotropy, and the amplitude of linear and circular anisotropy. The Jones matrices corresponding to these phenomena are well known [9,10], and are presented in Table 1.

The matrices contain six independent parameters: δ and azimuth α for linear phase anisotropy, ϕ for circular phase anisotropy; P and azimuth χ for linear amplitude anisotropy, and R for circular amplitude anisotropy. In [23] it was shown that the Jones matrix for an arbitrary object is in general a product of the four anisotropy matrices in the following order:

$$[Lin.Phase][Circ.Phase][Lin.Amp.][Circ.Amp.]$$

Table 1: The Jones matrices of amplitude and phase anisotropy

Anisotropy	The Jones matrix
Linear phase	$\begin{bmatrix} \cos^2(\alpha)+\sin^2(\alpha)\exp(-i\delta) & \cos(\alpha)\sin(\alpha)(1-\exp(-i\delta)) \\ \cos(\alpha)\sin(\alpha)(1-\exp(-i\delta)) & \cos^2(\alpha)\exp(-i\delta)+\sin^2(\alpha) \end{bmatrix}$
Circular phase	$\begin{bmatrix} \cos(\phi) & \sin(\phi) \\ -\sin(\phi) & \cos(\phi) \end{bmatrix}$
Linear amplitude	$\begin{bmatrix} \cos^2(\chi)+P\cdot\sin^2(\chi) & \cos(\chi)\cdot\sin(\chi)\cdot(1-P) \\ \cos(\chi)\cdot\sin(\chi)\cdot(1-P) & P\cdot\cos^2(\chi)+\sin^2(\chi) \end{bmatrix}$
Circular amplitude	$\begin{bmatrix} 1 & -i\cdot R \\ i\cdot R & 1 \end{bmatrix}$

Thus, we have the generalized Jones matrix. Its four elements are a function of six parameters: P, χ, R, δ, α, ϕ which have physical meanings. In our applications, the values for the anisotropy parameters P, χ, R, δ, α, ϕ were calculated numerically. In our applications we calculated the anisotropy parameters numerically from the deterministic Mueller matrix. In section 4 measurement of the Mueller matrix is described.

4. Measurement of the polarization of light and objects

In this section we present a general theory for the measurement of polarization characteristics of light, Stokes vector, Mueller matrix, and optimization procedures for minimizing time and errors of measurement. Stokes vector for a beam of light is determined by carrying out a series of measurements for the intensity of light transmitted through a set of polarization elements. Fig. 1. shows the polarization state analyzer (PSA).

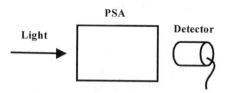

Fig. 1. Stokes vector measurement.

The measurement of a Stokes vector in most general form is described by the following equation:

$$G = W_S K .\qquad(4.1)$$

here, K is the Stokes vector to be measured, G is a vector of light intensity measured by a photodetector, and W_S is a matrix called measurement matrix whose rows are the first rows of the Mueller matrices for each configuration of polarization elements in PSA. The dimensions of the vectors and matrix in (4.1) depend on the number of measurements realized. Let the total number of measurements be N, then there are N Mueller matrices describing the corresponding configurations of polarization elements whose first rows form the $N \times 4$ matrix W_S:

$$W_S = \begin{bmatrix} w_1^1 & w_2^1 & w_3^1 & w_4^1 \\ w_1^2 & w_2^2 & w_3^2 & w_4^2 \\ \cdot & & & \\ \cdot & & & \\ \cdot & & & \\ w_1^N & w_2^N & w_3^N & w_4^N \end{bmatrix}\qquad(4.2)$$

Thus, if W_S is known, then the equation (4.2) can be solved by inverting for the incident Stokes vector:

$$K = W_S^{-1} G .\qquad(4.3)$$

The values of matrix elements W_S are determined during calibration of the Stokes polarimeter at each setting. There are three considerations for (4.3): existence, rank, and uniqueness of the inverse matrix W_S^{-1}. The first case is realized when four independent measurements are performed. In this case $N = 4$ and W_S is of rank four. The matrix W_S is nonsingular, that is, W_S^{-1} exists and is unique. The second case is realized if $N > 4$. When more than four measurements are made, the matrix W_S is not square and unique. This means that K is formally over determined by the measurements. If random measurement errors are absent, all possible W_S^{-1} result in the same values for elements of K. When there exists random measurement error, the best estimation (least square) for K is given by the pseudo-inverse of W_S: $W_S^{-1} = \left(W_S^T W_S \right)^{-1} W_S^T$. Thus, the best estimation for K in this case is:

$$K = \left(W_S^T W_S \right)^{-1} W_S^T G. \qquad (4.4)$$

The third case is realized when W_S is of rank three or less. As it was above, the optimal matrix inverse is the pseudo-inverse. However, in this case only three or less of the Stokes vector elements can be measured. The corresponding measurement apparatus is called an "incomplete Stokes polarimeter".

There are a considerable number of ways to construct the apparatus shown in Fig. 1 [24,25]. In many cases, PSA consists of a controllable retarder and fixed linear polarizer (RRFP), often with a quarter-wave plate as the retarder. In the case $N > 4$, the controllable retarder is rotated continuously and data reduction assumes Fourier transformable detector signal. In the case of $N \le 4$, the controllable retarder has four or less fixed orientations.

Thus, a PSA generally is characterized by two degree of freedom: the value of retardance and orientation of the controllable retarder. The choice of these parameters impacts measurement errors of Stokes vector elements, and is discussed in [26-29]. For example, in [26-27] an optimization with respect to the angular orientations of a (fixed value of retardance) quarter-wave plate in a PSA of RRFP scheme for case $N = 4$ is given. In [28], the retardance at equal orientations was included in optimization and, in particular, it was shown that a favorable configuration was the following: retardance of 132^0 and orientation angles of $\pm 51.7^0$ and $\pm 15.1^0$. Different approaches have been used as a figure of merit for assessing measurement errors: measurement matrix condition number, reciprocal of absolute determinant, equally weighted variance of measurement matrix, and signal-to-noise ratio [29].

Design of the apparatus for measurement of Mueller matrix elements (Mueller polarimeter) is more complicated than the design of the Stokes polarimeter described above. We initially proceed with the assumption that all 16 elements of the Mueller matrix are independent and that measurement of all 16 elements is desired. The general principle of Mueller matrix measurements consists in illuminating an object by light with controlled polarization and measurement of scattered polarization. Fig. 2 is a block scheme of the apparatus. The polarization state generator (PSG) generates polarized light. The polarization of light after linear interaction with an object (output light) is analyzed by the PSA.

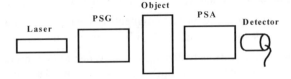

Fig. 2. Mueller matrix measurement.

Thus, the PSG forms either a sequence of definite polarizations or periodically changing polarization of input light. The output polarizations are analyzed by N configurations of polarization elements in the PSA. From the set of such measurements, a system of linear equations is developed that can be solved for the Mueller matrix elements. This yields:

$$G = QML = \sum_{i=1}^{4} \sum_{j=1}^{4} q_i^N m_{ij} l_j^N \qquad (4.5)$$

where, G is a vector of light intensity measurements by the photodetector, Q is a $N \times 4$ matrix whose rows are formed by the first rows of Mueller matrices which describes the configurations of polarization elements in PSA; M is the Mueller matrix to be measured, L is a $4 \times N$ matrix whose columns are Stokes vectors of input light.

The equation (4.5) is the most general equation for Mueller matrix measurement. This equation can be rewritten in a vector-vector product form. For that, the Mueller matrix to be measured is rewritten as a 16×1 Mueller vector: $\quad \vec{M} = \begin{bmatrix} m_{11} & m_{12} & m_{13} & m_{14} & \cdot & \cdot & \cdot & m_{43} & m_{44} \end{bmatrix}^T.$ Then, the $N \times 16$ measurement matrix W_M is introduced with elements defined as $w_{ij}^N = q_i^N t_j^N$. Equation (4.5) takes the form:

$$G = W_M \vec{M} = \begin{bmatrix} q_1^1 t_1^1 & q_1^1 t_2^1 & q_1^1 t_3^1 & \cdot & q_4^1 t_4^1 \\ q_1^2 t_1^2 & q_1^2 t_2^2 & q_1^2 t_3^2 & \cdot & q_4^2 t_4^2 \\ q_1^3 t_1^3 & q_1^3 t_2^3 & q_1^3 t_3^3 & \cdot & q_4^3 t_4^3 \\ \cdot & \cdot & \cdot & \cdot & \cdot \\ q_1^N t_1^N & q_1^N t_2^N & q_1^N t_3^N & \cdot & q_4^N t_4^N \end{bmatrix} \begin{bmatrix} m_{11} \\ m_{12} \\ m_{13} \\ \cdot \\ m_{44} \end{bmatrix}. \qquad (4.6)$$

If W_M is known, Mueller matrix m_{ij} can be solved from equation (4.6).

As in (4.3), the existence, rank, and uniqueness of inverse matrix W_M^{-1} plays a key role in the solution of equation (4.6). If W_M contains sixteen independent columns, then all sixteen elements of the Mueller matrix can be determined. When $N = 16$, then W_M^{-1} is unique. If $N > 16$, then W_M^{-1} is overdetermined. The optimal least squares estimation for m_{ij} is given by the pseudo-inverse of W_M:

$$\vec{M} = W_M^{-1} G = \left(W_M^T W_M \right)^{-1} W_M^T G. \qquad (4.7)$$

If $N < 16$, then the optimal matrix inverse is the pseudo-inverse as well. However, in this case only fifteen or less of the Mueller matrix elements can be measured. The corresponding measurement apparatus is called an "incomplete Mueller polarimeter". It is worth pointing out the following advantages of Mueller polarimetry:

(i)   no assumption was made that PSG and PSA had a particular configuration of polarization elements;

(ii)   no assumption was made on whether polarization elements were ideal or have imperfections; and,

(iii)   this approach assumes both overdeterimined and underdetermined measurement as providing least square estimation.

Eqn. (4.7) describes any existence scheme for Mueller polarimeters. Below we present examples to illustrate the application of equation (4.6) in complete and incomplete Mueller polarimeters. Consider a prevailing scheme of a Mueller polarimeter in which PSG forms four independent polarizations of input light, and PSA measures the complete Stokes vector. For this polarimeter it is convenient to rewrite (4.5) in the form:

$$Q^{-1}G = S = ML \qquad (4.8)$$

or in detailed form:

$$
\begin{bmatrix}
m_{11} & m_{12} & m_{13} & m_{14} \\
m_{21} & m_{22} & m_{23} & m_{24} \\
m_{31} & m_{32} & m_{33} & m_{34} \\
m_{41} & m_{42} & m_{43} & m_{44}
\end{bmatrix}
\begin{bmatrix}
l_1^1 & l_1^2 & l_1^3 & l_1^4 \\
l_2^1 & l_2^2 & l_2^3 & l_2^4 \\
l_3^1 & l_3^2 & l_3^3 & l_3^4 \\
l_4^1 & l_4^2 & l_4^3 & l_4^4
\end{bmatrix}
=
\begin{bmatrix}
s_1^1 & s_1^2 & s_1^3 & s_1^4 \\
s_2^1 & s_2^2 & s_2^3 & s_2^4 \\
s_3^1 & s_3^2 & s_3^3 & s_3^4 \\
s_4^1 & s_4^2 & s_4^3 & s_4^4
\end{bmatrix}
\qquad (4.9)
$$

where, l_i^k - are correspondingly arranged components of input Stokes vectors (i-th component of k-th input Stokes vector); M is the Mueller matrix to be measured; and s_i^k are components of the output Stokes vectors measured by PSA. By rewriting the Mueller matrix M in a 16×1 vector, the system (4.9) can be structured in the following way:

$$\sum_{i=1}^{4}\sum_{j=1}^{4} l_j^k m_{ij} = s_i^k; \quad k = \overline{1,4} \qquad (4.10)$$

Each of the structural parts (subsystems) of the system (4.10) have the characteristic matrix $V_{4\times4}$:

$$
V_{4\times4} =
\begin{bmatrix}
l_1^1 & l_2^1 & l_3^1 & l_4^1 \\
l_1^2 & l_2^2 & l_3^2 & l_4^2 \\
l_1^3 & l_2^3 & l_3^3 & l_4^3 \\
l_1^4 & l_2^4 & l_3^4 & l_4^4
\end{bmatrix}
\qquad (4.11)
$$

which is the matrix L which, in turn, is a 16×16 block diagonal matrix. Each ith subsystem of (4.10) can be solved independently for elements of the ith row of the Mueller matrix. Using four independent input polarizations it is possible to measure all 16 elements assuming all 16 elements of the Mueller matrix are independent. Such a polarimeter with four input polarizations is complete, and was demonstrated for light scatter measurement on English oak leaves [4].

In practice, all 16 elements very often may not be independent. Some may be zero and some may be identical to others depending on symmetry and other physical properties of the studied object [7,8]. A typical example is the deterministic class of objects which have less than seven independent elements. Hence, in measuring all 16 elements of the deterministic Mueller matrix, more than 50% are "uninformative" measurements. This problem is relevant to imaging polarimetry, since time and storage requirements are important considerations. The number of independent elements of the Mueller matrix can be taken into account in developing a polarimeter. If one takes only three input polarizations, (4.9) can be structured in the following way:

$$\sum_{i=1}^{4} \sum_{j=1}^{3} l_j^k m_{ij} = s_i^k; \quad k = \overline{1,3} \tag{4.12}$$

Each structural part (subsystem) has a characteristic matrix $V_{3\times3}$ of the following form:

$$V_{3\times3} = \begin{bmatrix} l_1^1 & l_2^1 & l_3^1 \\ l_1^2 & l_2^2 & l_3^2 \\ l_1^3 & l_2^3 & l_3^3 \end{bmatrix}, \tag{4.13}$$

which can be solved independently.

The demands are that the three input polarizations have to be linearly independent and form a well defined system of equations in transition from (4.9) to (4.12). The choice of $l_4^k = 0$, or $l_3^k = 0$ or $l_2^k = 0$, satisfies these demands. It is important to note that in this case, matrix $V_{3\times3}$ is composed of non-zero components corresponding to the set of input Stokes vectors. In the case of three input linear polarizations ($l_4^k = 0$), the matrix $V_{3\times3}$ is simply a principal minor of dimension (3×3) of the matrix L. In general, L is singular, whereas $V_{3\times3}$ for linear independent input Stokes vectors is invertible. Hence, three input polarizations can determine 12 elements of the Mueller matrix. Similarly, by setting $l_4^k = l_3^k = 0$, $l_4^k = l_2^k = 0$, or $l_3^k = l_2^k = 0$ eight elements of the Mueller matrix can be determined. Three and two input polarizations are incomplete [31]. The Mueller matrices measured by a polarimeter with three input polarizations

$$\begin{bmatrix} m_{11} & m_{12} & m_{13} \\ m_{21} & m_{22} & m_{23} \\ m_{31} & m_{32} & m_{33} \\ m_{41} & m_{42} & m_{43} \end{bmatrix} \begin{bmatrix} m_{11} & m_{12} & m_{14} \\ m_{21} & m_{22} & m_{24} \\ m_{31} & m_{32} & m_{34} \\ m_{41} & m_{42} & m_{44} \end{bmatrix} \begin{bmatrix} m_{11} & m_{13} & m_{14} \\ m_{21} & m_{23} & m_{24} \\ m_{31} & m_{33} & m_{34} \\ m_{41} & m_{43} & m_{44} \end{bmatrix} \tag{4.14}$$

are informationally complete for deterministic objects. In other words, any of these structures allow solving completely the inverse problem, that is to say that the values of parameters P, χ, R, δ, α, ϕ can be determined. The incomplete Mueller matrices measured by a polarimeter with two input polarizations:

$$
\begin{bmatrix} m_{11} & m_{12} \\ m_{21} & m_{22} \\ m_{31} & m_{32} \\ m_{41} & m_{42} \end{bmatrix}
\begin{bmatrix} m_{11} & m_{14} \\ m_{21} & m_{24} \\ m_{31} & m_{34} \\ m_{41} & m_{44} \end{bmatrix}
\begin{bmatrix} m_{11} & m_{13} \\ m_{21} & m_{23} \\ m_{31} & m_{33} \\ m_{41} & m_{43} \end{bmatrix}
\tag{4.15}
$$

in spite of having eight elements, are not complete for deterministic classes of objects, although measurement time is roughly reduced by half from the three input case.

Not every set of independent Stokes vectors in (4.11), (4.13), and lesser matrices can be used as a set of input polarizations [30]. In other words, in solving the systems (4.10) and (4.12), the corresponding inverse matrices $(V_{ixi})^{-1}$ can exist but the resulting measurement errors will be too high. This follows from the fact that when solving the system of linear equations, say (4.8), small changes (fluctuations) in components of S cause large changes in M. The magnitudes of the changes in M are connected with conditioning of the characteristic matrices $(V_{ixi})^{-1}$. We consider optimization of V_{ixi}. We take as a figure of merit the condition number $cond(V_{ixi})$ of the matrices V_{ixi} based on Frobhenius norm due to convenience and geometric interpretation. The condition number of a matrix can be calculated in the following way [18]:

$$
cond(V) = \left\| V^{-1} \right\| \cdot \left\| V \right\|;
\tag{4.16}
$$

where $\left\| V^{-1} \right\|, \left\| V \right\|$ are the norms of inverse and direct matrix V, respectively. The Frobhenius norm is

$$
\left\| V \right\| = \left(\sum_{i,j} \left| V_{ij} \right|^2 \right)^{1/2}.
\tag{4.17}
$$

Then the upper bound for the Mueller matrix element error δM can be estimated as

$$
\delta M \le \frac{2cond(V)\delta S}{1 - cond(V)\delta S}
\tag{4.18}
$$

where,

$$
\delta M = \frac{\left\| M_{exact} - M_{exp} \right\|}{\left\| M_{exact} \right\|}, \qquad \delta S = \frac{\left\| S_{exact} - S_{exp} \right\|}{\left\| S_{exact} \right\|}.
\tag{4.19}
$$

The quantity S_{exp} is the measured Stokes vector, S_{exact} is the corresponding exact Stokes vector, M_{exp} and M_{exact} are the measured and exact Mueller matrices of the object, respectively. The value of δS always can be obtained by measuring the Stokes vector of known polarization states. Here, the Frobhenius norm is taken for δM and the Euclidean norm is taken for δS. The explicit form of the condition numbers as a function of azimuths and ellipticities of four and three input polarizations is cumbersome, and best addressed through a geometric approach [32].

All possible polarization states in (2.3) form the sphere of unit radius known as the Poincaré sphere [33,10]. The longitude and latitude of a point on the sphere are respectively related to doubled azimuth θ and ellipticity ε of the polarization ellipse of the light. Polarizations, whose Stokes vectors form the matrices V_{ixi}, can be represented by unmatched points on the Poincaré sphere. These points form the vertices of a tetrahedron, triangle, and straight-line segment, corresponding to four-, three- and two input polarizations when their respective Stokes vectors are linearly independent. It can be shown [18] from vector algebra that the volume of the tetrahedron, area of the triangle, and length of the straight-line segment are proportional to the determinants of the matrices V_{ixi} for their respective input vectors. On the other hand, the condition numbers are inversely proportional to the determinants of these matrices.

Thus, the problem of minimizing the condition numbers narrows to the problem of maximizing the volume, area, and length of geometrical figures inscribed within the Poincaré sphere. From geometry, it can be shown that the inscribed tetrahedron with the largest volume and triangle with largest area are the regular tetrahedron and triangle. The straight-line segment has maximum length if it is a diameter of a Poincaré sphere in cross section. This geometrical interpretation allows defining the exact minimal values for condition numbers:

$$cond\left(V_{4\times4}\right) = 4.472$$

$$cond\left(V_{3\times3}\right) = 3.162 \qquad (4.20)$$

$$cond\left(V_{2\times2}\right) = 2.$$

In general, an infinite number of optimal sets of input polarizations exist that secure these condition numbers for four and three input polarizations. This corresponds to the rotation of the inscribed regular tetrahedron and triangle inside the Poincaré sphere. Whereas, for two input polarizations only three pairs of input polarizations, namely, linear polarizations with azimuths at either 0^0 and 90^0 or $\pm45^0$, and left and right circular polarizations are allowed. From (4.18) and (4.20) it can be seen that, all other circumstances being equal, we have a 25% reduction of the number of necessary measurements and an approximately 40% reduction of measurement error for two input polarizations in comparison to three input polarizations. In comparison with four input polarizations, we have a reduction by a factor of 2 of the number of necessary measurements, and approximately a factor of 2.3 reduction in measurement error. The combination of these two factors, the decrease of errors and decrease of number of necessary measurements, makes this approach to the measurement of Mueller matrix elements especially attractive in imaging polarimetry.

Similar analysis of (4.5) can be carried out for any other scheme of Mueller polarimeter. The dual rotating retarder polarimeter is discussed in [34,35].

5. Leaf Light Scatter

Polarization measurement on leaves and canopies offers the possibility of remotely recognizing leaf particle properties through the Mueller matrix and molecular aggregate features through the anisotropy parameters. Fluorescence (which is emitted as a small fraction of the energy balance during photosynthesis) is now commonly employed by botanist to determine a considerable amount of leaf level biochemical processes [36]. Fluorescence has also been detected in the derivative of leaf and canopy reflectance curves [37]. A limitation of fluorescence is that sensing is limited to areas with pigmentation, and cellular and molecular aggregate features can be difficult to detect. Most polarization measurements of vegetation at large distances have been made using radar, while laser lidar systems are making gains. In active radar sensing, the phase between transmitted and received waves have been used to identify urban and canopy features [38]. The back-scattered vertical polarization (VV) penetrates deeper into the canopy layer than horizontal polarization (HH), and comparison between the vertical and horizontal polarizations can be interpreted in terms of terrestrial features. In passive radar sensing, the brightness temperature is determined and compared to canopy features such as roughness, surface area, and volume [39]. All these methods take a "down-scale" approach, since remote sensing is related to terrestrial attributes through field campaigns.

Another approach is to "up-scale" to canopies from leaf-level light scattering matrices. To do this requires careful experimental characterization of the leaf. First, it must be determined whether the whole leaf is behaving as a simple single scatter system i.e., even though light multiple scatters within the leaf, the final "effect" when light is reflected or transmitted to a sensor may show single-scatter behavior. If the whole leaf were behaving as a single scatterer, then scaling to canopies can be accomplished by doubling methods. Single scatter versus multiple scatter is determined experimentally by examining photon flux as function of concentration. If the photon flux changes linearly with concentration, then single scattering occurs. If effective leaf light scatter were a multiple scattering phenomena, then up-scaling is more difficult since adding and doubling by each scattering layer within the leaf must be initiated starting with the single-scattering matrix. For the leaf, the analogous concept to "concentration" is particle density. Studies of leaf particle density can be accomplished through genetic mutations. For example, all chromosomes on *Arabidopsis thaliana* (a small mustard plant) have been mapped, and researchers have generated mutated varieties. At the Arabidopsis website (http://www.arabidopsis.org) one can input a desired mutation (such as chloroplast density, palisade density and so on) and germplasm identifier, and the mutated gene are given. This plant has a rapid growth period (a few weeks) making it an ideal vehicle for light-scattering studies for vegetation.

Fig. 3 shows the cross-sectional view of a typical leaf based on that given in [40]. Leaves can contain outer pubescence (hair) layers to enhance thermal dissipation. The cuticle wax serves to block loss of leaf moisture, and protect against invasion of pathogens. Within the epidermis are specialized cells called guard cells or stomata. The stomata regulate the flow of atmospheric gases (mainly CO_2 and O_2 are relevant), and water vapor into and out of the leaf. The stomata also contain chloroplasts with chlorophyll molecules. Specialized flavonoid compounds in the epidermal cells provide UV to the deeper layers. Even though the epidermis is compact, sub-micron spacing between cells can be found. Fig. 3 depicts only a few chloroplasts for drawing convenience, while in reality the mesophyll cells contain numerous chloroplast inclusions. The drawing of the cells also is also grossly simplified to standard shapes. Since light absorption by chlorophyll molecules initiates carbon fixation, we know that light has to reach nearly all internal parts of the leaf, to within chloroplast, to the still smaller grana, and specialized antenna structures (which we will generically call "harvesting complexes" here, keeping in mind that it is a simplification). The primary optical question of interest are: how does light get inside the leaf? what roles does the air spacing, and particle shapes and sizes play in either the random or preferential movement of light ? how can light scattering be non-invasively measured ? One theory [41] from geometric optics states that light at grazing angles is focused to the center of the leaf (as shown for a spherical particle by Hapke [40]).

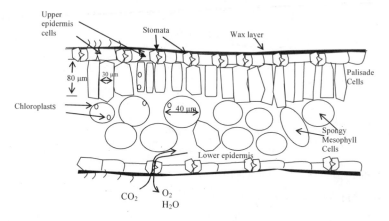

Fig. 3. Cross-sectional view of a typical leaf. In reality, all mesophyll cells and stomata contain numerous chloroplasts.

Because the cross-section of leaves is densely composed of particles of different shapes, sizes, pigmentation with diverse refractive indices, and variable particle spacing, a comprehensive development of a reflectance model is difficult. Many optical properties such as the complex refractive indices required to model a multi-particle system are unknown, except at the whole leaf level. Past attempts have included geometric optics [41], Fresnel reflection [43], and light absorption after accounting for leaf pigmentation [44]. Fig. 4 shows Fresnel reflectivity coefficients (at λ = 400 nm), for refractive index of m=1.34 + i0.13 obtained for stacked whole cotton leaves [43]. The parallel reflectivity reaches zero at the Brewster angle of about 55°. The phase shift of the parallel and perpendicular components also has the most dramatic change at the Brewster angle (Fig. 5).

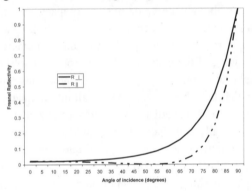

Fig. 4. Fresnel reflectivity coefficients for cotton leaves.

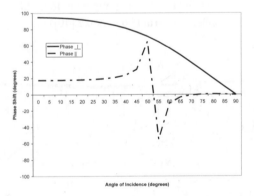

Fig. 5. Phase shift of the Fresnel coefficients.

Polarization defined as,

$$P = \frac{R_\perp - R_=}{R_\perp + R_=}$$
(5.1)

is shown in Fig. 6. Volten *et al.* [45] found that phytoplankton with and without gas vacuoles also displayed a bell-shaped polarization curve, consistent with Mie particle scatter. Therefore, light scatter observation through linear polarization suggests large sized particles ($> \lambda$) dominating light reflectance in leaves.

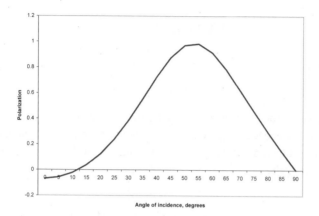

Fig. 6. Polarization curve for Fresnel reflection.

Experimentally, we obtained the following form of the Mueller matrix for transmitted and back scattered light for English oak leaves [4]:

$$
\begin{bmatrix}
m_{11}^{trans} & 0 & 0 & 0 \\
0 & m_{22}^{trans} & 0 & 0 \\
0 & 0 & m_{33}^{trans} & 0 \\
0 & 0 & 0 & m_{44}^{trans}
\end{bmatrix},
\tag{5.2}
$$

$$
\begin{bmatrix}
m_{11}^{bscatt} & m_{12}^{bscatt} & 0 & 0 \\
m_{21}^{bscatt} & m_{22}^{bscatt} & 0 & 0 \\
0 & 0 & m_{33}^{bscatt} & 0 \\
0 & 0 & 0 & m_{44}^{bscatt}
\end{bmatrix}.
\tag{5.3}
$$

These matrices correspond structurally with the Mueller matrix of so-called macroscopically isotropic and mirror-symmetric scattering media composed of randomly oriented particles with plane of symmetry and/or equal numbers of randomly oriented particles and their mirror-symmetric counterparts [46]:

$$
\begin{bmatrix}
m_{11} & m_{12} & 0 & 0 \\
m_{21} & m_{22} & 0 & 0 \\
0 & 0 & m_{33} & m_{34} \\
0 & 0 & m_{43} & m_{44}
\end{bmatrix}.
\tag{5.4}
$$

Non-zero elements of these matrices can be considered as the corresponding incomplete Mueller matrices. The Mueller matrix (5.4) contains eight non-zero elements. For this matrix, (4.10) yields:

$$
\begin{bmatrix}
m_{11}l_1^1 + m_{12}l_2^1 \\
m_{11}l_1^2 + m_{12}l_2^2 \\
m_{11}l_1^3 + m_{12}l_2^3 \\
m_{11}l_1^4 + m_{12}l_2^4 \\
\cdot \\
\cdot \\
m_{43}l_3^4 + m_{44}l_4^4
\end{bmatrix} .
\tag{5.5}
$$

It can be seen from (5.5) that the well-defined system can be obtained from two linear, independent polarizations with all four nonzero Stokes vector components. Equation (5.5) contains two independent systems of equations with following characteristics matrices:

$$
V_1 = \begin{bmatrix} l_1^1 & l_2^1 \\ l_1^2 & l_2^2 \end{bmatrix}, \text{ and } V_2 = \begin{bmatrix} l_3^1 & l_4^1 \\ l_3^2 & l_4^2 \end{bmatrix} .
\tag{5.6}
$$

Minimizing of the condition numbers of matrices (5.6) enables choosing optimal input polarizations for measuring the Mueller matrix (5.4). The same for the backscattered Mueller matrix (5.3), with some over determination that can be used to estimate elements m_{33} and m_{44}.

To measure all elements of the transmitted Mueller matrix (5.2), it is sufficient to use only one input polarization:

$$
\begin{bmatrix}
m_{11}l_1^1 \\
m_{22}l_2^1 \\
m_{33}l_3^1 \\
m_{44}l_4^1
\end{bmatrix} .
\tag{5.7}
$$

Optimal input polarization for (5.7) is the polarization with equal components: $|l_2| = |l_3| = |l_4|$. It is interesting to note that there is only one input polarization for studies of the upper layers of the atmosphere illuminated by unpolarized sunlight [47].

6. Multiple Scattering

Multiple scattering using the single scattering matrix has been modeled in the following way [48]:

1). With reflection matrix R and transmission matrix T,

$$Q_1 = R * R,$$
$$Qn = Q_1 Q_{n-1}$$
$$S = \sum_{n=1}^{\infty} Q_n$$
$$D = T + e^{-\tau/\mu_0} S + ST$$
$$U = e^{-\tau/\mu_0} R + RD$$

(6.1)

where, τ is optical depth, (μ', μ_0) is the incident and reflected cosine angle θ (about zenith).

2). The reflection and transmission matrices for layer thickness of 2τ are given by:

$$R(2\tau) = R + e^{-\tau/\mu_0} U + T * U$$
$$T(2\tau) = e^{-\tau/\mu} D + e^{-\tau/\mu_0} T + TD$$

(6.2)

where, the matrices $R*$ and $T*$ are the reflected and transmitted components from an illumination source placed at a distance τ within the media. The matrices D and U are the diffuse downward and upward radiation at the mid-level of combined thickness 2τ.

3). For homogeneous layers, $R* = R$, and $T* = T$.

4). The matrix multiplication of two phase matrices requires the Fourier series expansion of the azimuthally dependent functions, so that each term in the Fourier series expansion can be treated independently [48, 49].

Different types of rainbows form from water droplets, depending on the number of internal reflections: the primary rainbow is formed by rays that are internally reflected once, the secondary rainbow is formed by rays that are internally reflected twice, etc. Hansen [50] used steps (1)-(4) listed above to simulate the intensity of light (phase function), and the degree of polarization as functions of scattering angle for clouds. When the sun was at zenith ($\mu' = \mu_0$ = 1), and as the optical thickness of the clouds increased, the amplitude of the phase function decreased, and single-scattering effects on degree of polarization washed off. The zero point of the degree of polarization stayed the same for different optical thickness. Secondary rainbow effects did not appear in the polarization curves. For azimuth dependence of incident light, the single-scattering form of the degree of polarization was preserved, but the zero point was slightly shifted with optical thickness. In our experiments on leaves [4] with light source at zenith, a null result was obtained for transmission degree of polarization, and degree of polarization decreased with backscatter angles. Whether the periodic doubling method yields a similar degree of polarization has yet to be seen.

References

1. Y. I. Astrashevski, A.B. Sikorski, V.V. Sikorski, and G.F. Stelmakh, and V. I. Shuplyak, *J. Appl. Spec* **65**, 103 (1998) (in Russian).
2. Y.I. Astrashevski, A.B. Sikorski, V.V. Sikorski, and G.F. Stelmakh, *J. Appl. Spec.* **66**, 100 (1999) (in Russian).
3. P.N. Raven, D.L. Jordan, and C.E. Smith, Opt. Eng. **41**,1002 (2002).
4. S.N. Savenkov, R.S. Muttiah, and Ye.A. Oberemok, Appl. Opt. **42**, 4955 (2003).
5. G.G. Stokes, Trans. Cambridge Philos. Soc. **9**, 399 (1852).
6. M. Born, and E.Wolf, "Principals of optics" (Pergamon press, New York, 1968).
7. H.C. van de Hulst. "Light scattering by small particles" (Wiley. New York, 1957).
8. C. F. Bohren and D. R. Huffman, "Absorption and Scattering of Light by Small Particles" (Wiley, New York, 1983).
9. R.M. Azzam and N.M. Bashara, "Ellipsometry and polarized light" (North-Holland, Amsterdam, 1977).
10. C. Brosseau, "Fundamentals of polarized light. A statistical optics approach" (Wiley, New York, 1998).
11. A. Gerrard, and J.M. Burch, "Introduction to matrix methods in optics" (Wiley, New York, 1975).
12. Zhang-Fan Xing, J.Mod.Opt. **39**, 461 (1992).
13. J.W. Hovenier, Appl.Opt. **33**, 8318 (1994).
14. S.R. Cloud, Optik(Stuttgart), **7**, 26 (1986).
15. J.W. Hovenier, and C.V.M. van der Mee, J.Quant.Spectrosc.Radiat.Transfer. **55**, 649 (1996).
16. S.R. Cloude, and E. Pottier, Opt. Eng. **34**, 1599 (1995).
17. H. Hurwitz, and C.R. Jones, JOSA **31**, 493 (1941).
18. G.H. Golub, and C.F. van Loan, "Matrix computations" (John Hopkins Univ. Press, Baltimore, 1983).
19. C. Whitney, JOSA **61**, 1207 (1971).
20. J..J. Gil, and E. Bernabeu, Optik (Stuttgart), **76**, 67 (1987).
21. Shih-Ya Lu, and R.A. Chipman, JOSA A. **13**, 1106 (1996).
22. W.-M. Boerner (eds.), "Direct and inverse methods in radar polarimetry" (Kluwer, Dordrechet, 1992).
23. V.V. Mar'enko, and S.N. Savenkov, Optics and Spectroscopy **76**, 94 (1994).
24. P.S. Hauge, Surface Sci. **96**, 108 (1980).
25. R.M. Azzam, SPIE Proc. **3121**, 396 (1997).
26. A. Ambirajan and D.C. Look, Opt. Eng. **34**, 1651 (1995).
27. A. Ambirajan and D.C. Look, Opt. Eng. **34**, 1656 (1995).
28. D.S. Sabatke, M.R. Descour, E. Dereniak, W.C. Sweatt, S.A. Kemme, and G.S. Phipps, Opt. Lett. **25**, 802 (2000).
29. J.S. Tyo, Opt. Lett. **25**, 1198 (2000).
30. S.N. Savenkov, Opt. Eng. **41**, 965 (2002).
31. Ye.A. Oberemok, and S.N. Savenkov, J. Appl. Spectroscopy **70**, 203 (2003) (In Russian).
32. R. Azzam, I.M. Elminyawi, and A.M. El-Saba, J. Opt. Soc. Am. *A* **5**, 681 (1988).
33. H. Poincare, "Theorie mathematique de la lumiere" (Georges Carre, Paris. Vol.2. 1892).
34. S.N. Savenkov, and K.E. Yushtin, Radiotekhnika **124**, 111 (2002) (In Russian).
35. R.M. Azzam, Opt.Lett. **2**, 148 (1978).
36. G.C. Papageorgiou, and Govindjee, "Chlorophyll Fluorescence: A Signature of Photosynthesis" (Kluwer, Dordrecht, The Netherlands, 2004).
37. P.J. Zarco-Tejada, J.R. Miller, and G.H. Mohammed. "Remote sensing of solar-induced chlorophyll fluorescence from vegetation hyperspectral reflectance and radiative transfer simulation", in From Spectroscopy to Remotely Sensed Spectra of Terrestrial Ecosystems, R.S. Muttiah (Ed.). (Kluwer, Dordrecht, The Netherlands, 2002) 233-269.
38. D. L. Evans, T. G. Farr, J.J. Van Zykl, and H. A. Zebker, IEEE Trans. Geo. & Remote Sensing 26(6): 774-789 (1988).

39. A.K. Fung, "Microwave remote sensing of soil moisture" in From Spectroscopy to Remotely
     Spectra of Terrestrial Ecosystems, R.S. Muttiah (Ed.). (Kluwer, Dordrecht, 2002) 21-59.
40. P. S. Nobel. "Biophysical plant physiology and ecology" (W.H. Freeman and Company,
     New York, 1983) 3.
41. R.A. Bone, D.W. Lee, and J.M. Norman, Appl. Opt. **24**:1408 (1985).
42. B. Hapke, "Theory of reflectance and emittance spectroscopy," (Cambridge University
     Press, Cambridge, 1993).
43. W.A. Allen, H.W. Gausman, and A.J. Richardson, J. Opt. Soc. Am **60,** 542 (1970).
44. S. Jacquemoud, S.L. Ustin, J. Verdebout, G. Schmuck, G. Andreoli, and B. Hosgood,
     Remote Sens. Envr. **56,**194 (1998).
45. H. Volten, J. F. de Haan, J.W. Hovenier, R. Schreurs, W. Vassen, A. G. Dekker, H.J.
     Hoogenboom, F. Charlton, and R. Wouts. Limnol. Ocean. **43,** 1180-1197 (1998).
46. M.I. Mishchenko, L.D. Travis, and A.A. Lacis, "Scattering, absorption, and emission of light
     by small particles," (Cambridge Univ. Press, United Kingdom, 2002).
47. M.I. Mishchenko, personal communication.
48.   J. E. Hansen, J. Atmos. Sci. **28,**120-125 (1971).
49. J.F. de Haan, P. B. Bosma, and J.W. Hovenier, Astron. Astrophys. **183,** 371-391 (1987).
50. J.E. Hansen, J. Atmos. Sci. **28,** 1400-1426 (1971).

Ranjan Muttiah   *Sergey Savenkov*

NATO participants loitering at Chufut-Kale.

NATO participants demonstrate their keen sense of fashion at Chufut-Kale.

OPTICAL PROPERTIES AND BIOMEDICAL APPLICATIONS OF NANOSTRUCTURES BASED ON GOLD AND SILVER BIOCONJUGATES

N.G. KHLEBTSOV,[1,2] A.G. MELNIKOV,[1]
L.A. DYKMAN[1] AND V.A. BOGATYREV[1,2]

[1]*Institute of Biochemistry and Physiology
of Plants and Microorganisms,
Russian Academy of Sciences,
13 Pr. Entuziastov, Saratov 410049, Russia,*
[2]*Saratov State University, 155 Moskovskaya Str.,
Saratov 410026, Russia*

Abstract. We discuss optical properties of single and aggregated colloidal gold and silver conjugates that can be fabricated by adsorption of a biopolymer onto nanoparticle surfaces. We start with a discussion of two-layer and multilayer optical models for colloidal gold and silver nanoparticle conjugates that consist of a metal core and a polymer shell formed by recognizing and target molecules. The point at issue is the core-size optimization of conjugate-based nanosensors as elementary transducers of molecular binding events into optical signals. We present a detailed discussion of optical properties of various aggregated conjugate-based structures such as bispheres, linear chains, plane arrays on a rectangular lattice, compact and porous clusters embedded on a cubic body-centerd lattice, and random fractal aggregates. Our attention is focused on the following topics: (1) statistical and orientation averaging of optical observables; (2) dependence of extinction and scattering spectra on the optical binary coupling of conjugates; (3) optical effects related to the chain-like structures; (4) effects of polymer coating, interparticle spacing, and cluster structure; (5) simulation of kinetic changes in the optical properties of aggregated sols formed during biospecific binding. Finally, we discuss experimental data and biomedical applications of metal nanoparticles and their biospecific conjugates in various biomedical studies.

G. Videen et al. (eds.), Photopolarimetry in Remote Sensing, 265-308.

1. Introduction

Colloidal-gold (CG) and silver nanoparticles have been used widely during the past years as effective optical transducers of biospecific interactions [1, 2, 3, 4]. In particular, the resonance optical properties of nanometer-sized CG particles have been employed to design biochips and biosensors [1, 5, 6] used as analytical tools in biology (determination of DNA, RNA, proteins, and metabolites), medicine (drugs screening, antigen and antibody determination, virus and bacterial diagnostics), and chemistry (on-line environmental monitoring, quantitative analysis of solutions and disperse media). We consider polynucleotide detection, based on hybridization of conjugate linkers with complementary oligonucleotides [7, 8] as a special and important example.

It is well known [9, 10, 11, 12] that the localized plasmon resonance (LPR) peak magnitude and its spectral position are dependent upon the optical permittivity, size, shape, and interparticle spacing of the nanoparticles as well as upon their local environment, including adsorbate molecules, dispersion medium (solvent), and supporting substrate. This principle is behind all biomedical applications of noble-metal nanoparticles as elementary optical biosensors that are able to detect interactions between recognizing and target biomacromolecules near the particle surface [3, 4, 13]. Practical realization of this principle includes three basic steps: (i) formation of an ensemble of metal particles on a dielectric substrate by means of self-assembly process [14] or controlled nanolithography technology [15]; (ii) chemical modification of the particle surfaces to create functional chemical groups; and (iii) attachment of the recognizing molecules (oligonucleotides, antibodies, biotin, lectin, etc.) to the functionalized particle surface [2, 3, 4, 16].

Conjugation of gold particles with biopolymers leads to the formation of an adsorbed dielectric shell that does not change the spectral properties of conjugates significantly, as compared with bare particles [17]. Addition of complementary molecules to a bioconjugate probe results in two possible modes of biospecific interaction between the adsorbed recognizing molecules of conjugates and the complementary (target) molecules.

In the first mode, the interaction between target molecules and bioconjugate probe results in an aggregation of nanoparticles that can be monitored by extinction spectra. It is the key idea of the well-known sol particle immunoassay (SPIA) introduced by Leuvering et al. in 1980 [18]. The physical origin of pronounced changes in sol color and in extinction spectra is the strong electrodynamic interaction of gold particles caused by their close proximity due to biospecific or salt aggregation [10, 19, 20, 21, 22]. The same physical picture is applicable to CG particles with chemically attached oligonucleotides that can interact with complementary nucleotide chains [21, 23] or to the gold nanoparticle aggregation induced by a non-cross-linking

DNA hybridization [24]. This strategy (analogous to SPIA) can be applied to protein detection at nanogram level [25]. It should be emphasized that such types of interaction are possible when both complementary components (conjugate-conjugate or conjugate-molecule) possess several sites for cross-linked binding. Along with changes in extinction, the scattering of light is strongly dependent on particle-particle proximity, and the scattering technique has been proposed for optical sensing of biospecific interactions in suspension of colloidal gold bioconjugates [26, 27]. Resonance light scattering (RLS) has been used to find several promising applications in recent years. For example, Roll *et al.* [28] used RLS spectra excited by white LED radiation to detect CG aggregation induced by avidin-biotin interaction. Quite recently, RLS spectroscopy was applied to detect a trace amount of thiamazole that enhanced scattering intensity due to aggregation of gold colloid [29] and to detect Au nanoparticles inside single brome mosaic virus (BMV) capsids [30]. In addition, RLS spectroscopy was shown [31] to be useful in monitoring the preparation of gold nanoparticle-supported DNA probes.

In the second interaction mode, the binding of target molecules results in formation of a secondary polymer shell without any aggregation. Such a scenario is typical for interaction between the conjugates of CG to monoclonal antibodies (only one clone) and the corresponding target antigens, or between the conjugates of CG to polyclonal antibodies and the low-molecular antigens (haptens). The second interaction mode has been utilized for real-time monitoring of biomolecular protein-protein binding and high-throughput clinical screening [32, 33, 34]. The optical signal arises from the dependence of the extinction maximum and its position on the local dielectric properties near the particle surface, which are altered due to biospecific interaction [9, 12, 17].

It is possible to implement the second interaction mode not only with conjugates in solution but as planar nanoscale biosensors as well. These biosensors are based on spatially ordered gold or silver nanoparticle arrays placed on a substrate by using self-assembling [35], covalent attaching [36] or nanolythography [15, 37] techniques. The biospecific binding of the target molecules to the surface-attached recognizing molecules is monitored by changes in the extinction LPR magnitude [35, 36] or in its spectral location [37, 38]. Another approach is based on measuring the spectra of light scattering from a single nanoparticle placed on a dielectric substrate [3]. In the latter case, microscopic systems equipped with dark-field illumination and CCD cameras or micro-spectrographs are used [3]. Quite recently, two research groups [39, 40] simultaneously demonstrated a record sensitivity of such nanoscale biosensors utilizing single-particle LPR spectral shifts caused by adsorption of fewer-than-60,000 small molecules (about 100 zeptamoles). Moreover, this approach allows for the real-time kinetics of single-nanoparticle response. In line with quantum-dots technology [41], the single-particle resonant scattering approach opens the door to the detection of the single-molecule interactions.

It is clear that optimization of the nano-gold-markers methodology demands a deep insight into the optical properties of conjugates and aggregates built from these nanostructures. However, noticeable progress in the field has been achieved only very recently, because of the strong electrodynamic many-particle interaction of cluster nanoparticles when its spacing is less than their size [10, 22, 42]. This paper gives an overview of recent works related to optical properties of nanoscale structures based on colloidal gold and silver particles.

2.   Optical models for colloidal gold bioconjugates

Conjugation of biopolymers with noble-metal nanoparticles changes their local surrounding medium and, accordingly, alters their optical properties attributed to the plasmon resonance [9, 43, 44]. Experimental observations currently are treated in terms of a two-layer spherical Mie particle [45] with a gold core and a homogeneous polymer coating [12, 46]. However, it is known from theory and experiments (see, e.g., [47] and references in [17]) that adsorbed polymer shells are not homogeneous. Furthermore, the homogeneous-shell model is not suitable to describe biospecific-binding events between CG conjugates and target molecules. A conjugate, being prepared for further use, contains some "primary" inhomogeneous polymer coating with "recognizing" molecules (immunoglobulins, lectins, etc.). Bimolecular binding of target molecules results in the formation of an additional, "secondary," biopolymer shell that again is not homogeneous. To treat such complexes as two-layer particles, one needs an artificial recalculation of real-shell parameters into an average homogeneous dielectric phantom. Nevertheless, the simple two-layer model is useful as a starting point for understanding the bioconjugate optics.

2.1.   Two-layer conjugate model

A two-layer conjugate can be characterized by the metal core diameter $d = 2R$, the shell diameter $d_s = 2R_s = d + 2s$, and the corresponding refractive indices n and n_s, respectively. Throughout this chapter, the optical constants of gold and the surrounding medium (water) were calculated according to Ref. [48]; whereas, for silver particles we used the data of Ref. [49]. The extinction and light-scattering intensity of a conjugate suspension were calculated by equations [12]

$$A = \lg e \frac{3cl}{2\rho}\left(\frac{d_s}{d}\right)^2 \frac{Q_{ext}}{d},$$
(2.1)

$$I_{90}(\lambda) = \lg e \frac{3cl}{2\rho d}\left(\frac{d_s}{d}\right)^2\left[\frac{16F_{11}(90)}{3x_s^2}\right],$$ (2.2)

where e is the natural logarithm basis; c and ρ are the weight concentration and the density of metal, respectively; $l = 1\text{cm}$ is the cuvette thickness; $Q_{ext} = C_{ext}/\pi R_s^2$ is the extinction efficiency; $x_s = kR_s$ is the shell size parameter; and F_{11} is the first element of the Müller scattering matrix [50]. For colloidal gold and silver particles or conjugates, we used the mass-volume concentrations $c=57$ μg/ml and $c=5$ μg/ml, respectively.

Polymer adsorption on gold nanoparticles results in two effects: (1) it increases the extinction and scattering maxima and (2) it shifts the resonance to the red part of spectrum. Detailed calculations for the gold-core diameters $d = 10\text{-}160$ nm, the shell thickness $s = 0\text{-}10$ nm, and the shell refractive indices $n_s = 1.4$ and $n_s = 1.5$ can be found elsewhere [12]. Here we provide only two examples that illustrate the effect of polymer adsorption on the value of the extinction maximum (Fig. 1a) and on the extinction peak position (Fig. 1b).

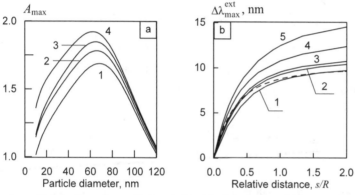

Figure 1. **a** – The size dependence of A_{max} calculated for bare gold particles (curve *1*) and for conjugates with $n_s = 1.4$, $s = 5$ nm (*2*), $n_s = 1.4$, $s = 10$ nm (*3*), and $n_s = 1.5$, $s = 5$ nm (*4*). **b** – The dependence of $\Delta\lambda_{max}^{ext}$ on the core/shell ratio s/R_g. Calculations for $d = 5$ (*1*), 10 (*2*), 15 (*3*), 30 (*4*), and 45 nm (*5*), $n_s = 1.5$. The dashed curve corresponds to Eq. (2.5).

Figure 1a shows that the maximal extinction increases with increasing adsorbed polymer shell thickness and shell refractive index. All curves have a maximum for gold core diameters 60-80 nm. This maximum is closely related to the optimization problems discussed in Refs. [12, 17]. For small particles, the shift of extinction plasmon resonance can be described by a universal dependence [12, 51]. In the first approximation, the optical properties of bioconjugates are well explained by a simple two-layer electrostatic model (previous analysis [43] contains some inaccuracies, see Ref. [12]). We shall consider the extinction spectra only. For small conjugates, the extinction is determined by the absorption efficiency

$$Q_{abs} = 4x_s \, \text{Im}\left(\frac{\varepsilon_{av} - \varepsilon_m}{\varepsilon_{av} + 2\varepsilon_m}\right), \tag{2.3}$$

$$\varepsilon_{av} = \varepsilon_s \frac{\varepsilon + 2\varepsilon_s + 2f(\varepsilon - \varepsilon_s)}{\varepsilon + 2\varepsilon_s - f(\varepsilon - \varepsilon_s)}, \tag{2.4}$$

where ε_m is the dielectric permittivity of the surrounding medium, ε_{av} is an equivalent ("average") dielectric permittivity of a two-layer sphere with dielectric parameters ε (core) and ε_s (shell), f and g are the volume fractions of core and shell, respectively

$$f \equiv 1 - g = 1/(1 + s/R_g)^3. \tag{2.5}$$

The well-known plasmon resonance condition $\varepsilon_{av} = -2\varepsilon_m$ [45] gives

$$\varepsilon = -2\varepsilon_s \frac{1 - f\alpha_s}{1 + 2f\alpha_s} = -2\varepsilon_s \frac{\varepsilon_s(1 - f) + \varepsilon_m(2 + f)}{\varepsilon_s(1 + 2f) + 2\varepsilon_m(1 - f)}, \tag{2.6}$$

$$\alpha_s = \frac{\varepsilon_s - \varepsilon_m}{\varepsilon_s + 2\varepsilon_m}. \tag{2.7}$$

In the case of small $f \ll 1$, Eq. (2.6) reduces to

$$\varepsilon = -2\varepsilon_s(1 - 3f\alpha_s). \tag{2.8}$$

For a thick shell ($f \ll 1$), the known conclusion on the "freezing" of the plasmon mode at $\varepsilon = -2\varepsilon_s$ [45] is valid. Let us rewrite the resonance condition (2.6) in an equivalent form

$$\varepsilon = -2(\varepsilon_m + \chi), \quad \chi = g\frac{\varepsilon_s - \varepsilon_m}{1 + 2\alpha_s(1 - g)}. \tag{2.9}$$

In the case of thick shells ($g \sim 1$), the parameter $\chi \approx \varepsilon_s - \varepsilon_m$, and we return again to Eq. (2.8), while for a thin shell ($g \ll 1$) we obtain the above resonance condition Eq.(2.9) with

$$\chi \cong g\frac{\varepsilon_s - \varepsilon_m}{1 + 2\alpha_s} = g\frac{\varepsilon_s - \varepsilon_m}{3}\left(1 + 2\frac{\varepsilon_m}{\varepsilon_s}\right). \tag{2.10}$$

To evaluate the spectral shift of the extinction maximum, we use the Drude formula [45] along with Eq. (2.9). After some simple algebra we get [12]

$$\Delta\lambda_{max} = \Delta\lambda_{max}^{\infty} \frac{g}{1 + 2\alpha_s(1 - g)} \approx \Delta\lambda_{max}^{\infty} g, \tag{2.11}$$

$$\Delta\lambda_{max}^{\infty} = \frac{\lambda_p^2}{\lambda_{max}^0}(\varepsilon_s - \varepsilon_m), \tag{2.12}$$

where for gold, $\lambda_p = 131$ nm [43], λ_{max}^0 corresponds to the extinction maximum of bare gold particles, $\Delta\lambda_{max}^{\infty}$ is the maximal spectral shift for a thick shell [12]. Thus, the plasmon extinction shift is proportional to the shell volume fraction and depends on the shell/core ratio in a universal fashion, as predicted by Eq. (2.5).

2.2. Multilayer conjugate model

In a recent work [17], we introduced a new model for CG bioconjugates. The model consists of a gold core (diameter d) and a primary inhomogeneous polymer shell formed by recognizing molecules. Additionally, the model includes a secondary shell formed by target molecules. Each of the inhomogeneous shells is modeled by an arbitrary number of discrete layers (N_1 and N_2) with layer thickness s_{1i} and s_{2i}, and corresponding refractive indices n_{1i} and n_{2i}. Equations (2.1) and (2.2) for the extinction and scattering remain valid, but now the parameters Q_{ext} and F_{11} are to be calculated by a multilayer Mie theory [52]. Detailed discussion of the relevant physical parameters and the calculated graphs has been presented in Ref. [17]. Here we again provide two illustrative examples only.

Curves 1 in Figure 2a show the differential extinction and light-scattering spectra corresponding to the formation of a primary polymer shell of conjugates (for example, $\Delta A_{10} = A_1 - A_0$; indices 0 and 1 denote the bare particles and the primary conjugates, respectively). In the same line, curves 2 in Fig. 2a show the differential spectra related to the target molecules binding or, in other words, to the formation of a secondary polymer shell (for example, $\Delta I_{21} = I_2 - I_1$; indices 1 and 2 denote the primary and secondary shell, respectively). These calculations were carried out for a simplified model with a gold core, a two-layer primary shell, and a two-layer secondary shell. The differential spectra $\Delta A(\lambda)$ and $\Delta I(\lambda)$ related to the adsorption of both recognizing and target molecules possess a characteristic resonance that is shifted to the red part of the spectrum as compared to the known plasmon resonance of bare gold particles. Keeping optimization of conjugate-nanosensors [12, 17] in mind, we consider the following problem: what particle size is optimal for the transduction of polymer adsorption events into variations of recorded optical signals? Our answer is shown in Fig. 2b: the maximal values of differential resonances ΔA_{max} and ΔI_{max} are observed for gold particles with diameters about 40–60 nm (extinction spectra) or 70–90 nm (scattering spectra).

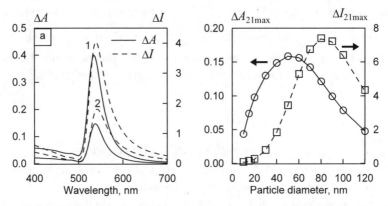

Figure 2. **(a)** - Differential extinction and scattering spectra corresponding to the formation of a primary polymer shell (curves *1*, $s_{11} = s_{12} = 2.5$ nm, $n_{11} = 1.50$, $n_{12} = 1.45$) and to the formation of a secondary shell (curves *2*, $s_{21} = 2$ nm, $s_{22} = 3$ nm, $n_{21} = 1.45$, $n_{22} = 1.40$). The gold-core diameter equals 40 nm. **(b)** – Dependence of the maximal changes in extinction ΔA_{21max} and scattering ΔI_{21max} on the gold-core diameter.

3.   Optical models for aggregated conjugates

3.1.   Computer models of aggregation

In our studies, we consider several types of aggregated structures such as bispheres, linear chains, plane arrays on a plane rectangular lattice, compact and porous body-centered clusters embedded on the cubic lattice (bcc clusters, the porosity was simulated by random elimination of monomers), and random fractal aggregates (RF clusters). To generate RF clusters, a three-dimensional lattice model with Brownian or linear trajectories of both single particles and intermediate clusters was employed for computer simulations of aggregation process. At the initial time moment, $N_0 = 50,000$ particles are generated at randomly selected points of a cubic lattice with size $L = 512$. When a particle moves to a lattice point adjacent to another particle or intermediate cluster, a combined cluster is formed. This model produces diffusion-limited (Brownian trajectories) or ballistic (linear trajectories) clusters with fractal dimension $d_f \approx 1.8$ and $d_f \approx 2.0$, respectively. A more detailed description of the model can be found elsewhere [53].

3.2.   Light scattering and extinction by a single aggregate

3.2.1. Coupled dipole method (CDM)

Consider a plane electromagnetic wave propagating in a dielectric medium with the refractive index n_0

$$\mathbf{E}_0 = \mathbf{e}_0 \exp(i\mathbf{kr}), \quad |\mathbf{e}_0| = 1, \quad |\mathbf{k}| = k = 2\pi n_0 / \lambda, \tag{3.1}$$

that encounters a cluster built from N small spherical nanoparticles with radius a and complex refractive index $n(\lambda)$ or dielectric permittivity $\varepsilon(\lambda)$. In the CDM method [54, 55], a real aggregate is replaced by a set of point dipoles $\mathbf{d}_i = \mathbf{d}(\mathbf{r}_i)$, $i = 1 - N$. If one uses the simplest physical version [56], then each dipole represents a real particle that is small enough to satisfy the dipole approximation. In a more sophisticated version [57], a real cluster is embedded on an appropriate 3-D lattice first. Then, the target dipole array is taken to consist of the lattice points located within the cluster volume. At present, this version of CDM is known as the discrete dipole approximation (DDA) [57].

The linear equations for interacting dipoles can be written in the form [20]

$$\sum_{jm=1}^{3N} A_{il,jm} d_{jm} = \alpha_i e_{0l} \exp(i\mathbf{kr}_i), \quad il = 1 - 3N, \tag{3.2}$$

where α_i is the polarizability of the ith dipole, the combined indices are $il = 3(i-1) + l$, $jm = 3(j-1) + m$; $i, j = 1 - N$; and indices $l, m = 1, 2, 3(x, y, z)$ correspond to the Cartesian components of vectors or tensors. The explicit form of the dipole interaction matrix $A_{il,jm}$ can be found in Ref. [20]. The solution of the linear systems of Eqs. (3.2) allows one to calculate the optical characteristics of the aggregate. For instance, the extinction cross section, related to the spectrum of optical density of a dilute suspension, can be calculated from the optical theorem [42, 45]

$$C_e = 4\pi k \, \mathrm{Im} \sum_i (\mathbf{e}_0 \mathbf{d}_i) \exp(-i\mathbf{kr}_i). \tag{3.3}$$

In the above sketch description we omitted some important questions related to the choice of optical polarizability and renormalization of interdipole spacing. The readers are referred to the corresponding discussion in review [42] and to references therein. In short, we use the interparticle spacing parameter $\gamma = d_L / a$ as a fitting parameter of a theory providing for the best agreement between theoretical predictions and experimental observations [58].

In practical applications, one usually needs average results for random orientations of clusters rather than calculations for a particular structure with a fixed orientation. In principle, such averaging can be carried out by numerical integration over Euler angles that define the orientation of a scatterer with respect to the incident wave. However, an analytical solution of such a problem is more effective. Examples of such analytical solutions are well known in the T-matrix method including its application to clusters [42, 50]. In works [59, 60], we derived an exact analytical solution for integral extinction, scattering, and absorption DDA cross sections averaged over random orientations of scatterers. Application of the solutions is illustrated by practical computations of average extinction cross sections for several examples of fractal clusters (soot in air and colloidal aggregates built from polystyrene, gold, and silver nanoparticles).

The interaction matrix in Eq. (3.2) does not depend on the incident wave orientation, so it is convenient to perform orientation averaging in the cluster coordinate frame. Using the inverse matrix $B = A^{-1}$ to solve Eq. (3.2), we obtain the following general equation for the extinction cross section

$$\langle C_e \rangle = 4\pi k \, \mathrm{Im} \left\{ \sum_{i,j=1}^{N} \sum_{p,q=1}^{3} \alpha B_{ip,jq} E_{pq}^{ij} \right\} , \tag{3.4}$$

$$E_{pq}^{ij} = \langle V_{pq} \exp(-i\mathbf{kr}_{ij}) \rangle , \tag{3.5}$$

where the angle brackets mean integration over Euler angles that define the orientation of the incident wave in the cluster frame, $V_{pq} = e_p e_q$ is the second-rank tensor, and e_p are the Cartesian components of the polarization vector in the cluster frame. The general scheme for calculation of average cross sections according to Eq. (3.4) consists of the following: First, we represent the tensor V_{pq} as a linear combination of irreducible spherical tensors in the incident wave coordinate frame. Such a transformation can be performed using Clebsh-Gordan coefficients and Wigner rotation functions [42]. Then, we expand the plane incident wave in a series over vector spherical harmonics (VSH) and perform orientation averaging using the orthogonality properties of VSH and Wigner functions. Omitting the technical details, we give the final result:

$$\langle C_e \rangle = 4\pi k \, \mathrm{Im} \left\{ \mathrm{Spur}(\hat{\mathrm{T}}) \right\} , \tag{3.6}$$

$$\langle C_a \rangle = 4\pi k \, \mathrm{Im} \left\{ \mathrm{Spur}(\hat{\mathrm{W}}) \right\} , \tag{3.7}$$

where Spur means the trace of a matrix, the matrices $\hat{\mathrm{T}}$ and $\hat{\mathrm{W}}$ are defined by equations

$$\hat{A}\hat{T} = \alpha \hat{E} , \tag{3.8}$$

$$\hat{W} = \frac{3 \, \mathrm{Im}(\varepsilon)}{a^3 |\varepsilon - 1|^2} |\alpha|^2 \, \hat{B}\hat{E}\hat{B}^+ , \tag{3.9}$$

and the modified polarizability α [55] is defined by relationship

$$\alpha = \frac{\alpha_0}{1 + \alpha_0 \varphi(ka)/a^3} , \tag{3.10}$$

$$\alpha_0 = a^3 \frac{\varepsilon - 1}{\varepsilon + 2} , \qquad \varphi(x) = 2 + 2(ix - 1)\exp(ix) \tag{3.11}$$

The explicit form of auxiliary matrix E_{pq}^{ij} [Eq. (3.5)] can be found in Ref. [60].

3.2.2. Generalized multiparticle Mie solution (GMM)

For large dielectric monomers or metal nanoparticles, the DDA model fails because of the multipole nature of electrodynamic particle coupling. An exact solution of the cluster-light-scattering problem can be formulated rather simply, using a generalized Mie theory for multisphere configurations [61, 62]. An incident electromagnetic field $\mathbf{E}_{inc}^{i}$ for the ith particle can be expanded over VSH $\mathbf{Y}_{mn1}^{(\tau)} = \mathbf{N}_{mn}^{(\tau)}(k\mathbf{r}_i)$ $\mathbf{Y}_{mn2}^{(\tau)} = \mathbf{M}_{mn}^{(\tau)}(k\mathbf{r}_i)$ of the first kind ($\tau =1$; spherical Bessel generating functions are used)

$$\mathbf{E}_{inc}^{i} = \sum_{n=1}^{\infty} \sum_{m=-n}^{n} \sum_{p=1}^{2} iE_{mn} \overline{p}_{mnp}^{i} \mathbf{Y}_{mnp}^{(1)}(k\mathbf{r}_i), \tag{3.12}$$

where E_{mn} are normalization coefficients, and $\overline{p}_{mnp}^{i}$ are the expansion coefficients.

The scattered field from the ith particle can be expanded over VSH of the third kind (Hankel generating functions are used) in the same manner as in Eq. (3.12)

$$\mathbf{E}_{s}^{i} = \sum_{n=1}^{\infty} \sum_{m=-n}^{n} \sum_{p=1}^{2} iE_{mn} a_{mnp}^{i} \mathbf{Y}_{mnp}^{(3)}(k\mathbf{r}_i). \tag{3.13}$$

The application of the usual boundary conditions leads to the following simple relations between the expansion coefficients

$$a_{mnp}^{i} = \overline{a}_{np}^{i} p_{mnp}^{i}, \tag{3.14}$$

where $\overline{a}_{n1}^{i} = a_n^{i}$, $\overline{a}_{n2}^{i} = b_n^{i}$ are the usual Mie coefficients for an isolated homogeneous sphere [45, 50]. Solution (3.14) is a crucial one in the GMM as it gives a simple and exact relation between p_{mnp}^{i} (exciting field) and a_{mnp}^{i} (scattered field). The unknown expansion coefficients of the exciting field can be found from the superposition principle that leads to the set of linear equations

$$\sum_{j=1}^{N} \sum_{v=1}^{\infty} \sum_{\mu=-v}^{v} \sum_{q=1}^{2} H_{mnp,\mu vq}^{ij} P_{\mu vq}^{j} = \overline{P}_{mnp}^{i}, \tag{3.15}$$

where $\overline{P}_{mnp}^{i}$ are the known expansion coefficients of the incident field in the ith coordinate frame. The interaction matrix $\hat{\mathbf{H}}$ is determined by the coefficients of translation of VSH based on spherical Hankel functions of the first kind (see explicit relations in Ref. [42]).

Once Eqs. (3.15) are solved and coefficients a_{mnp}^{i} are found, one can calculate all characteristics of light scattered by a cluster. For example, the extinction cross section is given by

$$C_e = \frac{4\pi}{k^2} \sum_{i=1}^{N} \sum_{n=1}^{L} \sum_{m=-n}^{n} \sum_{p=1}^{2} C_{mn} \, \mathrm{Re}[a_{mnp}^i (\overline{p}_{mnp}^i)^*], \qquad (3.16)$$

where coefficients C_{mn} depend on VSH normalization, and L is the maximal multipole order ensuring solution convergence.

The above scheme is applicable not only to the aggregates of homogeneous spheres but to aggregates built from multilayer spheres as well. To introduce such a generalization, it is sufficient to replace the usual Mie coefficients in Eqs. (3.14) by the corresponding VSH coefficients of multilayer spheres. Examples of such generalization have been demonstrated [63, 64] for the case of aggregated conjugates that are modeled by two-layer particles with a gold core and a polymer shell.

3.2.3. T-matrix formalism for cluster scattering

Usually, it is not just a calculation for an individual structure, but an averaged result over the statistical ensemble and orientation that is required. Formally, such an averaging can be carried out by simple summation of the results calculated for various orientations and polarizations of the incident light, but this is inefficient. It is more convenient to perform orientation averaging for clusters using T-matrix formalism [65], even for arbitrary nonspherical particles [66]. If we formally invert Eq. (3.15), we get the T-matrix of an individual cluster-particle [62]

$$a_{mnp}^i = \sum_j \sum_{\mu\nu q} T_{mnp,\mu\nu q}^{ij} P_{\mu\nu q}^j. \qquad (3.17)$$

Using theorems of VSH addition, one can combine these single-particle T-matrices into a common cluster T-matrix [62]

$$T_{a,b}^0 = \sum_{i,j} \sum_{c,d} J_{ac}^{oi} T_{cd}^{ij} J_{db}^{j0}, \qquad (3.18)$$

where for simplicity we use the multi-indices a,b,c,d to denote the order, degree, and mode (i.e. $a = mnp$, etc.), and J_{ab}^{oi} matrices describe the VSH translation to the common cluster center and are based on the spherical Bessel functions. Further calculations of the optical characteristics and their orientation averaging are performed according to the same scheme as that employed in the case of the usual nonspherical particles [50]. A more detailed consideration of orientation-averaged parameters of cluster scattering can be found in Ref. [66].

3.3. Light scattering and extinction by an ensemble of aggregates

A model for simulating the extinction spectra during the aggregation process was developed in our work [20]. We assume that a cluster suspension is dilute so that the single scattering approximation is valid. This means that the extinction (optical density or absorption) $A(\lambda)$ is directly proportional to the sum of the extinction cross sections of clusters per unit volume. It is convenient to normalize the extinction spectra to the monomer optical density $A_m(a,\lambda_{max})$ at the maximum of extinction $C_{em}(\lambda_{max})$

$$A_m(\lambda) \equiv A(a,\lambda)/A_m(a,\lambda_{max}) = C_e(a,\lambda)/C_{em}(a,\lambda_{max}).\qquad(3.19)$$

The cluster size distribution at an arbitrary stage of aggregation can be described by a set of number pairs (N_p, n_p) where N_p is the number of particles per aggregate from a given monodisperse ensemble ($N_p = const$) with different structures of aggregates, and n_p is the number of such aggregates. Light extinction by the (N_p, n_p) ensemble is given by the normalized extinction cross section

$$\overline{\langle Q_p \rangle} = \overline{\langle C_{ep}(R_m,\lambda) \rangle}/pC_{em}(R_m,\lambda),\qquad(3.20)$$

where the angle brackets denote averaging over random cluster orientations in a monodisperse ensemble of clusters, and the horizontal bar designates statistical averaging over cluster configurations of ensemble (N_p, n_p). The optical density of the suspension is given by the relationship

$$E(\lambda) = E_m(R_m,\lambda)\sum_p v_p \overline{\langle Q_p \rangle},\qquad(3.21)$$

where v_p is the fraction of N_p-cluster particles of the total number of initial monomers N_0

$$v_p = n_p N_p / \sum_p n_p N_p = n_p N_p / N_0.\qquad(3.22)$$

For calculation of the absolute values of extinction and scattering intensity, we used a cuvette thickness 1 cm and gold and silver concentrations of 57 μg/ml and 5 μg/ml, respectively.

3.4. Statistical and orientation averaging of light-scattering observables: some important simplifications

Equation (3.21) implies that the determination of an optical parameter at a given time consists of three steps: (1) *orientation* averaging for a cluster possessing a given particle number N_p, (2) *statistical* averaging over all possible configurations of monomers for a monodisperse ensemble $\{N_p = const\}$, and (3) *polydisperse* averaging over cluster size distribution. The calculation of $\overline{\langle C_{ep} \rangle}$ through the steps (1) and (2) is the most challenging task in the kinetic simulation. It has been shown that orientation averaging can be performed analytically using either the T-matrix [62, 66] or CDM [60] methods. However, the computational burden becomes impractical when

Table 1. Comparison of the statistically averaged optical parameters $\overline{A}$, $\overline{I_{90}}$ with those calculated with both orientation and statistical averaging, $\langle A \rangle$, $\langle I_{90} \rangle$. Calculations for ballistic clusters built from gold particles with diameters $d = 15$ and 30 nm, polymer shell thickness and refractive index $s = 2.5$ nm and $n_s = 1.4$, respectively. All data are averaged over 100 independent cluster configurations.

d, nm	N	$\lambda = 500$ nm				$\lambda = 600$ nm			
		$\overline{A}$	$\langle A \rangle$	$\overline{I_{90}}$	$\langle I_{90} \rangle$	$\overline{A}$	$\langle A \rangle$	$\overline{I_{90}}$	$\langle I_{90} \rangle$
15	10	0.955	0.954	0.176	0.174	0.213	0.213	0.163	0.161
	100	0.922	0.921	.364	0.368	0.296	0.298	0.589	0.603
30	10	0.894	0.893	0.779	0.766	0.551	0.545	0.201	0.195
	100	0.777	0.775	0.487	0.513	0.904	0.897	0.256	0.251

particle number is large, especially for T-matrix method. Recently [10] we proposed substituting statistically averaged optical parameters for those calculated with both orientation and statistical averaging. This allows us to skip step (1), simplifying the calculations. The approximation is

$$\langle C_{ep} \rangle \approx \overline{C}_{ep}, \quad N_p = const, \tag{3.23}$$

or in other words

$$\langle A \rangle \approx \overline{A}, \quad \langle I_{90} \rangle \approx \overline{I}_{90}, \quad N_p = const. \tag{3.24}$$

To test the proposed new procedure, we generate monodisperse statistical ensembles of clusters with $\{ N_p = const \}$ and calculate two sets of averaged parameters, for example $\langle A \rangle$ and $\overline{A}$, $\langle I_{90} \rangle$ and $\overline{I}_{90}$, etc. A portion of data is listed in Table 1 (see also Ref. [10]).

From Table 1, we draw two important conclusions. First, these data support the above hypothesis (3.24). Second, the scattered intensity is more sensitive to cluster orientations than extinction is [50]. Therefore, one has to use a large ensemble of monodisperse clusters to eliminate fluctuations in the scattered intensity. The first conclusion leads to great simplifications, as we can use effective and fast GMM codes [66] for a cluster with a *fixed* orientation instead of T-matrix codes that require large RAM and CPU-time.

4. Results of numerical simulations and discussion

4.1. Two-conjugate clusters

Strong interaction between small metal particles [60, 67] and the well known splitting of the extinction spectra can be illustrated by the simplest bisphere case: two bare 15-nm gold particles with separation distance $\Delta d = 0.15$ nm and random orientations in water (Fig. 3).

To stress the multipole nature of the binary coupling [68], the single-particle multipole order is specified as $L = 1 - 30$, while the cluster T-matrix order N_T was determined by convergence conditions. Satisfactory convergence was obtained at $L = 30$ and $N_T \approx 4$. The extinction spectrum calculated using the dipole approximation (i.e. with $L = 1$) is close to the single-particle spectrum (dashed curve) but differs dramatically from the exact result at $L = 30$.

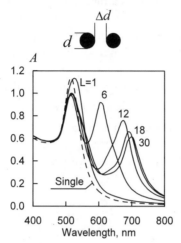

Figure 3. Extinction spectra of randomly oriented gold bispheres in water (diameter d = 15 nm, separation distance Δd = 0.15 nm). Calculations with different single-particle multipole order L = 1-30. The dashed lines shows the single-particle spectrum.

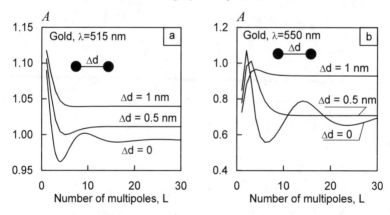

Figure 4. Dependences of the extinction A of gold bispheres in water on the number of multipoles included in the single-particle expansions. Calculations by the T-matrix method for randomly oriented bispheres with touching component spheres ($\Delta d = 0$) and those separated by distances $\Delta d = 0.5$ nm and $\Delta d = 1$ nm. Particle diameter equals $d = 15$ nm; the wavelength in vacuum equals $\lambda = 515$ nm (a) and 550 nm (b).

The interparticle distance Δd is a key parameter that determines the electrodynamic coupling of metal [10] or dielectric [69] monomers. It has been shown [20] that for clusters built from silver or gold nanospheres, the convergence of the GMM or T-matrix methods is too slow due to the multipole nature of the interparticle interaction. However, if the cluster particles are separated by a small distance, the number of multipole terms required in the single-particle expansions decreases to a tractable level [10, 70].

This effect is illustrated in Fig. 4 that shows the extinction of randomly oriented gold bispheres plotted as a function of single-particle multipole order L at different separations Δd between component particles. Calculations were performed for diameters of component gold particles $d = 15$ nm and wavelength in vacuum $\lambda = 515$ nm (a, plasmon resonance wavelength in water), and $\lambda = 550$ nm (b). In the case of touching particles, the convergence multipole order is greater than the total T-matrix order $N_T \sim 5$; whereas, for separated components, it is sufficient to use the single-particle order $L \approx N_T$ to ensure reasonable convergence of numerical results.

In our cluster models, the shell thickness (s) acts as a dielectric interparticle separation ($2s$) that is known [62, 67], to be responsible for the

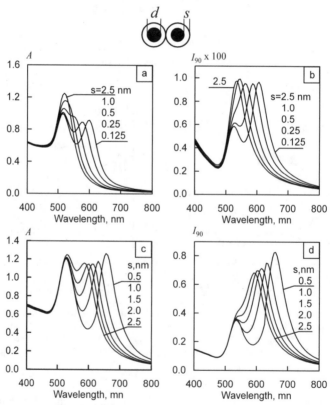

Figure 5. Extinction and scattering spectra of two-conjugate clusters with different shell thickness s. Parameters of conjugates: $d = 15$ (a,b) and 60 nm (c,d), the shell refractive index $n_s = 1.40$. Calculations for randomly orientated clusters in water.

strong reduction of binary optical coupling. The shell thickness effect is illustrated in Figure 5 for the case of randomly oriented clusters built from two-layer bispheres.

Calculations were carried out for particles having gold core diameters $d = 15$ and 60 nm with a variable shell thickness $s = 0.125 - 2.5$ nm. When s is close to $0.1d$ (i.e. the interparticle separation approaches $0.2d$), the extinction and scattering spectra become similar to those for monomers. For small bisphere conjugates with $d = 15$ nm (Fig. 5a,b), the extinction spectrum is governed by true absorption; whereas, for larger conjugates with $d = 60$ nm (Fig. 5d,e), the integral light scattering contributes about 50% to the second extinction peak. Both contributions, A_{sca} and A_{abs}, are determined by the usual 520-nm plasmon resonance peak and by the delocalized plasmon excitations [58] in the range of 600-700 nm.

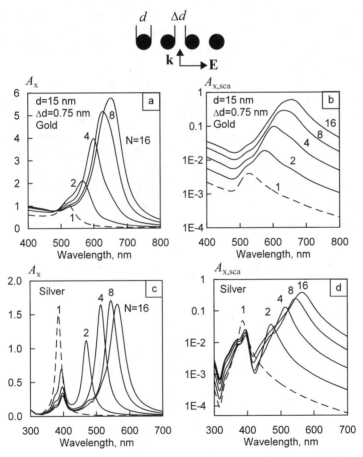

Figure 6. Extinction (a,c) and integral scattering (b,d) spectra of linear chains of gold (a,b) and silver (c,d) spheres with diameter $d = 15$-nm separated by $\Delta d = 0.75$ nm (in water). Calculations for TM plane wave at perpendicular incidence, the number of chain particles varies from 1 (dashed lines) to 16.

4.2.   Linear chains

Looking at the internal structure of real or simulated clusters, one can note numerous chain-like fragments. Therefore, it is desirable to understand possible optical effects related to the formation of such chain-like nanoparticle structures. We start from the simplest case of a linear chain of gold or silver spheres excited by TM electromagnetic wave at perpendicular incidence (electric vector is directed along the chain's axis). In a sense, the linear chain of metal nanospheres is analogous to typical nonspherical particles (like cylinders or spheroids), so we can expect the appearance of red shifted spectra corresponding to the longitudinal plasmon excitations [45]. Addition of a nanosphere results in a red-shifted spectra due to the increase in the chain axis ratio. Figure 6 illustrates these conclusions, showing the total extinction and the integral scattering spectra calculated for different particle numbers from 1 to 16. Note that there is a rapid saturation of magnitude and position of the longitudinal resonance with increasing length of chain. This means that the effective electrodynamic interaction involves only a finite number of particles, about $N = 16$-32 in our case [71].

Figure 7 shows extinction spectra calculated for randomly oriented linear chains of 13-nm gold spheres with interparticle spacing 1.1 nm. It is the same model that has been considered in Ref. [70]. We note two spectral resonances related to the transversal and longitudinal (red-shifted) plasmon excitations. Such properties are analogous to the randomly oriented metal spheroids [45, 72]. By contrast to Fig. 8 in Ref. [70], we conclude that random chain orientations do not eliminate the red-shifted longitudinal resonance. Perhaps, the spectra of Fig. 8 from [70] were calculated with an insufficiently large multipole expansion order.

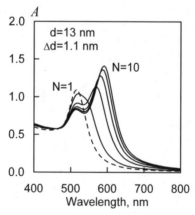

Figure 7. Extinction spectra of randomly oriented chains built from 13 nm gold particles separated by 1 nm distance in water. The number of spheres equals to 1,2,4,6,8, and 10.

Strong binary coupling also is observed for linear chains of two-layer conjugates. Figure 8 shows the dependence of extinction $A(\lambda)$ and scattering $I_{90}(\lambda)$ spectra of randomly oriented linear chains built from two-layer conjugates. The number of conjugates is the variable parameter of curves. It is worth noting a principal difference between the spectra of densely packed ($s = 1$ nm, strong binary coupling, Fig. 8c,d) and rare chains ($s = 5$ nm, weak binary coupling, Fig. 8a,b). The scattering intensity in Fig. 8b and Fig. 8d is expressed in the same scale and was calculated for the same gold concentration. In agreement with our previous observations [10], the extinction

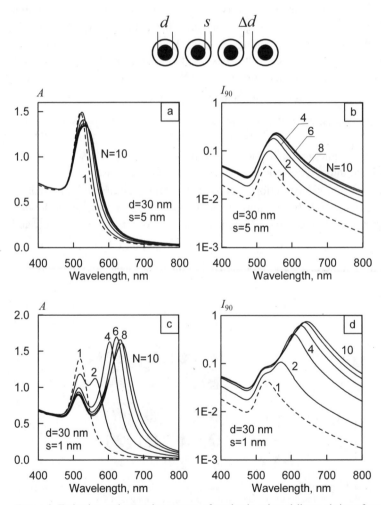

Figure 8. Extinction and scattering spectra of randomly oriented linear chains of two-layered conjugates with different conjugate number $N = 1-10$. Calculations for the core diameter $d = 30$ nm, shell thickness $s = 5$ (a,b) and 1 nm (c,d), and separation distance $\Delta d = 0$.

spectra do not change essentially for small and rare aggregates (Fig. 8a), while the plasmon resonance peak in the scattering spectra increases significantly due to constructive far-field interference. However, the peak position does not shift significantly. For a thin dielectric shell, the transformation of extinction and scattering spectra are due to both the electrodynamic coupling and the far-field interference. Again we observe a rapid saturation of particle-number effect. This can be interpreted as manifestation of an effective electrodynamic interaction between monomers, which belong to a finite conjugate group.

4.3. Lattice arrays

We consider a plane conjugate array embedded on a square or rectangular lattice. Bare silver-particle configurations have been studied recently [73, 74]. The spatial arrangement of conjugates (Fig. 9) is specified by the gold core diameter d, the polymer shell thickness s, the separation distances Δx and Δy, the number of monomers along x-axis N_x, and the number of monomers along y-axis N_y. In our model, a couple of touching conjugates along y-direction is used as a basic array element so the double conjugate chains are separated by Δy distance. This configuration allows one to simulate the electrodynamic interaction between nonspherical (bispherical) structures that are assembled into a linear chain with separation distance Δx.

Figure 10 shows extinction spectra for x and y-polarized light that is scattered and absorbed by a double chain of gold conjugates with the following parameters: $d = 15$ nm, $s = 0.125$ nm, $n_s = 1.40$, $\Delta x = 0$, $N_y = const = 1$, and number $N_x = 1, 2, 4, 8, 16, 32,$ and 64. The extinction efficiencies $\langle Q_{e,x(y)} \rangle$ are the extinction cross section normalized to the total surface of conjugates

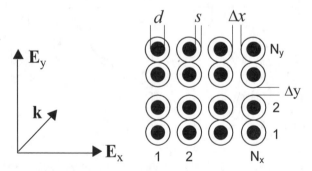

Figure 9. Scattering configuration for a plane array of conjugates embedded on a (x, y) rectangular lattice. The incident x, y-polarized light propagates along z-axis.

$$\langle Q_{e,x(y)} \rangle = \langle C_{e,x(y)} \rangle / N_x N_y \pi (a + s)^2 . \tag{4.1}$$

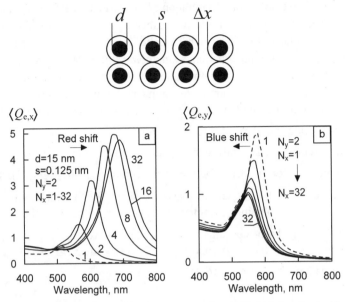

Figure 10. Extinction spectra for *x* (a) and *y*-polarized light (b). Calculations for double chain of gold conjugates with the following parameters: $d = 15$ nm, $s = 0.125$ nm, $n_s = 1.40$ $\Delta x = 0$, $N_y = const = 2$, and the number $N_x = 1, 2, 4, 8, 16,$ and 32.

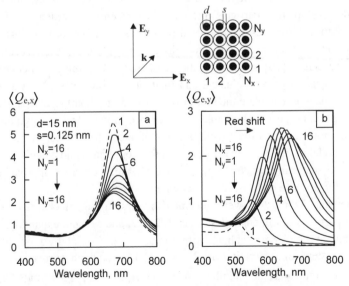

Figure 11. Extinction spectra for *x* (a) and *y*-polarized light (b). Calculations for rectangular array of gold conjugates with the following parameters: $d = 15$ nm, $s = 0.125$ nm, $n_s = 1.40$, $\Delta x = 0$, $\Delta y = 0$, $N_x = const = 16$, and the number $N_y = 1, 2, 4, ... 16$.

Generally, transformations of $\langle Q_{e,x} \rangle$ spectra are analogous to those shown in Fig. 6a. Again we note a rapid saturation of the extinction efficiency maximum and its position with the increase in number of conjugates, i.e. the length of chain. For x-polarized incident light, a noticeable *red* shift exists from 500 to 700 nm, whereas for y-polarization, a small *blue* shift is observed.

Figure 11 illustrates a transformation of extinction spectra for x and y-polarized light when the number of conjugates along the x-axis is constant ($N_x = 16$) while the number of doubled x-chains N_y increases from 1 to 16. With increasing number of linear x-chains N_y, the spectral position of $\langle Q_{e,x}(\lambda) \rangle$ remains essentially unchanged, but the maximum value decreases more than a factor of two. For the same conditions, the $\langle Q_{e,y}(\lambda) \rangle$ spectral maximum shifts to the red and increases in value. For the final configuration, $N_x = N_y = 16$, we obtain a symmetrical square array because the separation distances along both coordinates are equal to zero. Accordingly, both extinction spectra $\langle Q_{e,x}(\lambda) \rangle$ and $\langle Q_{e,y}(\lambda) \rangle$ are identical in the case of symmetrical array $N_x = N_y = 16$.

4.4. Optical properties of bcc-clusters

The aggregation of colloids results in the formation of low-dimension fractal structures such as those formed from underivatized Au and Ag nanoparticles that bind irreversibly after addition of a salt [75]. However, aggregation of biomolecule-dressed nanoparticles produce more compact structures. For example, compact structures are observed in the case of DNA-linked clusters [21] or clusters produced by antigen-antibody interactions [76]. Recently, Lazarides and Schatz [70] published a thorough study of correlations between the structural organization of clusters and their optical properties. They used the DDA approach to calculate the scatter from 13 nm gold particles separated by 6.5 nm and arranged on a simple cubic or bcc lattice. The gold volume fraction was adjusted by random elimination of cluster particles. Here, we extend this model to include a two-layer model of conjugates and calculate the extinction and scattering spectra using exact GMM and T-matrix approaches generalized for the case of two-layer spherical monomers [66].

The compact bcc-clusters were generated by first filling blocks of a bcc lattice with particles of dimensionless sizes 1, 2, 3... and second, excluding all particles that are outside a sphere with radius 1, 2, 3... and so on. We designate the steps of generation as $N_c = 1, 2, 3,...6$, which correspond to clusters with total particle number $N = 1, 9, 59, 169, 339,$ and 701, respectively.

Figure 12 shows the extinction and scattering spectra calculated for 15 nm and 60 nm gold particles covered by 2.5 nm polymer layer with refractive index 1.40. Evolution of extinction and scattering spectra is determined mainly by the overall cluster size [70]. For 15 nm particles, the calculated spectra display the two features known from experiment [10, 21, 76], namely, substantial red shifting and broadening of the plasmon peak. Extinction spectra for larger aggregates of 60-nm particles are more complicated and include several plasmon bands. In our experiments with colloidal gold particles of different sizes [10, 20], we observe analogous spectra at the final stages of aggregation.

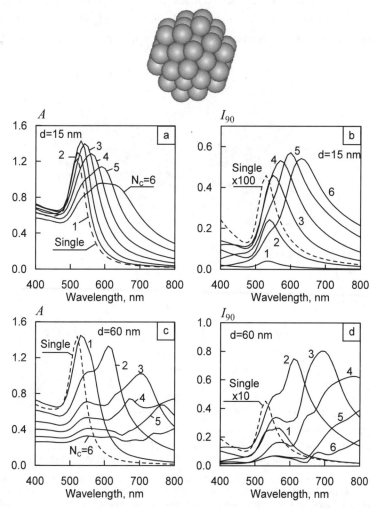

Figure 12. Extinction (a,c) and scattering (b,c) spectra of bcc-clusters with different number of monomers N. The numbers $N_c = 1$-6 correspond to $N = 1,9,59,169,339$, and 701. Calculations by GMM method gold conjugates with the following parameters: $d = 15$ nm (a,b) and 60 nm (c,d), $s = 2.5$ nm, $n_s = 1.40$. The upper picture is a bcc-cluster with $N = 59$.

We also have studied the optical properties of the porous bcc-clusters with variable conjugate volume fraction. To this end, some part of the occupied sites was removed randomly to obtain aggregates with a desired conjugate volume fraction. This is exactly the same procedure that has been introduced by Lazarides and Schatz [70], but for homogeneous metal monomers.

The initial dense clusters consist of $N = 9$, 59, ...to 1243 conjugates, and the porous aggregates are obtained by random elimination of 10-70% N (in steps of 10%) from the initial cluster. Each conjugate consists of the gold core (diameters of $d = 15$ nm or 30 nm) and a polymer shell with thickness $s = 0.5$, 1.0, or 2.5 nm; the shell refractive index equals $n_s = 1.40$. Figure 13 is an illustration for the case $N = 701$ (initial cluster), 561, 421, and 281. The corresponding volume fractions of cluster conjugates are equal to 0.53

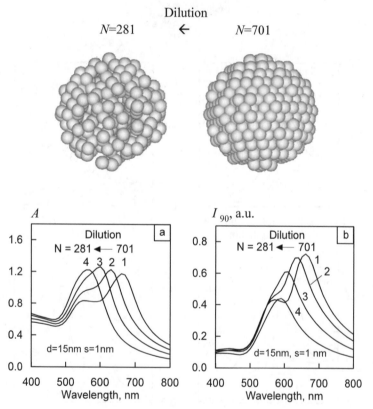

Figure 13. Extinction and scattering spectra of bbc-clusters with the number of conjugates N=701 (1, maximal volume density), 561(2), 421 (3), and 281 (4). "Diluted" clusters were obtained by random elimination of cluster monomers. The volume fractions of conjugates equal 0.53 (maximal density), 0.42, 0.316, and 0.21, respectively. Calculations were carried out by GMM method [66] using the following parameters of conjugates: gold core – $d = 15$ nm, polymer shell – $s = 1.0$, $n_s = 1.40$.

(maximal initial density), 0.43, 0.32, and 0.21, respectively. On the top of Fig. 13, the initial ($N = 701$) and the final ($N = 281$) clusters are shown. Figure 13 illustrates how the extinction and scattering spectra are transformed with cluster "dilution." For densely packed aggregates ($N = 701$), the extinction and scattering spectra possess two maxima; whereas, only one maximum is observed in the case of the final porous aggregates ($N = 281$).

4.5. Random fractal aggregates

An algorithm for generation of random fractal (RF) clusters has been described in Section 3.1.1. The major computer limitations were imposed by the number of cluster particles N, and their proximity in terms of separation distance between gold cores. Our calculations included both DLCA and ballistic RF

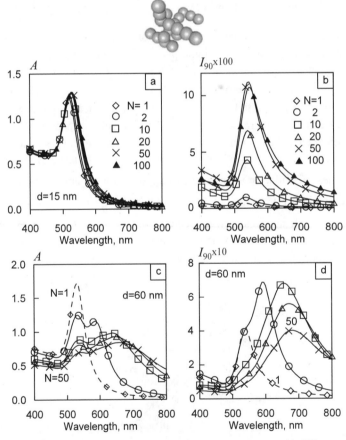

Figure 14. Extinction and scattering spectra for ballistic RF aggregates with different conjugate numbers $N = 1$-100. All data are averaged over random orientations (T-matrix method) without statistical averaging. Parameters of conjugates are the shell thickness and refractive index $s = 2.5$ nm, $n_s = 1.40$, respectively, the gold core diameter $d = 15$ nm (a,b) and 60 nm (c,d).

aggregates with different conjugate parameters and particle number limit $N = 100$. With all other conditions being equal, the extinction spectra are determined mainly by the core size and shell thickness. The core-size effect is illustrated in Fig. 14 for the case of RF aggregates that comprise of small numbers of conjugates $N = 1-100$ with $s = 2.5$ nm, $d_g = 15$ and 60 nm. With the same particle number N, the spectra are significantly different for 15-nm (Fig.3a,b) and 60-nm (Fig. 3c,d) conjugates. For 15-nm aggregated conjugates, there is essentially no near-field interaction between monomers and the extinction spectra are the same as those for a single conjugate, while the scattering spectra differ due to the constructive far-field interference. By contrast, for 60-nm conjugates, the near-field optical interaction significantly affects both the extinction and the scattering spectra.

4.6. Dynamic simulation of optical effects caused by particle aggregation

A three-dimensional lattice model (section 3.1.1) was used to simulate aggregation kinetics, in which single particles and intermediate clusters move on Brownian or linear trajectories. Initially, $N_0 = 50,000$ particles are placed randomly in a cubic lattice of size $L = 215 \times 215 \times 215$. A combined cluster is formed whenever a particle or a cluster moves to a lattice point adjacent to another particle or intermediate cluster. This model produces DLCA clusters [77] with fractal dimension around 1.8 (Brownian trajectories) and 2 (linear trajectories). A sequence of two integers (N_p, n_p) is used to describe the cluster-mass distribution at any stage of aggregation, where N_p is the number of particles in a given monodisperse ensemble $\{ N_p = const \}$ and n_p is the number of such aggregates (see Section 3.3.). Figure 15 is an example to illustrate the simulation of kinetic evolution of cluster-mass distribution for different fraction of aggregated particles

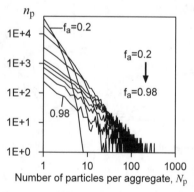

Figure 15. Evolution of the cluster-mass distribution for different time moments corresponding to the aggregation parameter $f_a = 0.2-0.8(0.2), 0.9, 0.92-0.98(0.02)$

$$f_a = \sum n_p N_p / N_0 (p \geq 2).$$ (4.2)

We see notable oscillations of the cluster-mass distribution curves in simulations without statistical averaging [77]. In our dynamic model, the "aggregation time" is expressed in terms of the aggregation parameter f_a. Clearly, the value $f_a = 0$ corresponds to zero aggregation time, and $f_a = 1$ means that there are no single particles in our "lattice colloid". Other approaches to the simulation of colloid aggregation dynamics can be found elsewhere [19, 20, 77].

Figure 16 shows the temporal evolution of extinction and scattering spectra of 30-nm and 60-nm gold conjugates. For 30-nm monomers and aggregation parameter $f_a \leq 0.9$, there are no significant changes in the extinction spectra with aggregation time. This is because interaction between monomers is weak due to large interparticle separation. Furthermore, the multiple scattering inside clusters is not sufficiently strong for a significant renormalization of the exciting internal field because the average cluster size is small.

On the other hand, the static light-scattering spectra allows for concise

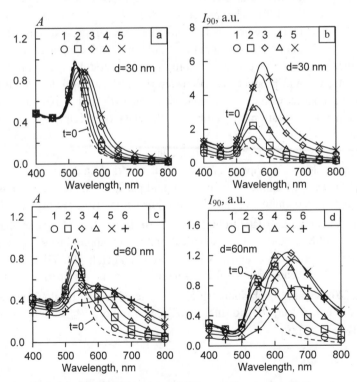

Figure 16. Evolution of the extinction and static light scattering spectra with time expressed in terms of the aggregation parameter $f_a = 0, 0.2(1), 0.4(2), 0.6(3), 0.8(4), 0.9(5), 0.98 (6)$

aggregation monitoring. The significant changes in scattering spectra are due to constructive far-field interference of fields that are scattered by monomers of a small cluster. In the case of 60-nm conjugates, both the extinction and the scattering spectra are highly time-dependent. In this case, the relative interparticle separation $2s/d = 0.08$ is only half of that for 30-nm conjugates. Also, the average cluster size is larger by a factor 2. These structural features result in stronger binary interparticle interaction and multiple scattering inside a cluster. Both these physical mechanisms are thought to be responsible for the significant differences between spectra shown in Figures 16a,b and 16c,d.

5. Experimental

5.1. Materials and methods

In this Section, we give a short review of our experimental results related to the optics of colloidal-gold conjugates. Colloidal gold particles are synthesized according to procedures described in Refs. [20, 48], by reducing tetrachloroauric acid with sodium citrate (Frens "citrate" method [78]). For smallest particles with diameter of 5 nm, we propose a new procedure as described in Ref. [79]. The protocol for preparing conjugates of CG to biospecific macromolecules, which involves preparing and purifying an aqueous probe solution, determining the "gold number," coupling the probe to the label, adding a secondary stabilizer, concentrating the marker, and optimizing the end product, has been described elsewhere [80]. We use the theoretical concept of the size dependence of the extinction maximum position to obtain the calibration by combining the data of spectrophotometry and electron microscopy [20, 48]. This method has proven to be convenient for expeditious evaluation of the mean size and has already found a number of applications [81, 82]. Figure 17 shows our working calibration dependence [48] of the average diameter of colloidal gold particles on the spectral position of extinction maximum, $\lambda_{max} - 500$ (nm). For comparison, we provide also in Fig. 17 experimental data from Tables 1 of Refs. [83, 84]. Note that the colloidal gold particles are not spherical but can be approximated by slightly elongated spheroids or cylinders with semispherical ends [48, 83]. That is why we use the diameter of equivolume sphere d_{ev} as the size parameter in our works [20, 48] and calibrations (Fig. 17). Therefore, we recalculate all data from Table 1 of Ref. [83] for major (d_{max}) and minor (d_{min}) diameters using the relationship [48]

$$d_{ev} = \sqrt[3]{d_{max} d_{min}^2} \ . \tag{5.1}$$

The solid line in Fig. 17 is a polynomial best fit according to Ref. [48]

$$d_{ev}(\text{nm}) = \sum_{n=0}^{3} A_n (\lambda - 500)^n \,, \tag{5.2}$$

where $A_0 = 12.558$, $A_1 = -2.593$, $A_2 = 0.1921$, $A_4 = -0.00253$. Equation (5.2) is valid up to particle diameters $d_{ev} = 60$ nm.

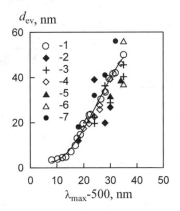

Figure 17. Dependence of the of the average diameter of colloidal gold particles d_{ev} on the spectral position of extinction maximum, $\lambda_{max} - 500$, nm. 1 – Ref. [48] (citrate Frens method, FM), 2 – Ref. [83] (FM), 3 – Ref. [20] from Ref. [83] (FM), 4 – Ref. [83] (seeding by 12 nm gold particles), 5 – Ref. [83] (seeding by 2.6 nm gold particles), 6 – Ref. [84] (FM), and 7 – Ref. [84] (seeded by NH_2OH). The solid line is the polynomial best fit Eq.(5.2).

5.2. Optical properties of clusters formed during slow and fast aggregation

Optical properties of aggregated gold and silver sols with different average particle size were studied in our Ref. [20] for two basic aggregation regimes, "fast" and "slow," which are known also as diffusion-limited-cluster-cluster-aggregation (DLCA) and reaction-limited-cluster-cluster-aggregation (RLCA) regimes, respectively [75, 77]. Aggregation of stable sols is initiated by addition of NaCl or K_2CO_3 salts. The extinction spectra (400-800 nm) of aggregates of colloidal-gold particles with diameters 5, 15 and 30 nm, and silver 20 nm particles are recorded using a Specord M-40 spectrophotometer. We have found that during fast aggregation, corresponding to the formation of the diffusion-limited (DLCA) clusters, the spectra are dependent on the size of the primary particles. For aggregates of 15- and 30-nm gold particles and for 20-nm silver particles, we record strongly broadened spectra with an additional red extinction maximum (Fig. 18a); whereas, the extinction spectra for aggregates of 5-nm particles has a single red-shifted and broadened extinction maximum [20]. In the case of slow aggregation, we observe a decrease in the plasmon extinction peak without an essential red shift and broadening.

According to the TEM data, the fast aggregation give typical [75, 77] ramified DLCA aggregates with chain-like branches (Fig. 18b); whereas, the

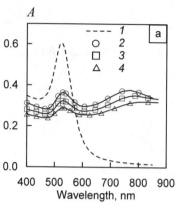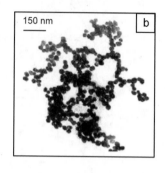

Figure 18. (a) – Extinction spectra of a 30-nm gold sol (dashed line 1) and its DLCA aggregates formed after addition of NaCl salt at a final concentration of 0.046 M. Aggregation time equals 0.5 (2), 12 (3), and 50 (4) min. (b) – TEM picture of a typical DLCA gold cluster after 60 min of aggregation.

slow aggregation leads to small compact structures along with an appreciable number of single (not aggregated) particles.

To explain these findings, we use a computer diffusion-limited cluster-cluster aggregation model, as described in Section 3.1.1. The optical properties of the aggregates are computed by the coupled dipole method and by a rigorous GMM and T-matrix methods. The bulk optical constants of metals are modified by the size-limiting effect of nanoparticles [20]. It was shown that a modified version of DDA [58] allows one to explain the shape of the experimental spectra for DLCA aggregates and the dependence of the spectra on the particle size.

5.3. Biospecific aggregation. Correlations between the extinction spectra and the cluster structure

Earlier, we reported on the optical properties of aggregates formed by biospecific interactions like antigen/antibody involving one or both reaction components immobilized on gold particles [19, 76]. The dashed line in Figure 19a shows the extinction spectra of conjugates of 30-nm gold particles with Protein A. After addition of human IgG (hIgG), biospecific aggregation occurs that easily can be monitored by extinction spectra in the visible region. It is evident that the temporal changes in the absorption spectra differ from those recorded during rapid and slow salt aggregations [20]. As in the case of rapid salt aggregation, the absorption peak decreases and shifts to the red part of the spectrum with simultaneous broadening (Fig. 19a). However, we do not observe the second red peak of the optical density. According to the TEM data, the biospecific aggregates are rather small and their monomers are separated by biopolymer substance (Fig. 19b). As in the case of DNA-linked clusters [21], we do not observe the low-dimensional branching fractal-like aggregates

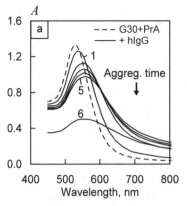
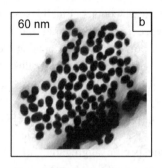

Figure 19. (a) – Extinction spectra of colloidal gold (d = 30 nm) conjugates to Protein A (PrA, dashed line) and its biospecific aggregates formed after addition of hIgG. Aggregation time equals 0.5 (2), 20 (3), 40 (3), 60 (4), 120 (5) 2280 min (6) min. (b) – TEM picture of a typical biospecific aggregate. Polymer surrounding is seen clearly near gold particles.

or small compact aggregates with particles in contact that are typical for RLCA clusters formed by salt aggregation.

To summarize, we recorded three types of cluster structures during the aggregation of colloidal-gold particles or bioconjugates, and also three types of corresponding temporal dynamics of absorption spectra. In the slow salt aggregation, relatively small, compact aggregates form that have fractal dimension $d_f > 2$. Such aggregation is accompanied by small decreases in the plasmon absorption-peak of monomers and by non-uniform widening of the long-wave wing. Rapid salt aggregation leads to the formation of fairly loose aggregates with characteristic branching DLCA [75, 77] structures and a fractal dimension of about 1.8 (Fig.18b). Contrary to the data of Ref. [85], we

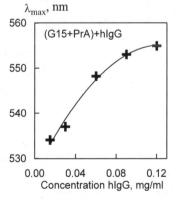

Figure 20. Dependence of extinction maximum position λ_{max} on the concentration of hIgG. The IgG molecules are added to a solution of biospecific conjugates of 15-nm gold particles with Protein A.

record the presence of a second long-wave absorption peak for gold particles with direct contact. Finally, a characteristic of biospecific aggregates is the presence of a biopolymer interlayer among the aggregate's gold particles, which prevents a direct conductive contact. The absorption-spectrum peak of such aggregates is reduced substantially and is shifted toward the red region, with the value of the peak decrease correlating with the concentration of the component initiating the aggregation of conjugates.

Figure 20 shows a calibration curve for a sufficiently rapid and technically simple quantitative sol-particle immunoassay (SPIA [18]). The assay is based on biospecific aggregation of 15-nm gold conjugates to Protein A due to interaction with hIgG molecules. We have found a direct correlation between the amount of the second added protein initiating aggregation and the spectral position of the extinction maximum.

5.4. Differential light scattering spectroscopy (DLSS)

Adsorption of a polymer onto the gold-particle surface results in relatively small changes in the optical density. The same is true for the initial stages of aggregation. Simple speculations [86] suggest that light-scattering spectra can be more informative for adsorption of a polymer onto a particle surface, as well as for the initial stages of aggregation as compared to the absorption technique. Recently, we proposed a new method [27] to study biospecific interactions in systems of conjugates of colloidal gold nanoparticles.

To explain our basic motivation, let us first consider qualitatively the simplest case where two isolated nanoparticles with diameters about 15-30 nm are combined into a trivial cluster (a bisphere). If the average interparticle spacing is around 1–3 nm, a weak electrodynamic interaction of particles does not change the extinction spectrum after particle clustering. Because the particle size and interparticle distance are much less than the incident light wavelength, the particles scatter light coherently and the scattered fields interfere constructively. This means that the coupling of particles into a bisphere results in a twofold increase in the scattered intensity. We can formulate this conjecture in terms of the well-known properties of dipolar Rayleigh scattering: the extinction and scattering cross-sections are proportional to the particle volume and to the square of volume, respectively. So, after particle coupling, their extinction does not change; whereas, the scattering increases twofold.

The DLSS method is based on measuring the differential spectra of light scattered at 90^0 within the wavelength range 350-800 nm. The above qualitative speculations have been confirmed by the cluster T-matrix method generalized to include two-layer spherical cluster particles (see Section 3.2.3).

For DLSS measurements, we use an attachment to a Specord M-40 spectrophotometer (Fig. 21). A detailed description of the attachment and the

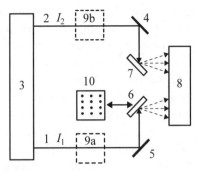

Figure 21. Scheme of DLSS measurements: *1* and *2* – sample and blank light beams, *3* – monochromator, *4* and *5* – mirrors, *6* and *7* – diffusive reflectors, *8* – photodetector (PM). For extinction measurements, the sample and blank cuvettes are placed in positions *9a* and *9b,* respectively. For scattering measurements, the diffusive reflector *6* is replaced by a sample cuvette *10,* and the other sample cuvette is placed in position *9b*.

corresponding procedure of spectral measurements have been discussed in Refs [27, 87], so we limit ourselves here to a short summary. During calibration of the device, the diffusive reflectors *4* and *5* are placed in both channels, and the calibration transmittances T_0 are stored in device memory

$$T_0(\lambda) = \frac{I_1(\lambda)r_1(\lambda)}{I_2(\lambda)r_2(\lambda)}, \tag{5.3}$$

where $I_{1,2}(\lambda)$ are the intensities of sample and blank monochromator beams, and $r_{1,2}(\lambda)$ are the reflection coefficients of diffusive reflectors. For extinction measurements, the sample and blank cuvettes are placed in positions *9a* and *9b*, respectively, and the measured transmittance is equal to

$$T_A(\lambda) = \frac{I_1(\lambda)r_1(\lambda)T_b(\lambda)10^{-A(\lambda)}}{I_2(\lambda)r_2(\lambda)T_b(\lambda)T_0(\lambda)} = 10^{-A(\lambda)}, \tag{5.4}$$

where T_b is the blank transmittance, and A is the measured sol extinction. In the case of scattering spectrum measurements, one of two identical sample cuvettes is placed in position *9b* and the other (*10*) is replaced with reflector *6* (Fig. 21). It can be shown that the light-scattering flux S_{90} is given by [87]

$$S_{90}(\lambda) = \beta I_1(\lambda)T_b(\lambda)10^{-A(\lambda)}I_{90}(\lambda), \tag{5.5}$$

where β is a constant coefficient. Hence, the measured transmission T_s equals

$$T_s(\lambda) = \frac{S_{90}(\lambda)}{I_2(\lambda)r_2(\lambda)T_b(\lambda)10^{-A(\lambda)}T_0(\lambda)} = \frac{\beta}{r_1(\lambda)}I_{90}(\lambda). \tag{5.6}$$

For experiments described in this work, glass plates coated with barium sulfate are used as scattering reflectors. These plates possess an almost neutral reflection spectrum within the range 400-700 nm, so that T_s is directly proportional to the single-scattering quantity $I_{90}(\lambda)$.

By contrast to known studies, in [27] we present experimental data on kinetic changes in extinction and scattering spectra caused by non-specific or biospecific aggregation of colloidal-gold conjugates. In both cases, already at 1-2 minutes after mixing the reagents we observe an essential increase in the resonance scattering maximum (up to 20 and even 400 times). Simultaneously, we record weaker changes in the extinction spectra. This observation allows one to assume that the developed light-scattering technique can be used as a sensitive analytical test. Figure 22 shows an example of kinetic measurements of extinction and scattering spectra during biospecific aggregation. As a model system, conjugates of 15-nm particles to Protein A (Sigma, USA) are used. Aggregation is initiated by addition of human IgG (hIgG) solution (Serva, Germany). The Protein A molecule has two or more sites for biospecific binding of IgG molecules. Data in Figure 23 correspond to an equimolar amount of reagent binding sites. Note how strongly the resonance scattering peak is increased in comparison with extinction spectral changes.

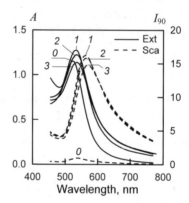

Figure 22. Evolution of extinction and scattering spectra with time: t = 0 (*0*, initial sol), 1 (*1*), 2(*1*), and 30 min (*3*). Experimental example for aggregation of conjugates CG-15+Protein A after addition of hIgG.

5.5. Adsorption of biopolymers on gold nanoparticles

Adsorption of polymers onto colloidal particles is of great interest for the chemical industry [88] and physical chemistry of disperse systems in connection with the fundamental problem of colloidal stability [89]. The structure of an adsorbed polymer layer can be probed by neutron [90] or X-ray [91] scattering, while the kinetics of adsorption can be followed by absorption spectroscopy [44].

To our knowledge, work [17] is the first instance where the adsorption of differently structured polymers (globular and random coil) onto gold nanoparticles of different sizes (18 and 34 nm) has been studied using extinction and differential light-scattering spectroscopy in combination with dynamic light scattering (DLS) particle sizing [92]. Rabbit antibodies against human immunoglobulins (designated as anti-hIgG), trypsin (Tr) and random coil (gelatin, Gel) polymers are used in our experiments that were treated in terms of homogeneous [12] and multilayer [17] optical models. All experimental details can be found in Ref. [17].

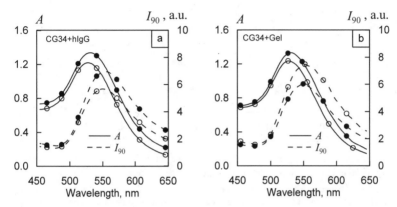

Figure 23. Extinction and scattering spectra measured for 34-nm colloidal gold conjugates with hIgG (a) and gelatin (b). The open and solid circles correspond to the spectra of bare gold particles and conjugates, respectively.

Table 2. Dynamic light-scattering data for gold particles, their bioconjugates, and the adsorbed polymer shells.

Polymer sample	Particle diameter, nm	Conjugate diameter, nm	Shell thickness, nm
hIgG	18 ± 2*	30.5 ± 3	6.2 ± 2.5
	34 ± 2	45 ± 3	5.5 ± 2.5
Tr	18 ± 2	27 ± 2	4.5 ± 2
Gel	18 ± 2	50 ± 3	16 ± 2.5
	34 ± 2	70 ± 3	18 ± 2.5

* The limits ±2 nm mean our estimation of a maximal error.

Figure 23a shows the extinction and scattering spectra measured for a 34-nm gold sol and CG34+hIgG conjugates. Adsorption of hIgG onto gold particles increases the extinction and scattering maxima by 10% and 20%, respectively, as well as slightly shifting peak positions to the red (3 nm for extinction and 4-5 nm for scattering). A theoretical simulation of the extinction and scattering spectra [12, 17] allows these changes to be attributed to the formation of an adsorbed homogeneous polymer shell with thickness of about $s = 5$ nm and average refractive index about $n_s = 1.40$-1.45. Using the DLS data (see Table 2), we find that the hydrodynamic radius of CG34+hIgG conjugates $d_1 = d + 2s = 45$ nm; hence, the average shell thickness is about 5.5 nm, in excellent agreement with spectral measurements and simulations, as well as in good agreement with data of Ref. [93] for 40-nm and 60-nm

CG+IgG conjugates. Similar results are obtained for CG18+hIgG conjugates ($d = 18$ nm, the shell thickness $s \approx 6$ nm), as well as for CG18+Tr conjugates ($d = 18$ nm, the shell thickness $s \approx 4.5$ nm).

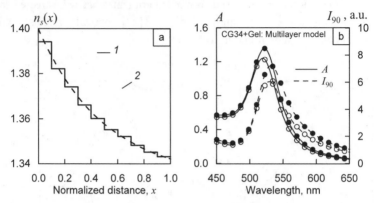

Figure 24. a - Inhomogeneous spatial profile of the refractive index (1) and its discrete approximation (2). The shell thickness of the adsorbed gelatin $s = 18$ nm, $x = r_s / s$ is the normalized distance from the gold particle surface. Panel b shows the corresponding theoretical extinction and light scattering spectra for 34-nm gold particles (open circles) and CG34+Gel conjugates (solid circles).

In the case of the random coil polymer (gelatin, Fig. 23b), the recorded changes in extinction and scattering spectra are close to those shown in Fig. 23a. This spectral analogy would imply that the adsorbed gelatin layer could be characterized by the same average thickness and density as for globular proteins. However, from DLS measurements, the adsorbed gelatin shell is found to be 15–18 nm, i.e., threefold as compared with the shell thickness for CG34+hIgG and CG18+Tr conjugates. Thus, we encounter a contradiction between quite similar optical properties and completely different shell thickness for the cases of globular proteins hIgG and Tr on the one hand, and for the case of random coil gelatin, on the other. Our attempts to reproduce the experimental extinction and light-scattering spectra with a homogeneous adsorbed gelatin layer are unsuccessful [12] when we use one and the same model for both spectra. To fix the problem, we use a new model with an inhomogeneous polymer coating [17]. The adsorbed shell is modelled by 10 discrete layers with 16–18 nm total thickness and exponential spatial profile of the shell refractive index (Fig. 24a)

$$n_s = n_m + K_1 \exp(-K_2 x), \tag{5.7}$$

where $x = r_s / s$ is the normalized distance from the gold-particle surface, n_m is the refractive index of the surrounding medium, and $K_{1,2}$ are the model constants. The exponential form of Eq. (5.7) is consistent with numerous theoretical and experimental data related to the polymer adsorption on an interface [94]. A more detailed discussion of the model and the determination of $K_{1,2}$ constants can be found elsewhere [17]. Using the inhomogeneous multilayer model depicted in Fig. 24a, we are able to obtain theoretical light-scattering and extinction spectra that are very close to the experimental observations (Fig. 24b). On the other hand, the hydrodynamic fitting size of CG34+Gel conjugates agrees with DLS measurements (Table 2).

6. Biomedical applications of colloidal gold and its bioconjugates

The pioneering biomedical applications of colloidal gold were published by Green in 1925 [95] and by Maclagan in 1944 [96]. Since 1971 [97], colloidal gold conjugates have been used traditionally in immunocyto- and histochemical studies as markers for electron microscopy. Gradually, the scope of use of gold markers was broadened. Currently, they are used in light microscopy methods and in various versions of force microscopy [98]. In addition, colloidal-gold conjugates are used in solid-phase-assay systems, such as dot-blot analysis [99], immunochromatographic test strips [100, 101], and as a carrier for antigens at immunization [102]. Readers are referred to the three-volume book [103] that describes numerous routine applications of immuno-gold conjugates in biology and medicine.

The up-to-date trends in using colloidal-gold markers for bioanalytical studies are as follows: (i) to minimize the analyte substance amount (antigens, oligonucleotides, lectins, etc.); and (ii) to enhance the sensitivity of analysis. For instance, a sensitivity up to 60-70 ng L^{-1} has been reported [104] for an assay based on a simple replacing of the nitrocellulose membranes by microarrays combined with a photometrical "reader." It should be emphasized that the microarray assay utilizes microliter volumes both for analyte probes and for labeled markers. Besides, the photometrical "reader," together with the accessory equipment, allows for 384 probes to be analyzed within a short time period.

The sensitivity of nanoparticle-based assays can be increased by the application of new detection techniques such as laser-based double beam absorption spectroscopy [105] and hyper-Rayleight scattering [106]. It is well known that the local electromagnetic fields are abnormally increased near metal rough surfaces or near a metal nanoparticle surface [107]. This effect has been intensively used in such fields as surface-enhanced Raman scattering [108] and surface-enhanced infrared absorption spectroscopy [109]. Also, such important and growing topics as targeted drug delivery [110] or genetic material [111, 112] delivery should be noted. The latter is used for genetic modifications of cells and novel DNA-vaccines.

Immobilized colloidal gold on electrode surfaces provides a microenvironment similar to that of the redox protein in native systems and gives the protein molecules more freedom in orientation, thus reducing the insulating property of the protein shell for the direct electron transfer and facilitating the electron transfer through the conducting tunnels of colloidal gold [113].

One of the important trends in biomedical application of gold and silver nanoparticles concern the study of DNA, RNA, and oligonucleotide molecules. For example, several papers [8, 114, 115] report on application of plasmon resonance nanoparticles for use with DNA microarrays. The reported sensitivity of microarray technologies in nucleic acid detection suggests that nanoparticle labels may have broad application in macroarray-based experiments. In the introduction, we mentioned some examples of novel nanoscale biosensors based on single-particle detection [39, 40]. Schultz *et al.* [3] demonstrated successful applications of single-particle detection for *in situ* labeling of a targeted DNA site on *Drosofila* chromosomes. Oldenburg *et al.* [116] extended this approach to ultrasensitive base pair mismatch recognition. It is expected [3], that the automation of nanoparticle identification, discrimination, and counting, may results in sensitive and high-throughput applications of gold and silver bioconjugates in the life science field. Quite recently, Storhoff *et al.* [117] reported on the technology that utilizes gold nanoparticles derivatized with thiol-modified oligonucleotides that are designed to bind the complementary DNA targets. This methodology is used for sensitive and specific DNA-based diagnostic. It has been shown that silver-amplified gold nanoparicle probes provide for a 1000-fold increase in sensitivity as compared to Cy3-based fluorescence.

In our research, fabricated colloidal-gold conjugates are used in studying the surface of nitrogen-fixing soil bacteria by TEM [118, 119]. The first application of gold markers to the dot-blot analysis of the surface antigens of soil bacteria is described in Refs. [120, 121]. Figure 25 shows a TEM image of an *Azospirillum brasilense* bacterium labeled by colloidal gold conjugates with whole-cell-antibodies (Fig. 25a) and a dot-blot analysis of Troponin I peptide (Fig. 25b).

Additionally, our markers and experimental procedures are used to develop an assay for rapid diagnosis of acute enteric infections [122], as well as in studies of a proliferative antigen of the initial cells of a wheat stem meristem [123] and plant cytoskeleton [124]. Recently, we recorded changes in the infrared spectra of Protein A–colloidal-gold conjugate after its interaction with immunoglobulin [125]. This result may serve as a basis for the development of new assay systems to detect biospecific interactions of the antigen-antibody type at the single-molecule level. Finally, one of the promising fields is the application of colloidal gold markers to preparation of antibodies both *in vitro* (by combinatorial phage display approach [126]), and *in vivo* (for enhancing of the immune response [127]). Note that this intriguing enhancing effect or, in other words, the adjuvant properties of gold sols have yet to be explained.

a

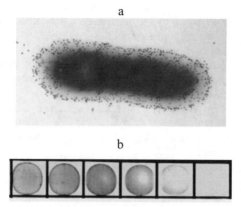

b

Figure 25. a – TEM image of an *Azospirillum brasilense* bacterium labeled by colloidal gold conjugates with whole-cell-antibodies; b – dot-blot staining of Troponin I peptide molecules using conjugates of 15-nm gold particles with anti-peptide antibodies. Details of the dot-blot technology can be found elsewhere [80].

Acknowledgements

This work was partially supported by grants from RFBR 04-04-48224, Education Ministry of Russian Federation 2.11.03, and by the President of Russian Federation grant NSH-25.2003.2. We thank D. N. Tychinin (IBPPM RAS) for help in preparation of the chapter, Yu-lin Xu (Univ. of Florida, USA) and Daniel Mackowski (Auburn Univ., USA) for help with implementation of computer codes, and Gorden Videen for thoroughly editing this chapter.

References

1. Th. Schalkhammer, Chemical Monthly, **129**, 1067 (1998).
2. K. K. Jain, Expert Rev. Mol. Diagn. **3**, 153 (2003).
3. D. A. Schultz, Curr. Opin. Biotechnol. **14**, 13 (2003).
4. W. J. Parak, D. Gerion, T. Pellegrino, D. Zanchet, C. Micheel, S. C. Williams, R. Boudreau, M. A. Le Gros, C. A. Larabell, and A. P. Alivisatos, Nanotechnology **14**, R15 (2003).
5. L. A. Lyon, M. D. Musick, and M. J. Natan, Anal. Chem. **70**, 5177 (1998).
6. N. Stich, A. Gandhum, V. Matyushin, J. Raats, C. Mayer, Y. Alguel, and T. Schalkhammer, J. Nanosci. Nanotechnol. **2**, 375 (2002).
7. C. A. Mirkin, Inorg. Chem. **39**, 2258 (2000).
8. P. Bao, A. G. Frutos, Ch. Greef, J. Lahiri, U. Muller, T. C. Peterson, L. Warden, and X. Xie, Anal. Chem. **74**, 1792 (2002).
9. U. Kreibig and M. Volmer, "Optical Properties of Metal Clusters" (Springer-Verlag, Berlin, 1995).
10. N. G. Khlebtsov, V. A. Bogatyrev, L. A. Dykman, Ya. M. Krasnov, and A. G. Melnikov, Izv. Vuz. Applied Nonlinear Dynamics **10** (Special English Issue No. 3), 172 (2002).
11. L. Kelly, E. Coronado, L. L. Zhao, and G. C. Schatz, J. Phys. Chem. B **107**, 668 (2003).

12. N. G. Khlebtsov, L. A. Dykman, V. A. Bogatyrev, and B. N. Khlebtsov, Colloid J. **65**, 552 (2003).

13. J. J. Mock, D. R. Smith, and S. Schultz, Nano Lett. **3**, 485(2003).

14. K. C. Grabar, R. G. Freeman, M. B. Hommer, and M. J. Natan, Anal. Chem. **67**, 735 (1995).

15. C. L. Haynes and R. P. Van Duyne, J. Phys. Chem. B **105**, 5599 (2001).

16. F. Caruso, Adv. Mater. **13**, 11 (2001).

17. N. G. Khlebtsov, V. A. Bogatyrev, B. N. Khlebtsov, L. A. Dykman, and P. Englebienne, Colloid J. **65**, 622 (2003).

18. J. H. W. Leuvering, P. J. H. M. Thal, M. van der Waart, and A. H. W. M. Schuurs, J. Immunoassay **1**, 77 (1980).

19. N. G. Khlebtsov, L. A. Dykman, Ya. M. Krasnov, and A. G. Melnikov, In: "Electromagnetic and Light Scattering by Nonspherical Particles: Theory and Applications," F. Obelleiro, J. L. Rodriguez, and Th. Wriedt (Eds.), p. 43 (Vigo Univ. Press, Vigo, Spain, 1999).

20. N. G. Khlebtsov, L. A. Dykman, Ya. M. Krasnov, and A. G. Melnikov, Colloid. J. **62**, 765 (2000).

21. J. J. Storhoff, A. A. Lazarides , R. C. Mucic, C. A. Mirkin, R. L. Letsinger, and G. C. Schatz, J. Am. Chem. Soc. **122**, 4640 (2000).

22. G. S. Schatz, Theochem. **573**, 73 (2001).

23. C. A. Mirkin, R. L. Letsinger, R. C. Mucic, and J. J. Storhoff, Nature **382**, 607 (1996).

24. K. Sato, K. Hosokawa, and M. Maeda, J. Am. Chem. Soc. **125**, 8102 (2003).

25. T. Ciesiolka and H.-J. Gabius, Anal. Biochem. **168**, 280 (1988).

26. V. A. Bogatyrev, L. A. Dykman, Ya. M. Krasnov, V. K. Plotnikov, and N. G. Khlebtsov, In: "Optical Technologies in Biophysics and Medicine III," V. V. Tuchin (Ed.), Proc. SPIE, Vol. 4707, p. 266 (SPIE, Bellingham, WA, 2002).

27. V. A. Bogatyrev, L. A. Dykman, Ya. M. Krasnov, V. K. Plotnikov, and N. G. Khlebtsov, Colloid. J. **64**, 671 (2002).

28. D. Roll, J. Malicka, I. Gryczynski, Z. Gryczynski, and J. R. Lakowicz, Anal. Chem. **75**, 3440 (2003).

29. X. Liu, H. Yuan, D. Pang, and R. Cai, Spectrochim. Acta Part A **60**, 385 (2004).

30. B. Dragnea, Ch. Chen, E.-S. Kwak, B. Stein, and C. Ch. Kao, J. Am. Chem. Soc. **125**, 6374 (2003).

31. Y.-F. Wang, D.-W. Pang , Z.-L. Zhang, H.-Z. Zheng, J.-P. Cao, and J.-T. Shen, J. Med. Virol. **70**, 205 (2003).

32. P. Englebienne, A. van Hoonacker, and J. Valsamis, Clin. Chem. **46**, 2000 (2000).

33. P. Englebienne, A. van Hoonacker, and M. Verhas, Analyst **126**, 1645 (2001).

34. P. Englebienne, A. van Hoonacker, M. Verhas, and N. G. Khlebtsov, Combinatorial Chemistry & High Throughput Screening **6**, 777 (2003).

35. N. Nath and A. Chilkoti, Anal. Chem. **74**, 504 (2002).

36. F. Frederix, J. M. Friedt, K. H. Choi, W. Laureyn, A. Campitelli, D. Mondelaers, G. Maes, and G. Borghs, Anal. Chem. **75**, 6894 (2003).

37. J. Haes and R. P. Van Duyne, J. Am. Chem. Soc. **124**, 10596 (2002).

38. J. C. Riboh, A. J. Haes, A. D. McFarland, C. Ranjit, and R. P. Van Duyne, J. Phys. Chem. B **107**, 1772 (2003).

39. G. Raschke, S. Kowarik, T. Franzl, C. Sönnichsen, T. A. Klar, J. Feldmann, A. Nichtl, and K. Kürzinger, Nano Lett. **3**, 935 (2003).

40. A. D. McFarland and R. P. Van Duyne, Nano Lett. **3**, 1057 (2003).

41. W. C. W. Chan, D. J. Maxwell, X. Gao, R. E. Bailey, M. Han, and S. Nie, Curr. Opin. Biotechnol. **13**, 40 (2002).

42. N. G. Khlebtsov, I. L. Maksimova, V. V. Tuchin, and L. Wang, In: "Handbook of Optical Biomedical Diagnostics," V. V. Tuchin (Ed.), p. 31 (SPIE, Bellingham, WA, 2002).

43. A. C. Templeton, J. J. Pietron, R. W. Murray, and P. Mulvaney, J. Phys. Chem. B **104**, 564 (2000).

44. D. Eck, C. A. Helm, N. J. Wagner, and K. A. Vaynberg, Langmuir **17**, 957 (2001).

45. C. F. Bohren and D. R. Huffman, "Absorption and Scattering of Light by Small Particles" (John Wiley&Sons, New York, 1983)

46. M. D. Malinsky, K. L. Kelly, G. C. Schatz, and R. P. Van Duyne, J. Am. Chem. Soc. **123**, 1471 (2001).

47. G. J. Fleer, M. A. Cohen Stuart, J. M. H. M. T. Scheutjens Cosgrove, and B. Vincent, "Polymers at Interfaces" (Chapman&Hall, London, 1993).

48. N. G. Khlebtsov, V. A. Bogatyrev, L. A. Dykman, and A. G. Melnikov, J. Colloid Interface Sci. **180**, 436 (1996).

49. P. B. Johnson and R. W. Christy, Phys. Rev. B **12**, 4370 (1973).

50. M. I. Mishchenko, L. D. Travis, and A. A. Lacis, "Scattering, Absorption, and Emission of Light by Small Particles" (Cambridge University Press, Cambridge, 2002).

51. H. Xu and M. Käll, Sens. Actuators B Chem. **87**, 244 (2002).

52. Z. C. Wu and Y. P. Wang, Radio Sci. **26**, 1393 (1991).

53. N. G. Khlebtsov and A. G. Melnikov, Colloid. J. **60**, 781 (1998).

54. E. M. Purcell and C. R. Pennypacker, Astrophys. J. **186**, 705 (1973).

55. A. Lakhtakia and G. W. Mulholland, J. Res. Natl. Inst. Stand. Technol. **98**, 699 (1993).

56. J. C. Ku, J. Opt. Soc. Am. A **10**, 336 (1993).

57. B. T. Draine, In: "Light scattering by Nonspherical Particles," M. I. Mishchenko, J. W. Hovenier, L. D. Travis (Eds.), p. 131 (Academic Press, San Diego, 2000).

58. V. A. Markel, V. M. Shalaev, E. B. Stechel, W. Kim, and R. L. Armstrong, Phys. Rev. B **53**, 2425 (1996).

59. N. G. Khlebtsov, In: "Light Scattering by Nonspherical Particles: Halifax Contributions," G. Videen, Q. Fu, and P. Chylek (Eds.), p. 123 (Army Research Laboratory, Adelphy MD, 2000).

60. N. G. Khlebtsov, Opt. Spectrosc. **90**, 408 (2001).

61. Y.-l. Xu, Appl. Opt. **34**, 4573 (1995).

62. K. A. Fuller and D. W. Mackowski, In: "Light Scattering by Nonspherical Particles: Theory, Measurements, and Applications," M. I. Mishchenko, J. W. Hovenier, and L. D. Travis (Eds.), p. 225 (Academic Press, San Diego, 2000).

63. N. G. Khlebtsov, A. G. Melnikov, and Y.-l. Xu, In: "Electromagnetic and Light Scattering - Theory and Applications VII," Th. Wriedt (Ed.), p. 147 (Universität Bremen, Bremen, 2003).

64. N. G. Khlebtsov, A. G. Melnikov, and Y.-l. Xu, *ibid.* p. 143.

65. D. W. Mackowski and M. I. Mishchenko, J. Opt. Soc. Am. A **13**, 2266 (1996).

66. Y.-l. Xu and N. G. Khlebtsov, J. Quant. Spectr. Radiat. Ttransfer **79-80**, 1121 (2003).

67. K. L. Kelly, A. A. Lazarides, and G. C. Schatz, Comp. Sci. Eng. **3**, 67 (2001).

68. D. W. Mackowski, J. Opt. Soc. Am. A **11**, 2851 (1994).

69. M. I. Mishchenko, D. W. Mackowski, and L. D. Travis, Appl. Opt. **34**, 4589 (1995).

70. A. A. Lazarides and G. C. Schatz, J. Phys. Chem. B **104**, 460 (2000).

71. N. G. Khlebtsov, V. A. Bogatyrev, L. A. Dykman, B. N. Khlebtsov, and A. G. Melnikov, In: Abstracts of NATO Advanced Study Institute on "Photopolarimetry in Remote Sensing", G. Videen, Ya. Yatskiv, A. Vid'machenko, V. Rosenbush, and M. Mishchenko (Eds.), p. 46 (Army Research Laboratory, Adelphy MD, 2003).

72. N. G. Khlebtsov, L. A. Trachuk, and A. G. Melnikov, Opt. Spectrosc. **94** (6), (2004, in press).

73. L. L. Zhao, K. L. Kelly, and G. C. Schatz, J. Phys. Chem. B **107**, 7343 (2003).

74. C. L. Haynes, A. D. McFarland, L. L. Zhao, G. C. Schatz, R. P. Van Duyne, L. Gunnarsson, J. Prikulis, B. Kasemo, and M. Käll, J. Phys. Chem. B **107**, 7337 (2003).

75. M. Y. Lin, H. M. Lindsay, D. A. Weitz, R. C. Ball, R. Klein, and P. Meakin, Nature **339**, 360 (1989).

76. L. A. Dykman, Ya. M. Krasnov, V. A. Bogatyrev, and N. G. Khlebtsov, In: "Optical Technologies in Biophysics and Medicine II," V. V. Tuchin (Ed.), Proc. SPIE, Vol. 4241, p. 371 (SPIE, Bellingham, WA, 2001).

77. P. Meakin, Ann. Rev. Phys. Chem. **39**, 237 (1988).

78. G. Frens, Nature Phys. Sci. **241**, 20 (1973).

79. L. A. Dykman, A. A Lyakhov, V. A. Bogatyrev, and S. Yu. Shchyogolev, Colloid J. **60**, 700 (1998).

80. L.A. Dykman and V. A. Bogatyrev, Biochemistry (Moscow) **62**, 350 (1997).

81. W. Nowicki, Colloids and Surfaces A **194**, 159 (2001).

82. A. Doron, E. Joselevich, A. Schlitter, and I. Willner, Thin Solid Films **340**, 183 (1999).

83. K. R. Brown, D. G. Walter, and M. J. Natan, Chem. Mater. **12**, 306 (2000).

84. K. R. Brown and M. J. Natan, Langmuir **14**, 726 (1998).

85. Yu. E. Danilova, "Localization of Optical Excitations in Colloidal Silver Aggregates'" PhD Thesis (Institute of Automatics and Electrometry RAS, Novosibirsk, 1999).

86. N. G. Khlebtsov, V. A. Bogatyrev, L. A. Dykman, B. N. Khlebtsov, and Ya. M. Krasnov, Th. Wriedt (Ed.), p. 135 (Universität Bremen, Bremen, 2003).

87. V. A. Bogatyrev, L. A. Dykman, B. N. Khlebtsov, and N. G. Khlebtsov, Opt. Spectrosc. **96**, 128 (2004).

88. A. Hubbard (Ed.), "Encyclopedia of Surface and Colloid Science" (Marcell Dekker, New York, 2002).

89. M. Faraday, Philos. Trans. R. Soc. London **147**, 145 (1857).

90. C. N. Likos, K. A. Vaynberg, H. Löven, and N. J. Wagner, Langmuir **16**, 4100 (2000).

91. S. Seelenmeyer and M. Ballauff, Macromol. Symp. 145, 9 (1999).

92. B. J. Berne and R. Pecora, "Dynamic Light Scattering with Applications to Chemistry, Biology, and Physics" (Dover Publications, Mineola NY, 2000).

93. J. M. C. Martin, M. Pâques, T. A. M. van der Velden-de Groot, and E. C. Beuvery, J. Immunoassay **11**, 31 (1990).

94. G. D. Parfitt and C. H. Rochester (Eds.), "Adsorption from Solution at the Solid/Liquid Interface" (Academic Press, New York, 1983).

95. F. Green, "The Colloidal Gold Reaction of the Cerebrospinal Fluid" (Medizin Fritz-Dieter Söhn, Berlin, 1925).

96. N. F. Maclagan, Brit. J. Exp. Pathol. **25**, 15 (1944).

97. W. Faulk and G. Taylor, Immunochemistry **8**, 1081 (1971).

98. C. Neagu, K. O. van der Werf, C. A. J. Putman, Y. M. Kraan, B. G. de Grooth, N. F. van Hulst, and J. Greve, J. Struct. Biol. **112**, 32 (1994).

99. M. Moeremans, G. Daneels, A. van Dijck, G. Langanger, and J. De Mey, J. Immunol. Meth. **74**, 353 (1984).

100. J.-H. Cho and S.-H. Paek, Biotechnology and Bioengineering **75**, 725 (2001).

101. K. Glynou, P. C. Ioannou, T. K. Christopoulos, and V. Syriopoulou, Anal. Chem. **75**, 4155 (2003).

102. S. Shiosaka, H. Kiyama, A. Wanaka, and M. Tohyama, Brain Research **382**, 399 (1986).

103. M. A. Hayat, (Ed.), "Colloidal Gold: Principles, Methods, and Applications" (Academic Press, San Diego, 1989).

104. A. Han, M. Dufva, E. Belleville, and C. B. V. Christensen, Lab. Chip. **3**, 329 (2003).

105. N. T. K. Thanh, J. H. Rees, and Z. Rosenzweig, Anal. Bioanal. Chem. **374**, 1174 (2002).

106. C. X. Zhang, Y. Zhang, X. Wang, Z. M. Tang, and Z. H. Lu, Anal. Biochem. **320**, 136 (2003).

107. V. M. Shalaev (Ed.), "Topics in Applied Physics. Optical Properties of Nanostructured Random Media," (Springer-Verlag, Berlin-Heidelberg, 2002).

108. J. Ni, R. J. Lipert, G. B. Dawson, and M. D. Porter, Anal. Chem. **71**, 4903 (1999).

109. C. W. Brown, Y. Li, J. A. Seelenbinder, P. Pivarnik, A. G. Rand, S. V. Letcher, O. J. Gregory, and M. J. Platek, Anal. Chem. **70**, 2991 (1998).

110. S. P. Vyas and V. Sihorkar, Advanced Drug Delivery Reviews **43**, 101 (2000).

111. D. Chen and L. G. Payne, Cell Research **12**, 97 (2002).

112. Z. Zhao, T. Wakita, and K. Yasui, J. Virol. **77**, 4248 (2003).

113. S. Liu, D. Leech, and H. Ju, Analytical Letters **36**, 1 (2003).

114. T. A. Taton, C. A. Mirkin, and R. L. Letsinger, Science **289**, 1757 (2000).

115. A. Csáki, G. Maubach, D. Born, J. Reichert, and W. Fritzsche, Single Mol. **3**, 275 (2002).

116. S. J. Oldenburg, Ch. C. Genick, K. A. Clark, and D. A. Schultz, Anal. Biochem. **309**, 109 (2002).

117. J. J. Storhoff, S. S. Marla, P. Bao, S. Hagenow, H. Mehta, A. Lucas, V. Garimella, T. Patno, W. Buckingham, W. Cork, and U. R. Müller, Biosensors Bioelectronics **19**, 875 (2004).

118. E. V. Egorenkova, S. A. Konnova, Yu. P. Fedonenko, L. A. Dykman, and V. V. Ignatov, Microbiology **70**, 36 (2001).

119. M. I. Chumakov, L. A. Dykman, V. A. Bogatyrev, and I. V. Kurbanova, Microbiology **70**, 232 (2001).

120. V. A. Bogatyrev, L. Yu. Ivanova, B. I. Schwartsburd, and N. G. Khlebtsov, In: 19th Meeting FEBS, July 2-7, 1989, Abstr. Book, p. 30 (FEBS, Rome, 1989).

121. V. A. Bogatyrev, L. A. Dykman, L. Yu. Matora, and B. I. Schwartsburd, FEMS Microbiol. Lett. **96**, 115 (1992).

122. L. A. Dykman and V. A. Bogatyrev, FEMS Immunol. Med. Microbiol. **27**, 135 (2000).

123. M. V. Sumaroka, L. A. Dykman, V. A. Bogatyrev, N. V. Evseeva, I. S. Zaitseva, S. Yu. Shchyogolev, and A. D. Volodarsky, J. Immunoassay **21**, 401 (2000).

124. L. A. Dykman, V. A. Bogatyrev, I. S. Zaitseva, M. K. Sokolova, V. V. Ivanov, and O. I. Sokolov, Biophysics **47**, 587 (2002).

125. A. A. Kamnev, L. A. Dykman, P. A. Tarantilis, and M. G. Polissiou, Bioscience Reports **22**, 541 (2002).

126. M. V. Sumaroka, L. A. Dykman, V. A. Bogatyrev, I. S. Zaitseva, O. I. Sokolov, S. Yu. Shchyogolev, and W. J. Harris, Allergology and Immunology **1**, 134 (2000).

127. L. A. Dykman, M. V. Sumaroka, S. A. Staroverov, I. S. Zaitseva, and V. A. Bogatyrev, Biol. Bull. **31** (2004) (in press).

N. G. Khlebtsov, V. A. Bogatyrev, L.A. Dykman, and A. G. Melnikov

Sweltering at Chufut-Kale.

POLARIZATION OF LIGHT BY PRE-MAIN-SEQUENCE STARS IN THE VISUAL WAVELENGTHS

V. P. GRININ

Pulkovo Astronomical Observatory, S. Petersburg, Russia;
Crimean Astrophysical Observatory, Crimea, Nauchny, Ukraine;
The Sobolev Astronomical Institute of S. Petersburg University,
S. Petersburg, Russia;

Abstract. The intrinsic polarization of young stars contains important information about the physical state of the objects and their surroundings. In particularly, the investigations of polarimetric activity of the variable stars can throw light upon variability mechanisms. In stars surrounded by circumstellar disks unresolved with a telescope, the position angle of the intrinsic linear polarization can indicate disk orientation. The wavelength dependence of the linear polarization is a sensitive function of the optical parameters of the dust particles and can be used for their diagnostics.

1. Introduction

The intrinsic polarization of the pre-main-sequence (PMS) stars is the result of interaction of the stellar radiation with their circumstellar (CS) surroundings. It can provide us with valuable information about the physical properties and geometry of CS matter. In this chapter I present mostly polarimetric observations of young stars in the optical region of spectrum. The current state of polarimetric investigations in the infrared and submillimeter regions of spectrum is discussed in the papers by Hough & Aitken (this book), Tamura (2001) and Wolf *et al.* (2003). The polarimetric observations of the T Tauri stars (TTSs) and related objects obtained up to the end of the eighties are summarized by Bastien (1988). The nature and properties of photopolarimetric activity of the PMS stars are discussed in review papers by Grinin (1994; 2000). The comprehensive review of the polarization by radiation of star-forming regions is given in the paper by

G. Videen et al. (eds.), Photopolarimetry in Remote Sensing, 309-324.
© *2004 Kluwer Academic Publishers. Printed in the Netherlands.*

Weintraub, Goodman & Akeson (2000). It is useful to present a short review
of the primary mechanisms of polarization of light by young stars.

In general, polarized radiation can be described by four Stokes parameters (I,Q,U,V): the linear polarization is a pseudo-vector: $P = (Q^2 + U^2)/I$;
its position angle $\theta = (1/2) \arctan(U/Q)$; and the ratio V/I gives the circular polarization. The observed polarization usually includes three components:
the instrumental polarization P_{inst}, the interstellar (IS) polarization P_{is}
and the intrinsic polarization of the object P_{in}:

$$\mathbf{P}_{obs} = \mathbf{P}_{inst} + \mathbf{P}_{is} + \mathbf{P}_{in} \tag{1}$$

Therefore in order to obtain the intrinsic polarization we have to correct
$\mathbf{P}_{obs}$ for the interstellar and the instrumental components. The classical
method of investigating interstellar polarization is through polarimetric
observations of neighboring stars. The instrumental polarization depends
on the optics of the telescope and the polarimeter and usually does not
exceed a few tenths of a percent. The correction for P_{inst} demands regular
observations of polarimetric standards collected in several papers (see e.g.
Coyne *et al.* 1974; Tinbergen 1979; Whittet 1992).

2. Mechanisms of light polarization in PMS stars

Most young stars are surrounded by CS disks consisting of gas and dust
(Strom *et al.* 1993). The inner parts of CS disks are ionized and some part of
the polarized radiation of the Herbig AeBe stars can arise due to Thomson
scattering of photons by free electrons.

2.1. THOMSON SCATTERING IN CS ENVELOPES AND DISKS

The optical thickness of the CS gas envelopes due to Thomson scattering τ_s
is usually small and the observed polarization is the result of single scattering. A well known example of such objects are the classical Be stars. Due
to the rapid rotation of these stars their CS envelopes are flattened and
the Thomson scattering of stellar radiation gives intrinsic polarization up
to about 1-2% (see e.g. Poeckert & Marlborough 1976; Brown & McLean
1977). The position angle of the linear polarization is parallel to the symmetry axis of the envelope. The degree of polarization is proportional to
$\sin^2 i$, were i is the angle between the line-of-sight and the symmetry axis
of the envelope, and depends also on its geometry. For example, in the
case of the point source surrounded by the flattened ellipsoidal envelope
with a uniform distribution of matter, the maximal degree of polarization

$p_{max}(\%) = 7\,\tau_s$ takes place when the ratio of major semi-axis to the minor one $A/B = 4$ (Dolginov *et al.* 1979).

In the case of Thomson scattering, the intensity of the scattered radiation does not depend on the wavelength of radiation λ. However, the thermal radiation of the gas envelope (which depends on λ) leads to a wavelength dependence of the intrinsic linear polarization. The most prominent details of p_λ in the visual wavelengths are the Balmer and Paschen jumps (Poeckert *et al.* 1979).

In the presence of a magnetic field, the scattered radiation undergoes Faraday rotation. The angle of the Faraday rotation of the electric vector of the electromagnetic wave

$$\chi \approx 0.8\,\lambda^2\,B\,\tau_s\,\cos\Theta\,, \tag{2}$$

where λ is the wavelength in microns, B is the magnetic strength in Gauss, Θ is the angle between the magnetic field and the light propagation vector (see e.g. Gnedin & Silant'ev 1997). If $\chi \gg 1$ then the depolarization effect is strong. In the opposite case the magnetic field cannot influence light propagation. Gnedin & Silant'ev (1997) considered different effects of Faraday rotation on the scattered envelopes due to the magnetic fields. Because of the wavelength dependence of χ the Faraday rotation is more important at radio wavelengths.

2.2. THOMSON SCATTERING AND POLARIZATION OF EMISSION LINES

If an emission line is formed due to the excitation of atoms by electron impacts or due to recombination, and the role of the resonance scattering is small, then the line emission is non-polarized even if the emitting region is non-spherical. The scattering of photons by free electrons can give different polarization effects within the spectral line. For example, the combination of the polarized continuum radiation and the non-polarized radiation in the line frequencies leads to depolarization effects within the line profile. This is possible if the polarized continuum is formed in a more compact region than the emission line. Such an effect is observed in the H_α line in the classical Be stars (Clarke & McLean 1974; Schulte-Ladbeck *et al.* 1994), in Herbig Be stars (Oudmajer & Drew 1999; Vink *et al.* 2002) and some other objects. A more complex situation takes place in the case of Thomson scattering of the emission line radiation. In this case a change of the linear polarization across the line profile is sensitive to the geometry and kinematic conditions of the emitting regions and is an important tool for their diagnostics (McLean 1979; Wood *et al.* 1993; Vink *et al.* 2002).

2.3. SCATTERING BY DUST

In stars surrounded by CS disks of gas and dust there are two possible different mechanisms of light polarization connected with the dust component: a) the light scattering in the non-spherical CS dust surroundings[1]; and b) the dichroic absorption and emission by non-spherical aligned grains. The light scattering depends on the sort of dust and the ratio of the characteristic grain sizes a to the wavelengths λ (see the chapter by Mishchenko & Travis, this book). In the case of spherical or misaligned non-spherical grains, the polarization vector is orthogonal to the radius vector toward the light source in the optically thin media (see e.g. Piirola *et al.* 1994). The optical parameters of grains are the scattering asymmetry factor g, the maximal polarization p_m (at a $90°$ scattering angle) and albedo $\omega = Q_{sc}/Q_{ext}$, where Q_{sc} and Q_{ext} are the scattering and extinction efficiency. In the case of Rayleigh scattering $g = 0$ and $p_m = 1$ (100%). Detailed calculations of these parameters for different types of dust particles were made by Daniel (1978), Draine & Lee (1984), Simmons (1983) and many others.

2.4. OPTICAL DICHROISM

Under certain conditions, non-spherical grains can be aligned in the CS magnetic fields (e.g. Dolginov & Mitrofanov 1978; Lazarian 2003 and references there). This leads to optical dichroism of CS dust and, as a result, to linear polarization of stars. The emission by aligned grains as a source of polarized radiation is important in the infrared and millimeter regions of spectrum (Tamura 2000; Wolf *et al.* 2003; Hough & Aitken, this book). The dichroic extinction can be important at all wavelengths where dust extinction takes place. In this case

$$P_{di} = \frac{I_\perp - I_\parallel}{I_\perp + I_\parallel} \tag{3}$$

Here

$$I_\perp = I_* \, e^{-\tau_\perp}; \; I_\parallel = I_* \, e^{-\tau_\parallel} \tag{4}$$

where $\tau_\perp$ and $\tau_\parallel$ are the optical thickness of the envelope for electromagnetic waves with electric vector orthogonal and parallel to the direction of the grains alignment.

---

[1]Recent observations by Ménard *et al.* (2002) of the intrinsic linear polarization of brown dwarfs show that dust scattering can take place immediately in the atmospheres of these objects.

2.5. CIRCULAR POLARIZATION

If the CS envelope contains spherical or non-spherical misaligned grains that are illuminated by non-polarized radiation then the single scattered radiation has zero circular polarization. Non-zero circular polarization can appear in the following situations:

1) When the spherical or misaligned non-spherical grains are illuminated by polarized light. This situation is realized in the case of multiple scattering of the initially non-polarized radiation (Bastien 1988). Two scattering events are enough and the degree of circular polarization $P_V \leq P_l^2$, where P_l is the linear polarization (Dolginov et al. 1979).

2) When the grains are non-spherical and aligned. In this case a single scattering event is enough (Bandermann & Kemp 1973; Shakhovskoj et al. 2001).

Besides these two cases, circular polarization can appear in the propagation of radiation in a media consisting of particles (or molecules) with chirality (Dolginov et al. 1979). Thus, the origin of circular polarization needs more specific conditions in the comparison with linear polarization. For example, if we observe an axially symmetric CS envelope in a diaphragm centered on the star, its scattered radiation is linearly polarized excepting the pole-on orientation. At the same time the circular polarization of this object is equal to zero for all orientations (Bastien 1988).

3. Polarization of young stars

Let us consider now in more detail the polarimetric properties of young stars. Many of them are surrounded by accretion disks and (or) disk-like envelopes of gas and dust. Near the young star (where its emission spectrum is formed) the gas is ionized and the Thomson scattering can provide some amount of polarized radiation. Recent spectropolarimetric observations of the Herbig AeBe stars in the H_α line have shown (Oudmajer & Drew 1999; Vink et al. 2002) that this mechanism can provide a sizable (a few tenths of a percent) contribution to the observed polarization of these stars. Vink et al. have found different behavior of the Stokes parameters within the Hα line profile in the Herbig Ae and Herbig Be stars. This difference is quite natural, since the more luminous Herbig Be stars can ionize the more extended region of their CS disks.

The broad-band polarimetric observations of a large number of PMS stars show, however, that the main source of their intrinsic polarization is the scattering of the stellar radiation by CS dust (Bastien 1985; Grinin et al. 1991). This conclusion follows from the wavelength dependence of the intrinsic polarization: it is far from that observed in the classical Be stars where Thomson scattering dominates. Another argument in favor of

CS dust is the correlation between the degree of linear polarization of the young stars p and their IR excesses (Bastien 1985; Yudin 2000).

The highest linear polarization (up to 30%) is observed in the youngest stars having the largest IR excesses. The CS disk around such stars are so powerful that the direct (non-polarized) stellar radiation is absorbed almost fully by CS dust and only the highly polarized scattered radiation from the polar regions above and beneath the disk plane (which are more transparent for the stellar radiation) can be observed (Elsässer & Staude 1978). According to Bastien (1988), a similar approach can be used for the interpretation of the polarization reversal that is observed in some TTSs: a change of the position angle θ of the linear polarization by 90° with the wavelength (Hough et al. 1981). The point is that the extinction by dust decreases with increasing λ and the situation is possible when the scattered radiation in the optics and near IR emerges from the polar regions or from the disk itself. The position angles of the linear polarizations in these two cases are orthogonal to each other.

The dependence of the position angle (PA) of the intrinsic linear polarization on the geometry of the scattered region frequently is used for the investigation of the orientation of symmetry axes of CS disks (which are unresolved with a telescope), both in single young stars (Grinin et al. 1991; Rostopchina et al. 1997; Rostopchina et al. 2001) and in the young binaries (Jensen et al. 2000; Monin et al. 1998, 2002; Maheswar et al. 2002).

Whitney & Hartmann (1992) investigated the polarimetric properties resulting from different models of accretion disks of TTSs and have shown that the flared disk gives much more polarized radiation then the flat one. The maximum polarization ($p \approx 13\%$) takes place when the flared disk is slightly inclined to the line-of-sight and the direct stellar radiation largely is obscured by the peripherical part of the disk. In both cases (flat or flared disks), the position angle θ of the linear polarization is parallel to the disk axis ($\theta = 0$) for any inclination of the disk to the line-of-sight. In the model of the disk surrounded by a dusty infalling envelope PA depends on the model parameters and situations are possible when the linear polarization has opposite sign at different wavelengths (Whitney & Hartmann 1993). Thus, this model can explain the changes of PA by 90° with wavelength mentioned above as well as the observed correlation between p and IR excesses of the young stars.

Repeated polarimetric observations of young stars show that in many of them the observed polarization is variable (Vardanian 1964; Serkowski 1969). The variability is observed in the position angle, degree of polarization and the wavelength dependence of P and θ. Several mechanisms have been suggested for explanation of this variability. One of them is the surface inhomogeneity of the young stars.

3.1. POLARIZATION OF SPOTTED T TAURI STARS

Most TTSs are cool stars. Therefore, their radiation is sensitive to temperature variations on their surface, especially at short wavelengths. The existence of surface inhomogeneities (cool and hot spots) of rotating stars leads to a periodic modulation of the flux illuminating the CS dust and, as a result, to the periodic modulation of the intrinsic polarization of stars. Its amplitude depends on the amplitude of the light modulation as well as on the CS disk parameters including its inclination to the line-of-sight and the optical parameters of CS dust.

In most cases the cool magnetic spots yield low-amplitude photometric waves in the light curves of TTSs. The amplitude maximum ($\Delta V \approx 0.6$) is observed in the weak T Tauri star V410 Tau (Vrba *et al.* 1988). Stars of such type have weak accretion disks. In this case it is difficult to expect a sizable amount of polarized radiation. The hot spots on the classical T Tauri stars (which are surrounded by powerful accretion disks) are more important. Such spots are caused by magnetically tunnelled gas accretion onto the stars and can provide a large, time-dependent contribution to the luminosity of TTSs (see, e.g. Bouvier & Bertout 1989).

Wood *et al.* (1996) and Stassun and Wood (1999) investigated the expected photopolarimetric variability of TTSs in the magnetic accretion disk model. Their model includes two diametrically opposed hot spots with radii of 20° at some fixed latitudes ϕ_s, a spot temperature $T_s = 10^4$ K and effective stellar temperature $T_{ef} = 4000$ K. The spots with such parameters are ≈ 3 times more luminous than the stellar hemisphere in the V band. Calculations have shown that the polarization effect depends strongly on the spot latitudes ϕ_s. The strongest effect is expected at small ϕ_s. In this case the maximum polarization (about 2%) takes place near 90° and 270° phases when the spots are close to the stellar limb. At these phases the direct (non-polarized) radiation of one of the spots is occulted by the stellar disk and the scattered radiation gives maximal polarization. The maximum p is expected at small inclinations of the CS disk to the line of sight. In the Q-U plane the theoretical tracks form loops that become more elongated with decreasing inclination.

Polarimetric effects predicted by this model have not been found up to now. We note, however, that the accretion of CS matter onto young stars is essentially the non-stationary process. Therefore the hot-spot parameters can be highly variable. Besides, in the young stars surrounded by a CS disks there is a stronger mechanism of light polarization that can mask the thin effect discussed above (see below).

The variable illumination of the CS dust also can be caused by the variable extinction of the stellar radiation due to the optically thick fragments of the CS disks themselves. An example of such an object is the T Tauri star AA Tau whose variability is caused by the large-scale dust "clump" orbiting the star with the same angular velocity as the star (Bouvier *et al.* 1999; Meńard *et al.* 2003). It is possible that the "clump" is the part of the inner region of the accretion disk perturbed by the stellar magnetosphere. It blocks the stellar radiation within the sizable solid angle that provides the time-dependent illumination of the CS disk. However, the observed polarimetric effect in this star is caused by the so-called coronagraphic effect (Meńard *et al.* 2003), which is similar to that observed in the UX Ori type stars, which we discuss in the next section.

3.2. CS CLOUDS, CORONAGRAPHIC EFFECT, UX ORI STARS

There is numerous observational evidence that the CS matter around young stars is distributed in an extremely inhomogeneous fashion. The clouds of dust fragments can appear in the neighborhoods of stars and lead to strong fluctuations of scattered light (Grinin 1994). A single large cloud of dust moving in the neighborhood of a star can give a sizable contribution to the scattering light. In this case the Stokes parameters of the scattered radiation is shifted on the Q-U plane along loops similar to that observed in some binary stars (Bastien 1988). More important, however, the CS dust clouds can be opaque to the stellar radiation. When intersecting the line-of-sight such clouds can screen a star from an observer. In these moments, the scattered radiation of the CS disk (which is usually weak) becomes a dominant source of the observed radiation and the intrinsic polarization of the young star. In other words, the CS clouds of gas and dust can play a role of the natural coronagraph and make possible the observations of the scattered radiation from CS disks.

This mechanism of photometric and polarimetric activity is realized in most spectacular form in young stars with non-periodic algol-type minima, so called UX Ori type stars, or UXORs (Grinin *et al.* 1991). The synchronous polarimetric and photometric observations of these stars started at CrAO in 1986[2] and lead us to the following conclusions:

i) The linear polarization of UXORs anti-correlates with brightness changes reaching 5-10% in the very deep minima (see Figs. 1 and 2), and the dependence of P_{obs} on the amplitude Δm agrees with that predicted by the variable extinction model (Figure 3):

$$\mathbf{P}_{obs}(\Delta m) = \mathbf{P}_{is} + \mathbf{P}_{in}(0) \cdot 10^{0.4\,\Delta m} \qquad (5)$$

---

[2]Part of these observations were made in cooperation with N. Kiselev and his team at the Sunglok Observatory (Tadjikistan), and with A. Okazaki and S. Kikuchi at the Dodajra Astronomical Observatory (Japan).

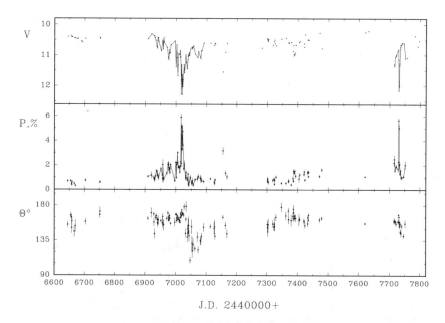

Figure 1. Behaviour of the brightness, degree and position angle of the linear polarization of the UX Ori star WW Vul in the V-band from Grinin *et al.* (1988) and Berdjugin *et al.* (1992). The complex structure of the long lasting minimum of 1987 was caused by the complex structure of the CS cloud of gas and dust that intersected the line-of-sight.

where $\mathbf{P}_{in}(0)$ is the intrinsic polarization of a star in the bright state.

This dependence follows from the assumption that the observed radiation of the young star is a sum of the direct (non-polarized) stellar radiation and the scattered radiation of the CS disk:

$$I_{obs} = I_* e^{-\tau_\lambda} + I_{sc} \qquad (6)$$

The first is variable due to the variable CS extinction on the line-of-sight. The second one is assumed to be constant.

ii) This model gives a natural explanation of the unusual photometric effect, observed in the UX Ori stars (Grinin 1986): it is the so-called "blueing" effect: a star becomes redder when fainter due to the selective extinction in the CS dust cloud intersecting the line-of-sight, but in the deep minima its color is determined by the scattered light[3] and becomes

[3]Worthy of mention is that in the middle of eighties it an idea was widespread that the photometric variability of the UX Ori stars is caused by the surface magnetic activity and the "blueing" effect is a result of increasing deposits of gas emission (Herbst 1986; see also Herbst & Shevchenko 1999).

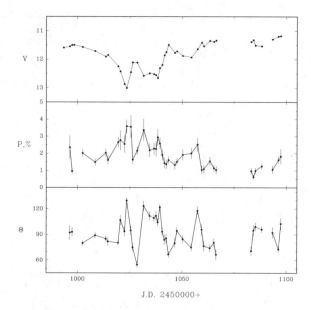

J.D. 2450000+

Figure 2.   The short fragment of the multi-year photopolarimetric monitoring of the
UX Ori star VX Cas from Shakhovskoj *et al.* (2003). The clear anti-correlation between
the brightness and linear polarization variations shows that the complex structure of the
minimum was caused (as in the case of WW Vul) by the highly inhomogeneous structure
of the CS cloud intersecting the line-of-sight. The strong variability of PA is a result of
the competition between the intrinsic and IS polarization.

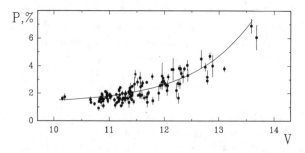

Figure 3.   The dependence of the degree of linear polarization of the UX Ori star RR
Tau in the V band on the stellar magnitude from Rostopchina *et al.* (1997). The solid
line is the model solution of Equation (5).

bluer again. So, the photopolarimetric observations of UXORs in the deep
minima were the key observational test of the mechanism of variable CS extinction as the source of the large-amplitude photometric activity of young
stars.

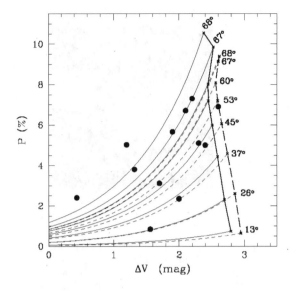

Figure 4. The degree of linear polarization in the V band in the model by Natta & Whitney (2000) for different inclination angles as function of brightness variations ΔV caused by screening of a star by CS cloud. The disk masses are 0.1 M$_\odot$ (solid line) and 0.01 (dashed line). The dots show the observed points from the Crimean data base.

iii) The high linear polarization systematically observed in the deep minima of UXORs led to the important conclusion about the nature of the UXORs phenomenon (Grinin et al. 1991; Grinin 1994): it is caused by the "optimal" orientation of CS disks surrounding these stars relative to the line-of-sight: edge-on or under a small inclination angle. This conclusion was slightly modified by Natta & Whitney (2001): using the model of the flared accretion disk (Fig. 4) they have shown that a small inclination angle must exist between the line-of sight and the disk "surface" (Fig. 5).

iv) The photopolarimetric observations of UXORs provide us with immediate information about the changes of the column density of CS dust on the line-of-sight and create the basis for diagnostics of the CS dust. Using Equation (5) one can isolate the intrinsic component of the observed polarization with high accuracy (see e.g. Grinin et al. 1991; Rostopchina et al. 1997; Shakhovskoj et al. 2001) and use $P_{in}(\lambda)$ for the theoretical modelling. It was shown (see e.g. Voshchinnikov et al. 1995; Rostopchina et al. 1997; Natta & Whitney 2001) that CS dust around UXORs differs slightly from interstellar dust: the minimal characteristic sizes of CS dust are larger by about ten times in comparison with interstellar dust.

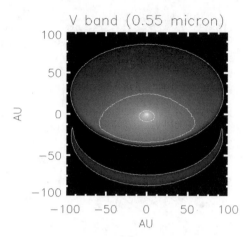

Figure 5. The predicted image of the CS disk of UX Ori type star in the V band during the deep minimum from Natta & Whitney (2000).

v) The results of the long-term photometric and polarimetric monitoring of UXORs (Grinin *et al.* 1998; Rostopchina *et al.* 1999; 2000; Bertout 2000) show that the azimuthal distribution of CS matter in the neighborhoods of these stars deviates strongly from the axially symmetric case. Such deviations can be caused by the dusty disk winds of the low-mass companions (Grinin & Tambovtseva 2002) or the large-scale vortices (Klahr *et al.* 2003).

Finally, the position angle of the linear polarization observed in deep minima of UXORs permits us to find the orientation of the symmetry axis of the CS disks on the sky plane unresolved with a telescope. It was found that the disk axes are parallel in many cases (sometimes with a high accuracy) to the direction of the local IS magnetic field (Grinin *et al.* 1991; Rostopchina *et al.* 1997; 2001). Such an alignment tells us about the important role of the IS magnetic field at the earlier stages of the gravitational collapse of the protostellar clouds from which these stars were formed.

The importance of the synchronous polarimetric and photometric observations of the young stars for the investigation of their variability mechanism and study of CS disks stimulated further activity in this field (see e.g. Kardopolov & Rspaev 1989; 1990; Matsumura *et al.* 1991; Oudmajer *et al.* 2000). As a result the number of UXOrs for which the anti-correlation between optical fluxes and degree of linear polarization was confirmed by the observations is about twenty at present.

3.3. POLARIZATION OF YOUNG BINARY STARS

Numerical simulations by Artymowics & Lubow (1996) show that young binary systems (YBS) generally consist of a circumbinary disk surrounding the binary components with circumstellar disks surrounding each of them. All components can be sources of intrinsic polarization of YBSs. Recently, on the basis of Mie theory in the single-scattering approximation, Manset & Bastien (2001a) have calculated polarimetric effects expected in the YBSs and found that the orbital motions of the binary components in the circumbinary (CB) dusty disks can produce weak (about 0.1%) periodic changes of linear polarization.

In a series of subsequent papers, Manset & Bastien (2000b, 2001, 2003) investigated several young binaries and found a weak periodic modulation of the intrinsic linear polarization in a few of them. In most cases, however, the stochastic variability dominates and masks the periodic low-amplitude variations. In the eclipsing binary HAEBE star MWC 1080, the observed periodic variations can be due to the eccentricity of the orbit and/or the asymmetric configuration of CS or CB material. The maximum of the observed polarization coincides (as in the case of UXORs) with the eclipse of the primary in this system.

Another interesting result recently was obtained by Jensen et al. (2000), Monin et al. (1998, 2002) and Maheswar et al. (2002). Jensen et al. investigated 18 T Tauri binaries with separations of about 1" and more and detected the linear polarization from both components in 16 of them. The analysis of the position angles of the observed polarization have shown that in approximately 70% of the wide (> 200 AU) binaries, the CS disks tend to be aligned. This result is similar to that obtained by Monin et al. (1998, 2002) and Maherswar et al. (2002) and is important for understanding the binary formation process.

4. Conclusion

Thus, photopolarimetric observations of PMS stars are an important tool for studying physical properties of their nearest circumstellar surroundings as well as the mechanisms of their variability. In particular, the spectropolarimetry of the emission lines, such as the H_α line is a valuable source of information about the geometry and kinematics of CS disks or envelopes in the regions nearest to the stars. Polarimetric observations of young variable objects are the most informative when they are accompanied by synchronous photometry. This is especially important in the case of the highly variable young stars, so called UXORs whose violent activity is a result of the "optimal" orientation of their CS disks relative to the line-of-sight. Dense in time monitoring of these objects is one the most effective ways to investigate a thin structure of the CS disks around young stars and search for low-mass companions in their nearest neighborhoods.

5. Acknowledgments

I would like to thank Yu. N. Gnedin and O. S. Shulov for useful discussions and comments. This work was supported in part by the grant: "Nonstationary phenomena in astronomy" and grant "Scientific School", N 1088. 2003.2.

References

1. P. Artymowicz & S.J. Lubow, Astrophys. J., **467**, L77 (1996)
2. L.W. Bandermann & J.C. Kemp, Mon. Not. Roy. Ast. Soc. **162**, 367 (1973)
3. P. Bastien, in G.V. Coyne et al. (eds.), Vatican Conference: "Polarized radiation of circumstellar origin", p. 303 (1988)
4. P. Bastien, Astrophys. J. S. **59**, 277 (1985)
5. A.V. Berdjugin, V.P. Grinin & N.Kh. Minikulov, Bull. Crimean Astrophys. Obs. **86**, 69 (1992)
6. J. Bouvier & C. Bertout, Astron. Astrophys. **211**, 99 (1989)
7. J. Bouvier, A. Chelli, S. Allain, et al. Astron. Astrophys. **349**, 619 (1999)
8. J.S. Brown, I.S. McLean, Astron. Astrophys. **57**, 141 (1977)
9. J.S. Brown, I.S. McLean, A.G. Emslie, Astron. Astrophys. **68**, 415 (1978)
10. G.V. Coyne, T. Gehrels & K. Serkowsky, Astron. J., **79**, 581 (1974)
11. J.-Y. Daniel, Astron. Astrophys. **67**, 345 (1978)
12. J.-Y. Daniel, Astron. Astrophys. **87**, 204 (1980)
13. A.Z. Dolginov, Yu.N. Gnedin, & N.A. Silant'ev, "Propagation and polarization of radiation in cosmic media", Gordon & Breach Pub. (1979)
14. A.Z. Dolginov & I.G. Mitrofanov, Astr. Ap. Sp. Sci. **69**, 421 (1978)
15. B.T. Draine & H.M. Lee, Astrophys. J. **285**, 89 (1984)
16. H. Elsässer & H.J. Staude, Astron. Astrophys. **70**, L3 (1978)
17. Yu.N. Gnedin & N.A. Silant'ev, Astrophys. Sp. Phys. Rev. **10**, 1 (1997)
18. V.P. Grinin, in "The nature and evolutionary status of Herbig Ae/Be stars", Eds. P.S. Thé, M.R. Pérez & E.P.J. van den Heuvel, ASP Conf. Ser. 62, 63 (1994)
19. V.P. Grinin, Unpublished poster paper presented at IAU Coll. 122 "Circumstellar matter" (1986), (see Sov. Astron. Letters, **14**, 27 (1988))
20. V.P. Grinin, N.N. Kiselev, N.Kh. Minikhulov, & G.P. Chernova, Pis'ma in Astron. Zn., **14**, 514 (1988)
21. V.P. Grinin, N.N. Kiselev, N.Kh. Minikhulov, G.P. Chernova, & N.V. Voshchinnikov, Astrophys. Sp. Sci. **186**, 283 (1991)
22. V.P. Grinin, in "Disks, Planetesimals and Planets", eds. F.Garzón, et al. ASP. Conf. 219, 216 (2000)
23. V.P. Grinin, & L.V.Tambovtseva, Astron. Letters, **28**, 592 (2002)
24. W. Herbst, PASP, **98**, 1088 (1986)
25. W. Herbst & V.S. Shevchenko, Astron. J., **118**, 1043 (1999)
26. J.H. Hough, J. Bailey, E.C. Cunningnam et al. Mon. Not. Roy. Ast. Soc. **195**, 429 (1981)
27. J.L.N. Jensen, A.X. Donar, R.D. Mathieu, in "Birth and Evolution of Binary Stars", Eds. Bo Reipurth and H. Zinnecker, p. 85, (2000)
28. V.I. Kardopolov & F. Rspaev, Kinem. Phys. Neb. Tel, **5**, 50 (1989)
29. V.I. Kardopolov & F. Rspaev, Astron. Zn. **76**, 1253 (1990)

30. H.H. Klahr, & P. Bodenheimer, Astrophys. J., **582**, 869 (2003)
31. A. Lazarian, J. Quant. Spectrosc. Rad. Trans. **79**, 881 (2003)
32. G. Maheswar, P. Manoj, & H.C. Bhatt, 2002, Astron. Astrophys. **387**, 1003 (2002)
33. N. Manset, & P. Bastien, Astron. J., **122**, 2692 (2001)
34. N. Manset, & P. Bastien, Astron. J., **122**, 3453 (2001)
35. N. Manset, & P. Bastien, Astron. J., **124**, 1089 (2002)
36. N. Manset, & P. Bastien, Astron. J., **125**, 3274 (2003)
37. M. Matsumura, M. Seki, & K.S. Kawabata, Astron. J., **117**, 429 (1999)
38. I.S. McLean, Mon. Not. Roy. Ast. Soc. **186**, 265 (1979)
39. F. Ménard, X. Delfosse, & J.-L. Monin, Astron. Astrophys. **396**, L35 (2002)
40. F. Ménard, J. Bouvier, C. Dougados *et al.* preprint, astro-ph/0306552 (2003)
41. J.-L. Monin, F. Menard, & G. Duchêne, Astron. Astrophys. **339**, 113 (1998)
42. J.-L. Monin, M. Menard, & N. Peretto, in "The origin of stars and planets", Proceedings of the Workshop, Garching, p. 121 (2002)
43. A. Natta, & B. Whitney, Astron. Astrophys. **364**, 633 (2000)
44. A. Natta, V. Grinin, V. Mannings V. in "Protostars and Planets" IV. eds. by V. Mannings, A.P. Boss, S.S. Russel, University of Arizona Press, (2000)
45. R.D. Oudmaijer, & J.E. Drew, Mon. Not. Roy. Ast. Soc. **305**, 166 (1999)
46. R.D. Oudmaijer, J. Palacios, C. Eiroa *et al.* , Astron. Astrophys. **379**, 564 (2001)
47. V. Piirola, F. Scaltriti & G.V. Coyne, in "The Nature and Evolutionary Status of Herbig Ae/Be Stars", Eds. P.S. Thé, M.R. Pérez & Ed P.J. van den Heuvel, PASC, **62**, p. 78 (1994)
48. R. Poeckert, P. Bastien, & J.D. Landstreet, Astron. J. **84**, 812 (1979)
49. R. Poeckert, J.M. Marlborough, Astrophys. J., **206**, 182 (1976)
50. A.N. Rostopchina, V.P. Grinin, & D.N. Shakhovskoj, Astron. Rep. **45**, 51 (2001)
51. A.N. Rostopchina, V.P. Grinin, A. Okazaki, *et al.* Astron. Astrophys. **327**, 145 (1997)
52. A.N. Rostopchina, V.P. Grinin, D.N. Shakhovskoi, P.S. The, N.Kh. Minikulov, Astron. Rep. **44**, 365 (2000)
53. G.D. Schmidt, L. Ferrario, D.T. Wickramasinghe, & P.S. Smith, Astrophys. J. **553**, 823 (2001)
54. R.E. Schulte-Ladbeck, G.C. Clayton, D.J. Hillier *et al.* Astrophys. J. **429**, 846 (1994)
55. K. Serkowski, Astrophys. J. **156**, L55 (1969)
56. J.F.L. Simmons, Mon. Not. Roy. Ast. Soc. **205**, 153 (1983)
57. D.N. Shakhovskoj, O.S. Shulov, E.N. Kopatskaja, Astron. Letters, **27**, 438 (2001)
58. D.N. Shakhovskoj, A.N. Rostopchina, V.P. Grinin, N.Kh. Minikulov, Astron. Rep. **47**, 301 (2003)
59. K. Stassun, & K. Wood, Astrophys. J. **510**, 892 (1999)
60. S.E. Strom, S. Edwards, & M.F. Skrutskie, in "Protostars and Planets III", Eds. E.H. Levy & J.I. Lunine, (The Univ. Arizona Press, Tucson), p. 837 (1993)
61. M. Tamura, in "Star Formation 1999", (2000)
62. J. Tinbergen, Astron. Astrophys. Suppl., **35**, 325 (1979)
63. R.A. Vardanian, Soobshch. Bjurak. Obs. **35**, 3 (1964)
64. J.S. Vink, J.E. Drew, T.J. Harries, & D. Oudmaijer, Mon. Not. Roy. Ast. Soc. (2002)
65. N.V. Voshchinnikov, V.P. Grinin & V.V. Karjukin, Astron. & Astrophys. **294**, 547 (1995)
66. F. Vrba, W. Herbst & J.L. Booth, Astron. J. **96**, 1032 (1988)
67. D.A. Weintraub, A.A. Goodman, R.L. Akeson, in "Protostars and Planets IV", Eds. by V. Mannings, A.P. Boss and S.S. Russel, (The Univ. Arizona Press), p. 247, (2000)
68. B.A. Whitney, & L. Hartmann, Astrophys. J. **395**, 529 (1992)
69. B.A. Whitney, & L. Hartmann, Astrophys. J., **402**, 605 (1993)
70. D.C. Whittet, P.G. Martin, J.H. Hough, *et al.* Astrophys. J., **386**, 562 (1992)
71. S. Wolf, B. Stecklum, Th. Henning *et al.* in "Polarimetry in Astronomy", Proc. of the SPIE, Ed. by S. Fineschi, v. 4843, p. 533, (2003)

72. K. Wood, J. S. Brown, & G. K. Fox, Astron. & Astrophys. **271**, 492 (1993).
73. K. Wood, S. J. Kenyon, B. Whitney & J. E. Bjorkman, Astrophys. J. **458**, L79 (1996).
74. R. V. Yudin, , Astron. & Astrophys. S. **144**, 285 (2000).

Khersonesus Tavricheskiy near Sevastopol

Crimean Astrophysical Observatory

INFRARED POLARIMETRY OF INTERSTELLAR DUST

J. H. HOUGH AND D. K. AITKEN

Department of Physics, Astronomy & Mathematics,
University of Hertfordshire, Hatfield AL10 9AB, UK

Abstract. Grains in the diffuse interstellar medium, and in at least some parts of dense molecular clouds, are aligned with their short axis preferentially along the direction of the local magnetic field. At longer infrared wavelengths radiation can be linearly polarized through the emission from such dust grains (dichroic emission), whilst at shorter wavelengths the radiation becomes polarized through dichroic absorption. The linear birefringence of dust grains can lead to the production of circular polarization. Enhanced linear polarization, relative to the continuum, occurs across solid-state features at ~3-5μm, attributed to ices, and at ~10μm and ~20μm, attributed to silicate grains. Continuum polarimetry and spectropolarimetry of the features can provide important diagnostics of the grain properties with significant advantages over total flux photometry or spectroscopy. Polarization produced by dichroic absorption and emission also can be used to determine the geometry of magnetic fields. In the near-infrared, typically for wavelengths up to a few microns, scattering of radiation from dust grains close to a radiation source can produce both linear and circular polarization. Polarimetry, especially over a wide wavelength range, can help to determine the properties of dust grains in the environs of stars ranging from young stellar objects that are undergoing accretion and mass outflow, to evolved stars that are losing mass. In the case of star-forming regions it is possible that dust grains play a key role in the origins of homochirality on the Earth.

1. Introduction

The study of dust grains is one of the key areas of research in astronomy. Dust is found in virtually all astrophysical situations, often obscuring our view of many objects and of large regions of galaxies. It is intimately involved in the

325

G. Videen et al. (eds.), Photopolarimetry in Remote Sensing, 325-350.
© 2004 *Kluwer Academic Publishers. Printed in the Netherlands.*

cycling of material from old to new stars and planets; it plays a critical role in the chemistry of the interstellar medium; and it might even play a role in the origin of homochirality of life on Earth. Lastly, because of the way non-spherical grains are aligned relative to the local magnetic field, polarization produced by the grains can be used to determine the geometry of the field.

At first sight there appears to be some disadvantages in using polarimetry to study dust grains in the interstellar medium. They are as follows: degrees of polarization are usually of order 1% or less and hence larger telescopes or longer integration times are required to get adequate signal to noise; instrumentation is more costly as the camera or spectrograph needs polarization optics, although these are usually a small fraction of the overall instrument cost; and lastly, corrections need to be made for the affect the telescope and the instrument can have on the polarization state of the observed radiation.

On the other hand, there are a number of very important advantages in polarimetry. As it is a differential measurement any effects of the terrestrial atmosphere can be excluded for most polarimeters and quantitative polarimetry can be carried out under poor observing, the only penalty being the loss of photons and hence longer integration times. Far more information on the physical and chemical properties of dust grains can be gained through polarimetry, than just through measurement of fluxes, as the scattering and or the absorption of light by dust grains changes the polarization state of the radiation, provided there is some overall asymmetry (note that intensity provides only a quarter of the total information carried by radiation, with the two components of linear polarization and the circular polarization needed for a complete description of the radiation). Other significant advantages of polarimetry in determining dust properties are noted in later sections.

The infrared, defined here as covering wavelengths between 1 and 20μm, is particularly important for studying dust grains (see [1,2,3] for excellent reviews of polarimetry at longer infrared wavelengths). Extinction is reduced substantially in the infrared for all lines of sight, being ~10 times less at 2μm compared to the visible (0.55μm) for lines of sight that do not include dense molecular clouds [4], and hence the interaction of radiation with dust grains can be studied in regions of high obscuration, as occurs for example in star-forming regions or towards the Galactic Centre. Second, there are several important spectral features associated with dust in this wavelength range.

The remainder of the chapter will first cover the alignment of dust grains; the fact that grains do align in many astrophysical situations is often a key to the production of polarized radiation. The use of polarimetry to study dust grains in the diffuse interstellar medium (ISM), in star-forming regions of dense molecular clouds, in the circumstellar environs of pre-main sequence stars such as T Tauri stars, of main sequence stars, and evolved stars such as proto-planetary nebulae is then presented. Figure 1 shows a cartoon of these different scenarios. In practice, it is only the use of field stars in the diffuse ISM that provides an unambiguous single mechanism (absorption) to study dust grains, although even for this situation there could be significant changes in the grain composition along the line of sight.

A very useful introduction to polarimetry, and the production of polarization by the interaction of radiation with dust, is given by Mishchenko & Travis [5], and Voshchinnikov [6], respectively.

2. Alignment of dust grains

Soon after the discovery of interstellar polarization along the line of sight to most stars [7,8], it was realised that the most likely explanation was differential extinction as the radiation passed through a medium of aligned grains. In the intervening years there have been many theoretical papers on the mechanisms by which grains may be aligned and it is probable that different mechanisms operate in different environments. Alignment mechanisms have been reviewed recently by Lazarian [9] and see Whittet [10]. The dust grains themselves are extremely small, $\sim 0.1\mu m$ in size, and in equipartition with the ambient gas they spin about the grain short axis with rates in the 10^5 Hz range. The ambient field induces rapid cycles of hysteresis in the grain that tend to damp components of spin that are orthogonal to the field. This mechanism, paramagnetic relaxation, was first proposed by Davis & Greenstein [11] and results in the short axis of grains becoming aligned with the field (that is the grains do not act like compass needles): dichroic absorption of radiation results in a net polarization of radiation along the field direction. However, when the strength of the field in the interstellar medium became known (a few μG), it was clear that collisions with gas molecules would dominate over this alignment process.

Modifications to the DG mechanism have been proposed (see [9] for more details and references), such as suprathermal spin of grains due to the formation and ejection of molecular H_2 at particular sites resulting in a *pinwheel* action, or the presence of ferromagnetic inclusions in grains. Other mechanisms include streaming of grains through gas in which collisions tend to align the spin axis normal to the direction of streaming, and similar effects with anisotropic radiation fields. Fortunately there is an over-riding effect by which spinning grains always acquire a magnetic moment (the Barnett effect) and each grain precesses about the ambient field, even if the field is weak. This does not constitute alignment in itself, since an isotropic distribution of grain spins remains isotropic, but it does ensure that any disturbance to isotropy is oriented by the magnetic field. In this way virtually all mechanisms lead to a similar outcome with the short axis of the grain aligned with the local magnetic field. This allows charting of the magnetic field structure from observations of polarization, important in many astrophysical situations (see section 3.5).

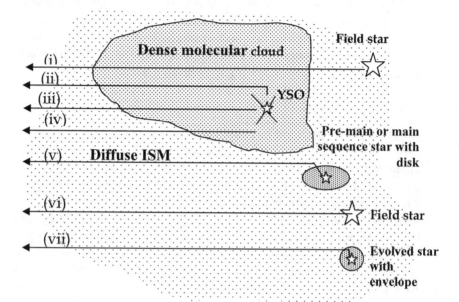

Figure 1. Radiation paths for the study of dust grains: (i) grains in dense clouds using a background field star; (ii) grains in the circumstellar environment of a young stellar object (YSO) scattering light into the line of sight (los); (iii) grains in a dense cloud using the direct light from a young star (although some scattering may be unavoidable at shorter wavelengths); (iv) radiation from hot dust grains; (v) grains in the circumstellar environs of a pre-main or main sequence star, scattering light into the los; (vi) grains in the diffuse ISM using a field star; (vii) grains in the circumstellar envelope of an evolved star scattering light into the los.

A requirement of most alignment mechanisms is that the rotational energy of the grain should exceed its thermal energy, otherwise thermal fluctuations significantly disturb the spin angle. It normally is sufficient to require that the gas temperature exceeds the grain temperature, but this introduces difficulties in explaining alignment in dense molecular clouds where the gas and grain temperatures should be tightly coupled and atomic hydrogen is not available to cause *pinwheel* spin-up. Indeed it seems that in the interior of quiescent cold clouds, grains are not aligned [12]; whereas by contrast, in dense molecular clouds, where active star formation is taking place, polarization by emission and absorption is observed frequently: here the alignment mechanism is uncertain but presumably is associated with dynamical activity in some way, possibly ambipolar diffusion, in which the charged grains see neutral gas streaming normal to the field.

Because of the uncertainty about the alignment mechanisms, it is not possible to relate the observed polarization fraction to the strength of the magnetic field. Indirect estimates can be made, however, by measuring the dispersion of the polarization vectors which is related to the magnetic field strength and the gas density and turbulent velocity [13,14], but the field strength can be overestimated unless the polarization structure is resolved spatially.

3.   Absorption and emission

A medium in which the grains are aligned has two optical depths, $\tau_x(\lambda)$ and $\tau_y(\lambda)$, and shows linear dichroism (here x and y are axes on the plane of the sky such that the x direction is the direction in which the projection of the long dimension of grains are aligned and typically the y direction is that of the projection of **B** on the plane of the sky). For clarity the dependences of the optical depths τ_x and τ_y on wavelength are omitted, and the polarization in absorption and in emission is given by

$$P_{abs} = \left(e^{-\tau_x} - e^{-\tau_y}\right)/\left(e^{-\tau_x} + e^{-\tau_y}\right) = -\tanh\{(\tau_x - \tau_y)/2\} \approx -(\tau_x - \tau_y)/2$$

$$P_{em} = \{(1 - e^{-\tau_x}) - (1 - e^{-\tau_y})\}/\{(1 - e^{-\tau_x}) + (1 - e^{-\tau_y})\} \approx (\tau_x - \tau_y)/(\tau_x + \tau_y)$$

The negative sign for P_{abs} indicates that the direction of polarization is along the y direction and is orthogonal to that for P_{em}. For optically thin emission,

$$P_{em}(\lambda) = -P_{abs}(\lambda)/\tau(\lambda), \quad \tau(\lambda) = \tfrac{1}{2}(\tau_x + \tau_y)$$

where $\tau(\lambda)$ is the optical thickness of a layer with aligned particles and the incident radiation is non-polarized. Note that P_{abs} requires a significant optical depth; whereas, P_{em} is independent of optical depth and is often referred to as the *specific polarization*. Because the spectral profiles of P_{abs} and P_{em} differ, it is possible to identify which process is occurring.

Some useful points can be made about the absorptive and emissive polarization:

1. The absorptive polarization, unlike the flux spectrum, is independent of the source spectrum, provided it is not polarised – often a reasonable assumption, particularly at wavelengths longer than a few microns.   This is a significant advantage of polarimetry;

2. if the position angle (hereafter PA) of grain alignment changes along the line of sight then circular polarization can be produced (see later), but the PA of linear polarization does not change with wavelength unless the optical properties of the grains also vary along the line of sight [15];

3.   in the mid-infrared, polarization can occur through emission as
     well as absorption and unless the warm and cold regions have
     precisely similar or orthogonal alignments (highly unlikely), their
     presence is revealed. At wavelengths where emission from dust is
     important, a wavelength-dependent polarisation PA usually
     indicates a medium of cold absorbing grains overlying much
     warmer emitting grains and a twisting magnetic field direction.
     The change in spectral profile, together with that of PA, can be
     used to separate out the contributions to the polarization of the
     warm and cold grains and find the average field directions in these
     regions, again a distinctive advantage of polarimetry (see section
     3.3.2).

It is useful to express the dichroic extinction in terms of elements of the
extinction matrix [15,16,17], with the Stokes parameters I, Q, U & V of the
radiation being given for axisymmetric aligned particles by

$$I = 0.5\{(I_i + Q_i)\exp[-n(K_{11}+K_{12})s] + (I_i-Q_i)\exp[-n(K_{11}-K_{12})s]\}$$

$$Q = 0.5\{(I_i + Q_i)\exp[-n(K_{11}+K_{12})s] - (I_i-Q_i)\exp[-n(K_{11}-K_{12})s]\}$$

$$U = \exp[-nK_{11}s].\{U_i\cos(nK_{34}s) - V_i\sin(nK_{34}s)\}$$

$$V = \exp[-nK_{11}s].\{V_i\cos(nK_{34}s) + U_i\sin(nK_{34}s)\},$$

where K_{11}, K_{12} and K_{34} are the 3 independent elements of the 4x4 extinction
matrix acting on the Stokes vector, s is the distance through a uniform cloud
with grain number density n, and subscript i denotes incident radiation. The
Q and U planes in the dichroic extinction equations are defined by the axis of
grain alignment with Q positive for polarization parallel to **B**, and negative for
polarization perpendicular to **B**. Note that it is possible to write $\tau = nK_{11}s$, and
all the above expressions for the Stokes parameters reduce to those given
earlier in the section for the case of non-polarized incident radiation.

These transfer equations show that it is possible to generate circular
polarization from initially linearly polarised radiation by conversion of Stokes
U to Stokes V due to the birefringence term K_{34}. The (prior) linear polarization
can be produced by dichroic extinction, with the circular polarization then
produced by the light passing through a medium of grains with a different
alignment direction, or by polarized sources (e.g. by scattering from a
reflection nebula), followed by passage of the light through a medium of
aligned grains whose axes are inclined relative to the **E**-vector of the linearly
polarized light.

Usually the degree of circular polarization is small, e.g. in the diffuse ISM
where the degrees of linear polarization from linear dichroism are usually only
a few per cent at most. However, degrees of linear polarization can be as high

as 80% in a reflection nebula making it possible to get very high degrees of circular polarization through linear birefringence (see section 5.2).

3.1 Wavelengths short-ward of 2μm

Although this chapter mainly is concerned with infrared polarimetry, it is useful here to consider the polarization in the optical and the UV. Linear polarization along a typical line of sight in the diffuse ISM shows a remarkably smooth dependence with wavelength, peaking in the visible. All the early observations, typically covering 0.35 to 0.9μm, are well described by the empirical formula of the form $P_\lambda = P_{max} \exp\{-K\ln^2(\lambda_{max}/\lambda)\}$, with K a measure of the width of the curve initially set to 1.15 for all stars [18]. P_{max} is the maximum polarization occurring at a wavelength λ_{max}, with λ_{max} covering the range 0.40-0.75μm for various lines of sight. As observations were extended into the space UV (~0.12μm) and the near-IR (~1-2.5μm), it became apparent that a better fit to data is obtained if K is treated as a free parameter, and fits to a large number of stars showed that K and λ_{max} are linearly correlated with $K = c_1\lambda_{max} + c_2$; $c_1 = 1.66 \pm 0.09$ and $c_2 = 0.01 \pm 0.05$ [19]. This implies a reduction in the width of the curve with increasing λ_{max}. It should be noted that the best values for c_1 and c_2 depend to some extent on the value of λ_{max}, particularly for low values of λ_{max}, perhaps not too surprising as the fit is entirely empirical with no physical basis.

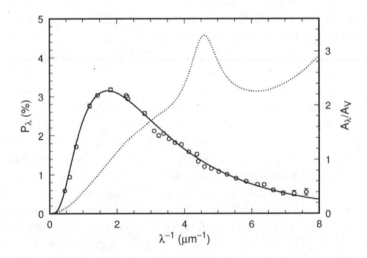

Figure 2. Comparison of the linear polarization (solid curve) and extinction (dotted curve) for the reddened star HD 99872. The points plot the observed polarization, and an empirical fit based on the Serkowski formula with $\lambda_{max} = 0.58$μm (continuous curve) also is shown. The extinction (dotted curve) is represented by a fit to observational data (taken from Whittet [20]).

The following can be deduced:

(i)   The extinction in the UV is high but the polarization is low implying that the small grains responsible for the extinction are very difficult to align. This is either because small grains are spherical and/or the grain composition is not conducive to the alignment mechanisms outlined in section 2.0;

(ii) maximum polarization for a dielectric cylinder with radius a (cylinder length $>> a$) and refractive index n, occurs when $2\pi a(n-1)/\lambda \sim 1$ ([10], p.113). Assuming n ~ 1.6, appropriate for silicates, then $a \sim 0.15\mu m$ for $\lambda_{max} \sim 0.55\mu m$, typical of most lines of sight;

(iii) that the polarization does not increase across the UV extinction bump at $\sim 2175\text{Å}$ ($4.6\mu m^{-1}$), usually attributed to graphitic carbon, shows that the carriers of the feature are not themselves aligned nor are they physically part of the general silicate grain population that is aligned. This behaviour is typical for virtually all lines of sight, the exceptions being HD197770 and ρOph [21,22], but even for these the amplitude of excess polarization is very small relative to the increased extinction.

Circular polarization in the diffuse ISM first was discovered along the line of sight to the Crab Nebula [23] and several stars [24], and Martin described the mechanism for its production [15]. The circular polarization is linked to the linear polarization by the Kramers-Krönig relationships and thus in itself does not provide any independent information about the dust grain properties. Martin found that there was a reversal in the sign of circular polarization, λ_c, at the wavelength of peak linear polarization (λ_{max}) and this was interpreted as evidence that the grain material responsible for the polarization is a dielectric. Grains with increasing absorption were predicted to have a change in circular polarization at wavelengths longer than λ_{max} [15,25], although Shapiro [26] calculated that λ_c and λ_{max} would occur at the same wavelength even for strongly absorbing conductors, and Chlewicki & Greenberg [27] suggested λ_c and λ_{max} always would be at the same wavelength if the adopted optical constants satisfy the Kramers-Krönig relation (see also Voshchinnikov [6]).

The interstellar polarization curve can be modelled by assuming oblate grains similar to "astronomical silicates" (the optical constants of astronomical silicates are constructed from laboratory and astronomical data [28]), axial ratio $\sim 2:1$, with no polarization contribution from other absorbers such as graphite [29,30]. A power-law distribution of grain sizes, a, is assumed, as used to explain the interstellar extinction curve, with $n(a) \propto a^{-q}$ (q = 3.5 [31], with the grain sizes in the range $a_{min} = 0.005\mu m$ and $a_{max} = 0.25\mu m$), but with no contribution to the polarization from the small grains (that is they are not aligned). A model in which the axial ratio is less than 2:1 also can fit the data, provided the degree of alignment is increased. Li & Greenberg [32] modelled the interstellar polarization using silicate grains but with organic refractory mantles produced by UV photo-processing of ice-mantled silicate cores.

3.2 Wavelengths long-ward of 2μm

Beyond a wavelength of ~2μm a power–law of the form $P_\lambda \propto \lambda^{-\beta}$, with β typically in the range 1.5-2.0 [33,34] provides a much better fit to polarization observations of the interstellar medium than the Serkowski formula, even if K is allowed to vary (see the previous section). At these wavelengths the polarization shows the same behaviour as the extinction, as shown in Figure 3, taken from Whittet [20] for lines of sight in the diffuse ISM.

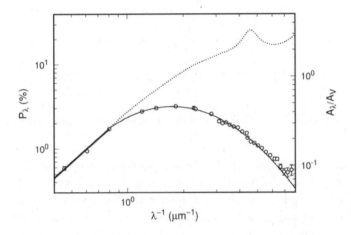

Figure 3. The same data as in Figure 2, but here shown on a log-log plot with polarization and extinction scales adjusted so that the curves overlap in the near-infrared. The figure is taken from Whittet [20].

The common (power-law) behaviour of the extinction and the polarization in the near-infrared implies that the same grains producing the extinction are responsible for the polarization. The extinction has a power-law form when the wavelength is large compared to the grain size, implying that large grains are not common (consistent with the steep power-law distribution of dust grain sizes noted in section 3.1).

3.3   Spectral features

For most lines of sight that include dense molecular clouds, the polarization spectrum beyond 2μm shows a number of features that most clearly are identified in the spectropolarimetry of the BN object in the Orion Molecular Cloud, OMC-1, and shown in Figure 4. The main features, clearly extending above the power-law fit to the near-infrared data, are the H_2O-ice feature at

$0.32\mu m^{-1}$ (3.1μm O-H stretching mode), the features associated with silicate materials at $0.10\mu m^{-1}$ (10μm Si-O stretching mode) and at $0.055\mu m^{-1}$ (19μm O-Si-O bending mode). Fortunately atmospheric windows allow all the features to be observed from ground-based observatories although even at the best sites the 20μm window is poor.

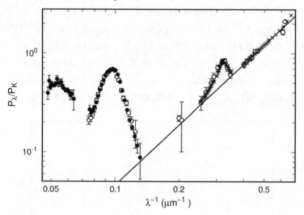

Figure 4. Infrared spectropolarimetry of the BN object in the Orion Nebula. The figure is taken from Whittet [20].

Data at sufficiently high spectral resolution show that the peak of polarization of the features occurs at a slightly longer wavelength than the peak of absorption (see Figure 5 for the spectropolarimetry of the BN ice feature). This is an important diagnostic showing that the polarization is indeed largely due to dichroic absorption by aligned grains [35] rather than by a directly viewed, but relatively obscured, unpolarized source with indirectly viewed scattered light [36]. The shift for dichroic absorption occurs because the absorption is slightly different for radiation with the **E**-vector parallel and perpendicular to a symmetry axis of the grain.

3.3.1   Ices

These are only observed along the lines of sight through dense molecular clouds with the radiation sources usually young stellar objects (one exception is the water-ice feature for the field star Elias 16 in the Taurus dark cloud [37]). Presumably the ices cannot exist in the UV radiation field of the diffuse ISM.

Figure 5 shows the excess optical depth and the excess polarization over the continuum for the BN object in OMC-1 [38]. Polarization structure in the long-wavelength wing of the ice profile is apparent, including a feature at 3.47μm which matches closely the spectroscopic feature discovered in several protostars and attributed to carbonaceous material with diamond-like structure

[39]. Modelling of the ice feature [38] gives a best fit with the following parameters: a mix of ices with $H_2O:CH_3OH:CO:NH_3$ in the ratio 100:50:1:1, on silicate cores with axial ratio $b/a \sim 1.5$, with $b \sim 0.02\mu m$ and mantle thickness $\sim 0.008\mu m$.

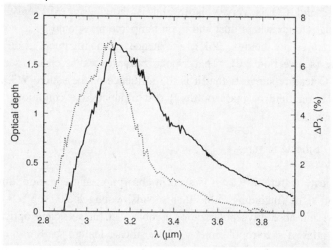

Figure 5. Spectropolarimetry of the ice-feature in the BN object in Orion [38]. The continuous curve is the excess polarization and the broken curve the excess optical depth, relative to the continuum. Note that the peak polarization occurs at a slightly longer wavelength than the overall extinction, indicative of polarization by dichroic absorption.

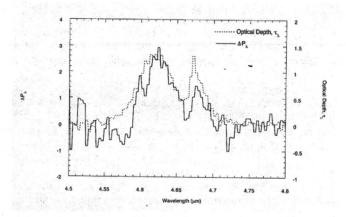

Figure 6. Excess polarization (solid curve) and optical depth (broken curve) against wavelength for the XCN feature at $4.62\mu m$ and the solid CO feature at $4.67\mu m$, along the line of sight to W33A [39].

Spectropolarimetry of other ice features is very sparse with observations only for the solid CO feature near $4.67\mu m$ and the XCN feature at $4.62\mu m$

observed towards W33A (Figure 6), a compact infrared source generally assumed to be a young stellar object, with a particularly rich solid state molecular spectrum [40]. The XCN feature is thought to arise from a cyanate produced by irradiation of ices containing NH_3 or N_2.

The solid CO is very volatile and is thought to exist only in dense molecular clouds where dust and grain temperatures should be in equilibrium (temperatures are below ~25K), and hence grain alignment should not be possible (see section 2.0). That a polarization excess is clearly seen for the solid CO feature suggests that it is the influence of the source W33A (either the radiation field or mass outflow) that produces the conditions for grain alignment.

3.3.2 Silicate features

The silicate feature at 9.7μm is seen in absorption and emission along many lines of sight (including molecular clouds, young stellar objects and HII regions) and absorptive polarization is observed for sufficient optical depths (for an atlas of spectropolarimetry of the silicate feature see Smith *et al.* [41] and references therein). It is seen in emission from circumstellar shells of late-type supergiants and more generally in ionized regions. In the latter case there is nearly always an overlying region of cold dust producing the feature in absorption and only one such source in [41] is attributed to emission alone.

Figure 4 shows spectropolarimetry of the silicate features at ~10μm and ~20μm, for the BN object in OMC-1. This profile of the 10μm silicate feature is acceptably similar to that in many other sources (eg Figure 7a), being smooth with a small shift between the peak of extinction and polarization (typically ~0.5μm), suggesting that the band strength of the silicates (absorption cross section per gram) is weak, appropriate for amorphous rather than a crystalline silicate [42].

One source, AFGL2591 (Figure 7b), does show evidence for some crystalline material, with a relatively sharp feature at 11.2μm, and a slightly narrower overall profile than the standard silicate. The 11.2μm feature has been attributed to crystalline olivine [43] and is similar to a spectroscopic feature in the dust emission from Comet Halley [44,45]. Note that in AFGL2591, the 11.2μm feature is not seen in the flux spectrum; the polarization and the extinction spectrum are different functions of the optical constants of the grain material and the grain geometry, so spectropolarimetry gives additional, and in this case more explicit, information about grain properties.

Detailed analyses of some of the silicate features have been carried out by a few authors and Aitken *et al.* [46] present a prescription for this. Principally, the analysis separates out the emissive and absorptive components as the warm and cold grains, responsible for the two components respectively, usually have a different alignment, arising from a twisting of the magnetic field along the line of sight. The absorptive component has a profile that is similar to the optical depth of the feature whilst the emissive polarization has a flattish, tilde-like profile (see Section 3 and Figure 7a). The position and detailed shape of the feature depends on the physical structure of the grains as well as their chemical mix. In general the polarization profiles of the silicate features are quite similar and are typified by that of the BN object. AFGL2591 is an exception; apart from the sharp 11.2µm feature, the general shape of the band is different (Figure 7(b)). "Astronomical silicate" [28] underestimates the polarization in the 8-9µm region and slightly overestimates it at 12-13µm. Grain sizes are typically 0.1µm and the grain shapes needed to account for the typical specific polarizations of ~2% have modest axial ratios of 1.5 and there seems some preference for a preponderance of oblate grains [43].

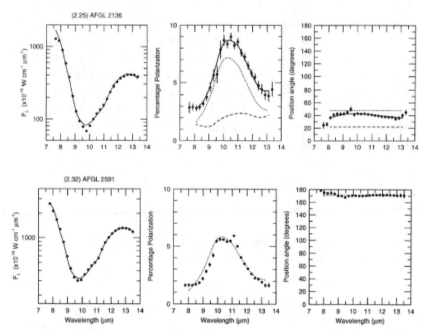

Figure 7. Spectropolarimetry of the 10µm silicate feature [41]: (a) upper panel; along the line of sight to the source AFGL 2136. The middle-box shows the polarization spectrum is composed of an emissive component (long-dashed line) and an absorptive component (short-dashed line). Note the change in PA of polarization across the feature (right-hand box); (b) lower panel; along the line of sight to the source AFGL 2591. Note that a purely absorptive component is sufficient to account for the polarization spectrum. Note also, the feature at 11.2µm attributed to crystalline olivine.

There is very little spectropolarimetry reported of the 19µm silicate feature; Smith *et al.* [41] list just 6 sources with both 10µm and 20µm spectropolarimetry.   Only two, BN and AFGL2591, show absorptive polarization without an additional emissive component, and the ratio of 20 to 10μm polarization in the former is much larger than expected from "astronomical silicate," while the unusual AFGL source conforms.   It has proven very difficult to fit a consistent model despite using a variety of grain types [47-50].

There is a strong correlation between the water-ice and silicate degrees of polarization and their PA of polarization [49,51], interpreted as evidence that the water-ices exist as mantles on the silicate grains with the core-mantle ratio similar for most objects.

3.4   The 3.4μm stretch C-H feature

The 3.4μm aliphatic C-H stretch feature generally is attributed to carbonaceous dust in the diffuse interstellar medium.   Spectropolarimetric observations of two sources (Sgr A IRS7 [52] and IRAS 18511+0146 [53]) suggest that the feature is unpolarized (see Figure 8).   If the carrier of the feature is associated with silicate grains, which are usually aligned in the diffuse interstellar medium, then a polarization excess would be expected, as occurs for the water-ices in dense molecular clouds.   One implication is that the 3.4μm carrier resides in a population of small, non-polarizing, carbonaceous grains, and in that regard shares many of the characteristics of the unpolarized 217.5nm extinction bump (see section 3.1).

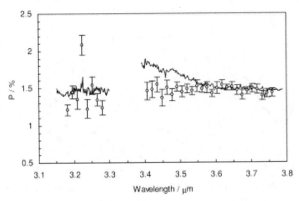

Figure 8. Polarization towards Sgr A IRS7 [52]. The data points are compared with the prediction (curve) for aligned silicate core/organic mantle grains. The gap in the data arises from absorption in the cement used in the Wollaston prism.

3.5   Determining magnetic field geometries

Because the polarization at mid-IR wavelengths can be produced by absorption or emission, or both, care has to be taken to identify the process(es) involved. The field direction derived from emissive and absorptive polarizations differs by 90 degrees and these processes are not identified by observations of polarization at a single wavelength. With spectropolarimetry it is possible to identify the components, and the same technique [46] can be applied to observations at two wavelengths. If both emission and absorption contribute to the polarization, it is possible to separate out the field direction and thus investigate field structure along the line of sight.

Whereas polarimetry of star-forming regions at far-infrared and sub-mm wavelengths often finds that magnetic fields play a dynamical role in early stages of star formation, the mid-infrared, which tends to look at compact central regions of young stellar objects seems to indicate a field which is largely passive and driven by dynamics. Here the polarization indicates a strong preference for fields that are in the direction of extended molecular structures (putative accretion disks?) and normal to outflow regions [54].

In our Galactic Centre, ionized filamentary structures are in motion about a candidate super-massive black hole. The grains are more strongly aligned here than other regions of the Galaxy and the emissive polarization is large (~8%) while its direction reveals fields directed along the filaments and probably ordered by tidal shearing within the filaments. It is thought that the alignment is probably saturated, that is to say the grains can be aligned no further, and differences of polarization fraction then indicate the out-of-sky plane component of the field, so that some three-dimensional information is retrievable [55].

4.   Summary of grain properties obtained by dichroic absorption and emission

There is strong evidence for the presence of amorphous silicates, with modelling of the silicate feature at 10μm suggesting grains with size up to a few tenths of a micron with modest asymmetry ($b/a \leq 2.0$). The interstellar polarization curve can be modelled assuming a power-law distribution of grain sizes, exponent 3.5, but there is no contribution to the polarization from the small particles responsible for extinction at UV wavelengths. There is no evidence for appreciable quantities of carbonaceous materials in grains that align; this is true for grains that contribute to the UV bump, and for the C-H stretch feature at 3.4μm. Various ices, water-ice, solid CO and XCN spectroscopic features show enhanced polarization and almost certainly arise from mantles on the silicate grains.

5.   Scattering

Although scattering contributes to the general extinction of radiation, its importance in studying the properties of dust grains mainly arises from the scattering of light into our line of light from the environs of young stellar objects, of pre-main and main sequence stars, and of evolved stars.

The polarization state of the radiation is described by the (I,Q,U,V) Stokes vector, defined relative to the scattering plane. This is modified in a single scatter by the scattering Mueller matrix shown below, with the matrix elements functions of the scattering angle and the grain properties including the grain orientation, with k the wave number of the radiation and r the distance to the observer.

$$
\begin{pmatrix} I_S \\ Q_S \\ U_S \\ V_S \end{pmatrix} = \frac{1}{k^2 r^2} \begin{pmatrix} S_{11} & S_{12} & S_{13} & S_{14} \\ S_{21} & S_{22} & S_{23} & S_{24} \\ S_{31} & S_{32} & S_{33} & S_{34} \\ S_{41} & S_{42} & S_{43} & S_{44} \end{pmatrix} \begin{pmatrix} I_i \\ Q_i \\ U_i \\ V_i \end{pmatrix}
$$

It is useful to identify some important cases:

(i)   Rayleigh scattering occurs when the size of the dust grain (radius a) is small compared to the wavelength of the radiation λ (more specifically when the size parameter $x = 2\pi a/\lambda \ll 1$), and abs$(mx) \ll 1$, where m is the grain refractive index. In this case the incident wave affects all interior regions of the particles simultaneously and an electrostatic approximation is valid (e.g. van de Hulst [56], p75). Matrix elements S_{13}, S_{14}, S_{23}, S_{24}, S_{31}, S_{32}, S_{34}, S_{41}, S_{42}, $S_{43} = 0$, and the linear polarization produced (LP) is given simply by $P = \sin^2\theta/(1+\cos^2\theta)$, where θ is the scattering angle and the polarization **E**-vector is normal to the scattering plane. Rayleigh scattering usually gives no information about the grain properties except that the grains are small relative to the wavelength;

(ii)   although not always considered, Rayleigh scattering is strictly for spherical particles, as scattering off a non-spherical dust grain can produce circular polarization (CP), even for a single scatter, as all matrix elements can be non-zero. Note that for very small size parameters (the dipole limit) the CP is zero for non-absorbing particles (i.e. particles with a zero imaginary component of the refractive index);

(iii) for an optically thin assembly of dust grains, CP will only be produced by single scatters if the grains have some overall 'alignment' (see section 2.0), otherwise equal amounts of left and right CP are produced which simply cancel out;

(iv)   for non-aligned grains, CP can only be produced by multiple scatters in which the LP produced by one scatter is modified by a second scatter with conversion of some of the LP into CP;

(v)   Mie scattering (S_{13}, S_{14}, S_{23}, S_{24}, S_{31}, S_{32}, S_{41}, S_{42}, = 0) applies strictly to the scattering from homogeneous spheres when the Rayleigh criterion does not apply. Analytical solutions exist for the scattering of radiation for such grains, and Mie-type solutions exist for some special cases such as infinitely long cylinders. Generally the degrees of polarization are less than for Rayleigh scattering, more radiation is scattered in the forward direction, and the **E**-vector of the scattered radiation can be either parallel to or perpendicular to the scattering plane, switching with wavelength and with scattering angle. This 'oscillation' in the direction of the **E**-vector increases with increasing size parameter;

(vi)   for scattering from non-spherical and non-homogeneous grains, numerical solutions are calculated using approaches such as the T-matrix method in which the electromagnetic wave is expanded as a series of spherical harmonics (see [6] for a number of references). A more general method is the discrete dipole approximation (DDA) in which a dust grain of arbitrary shape is represented by an array of dipoles with each dipole responding to the incident radiation and the electric fields produced by the other dipoles.

The penalty of dealing with non-spherical grains is the computation time involved, although there is ample evidence that many – if not all – grains are indeed non-spherical. Our present knowledge of interstellar dust grains, however, does not normally warrant dealing with shapes more complex than oblate and prolate spheroids (but see section 5.5).

5.1   Scattering Models

Studying the properties of dust grains from the scattering of radiation close to a star requires detailed models to be constructed. Monte-Carlo techniques usually are employed as these can most easily deal with a combination of processes, the production of linear and or circular polarization, and a range of dust density distributions ([17], and references therein). Such models only can provide useful information on grain properties if observations are made over as large a range of wavelengths as possible, although it is of course important that all the scattered radiation is coming from the same population of dust grains.

5.2   Young Stellar Objects

Figure 9(a) shows schematically the distribution of dust around a young star, formed within a dense cloud, with material flowing onto the star through an accretion disk around the star, surrounded by a much larger dusty envelope. An outflow along the poles of the accretion disk, believed to be in the same direction as the rotational axis of the star and the cloud magnetic field, clears a cavity within the cloud allowing photons from the star to escape from the close vicinity of the star. Photons then can be scattered into our line of sight, although they still suffer extinction in the dense parent cloud, and to a lesser extent in the diffuse ISM, before reaching Earth. It is generally only at wavelengths longer than ~1μm that photons originating close to the star can be observed. At wavelengths of ~10μm or longer, photons can be observed directly along the line of sight to the star with extinction almost entirely dominated by absorption.

Figure 9(b) shows the observed linear polarization at a wavelength of 2.2μm (K-band) for GSS30 [57], a low mass young stellar object in the star-forming region of the ρ Oph cloud. Light emerging along the outflow axis is scattered into our line of sight, producing a reflection nebula.

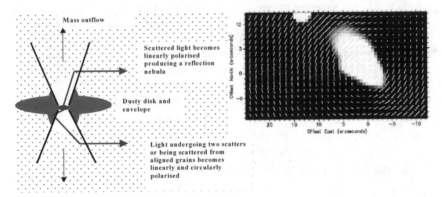

Figure 9. (a) Shows a schematic of the dust in the environs of a young stellar object; (b) shows a polarization image at 2.2μm for the young stellar object GSS30 [57]. The source is at position (0,0), the length of the lines represent the degrees of polarization, with 100% polarization equivalent to the length of 6arcsec on the axes, and the line direction represents the **E**-vector of the radiation. The image needs to be rotated through ~45 degrees clockwise to correspond to the geometry of the schematic shown in Figure 9(a).

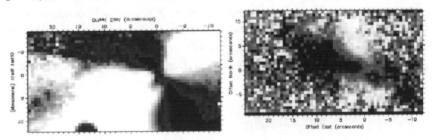

Figure 10. (a): as for Figure 9(b) except the degrees of linear polarization are represented by a grey scale; (b) a circular polarization image of GSS30 at 2.2μm [58]. The light and dark patches represent alternate signs of CP, with the maximum value ~1%.

In Figure 10(a) the degrees of polarization are represented by a grey scale which very clearly shows the geometry of the dusty envelope and outflow. Figure 10(b) is a circular-polarization (CP) image at the same wavelength [58]. The CP has alternate left and right handedness as the orientation of the scattering plane rotates about the source, with the maximum CP when the scattering plane is at 45 degrees to the linear polarization vector produced by scattering of radiation from the source. The CP, peak value ~1%, is interpreted as being produced by (secondary) scattering of light that is first linearly polarised by scattering that takes place close to the star [58].

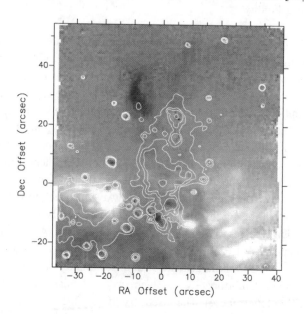

Figure 11. A circular polarization image at 2.2μm of the star-forming region in the Orion Molecular Cloud, OMC-1 [59]. Light and dark patches represent alternate signs of circular polarization. The main source (not visible in the image) is the IRC2 complex at position (0,0).

Circular polarimetry of the high-mass star-forming region in OMC-1 [59] (see Figure 11) shows much higher degrees of circular polarization, reaching ~20% in parts of the nebula. It is very difficult to produce such high degrees of CP by multiple scattering and it is suggested that scattering off aligned grains more readily produces the high degrees of CP [59,60], with the maximum CP occuring when the particle symmetry axis is at 45 degrees to the scattering plane. More recent detailed modelling of the linear and circular polarization produced by scattering close to a young stellar object [17] shows that another mechanism also might be responsible for the high CP in the OMC-1 star-forming region. Radiation with high linear polarization, produced in the reflection nebula, passing through grains that are aligned at an angle to the **E**-vector of the radiation, can convert the U Stokes vector into V (see section 3.0 for more details). In the case of OMC-1 both the LP and the CP can be fitted [17] with a power law size distribution of grains of the form $n(a) \propto a^{-3.5}$, with $a_{min} \sim 0.005$ and $a_{max} \sim 0.3 \mu m$ in the general cloud (barely larger than the diffuse ISM). In the circumstellar envelopes grains are a little larger with $a_{max} \sim 0.75$ μm. The observed CP is produced either by dichroic scattering or by dichroic extinction, although the latter requires only modest axial ratios for the dust grains (~ 1.5). "Astronomical silicates" (see Section 3.1) are able to produce the observed linear and circular polarization.

5.2.1   Homochirality

Amino acids (the basic building blocks of proteins) are chiral (the one exception being glycine), existing in two distinct forms with left and right-handed configurations (enantiomers) that are mirror images of each other and cannot be superimposed. In all living organisms only the left-handed amino acid exists, and it appears that self-replication is impossible if there is a mix of enantiomers. The origin of this homochirality has remained a mystery, since it was discovered by Pasteur over 150 years ago. In the laboratory the easiest way to produce an enantiomeric excess is through circular dichroism in which circularly polarised light (CPL) preferentially destroys one enantiomer, leaving an excess of the other. The problem always has been to find a natural source of CPL. On the Earth sunlight produces only a very small amount of CPL and the sign changes as the Sun sweeps across the sky. An extraterrestrial origin was proposed [61,62] in which circularly polarized UV radiation from supernova remnants or pulsars generates chiral asymmetry in molecules within interstellar clouds. Unfortunately the degrees of circular polarization from these sources are low (<0.2%). The discovery of very high degrees of circular polarization in star-, and presumably planet-forming regions, which contain many organic molecules, provides a natural and effective way of producing the homochirality. Added weight has been given to the idea of an extraterrestrial origin of homochirality through the discovery of a left-handed enantiomeric access in the Murchison and other meteorites. Although the enantiomeric excesses that are produced by circular dichroism are relatively low, without destroying all the molecules, non-linear autocatalytic reactions with back reactions can produce complete homochirality [63]. For more details on the origins of homochirality see [17,64,65,66].

5.3   Pre-main sequence stars

Before stars like the Sun move onto the main sequence they go through the T Tauri phase with a circumstellar or protoplanetary disk. Mid-infrared imaging of scattered light around the pre-main sequence star HK TauB, rather than polarimetry, has shown that there is a population of dust grains of order 1.5-3.0µm in size, significantly larger than in the diffuse interstellar medium, although it is not yet clear if the grain growth has occurred in the disk itself [67]. In fact much larger grains might be expected by the T Tauri stage of stellar evolution (~1Myr) but to detect these observations of scattered light at a wavelength of order the size of the particles are needed.

5.4   Main sequence stars

The most famous main sequence star with extensive observations of a circumstellar disk is β Pictoris, whose age of ~12Myr [68] makes it a young main-sequence star. Modelling polarization observations of the disk [69] show that a wide range of grain sizes is required but with grains smaller than a few microns being depleted by a factor of 10-100 compared to the usual power-law distribution with exponent 3.5; presumably the small grains are blown away by radiation pressure (see also [70] for a discussion on the grain properties of the disc of β Pictoris). Although not well constrained, the grain types are most likely compact, or just slightly porous, silicate particles. Mid-infrared spectroscopy shows both amorphous and crystalline silicate features in emission for the dust within 0.8 arcsec of the star, showing that small grains must exist in the disk [71]; whereas, imaging of the disk at 1.2mm provides evidence that large, millimetre-sized, grains also must exist [72]. That β Pictoris is young raises the possibility that it is in the later stages of planet formation, and warps and small scale asymmetries observed in the disk have been explained by the presence of one or more planets [73].

About 100 main sequence stars now have been shown to have orbiting planets [74]. Many are typically the size of Jupiter but with orbits very close to the star, having periods of just a few days, the so-called hot-Jupiters. Almost all are detected through radial velocity measurements of the star with a very small number observed through the change in brightness of the star as the planet transits across the stellar disk. In both cases light from the planet itself is not separated from that of the star. Hough *et al.* [75] describe a polarimeter that detects the scattered, polarized, light from hot-Jupiters. Models of the expected polarization signal (e.g. [76]) predict fractional polarizations of a few $\times 10^{-6}$. The polarization signal provides the inclination i of the planet orbit and hence the planet mass M (radial velocity measurements give only *Msini*), the albedo of the planet and the type of dust particles in any planetary cloud deck.

5.5   Evolved stars

Here the dust is produced by condensation of gas outflows from the evolved star, enriched in heavy elements by nuclear processes taking place throughout the life of the star, and eventually will be recycled to form new stars and planets.

Figure 12 provides a good example of the usefulness of imaging polarimetry at near-infared wavelengths, in which the degrees of polarization and the polarised flux for the proto-planetary nebula IRAS 17436+5003 are shown [77]. In polarised flux the central star, which is unpolarized, effectively is masked out, revealing the structure of the circumstellar envelope. A small grain (~0.01μm) model with a steeper power-law, $n(a) \propto a^{-6}$, than for the diffuse interstellar medium, fits best the near-IR polarimetry and the mid-infrared spectral energy distribution of the envelope [78]. There is some evidence that a relatively complicated grain structure involving highly non-spherical and/or non-homogeneous (layered or fluffy) grains is required. This is to account for the high polarization, indicative of small grains, and the relatively neutral colours of the reflected light (indicative of large grains).

IRAS 17436+5003

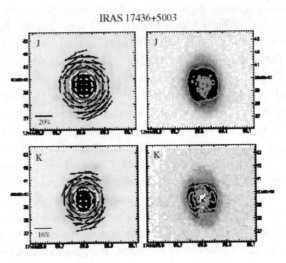

Figure 12. Degrees of polarization and the polarised flux in the J (1.2μm) and K (2.2μm) bands for IRAS 17436+5003 [77].

6. Summary of grain properties obtained by scattering

The maximum grain size in dense molecular clouds is little more than in the diffuse interstellar medium, with $a_{max} \sim 0.3$μm, and approaches ~1μm in the circumstellar envelopes of young stellar objects. There is strong evidence that aligned grains are responsible for producing circular polarization. In pre-main sequence stars, a_{max} can be several microns, and in main sequence stars with disks there is evidence for millimetre-size grains, and an increasing number of main-sequence stars are found to have at least one orbiting planet. In order to account for the large linear polarizations observed in the scattering envelopes around evolved stars, a population of grains smaller than those typically assumed for the diffuse interstellar medium is required.

7. Acknowledgements

We thank Tim Gledhill, Phil Lucas and Alan McCall for many useful comments, and Nikolai Voshchinnikov for a thorough reading of the text and a number of very useful suggestions which helped clarify parts of the text.

References

1. R. H. Hildebrand, in "Polarimetry of the Interstellar Medium", Ed. by W. G. Roberge and D. C. B. Whittet (Astron. Soc. of the Pacific Conference Series, 1996, Vol. 97) 254.

2. R. H. Hildebrand, J. A. Davidson, J. L. Dotson, C. D. Dowell, G. Novak and J. E. Vaillancourt, PASP, **112**, 1215 (2000).

3. R. H. Hildebrand, in "Astrophysical Spectropolarimetery", Ed. by J. Trujillo-Bueno, F. Moreno-Insertis and F. Sanchez (Cambridge University Press, 2002) 265.

4. G. H. Reike and M. J. Lebofsky, Astrophys. J. **288**, 618 (1985).

5. M. I. Mishchenko and L. D. Travis, Chapter 1, this book.

6. N. V. Voshchinnikov, Astrophysics and Space Science Reviews, **13**, 1 (2003).

7. J. S. Hall, Science, **109**, 166 (1949).

8. W. A. Hiltner, Science, **109**, 165 (1949).

9. A. Lazarian, J. Quant. Spectrosc. Rad. Trans. **79-80**, 881 (2003).

10. D. C. B. Whittet, "Dust in the Galactic Environment", (Second Edition, Institute of Physics, 2003).

11. L. Davis and J. L. Greenstein, Astrophys. J. **114**, 206 (1951).

12. A. A. Goodman, T. J. Jones, E. A. Lada and P. C. Myers, Astrophys. J. **448**, 748 (1995).

13. S. Chandrasekhar and E. Fermi, Astrophys. J. **118**, 113 (1953).

14. E. G. Zweibel, Astrophys. J. **362**, 545 (1990).

15. P. G. Martin, Astrophys. J. **187**, 461 (1974).

16. B. A. Whitney and M. Wolff, Astrophys. J. **574**, 205 (2002).

17. P. W. Lucas, J. H. Hough, J. A. Bailey, A. C. Chrysostomou, T. M. Gledhill and A. McCall, Origins of Life and Evolution of the Biosphere, in press (2003).

18. K. Serkowski, in "Interstellar Dust and Related Topics", Ed. by J.M. Greenberg and H.C. van de Hulst (IAU Symp. No.52, Dordrecht: Reidel, 1973), p.145.

19. D. C. B. Whittet, P. G. Martin, J. H. Hough, M. F. Rouse, J. A. Bailey and D. J. Axon, Astrophys. J. **386**, 562 (1992).

20. D. C. B. Whittet, in "Astrophysics of Dust", Ed. By A. Witt, B. Draine and G. Clayton (Astronomical Society of the Pacific Conference Series, 2003) in press.

21. M. J. Wolff, G. C. Clayton and M. R. Meade, Astrophys. J. **403**, 722 (1993).

22. M. J. Wolff, G. C. Clayton, S.-H Kim, P. G. Martin and C. M. Anderson, Astrophys. J. **478**, 395 (1997).

23. P. G. Martin, R. Illing and J. R. P. Angel, Mon. Not. R. Soc. **159**, 191 (1972).

24. J. C. Kemp and R. D. Wolstencroft, Astrophys. J. **176**, L115 (1972).

25. P. A. Aannestad and J. M. Greenberg, Astrophys. J. **272**, 551 (1983).

26. P. R. Shapiro, Astrophys. J. **201**, 151 (1975).

27. G. Chlewicki and J. M. Greenberg, Mon. Not. R. Soc. **210**, 791 (1984).

28. B. T. Draine and H. M. Lee, Astrophys. J. **285**, 89 (1984).

29. S. H. Kim and P. G. Martin, Astrophys. J. **442**, 172 (1995).

30. S. H. Kim and P. G. Martin, Astrophys. J. **444**, 293 (1995).

31. J. S. Mathis and S. G. Wallenhorst, Astrophys. J. **244**, 483 (1981).

32. A. Li and J. M. Greenberg, Astr. Astrophys. **323**, 566 (1997).

33. P. G. Martin and D. C. B. Whittet, Astrophys. J. **357**, 113 (1990).

34. P. G. Martin, A. J. Adamson, D. C. B. Whittet, J. H. Hough, J. A. Bailey, S.-H. Kim, S. Sato, M. Tamura and T. Yamashita, Astrophys. J. **392**, 691 (1992).

35. P. G. Martin, Astrophys. J. **202**, 393 (1975).

36. H. Elsasser and H. J. Staude, Astr. Astrophys. **70**, L3 (1978).

37. J. H. Hough, S. Sato, M. Tamura, T. Yamashita, A. D. McFadzean, M. F. Rouse, D. C. B. Whittet, N. Kaifu, H. Suzuki, T. Nagata, I. Gatley and J. Bailey, Mon. Not. R. Soc. **230**, 107, (1988).

38. J. H. Hough, A. Chrysostomou, D. W. Messinger, D. C. B. Whittet, D. K. Aitken and P. F. Roche, Astrophys. J. **461**, 902 (1996).

39. L. J. Allamandola, S. A. Sandford, A. G. G. M. Tielens and T. M. Herbst, Astrophys. J. **399**, 134 (1992).

40. A. Chrysostomou, J. H. Hough, D. C. B. Whittet. D. K. Aitken, P. F. Roche and A. Lazarian, Astrophys. J. **465**, L61 (1996).

41. C. H. Smith, C. M. Wright, D. K. Aitken, P. F. Roche and J. H. Hough, Mon. Not. R. Soc. **312**, 327 (2000).

42. R. W. Capps and R. F. Knacke, Astrophys. J. **210**, 76 (1976).

43. D. K. Aitken in "Polarimetry of the Interstellar Medium", Ed. by W. G. Roberge and D. C. B. Whittet (Astron. Soc. of the Pacific Conference Series, 1996, Vol. 97) 225.

44. J. D. Bregman, H. Campins, F. C. Witteborn, D. H. Wooden, D. M. Rank, L. J. Allamandola, M. Cohen and A. G. G. M. Tielens, Astr. Astrophys. **187**, 616 (1987).

45. H. Campins and E. V. Ryan, Astrophys. J. **341**, 1059 (1989).

46. D. K. Aitken, J. H. Hough, P. F. Roche, C. H. Smith and C. M. Wright, Mon. Not. R. Soc. in press (2003).

47. D. K. Aitken, C. H. Smith and P. F. Roche, Mon. Not. R. Soc. **236**, 919 (1989).

48. C. M. Wright, D. K. Aitken, C. H. Smith, P. F. Roche and R. J. Laureijs, in "The Origins of Stars and Planets: The VLT View," Ed. by J. Alves and M McCaughrean (New York: Springer, 2002).

49. R. P. Holloway, "Infrared spectropolarimetry of Young Stellar Objects", PhD Thesis, University of Hertfordshire (2003).

50. J. M. Greenberg and A. Li, Astr. Astrophys. **309**, 258 (1996).

51. R. P. Holloway, A. Chrysostomou, D. K. Aitken, J. H. Hough and A. McCall, Mon. Not. R. Soc. **336**, 425 (2002).

52. A. J. Adamson, D. C. B. Whittet, A. Chrysostomou, J. H. Hough, D. K. Aitken, G. S Wright and P. F Roche, Astrophys. J. **512**, 224 (1999).

53. M. Ishii, T. Nagata, A. Chrysostomou and J. H. Hough, Astron. J. **124**, 2790 (2002).

54. D. K. Aitken, C. M. Wright, C. H. Smith, P. F. Roche, Mon. Not. R. Soc. **262**, 456 (1993).

55. D. K. Aitken, C. H. Smith, T. J. T. Moore, P. F. Roche, Mon. Not. R. Soc. **299**, 743 (1998).

56. H. C. van de Hulst, in "Light Scattering by Small Particles" (John Wiley, New York, 1957).

57. A. Chrysostomou, S. G. Clark, J. H. Hough, T. M. Gledhill, A. McCall and M. Tamura, Mon. Not. R. Soc. **278**, 449 (1996).

58. A. Chrysostomou, F. Menard, T. M. Gledhill, S. Clark, J. H. Hough, A. McCall and M. Tamura, Mon. Not. R. Soc. **285**, 750 (1997).

59. A. Chrysostomou, T. M. Gledhill, F. Menard, J. H. Hough, M. Tamura and J. Bailey, Mon. Not. R. Soc. **312**, 103 (2000).

60. T. M. Gledhill and A. McCall, Mon. Not. R. Soc. **314**, 123 (2000).

61. W. A. Bonner and E. Rubenstein, Biosystems **20**, 99 (1987).

62. J. M. Greenberg, A. Kouchi, W. Niessen, H. Irth, J. van Paradijs, M. de Groot and W. Hermsen, J. Biol. Phys. **20**, 61 (1994).

63. Y. Saito and H. Hyuga, arXiv:physics/0310142 (2003).

64. J. A. Bailey, A. Chrysostomou, J. H. Hough, T. M. Gledhill, A. McCall, S. Clark, F. Menard and M. Tamura, Science **281**, 672 (1998).

65. J. A. Bailey, Origins of Life and Evolution of the Biosphere, **31**, 167 (2001).

66. J. H. Hough, J. A. Bailey, A. Chrysostomou, T. M. Gledhill, P. W. Lucas, M. Tamura, S. Clark, J. A. Yates and F. Menard, Adv. Space Res. **27**, 313 (2001).

67. A. McCabe, G. Duchêne and A. M. Ghez, Astrophys. J. **588**, L113 (2003).

68. B. Zuckerman, I. Song, M. S. Bessell and R. A. Webb, Astrophys. J. **562**, L90 (2001).

69. N. A. Krivova, A. V. Krivov, I. Mann, Astrophys. J. **539**, 424 (2000).

70. N. V. Voshchinnikov and E. Krugel, Astr. Astrophys. **352**, 508 (1999).

71. A. J. Weinberger, E. E. Becklin and B. Zuckerman, Astrophys. J. **584**, L33 (2003).

72. R. Liseau, A. Brandeker, M. Fridlund, G. Olofsson, T. Takeuchi and P. Artymowicz, Astr. Astrophys. **402**, 183 (2003).

73. J. C. Augereau, R. P. Nelson, A. M. Lagrange, J. C. B. Papaloizou and D. Mouillet, Astr. Astrophys. **370**, 447 (2001).

74. J. Schneider, www.obspm.fr/encycl/encycl.html.

75. J. H. Hough, P. W. Lucas, J. A. Bailey and M. Tamura in "Polarimetry in Astronomy", Ed. by S. Fineschi (Proceedings of SPIE Volume 4843 (SPIE, Bellingham, WA, 2003), p. 517.

76. S. Seager, B. A. Whitney and D. D. Sasselov, Astrophys. J. **540**, 504 (2000).

77. T. M. Gledhill, A. Chrysostomou, J. H. Hough and J. A. Yates, Mon. Not. R. Soc. **322**, 321 (2001).

78. T. M. Gledhill and J. A. Yates, Mon. Not. R. Soc. **343**, 880 (2003).

*Gudrun Jockers and Jim Hough
discuss merits of Crimean wine at
Livadia Palace.*

*Tracy Smith and Mark Thoreson
enjoy the beach in front of the
Druzhba.*

*Yuriy Shkuratov, Klaus Jockers, and Leonid Shulman scouting out the
poster session.*

MEASUREMENTS OF GENERAL MAGNETIC FIELDS ON STARS WITH VIGOROUS CONVECTIVE ZONES USING HIGH-ACCURACY SPECTROPOLARIMETRY

S. I. PLACHINDA
Crimean Astrophysical Observatory,
p/o Nauchny, Crimea, 98409, The Ukraine

Abstract. Magnetic fields are studied to understand the nature of activity on stars with convective envelopes. Results of high-accuracy General Magnetic Field (GMF) measurements of different luminosity stars on the right hand side of the Cepheid instability strip of the H-R diagram are reviewed: the presence of a weak general magnetic field for 21 stars with vigorous convection is detected (F9-M3 spectral types and I-V luminosity classes). A substantial (up to some dozens) GMF value was detected on two supergiant stars, three bright giants, twelve giant stars, one subgiant, and three solar-like dwarfs. Furthermore, the variation of global nonaxisymmetric magnetic fields as a function of the stellar rotation is determined for two solar-like stars other than the Sun: the magnetic field of the young solar-like star ξ Boo A shows periodic variations from -10 G to +30 G, and the magnetic field of the old solar-like star 61 Cyg A varies from -10 G up to +4 G. Currently, the nature of this field is unknown. The nonaxisymmetric GMF as a phenomenon is absent in the Babcock' and Leighton' phenomenological magneto-kinematic model of the solar cycle. In terms of standard α-Ω dynamo theory, GMF is absent also. There are only two main components of large-scale magnetic fields on the Sun: the toroidal magnetic field and the axisymmetric poloidal field. The coincidence of theoretical conclusions of different authors as well as results of their numerical simulations and new data on the observed magnetic field for solar-like stars (i.e., the presence of a nonaxisymmetric large-scale field) leads to a working hypothesis that GMF reflects properties of a stationary global magnetic field of the Sun's (or convective star's) radiative interior onto its surface. There appears to be a third nonaxisymmetric, large-scale component of the magnetic field (Origin Magnetic Field) – the initial magnetic field for dynamo mechanisms; a global magnetic field of radiative interior penetrates into surface of the Sun, and the observed GMF is a time-dependent superposition of all large-scale components: axisymmetric poloidal, nonaxisymmetric toroidal and nonaxisymmetric origin fields.

G. Videen et al. (eds.), Photopolarimetry in Remote Sensing, 351-368.
© 2004 *Kluwer Academic Publishers. Printed in the Netherlands.*

1. Introduction

Investigations of solar activity have shown that almost all manifestations of solar activity (chromospheres and coronae, plages and spots, flashes, etc.) are related to magnetic fields. Active regions on the surface of the Sun appear as bright plages of emission and represent moderate concentrations of magnetic flux with fields on the order of 100 G. The most intense phase of an active region is characterized by the presence of sunspots in the photosphere. The magnetic field in the center of a spot has strength of about 2000 - 3000 G, reaching as high as 4000 G. These and other small-scale magnetic structures are simply one manifestation of the large-scale, organized magnetic field of the Sun.

In terms of standard α-Ω dynamo theory, there are two main components of large-scale magnetic field on the Sun: toroidal (azimuthal) magnetic field and axisymmetric poloidal field (essentially a dipole):

A). Toroidal magnetic field lies in the base of the convective zone and manifests itself when magnetic loops emerge on the surface in bipolar active regions. In the northern and southern hemispheres of the Sun, the directions of the azimuthal field components are reversed cyclically with a period of about 22 yr.

B). The axisymmetric poloidal field lies under the photosphere and changes its polarity with a period of 22 yr as well, reaching peak values of about 1-2 G on rotation poles during minima of spot activity.

It is believed that the underlying cause of the solar activity cycle is the interplay between poloidal magnetic field, differential rotation, and convection. This is illustrated by the most developed Babcock' [1] and Leighton' [2,3] phenomenological model of the solar cycle.

The General Magnetic Field (GMF) [4-6] is the surface-averaged value of the longitudinal component of small- and large-scale magnetic structures on the Sun, seen as a star. Observations of the Sun' GMF were obtained mainly at four observatories (Crimean Astrophysical Observatory, 1968-present; Mount Wilson Observatory, 1970-1982; Wilcox Solar Observatory of the Centre for Space Science and Astrophysics of Stanford University, 1975-present; and the Sayan Observatory of the Solar-Terrestrial Physics Institute, 1982-1993) [6-8]. The following general properties of GMF of the Sun as a star are derived:

A). GMF strength versus the synodic rotational period ($P_\odot = 26.93$ days) shows both sign and shape variations. This period has not varied over the last 34 years of direct measurements. Using information about measurements of the interplanetary magnetic field, $P_\odot$ has not varied over the 76-year time span [9-11]. Both dipole, as dominant, and quadrupole components of the field are detected in the observations (see Figure 3 and [8]).

B). The amplitude of variations of the GMF varies with the period of sunspots cycle: the GMF is strongest during peaks in spot activity, reaching values of about 1 - 2 G [6].

C). During three decades of direct observations, excess positive magnetic flux is concentrated on one side of the Sun, and excess negative flux is concentrated on the opposite side (see Figure 4 in [11]), therefore we cannot claim that the GMF of the Sun reverses its polarity with the 22 yr solar cycle period [12].

D). The ratio of the positive to negative magnetic flux $\Delta\Phi_+/\Delta\Phi_- = 0.99$ [12].

All above-mentioned properties tell us that the Sun is a star with real global nonaxisymmetric field. The nature of this field currently is unknown. The nonaxisymmetric GMF as a phenomenon is absent in the Babcock' and Leighton' phenomenological model of the solar cycle. GMF is absent also in the standard α-Ω dynamo theory. There are only two main components of large-scale magnetic fields on the Sun: toroidal magnetic field and axisymmetric poloidal field.

Our working hypothesis is the following: *GMF reflects properties of a stationary global magnetic field of Sun's radiative interior on its surface, and there appears to be third nonaxisymmetric, large-scale component of the magnetic field (Origin Magnetic Field – initial magnetic field for dynamo mechanisms). The GMF of the radiative interior penetrates into the surface of the Sun. The GMF is a time-dependent superposition of all large-scale components: axisymmetric poloidal, nonaxisymmetric toroidal and nonaxisymmetric origin fields.* As a rule, possible contributions to the GMF by strong local magnetic fields similar to solar active regions (toroidal field) is small because the mutual cancellation of opposite polarities typically are found in active regions. The expected contribution to the GMF from the net longitudinal component of the solar north and south polar fields is a long term drift with approximately an annual period, but this variation is negligible when averaging by years [5]. If a star surface were seen at angle $i < 90°$ to the axis of rotation, the observed GMF curve would be shifted along the y-direction because of the contribution to GMF from an axisymmetric poloidal field of the nearest polar region. In this case, the value $\Delta\Phi_+/\Delta\Phi_-$ cannot be equal to 1. For ξ Boo A, $\Delta\Phi_+/\Delta\Phi_- \sim 1.5$ [12].

The gas in the convective outer layers of the Sun rotates faster at the equator than at the poles, and gas rotates almost uniformly in the radiative zone. This structure (including the presence of the tachocline zone, located at the interface between the latitude-dependent rotation of the convective zone and the rigid radiative interior) has been measured seismologically [13-15]. Gough & McIntyre [16] argue that we must have a magnetic field in the radiative interior in order to explain the uniform rotation of the radiative zone.

Such an internal field of the Sun also is required in the magnetic models of Rüdiger & Kitchatinov [17] and MacGregor & Charbonneau [18].

As we know, the magnetic flux (primordial magnetic field) can be captured from a protostellar cloud by the forming star. The star then evolves through a fully-convective Hayashi-phase. Direct observations of magnetic fields [19,20] and magnetic activity (e.g., [21]) of T Tauri stars support the hypothesis that rotating pre-main-sequence convective stars can drive hydromagnetic dynamos. This dynamo-generating field can be incorporated into the growing radiative core [22,23]. Kitchatinov et al. [24] also argued that contemporary magnetic fields in radiative cores of solar-like stars are relics of hydromagnetic dynamous operating over the pre-main-sequence epoch when a core was formed. Their numerical simulations show that this internal field is largest for an orientation normal to the rotation axis of the star (*nonaxisymmetric internal field*).

The GMF in the Sun's radiative interior beneath the tachocline must be stationary: "...it seems unlikely that the rapidly oscillating field associated with the solar cycle would contribute significantly to the dynamics in the radiative zone, particularly in view of the 10^6 yr tachocline ventilation time" (Gough & McIntyre [16]) (*stationary internal field*).

The hypothesis about the effect of quasistationary primary field of the Sun on the behaviour of solar activity and background magnetic field during the 22-year solar cycle has been discussed by many authors (see [25-29]). Benevolenska and Pudovkin [29] have concluded: "Leighton's model with an internal magnetic field taken into account, explains the main regularities observed in the course of solar activity..." and "long-term variations of solar activity including prolonged periods of abnormally low and/or high activity, may be explained within the framework of Leighton's model by variations of the hypothetical quasi-stationary internal magnetic field of the Sun...." Plachinda and Tarasova [12] wrote: "Therefore, we think that the aforementioned general characteristics of GMF point to a presence of a global magnetic field in the Sun's radiative interior beneath the tachocline and reflect its properties as well."

The intricate and time-dependent magnetic structure that is directly observed in the solar atmosphere is attributed to the interaction of magnetic field, convection, and rotation in the solar envelope. This phenomenon is not expected to be unique to the Sun, and that is inferred to be present in other late-type stars with vigorous convection in the envelope below the atmosphere. Therefore we can expect the presence of the GMF in the radiative interior of these stars and penetration of it into the surface, where the origin magnetic field becomes the initial magnetic field for generating magnetic fields by magnetohydrodynamic dynamos.

There is a wealth of indirect evidence for the presence of magnetic field on late-type stars of all luminosity classes: spots, flashes, chromospheres, transition regions, coronae, winds, etc. Currently, we have a wealth of direct spectroscopic data indicating locally strong magnetic field (1000 – 4000 G) on the surface of main-sequence stars of F-G-K-M spectral classes (see, for example, [30,31]). The existence of strong local magnetic field on the surfaces of rapidly rotating RS CVn stars (K0 dwarfs AB Doradus and LQ Hydrae, and K1 subgiant HR 1099 (V711 Tauri)) was determined using the spectropolarimetric technique of Zeeman-Doppler imaging from observations collected at the Anglo-Australian Telescope [32].

Although the study of large-scale magnetic fields has revealed the main processes causing the activity of a star as a whole [33-35], little is known about the global magnetic configuration of stars with vigorous convective zones. This is a result of difficulties of spectropolarimetric magnetic field measurements with an accuracy of 1-5 G and better [36].

Because the measurement of accurate magnetic fields is a very difficult task we developed a data reduction procedure using four spectra obtained in two exposures with a turn of the quarter-wave plate by 90°. This procedure was labeled as "Flip-Flop" Zeeman Measurements (FFZM).

Zeeman-Doppler imaging technique [32,37] is the preferred method to measure magnetic fields on the surfaces of rapidly rotating stars, however, FFZM technique is more robust and effective for global magnetic field measurements on slowly rotating stars (especially, in the case of weak magnetic fields).

The program of systematic measurements of GMF on slowly rotating stars with convective envelopes was initiated at Crimea in 1989. The observations and data reduction were carried out using 2.6m Shajn telescope, Stokesmeter, coudé spectrograph and FFZM technique [36,38]. The spectrograms were taken in the spectral region 6130-6270 Å. The reciprocal linear dispersion was 3 Å mm^{-1} (0.066 Å pixel^{-1}), and the resolving power of spectra was approximately 3×10^4 (3.0 pixel). Signal-to-noise ratios of one polarized spectrum were typically 300-400.

2. "Flip-Flop" Zeeman Measurement Technique

The primary direct means of detecting a magnetic field on a star is an observation of Zeeman splitting of spectral lines. The displacement caused by the splitting of atomic energy levels in the magnetic field of a star produces a wavelength shift $\Delta\lambda_B$ equal to

$$\Delta\lambda_B = (e/4\pi m_e c^2) z \lambda^2 B = 4.67 \times 10^{-13} z \lambda^2 B \,, \qquad (1)$$

where e is the electronic charge, m_e is its mass, c is the velocity of light, z is the effective Lande factor, λ is the wavelength in Å, and B is the magnetic field strength in Gauss. In particular, if the magnetic field is parallel to the line of sight, the π components are not visible and the wavelength displacement of the circular polarized σ components from its zero-field wavelength is calculated using Eq. (1).

The Crimean polarization analyzer (Stokesmeter) is mounted in front of the entrance slit of the spectrograph. The device consists of an entrance-retarding achromatic quarter-wave plate (the working region is 4000-6800 Å), a plate of Iceland spar, and an exit achromatic quarter-wave plate for converting the resulting linearly polarized light after the Iceland spar plate into circularly polarized light. This last step is needed to balance differences in the reflectivity of the two light beams from the diffraction grating. After passing through the exit quarter-wave plate and entrance slit of the spectrograph, two beams are projected on the plane of the CCD detector as two spectra with different initial circular polarization. Since only circular polarization is recorded, we are sensitive only to a net longitudinal component of magnetic field, averaged over the visible disk of a star. This is the so-called Effective Magnetic Field (of hot stars) or the GMF (of convective stars).

As mentioned, stellar spectra with right circular polarization obtained in the first exposure and left circular polarization in the second exposure are projected in turn on the same section of the CCD detector. Thus, errors in the flat-fielding procedure for two spectra with opposite circular polarization are practically the same and do not affect the calculation of GMF in the case of weak magnetic fields. Additionally, this observational technique automatically allows us to rule out shifts of spectral lines caused by inaccurate adjustment of the CCD plane to the focal plane of the spectrograph and instrumental drift of contours of spectral lines during the second exposure relative to the first one.

The final value for the mean longitudinal field is the arithmetic mean of the two individual measurements, in which the instrumental effects are cancelled out to first order:

$$B_e = (B'_e + B''_e)/2,$$

where

$$B'_e = k(\lambda_{1rcp} - \lambda_{2lcp})/2$$

and

$$B''_e = k(\lambda_{1lcp} - \lambda_{2rcp})/2.$$

Here, λ_{1rcp} stands for the center of gravity of a line with right circular polarization obtained during the first exposure, and λ_{1lcp} stands for the center of gravity of a line with left circular polarization. The same designation for the second exposure are λ_{2rcp} and λ_{2lcp}. The constant $k = 1/(4.67 \times 10^{-13} z \lambda)$.

If the spectral line is sharp and occupies a small number of pixels, the pixels' asymmetrical displacement relative to the center of gravity of the contour will increase the errors of the calculation. Therefore, it is desirable to fit the line with a function that is used in the subsequent calculations. We fit an approximate cubic spline function to the lines.

3. Results of Measurements of GMF on Cool Stars

Table 1.   General Magnetic Field on Cool Stars

	Object	Sp	B_e(Gauss)	B_e/σ	Ref
1	ε Gem[a]	G8 Ib	+11.1 ± 2.7	4.1	SM [44]
2	ε Peg[*]	K2 Ib	-5.3 ± 0.9	5.9	SM [44]
3	ε Leo	G1 II	+49.2 ± 6.1	8.1	SM [40]
4	ζ Cyg	G8 II	+5.4 ± 1.7	3.2	SM [41]
5	ζ Hya	G9 II	-15.3 ± 2.9	5.3	SM [41]
6	η Psc	G7 III	+11.4 ± 3.9	2.9	SM [41]
7	κ Gem[*]	G8 III	+13.0 ± 3.8	3.4	SM [41]
8	μ Peg[*]	G8 III	-20.1 ± 3.3	6.1	SM [41]
9	ε Vir	G8 III	-10.8 ± 3.2	3.4	SM [41]
10	ξ Her	G8 III	-28.1 ± 4.5	6.2	SM [40]
11	γ Tau	K0 III	+19.8 ± 5.2	3.8	SM [40]
12	ε Cyg[*]	K0 III	+9.3 ± 2.5	3.7	SM [41]
13	ε Tau	K1 III	-22.3 ± 5.4	4.1	SM [40]
14	α Boo	K2 III	+3.3 ± 0.5	6.6	MM [39]
15	δ And[*]	K3 III	+8.5 ± 2.8	3.0	SM [41]
16	β And	M0 III	+12.6 ± 2.2	5.7	SM [41]
17	μ Gem	M3 III	+9.1 ± 2.0	4.6	MM [39]
18	ζ Her	F9 IV	-10.1 ± 3.1	3.3	SM [36]
19	ξ Boo A[b]	G8 V	-10, +30	---	MM [39,45] SM [12]
20	ε Eri[c]	K2 V	+21.3 ± 4.5	4.7	SM [42]
21	61 Cyg A[d]	K5 V	-10, +4	---	MM [39,45] SM [43,44]

[*] $B_e/\sigma > 3.0$ were registered twice; [a] $B_e/\sigma > 3.0$ were registered 5 times; [b] general magnetic field varies from -10 G up to +30 G as a function of stellar rotation phase of the period $P = 6.1455$ days; [c] $B_e/\sigma > 3.0$ has been registered 3 times; [d] general magnetic field varies from -10 G up to +4 G as a function of stellar rotation phase of the period $P = 36.617$ days.

A review of published GMF data of convective stars of all luminosities is listed in the Table 1. The first column contains the object name. In the second column the spectrum is given. Column three gives the GMF (B_e) and its observed error. In the fourth column, we give ratios of $B_e/\sigma > 3$. In the 5th column, devices and references are given. In this column SM indicates observations which were carried out using Stokesmeter and 2.6-m Shajn telescope at Crimean Astrophysical Observatory, MM indicates observations that carried out using the multislit magnetometer.

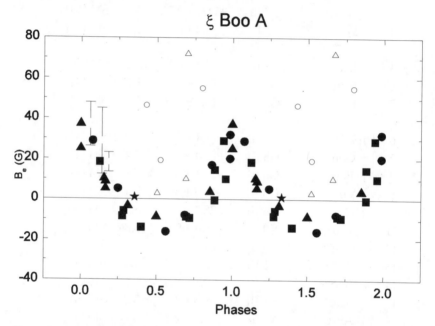

Figure 1. Variations of GMF of ξ Boo A phased with the rotation period, using measurements by different authors.

The presence of weak GMF is detected with high statistical assurance for 21 stars with convective envelopes (F9-M3 spectral types and I-V luminosity classes). For 19 stars only episodic measurements were carried out. For two solar-like stars, variation of the general magnetic field as a function of the stellar rotation has been determined. For ξ Boo A GMF variations have been measured from -10 G up to +30 G (see Figure 1) [12]. Magnetic field measurements are in Gauss and the total period is shown twice. Phases are calculated with a zero epoch at the peak value of the magnetic field:

$$HJD_{max} = 2\,445\,416.523 + 6.1455E \pm 0.0003,$$

where HJD - the Julian Date in the heliocentric coordinate system. The single measurement by Brown *et al.* [45] is shown by an asterisk "★"; it was obtained with a multislit magnetometer. Measurements obtained by Borra *et al.* [39] using a multislit magnetometer are shown with filled and open triangles. Measurements obtained at Crimean Astrophysical Observatory with Stokesmeter in 1990 [40] are shown with filled and small open circles. Results of analogous observations with Stokesmeter obtained in 1998 are shown by filled squares. Small open circles and triangles represent data that lie out of the supposed curve of magnetic field variations with rotational period. Bars are mean values of rms errors of every type of measurements. Six of the total 32 points are discrepant, and only 2 of them significantly deviate from the curve. The discrepancy may be a result of the experimental errors. More precise measurements obtained with Stokesmeter in 1998 have no data that significantly deviate from the curve. Apparently, this variability has remained phase constant during 16 years of observations (since 1982). ξ Boo A is a younger and more active solar-like dwarf than the Sun, with variability of Ca II emission (which indicates the chromosphere activity), but without clearly expressed solar-like periodicity of activity. This star is the first solar-like star with determined variations of the global magnetic field throughout a full rotation cycle.

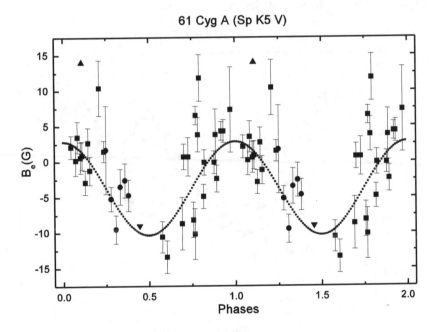

Figure 2. Variations of GMF of 61 Cyg A phased with the rotation period.

For 61 Cyg A, variations of the GMF have been measured from -10 G up to +4 G (see Figure 2). Measurements obtained with 2.6-m Shajn Telescope at Crimean Astrophysical Observatory are shown by filled squares (1998-1999) and filled circles (2002). The filled triangle down shows the single measurement by Borra *et al.* [39] obtained with multislit magnetometer. Single measurement obtained by Brown *et al.* [45] using multislit magnetometer is shown by triangle up. In both cases of multislit magnetometer measurements errors are big (40 G and 14 G, respectively). The total period is shown twice. Phases are calculated with a zero epoch at the peak value of the magnetic field:

$$HJD_{max} = 2450989.2 + 36.617 \pm 0.054.$$

This star has largely lost its rotational angular momentum, has the same level of activity as the Sun and clearly expressed solar-like Ca II emission periodicity 7.3 ± 0.1 yr [46]. 61 Cyg A is the second solar-like star for which variations of the global magnetic field have been determined throughout a full rotation cycle.

Variations of GMF of the Sun as a star with period of 26.92 days in 1991 during the spot activity maximum are shown in Figure 3, using data from Wilcox Solar Observatory [8]. The measurement uncertainty is about 0.05 G, which is less than the plotted symbol size. Open circles indicate GMF measurements for the first rotation period in 1991.

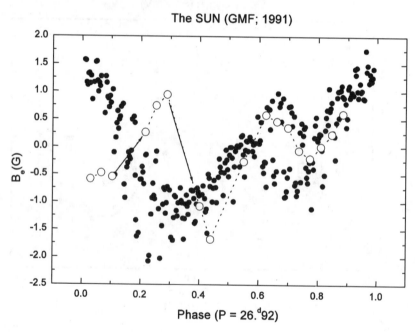

Figure 3. Variations of the general magnetic field of the Sun as a star with the synodic period of rotation in 1991.

The picture demonstrates complicated behavior of the GMF curve shape from period to period in spite of the phase stability of the main maximum

(0.9-1.0) and the main minimum (0.3-0.4) for a mean phase curve of 1968-2001 span (the total number of GMF measurements $N = 12428$) [7].

One can see, that in the case of GMF investigations of convective stars, there are not only difficulties in carrying out spectropolarimetric measurements with an accuracy of 1-5 G or better, but there is a phenomenon of nonstationary variability of an amplitude and curve shape of magnetic field against rotational period of the star as well. In such cases the reliability and accuracy of reproduction of results by using devices comes into question.

4.   The Reliability of the "Flip-Flop" Zeeman Measurement Technique.

Excluding references [39] and [45] in Table 1, all other measurements were carried out at Crimean Astrophysical Observatory using Stokesmeter and "Flip-Flop" Zeeman Measurement Technique. For these measurements a number of criteria for measurement reliability are examined. We discuss these in turn.

4.1   Systematic control of adjustment of the Stokesmeter.

Before each run of observations the adjustment of the Stokesmeter was examined using any bright star. The efficiency of the Stokesmeter is 94-95%, that includes inefficiency of the calibration device.

4.2   Reproduction of the known magnetic curve of a magnetic star for testing the value and sign of the magnetic field.

The value and sign of the magnetic field of the magnetic star β CrB are used for testing in order to reproduce the accepted curve shown in Figure 4. All curves are results of geometrical dipole least-squares fittings. This figure demonstrates the real discrepancy between magnetic field curves for different spectral lines. It is an old and well-known effect, as well as an important and deep problem for not only Ap-stars, which we must remember when high-accuracy measurements are analyzed. This phenomenon of discrepancy between different spectral line measurements also is present for measurements of the GMF of the Sun as a star [48]. The nature of this effect is unknown.

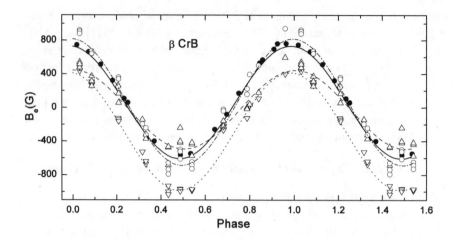

Figure 4. Mean longitudinal magnetic field strength (B_e) of β CrB vs. phases of stellar rotational period. Solid circles – B_e measurements by Wade [47] using the bulk of spectral lines. Open triangles up are Crimean B_e measurements obtained using Fe I 6136.615 Å spectral line. Open circles are our B_e measurements obtained using Fe I 6137.692 Å spectral line. Open triangles down are our B_e measurements obtained using Ca I 6162.173 Å spectral line.

4.3 Reproduction of the zero field of a nonmagnetic star and knowledge of the value of the systematic instrumental shift.

The comparison of the Crimean data for Procyon (B_e = -1.34 ± 1.0 G) [36] with those which have been obtained by Bedford *et al.* [49] using a magneto-optical filter (B_e = -1.86 ± 0.9, 0.49 ± 0.8 G) shows good agreement within the observational errors. Using 27 observing nights in 1989-1997, the systematic instrumental shift was determined as $<B_e>$ = -0.12 ± 0.99 G [36].

4.4 Knowledge of the statistical distribution characteristics of experimental values; i.e., whether the mean and standard deviation are unbiased estimates, signified by whether the statistical distribution is symmetric like a normal distribution.

Using magnetic-field measurements on supergiant ε Peg (Sp K2Ib), the statistics associated with polarization measurements are evaluated (number of measurements N = 971). Measurements used in the analysis include weak spectral lines with $z \times (r_0 - r_c) < 0.2$, where z - Landé factor, r_0 - contour restriction level, and r_c - central line depth in continuum units [36]. To fit a normal distribution, skewness is negligible, S = -0.154 ± 0.078, the positive kurtosis ("peakedness") is essential, E = 2.280 ± 0.157, and the probability of the normality P = 93.7% using Kolmogorov-Smirnov test. The total experimental distribution is symmetric: the mean and standard deviation are unbiased estimates.

By virtue of testing for normality, we eliminate spectral lines, for which $z \times (r_0 - r_c) < 0.2$, and then use the more homogeneous sample of observations with $10 < \sigma < 15$ G for four supergiants: β Aqr (Sp G0 Ib), α Aqr (Sp G2 Ib), ε Gem (Sp G8 Ib), ε Peg (Sp K2 Ib). For $N = 460$ values of B_e, we obtain the skewness $S = -0.115 \pm 0.114$ and kurtosis $E = -0.020 \pm 0.227$ with the probability of normality $P \sim 96.4\%$ using the Kolmogorov-Smirnov test. The possibility to use the mean and standard deviation for analysis as unbiased estimates is present because the experimental distribution is symmetric and normal.

4.5 The coincidence between the numerical simulation value (Monte-Carlo method) of the standard deviation and the experimental standard deviation using normal or the known experimental statistical distribution.

In order to test observational errors using a full sample of unblended spectral lines, the Monte-Carlo method with a generator of normally distributed numbers was used. For $N = 2545$ measurements of magnetic fields on four yellow supergiants (β Aqr, α Aqr, ε Gem, ε Peg), including weak unblended spectral lines, the relation between mean the Monte-Carlo simulated standard error $<\sigma_{(m-c)}>$ and the mean experimental standard error $<\sigma>$ was estimated as $<\sigma_{(m-c)}> = 1.033<\sigma>$. Further, weak spectral lines for which $z \times (r_0 - r_c) < 0.2$ were eliminated to strengthen the data uniformity. For $N = 1844$ measurements $<\sigma_{(m-c)}> = 0.968<\sigma>$. The discrepancy is 3.3 % in the first case and 3.2 % in the second case; both appear to be very small.

4.6 The use of the homogeneous sample of measurements for estimation of the mean and rms error.

As a rule, various spectral lines form in different physical conditions, so longitudinal magnetic-field measurements may bring us a variety of magnetic field strengths. For a solar-like spectrum with signal-to-noise ratio $S/N \sim 400$, errors lie from 4-5 G to 20-25 G, depending on magnetic sensitivity, half-widths and depths of spectral lines, therefore a uniform sample of spectral lines for magnetic field calculations must be used. The weighted values are proper if the statistical assurance of the discrepancy between experimental and Monte-Carlo standard errors is less than 95%. Otherwise, when $P > 95\%$, we must calculate the arithmetic mean B_e and σ, and analyze sources for discrepancies.

4.7   Reproduction of known magnetic curves of a star having a weak field.

The investigations were carried out using solar like stars ξ Boo A and 61 Cyg A (see Figure 1 and Figure 2). Because this is the first study of the general magnetic field as a function of rotation on a solar-like star other than the Sun, we do not have the possibility to reproduce a known magnetic curve of a star having a weak field. But Stokesmeter measurements have shown good agreement with multislit magnetometer measurements by Brown *et al.* [45] and Borra *et al* [39]. Moreover, rotational periods of these stars are well known from flux index S measurements for the Ca II H and K lines; they are in agreement with magnetic periods.

4.8   "Flip-Flop" Zeeman Measurement technique - internal Stokesmeter test for presence or absence of significant stochastic or spurious time-dependent Stokes signatures.

When using the "Flip-Flop" Zeeman Measurement technique an internal device possibility allows us to control for the presence or absence of significant stochastic or spurious time-dependent Stokes signatures. When using more than one pair of exposures (see Figure 5), spectra with identical circular polarization are projected on the same location of the CCD. Therefore, we can calculate the value of the spurious "magnetic field" that must be equal to zero if all spurious effects are negligible.

    In Table 2 the results of such testing of the magnetic field for two supergiants are shown. "Zero field" ($B_{test} \pm \sigma_{test}$) using pairs of exposures over one are evaluated. The first column contains the Heliocentric Julian Date. The second column contains the number of GMF measurements. The third column contains the GMF and its observed error. The fourth column contains the ratio of $k = B_e/\sigma > 3$. The fifth column contains B_{test} and its error σ_{test}, and the 6th column contains the ratio of $k_{test} = B_{test}/\sigma_{test}$.

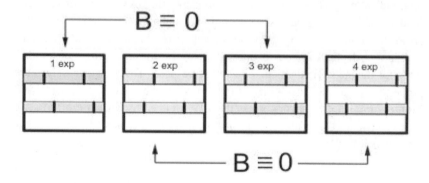

Figure 5. Internal Stokesmeter test on presence of spurious Stokes signatures.

Table 2. Test of "Spurious field"

	JDH (+2400000)	N	$B_e \pm \sigma$ (Gauss)	k	$B_{test} \pm \sigma_{test}$ (Gauss)	k_{test}
	ε Gem (Sp G8 Ib)					
1	51907.500	79	11.1 ± 2.7	4.1	3.8 ± 3.0	1.3
2	51912.219	44	9.8 ± 2.5	3.9	-4.4 ± 2.2	2.0
3	52217.516	108	-10.5 ± 3.0	3.5	7.2 ± 3.4	2.1
4	52306.310	13	38.1 ± 7.4	5.1	---	---
5	52309.356	77	5.3 ± 1.5	3.5	0.1 ± 1.6	0.1
	ε Peg (Sp K2 Ib)					
6	51035.468	178	-5.3 ± 0.9	5.9	0.3 ± 1.0	0.3
7	52509.493	206	-2.7 ± 0.8	3.4	-0.7 ± 0.9	0.9

For the date JDH 2452306.310 only two exposures were made. Therefore, the above-mentioned testing is not possible. The reliability of this result is argued by the insufficient statistical assurance ($P \sim 65\%$) and discrepancy between Monte Carlo and experimental errors. For the six other dates the ratio $k_{test} = B_{test}/\sigma_{test} < 3.0$, and the statistical assurance of registered GMFs is reliable.

Notes: The "Photometrics" CCD camera used at Crimean observatory, is cooled to $-113.7°$ C. The temperature stability is $\sim 0.1°$ C during 15 hours. The longest night lasts 12 hours. The size of a pixel is 2.4×10^{-05} m. The silicon coefficient of volume of the linear expansion is $\alpha \sim 3.58 \times 10^{-06}$. In such a case, the thermal expansion is 8.59×10^{-12} m pixel$^{-1} \sim 2.6 \, 10^{-08}$ Å. For Landé factor $Z = 1.000$ and wavelength 6000 Å, this expansion is ~ 0.0015 Gauss pixel^{-1}. We have three orders of margin safety.

5. Conclusions

The presence of weak GMF for 21 star with vigorous convection is detected (F9-M3 spectral types and I-V luminosity classes). Furthermore, the variation of global nonaxisymmetric magnetic field – general magnetic field - as a function of the stellar rotation has been first determined for two solar-like stars other than the Sun: ξ Boo A and 61 Cyg A.

Today we still do not know what the nature of this field. The nonaxisymmetric GMF as a phenomenon is absent in the Babcock' and Leighton' phenomenological model of the solar cycle. In terms of standard α-Ω dynamo theory GMF is absent also; there are only two main components of large-scale magnetic fields: toroidal magnetic field and axisymmetric poloidal field.

The working hypothesis is that GMF reflects properties of a stationary global magnetic field of Sun's and convective star's radiative interior on its surface, and there appears to be a third, nonaxisymmetric, large-scale component of magnetic field (Origin Magnetic Field – initial magnetic filed for dynamo mechanisms). A global magnetic field of the radiative interior penetrates into the surface of the Sun, and GMF is a time-dependent superposition of all large-scale components: axisymmetric poloidal, nonaxisymmetric toroidal and nonaxisymmetric origin fields.

6. Acknowledgments

I acknowledge with thanks Dr. G. Videen for his kind editing of the manuscript. This work has been supported in part by the Ukrainian SFFD grant No. 02.07/00300, by grants R2Q000 and U1C000 from the International Science Foundation, and by the grant A-05-067 from the ESO C&EE Programme.

References

1. H. W. Babcock, ApJ, **133**, 572 (1961)

2. R. B. Leighton, ApJ, **140**, 1559 (1964)

3. R. B. Leighton, ApJ, **156**, 1 (1969)

4. A. B. Severny, Nature, **224**, 53 (1969)

5. P. H. Scherrer, J. M. Wilkox, L. Svalgaard, T. L. Duvall, P. H. Dittmer, & E. K. Gustafson, Sol. Phys., **54**, 353 (1977)

6. V. A. Kotov, P. H. Scherrer, R. F. Howard, & V. I. Haneychuk, ApJS, **116**, 103 (1998)

7. V. I. Haneychuk, V. A. Kotov, & T. T. Tsap, A&A, **403**, 1115 (2003)

8. Solar Geophysical Data

9. V. A. Kotov, Bull. Crimean Astrophys. Obs., **77**, 39 (1987)

10. L. Svalgaard, & J. M. Wilcox, Sol. Phys., **41**, 461 (1975)

11. V. I. Haneychuk, V. A. Kotov, & T. T. Tsap, A&A, **403**, 1115 (2003)

12. S. I. Plachinda, & T. N. Tarasova, ApJ, **533**, 1016 (2000)

13. M. J. Thompson, et al., Science, **272**, 1400 (1996)

14. A. G. Kosovichev, et al., Sol. Phys., **170**, 43 (1997)

15. J. Schou, et al., ApJ, **505**, 390 (1998)

16. D. O. Gough, & M. E. McIntyre, Nature, **394**, 755 (1998)

17. G. Rüdiger, & L. L. Kitchatinov, Astronomische Nachrichten, **318**, 273 (1997)

18. K. B. MacGregor, & P. Charbonneau, ApJ, **519**, 911 (1999)

19. C. M. Johns-Krull, J. A. Valenti, A. P. Hatzes, & A. Kanaan, ApJ, **510**, L41 (1999)

20. C. M. Johns-Krull, J. A. Valenti, & C. Koresko, ApJ, **516**, 900 (1999)

21. G. Basri, G. W. Marcy, & J. A. Valenti, ApJ, **390**, 622 (1992)

22. E. N. Parker, Geophys. Astrophys. Fluid Dyn., **18**, 175 (1981)

23. A. E. Dudorov, V. N. Krivodubskij, T. V. Ruzmaikina, & A. A. Ruzmaikin, SvA, **33**, 420 (1989)

24. L. L. Kitchatinov, M. Jardine, & A. C. Cameron, A&A, **374**, 250 (2001)

25. M. I. Pudovkin, & E. E. Benevolenskaya, SvA, **28**, 458 (1984)

26. E. H. Levy, ASP Conference Series, **26**, 223, (1992)

27. E. H. Levy, & D. W. Boyer, ApJ, **254**, L19 (1982)

28. D. W. Boyer, & E. H. Levy, ApJ, **277**, 848 (1984)

29. E. E. Benevolenska, & M. I. Pudovkin, Sol. Phys., **95**, 381 (1985)

30. I. Rueedi, S. K. Solanki, G. Mathys, & S. H. Saar, A&A, **318**, 429 (1997)

31. C. M. Johns-Krull, J. A. Valenty, & C. Koresko, ApJ, **516**, 900 (1999)

32. J.-F. Donati, et al., MNRS, **345**, 1145 (2003)

33. E. N. Parker, "Cosmical Magnetic Fields: Their Origin and Their Activity" (Oxford: Calerdon Press, 1979)

34. F. Krause, & K.-H. Rädler, "Mean-Field Magnetohydrodynamics and Dynamo Theory" (Berlin: Verlag, 1980)

35. S. I. Vainshtein, Ya. B. Zel'dovich, & A. A. Ruzmaikin, "Turbulent Dynamo in Astrophysics" (Moscow: Nauka, 1980)

36. S. I. Plachinda, & T. N. Tarasova, ApJ, **514**, 402 (1999)

37. O. Kochukhov, et al., A&A, **414**, 613 (2004)

38. S. I. Plachinda, A. V. Jakuschechkin, & S. G. Sergeev, Izv. Krym. Astropfiz. Obs. **87**, 91 (1993)

39. E. F. Borra, G. Edwards, & M. Mayor, ApJ, **284**, 211 (1984)

40. S. Hubrig, S. I. Plachinda, M. Hunsch, & K.-P. Schröder, A&A, **291**, 890 (1994)

41. T. N. Tarasova, ARep, **46**, 474 (2002)

42. T. N. Tarasova, S. I. Plachinda, & V. V. Rumyantsev, ARep, **45**, 475 (2001)

43. S. I. Plachinda, C. M. Johns-Krull, & T. N. Tarasova, Odessa Astron. Publ., **14**, 219 (2001)

44. S. I. Plachinda, in press (2004)

45. D. N. Brown, & J. D. Landstreet, ApJ, **246**, 899 (1981)

46. S. L. Baliunas, et al., ApJ, **438**, 269 (1995)

47. G. A. Wade, J.-F. Donati, J. D. Landstreet, & S. L. S. Shorlin, MNRS, **313**, 851 (2000)

48. V. A. Kotov, & T. V. Setyaeva, ARep, **46**, 246 (2002)

49. D. K. Bedford, W. J. Chaplin, A. R. Davies, J. L. Innis, G. R. Isaak, & C. C. Speake, A&A, **293**, 377 (1995)

Sergei Plachinda

Performers at first NATO ASI dinner

Ben Veihelmann investigates dormitories at Chufut-Kale.

Igor Geogdzhayev with wife Elena and daughter Masha.

POLARIMETRY AND PHYSICS OF SOLAR SYSTEM BODIES

ALEXANDER MOROZHENKO AND
ANATOLIY VID'MACHENKO
*Main Astronomical Observatory of the Ukrainian
National Academy of Sciences, 27 Akademika
Zabolotnogo Street, Kyiv, 03680, Ukraine*

Abstract. We present a historical review of polarimetric observations of planetary atmospheres, comets, atmosphereless solar system bodies, and terrestrial materials. We highlight the study of physical and optical parameters of planetary atmospheres. Polarimetric observations of the atmospheres of Venus, Mars, Jupiter and Saturn have made it possible to determine the real part of the refractive index and the cumulative size distribution function for the constituent cloud layers. We describe a simple and reliable method of quantifying absorptive cloud layers of the giant planets and predict the vertical structure of aerosol layers of planetary atmospheres based on the analysis of observational spectropolarimetric data of contours of molecular absorption bands at the center of the planetary disk. The method is effective only when experimental data exist in a broad interval of phase angles. Using this method we can determine aerosol sizes in the atmospheres of Uranus and Neptune.

1. Introduction

Polarimetry is a powerful method for studying solar-system bodies. It has allowed the determination of such parameters as the complex refractive index of particles in planetary atmospheres, the size distribution functions of these particles, the methane concentrations, the atmospheric pressure values above the cloud layers, etc. Independent spectral analyses of linear (P) and circular (V) polarization observational data also may facilitate the determination of physical characteristics of particles at different heights in a planetary atmosphere. Polarimetry enables us to make qualitative conclusions about

G. Videen et al. (eds.), Photopolarimetry in Remote Sensing, 369-384.
© 2004 *Kluwer Academic Publishers. Printed in the Netherlands.*

physical characteristics and surface structure of atmosphereless celestial bodies as well as to estimate approximate values of asteroid reflectivity.

The degree of linear polarization (P) results primarily from first-order light-scattering events in the upper atmosphere. The degree of circular polarization (V) is predominantly the result of multiple scattering. Therefore independent analyses of P and V allow for the determination of some physical characteristics of particles at different heights in a planetary atmosphere. There have been fewer opportunities to determine physical characteristics of atmosphereless celestial bodies and comets. This is explained by the absence of a comprehensive theory of formation of the Stokes parameters of light reflected diffusely by rough surfaces and heterogeneous particles. In this chapter we summarize the most essential effects in observed polarization properties of diffusely reflected light by these bodies, and we supplement these effects by the data of laboratory investigations of particles made of terrestrial materials.

2.   Features of polarization properties

2.1. Atmosphereless bodies and terrestrial materials

In 1811 the French scientist Arago was the first to observe the Moon through a polariscope. He discovered that light reflected by dark lunar details (seas) is polarized more strongly than light reflected by bright details (continents). In 1905–1912 the Russian scientist Umov studied this effect in detail in the laboratory and consequently it was named in his honour. In the 1950s, Rosenberg [65] showed that the cause of this effect was depolarization due to multiple scattering by surface micro-roughness. He offered a semi-empirical formula that relates depolarization, single-scattering albedo and roughness of the particulate surface layer. The following effects have been observed: 1) In continuum, the degree of polarization decreases with increasing wavelength practically for all atmosphereless solar system bodies; 2) For Mars, there is a dependence of the degree of linear polarization on the longitude of the central meridian as observed from the Earth [44]; 3) For atmosphereless bodies, there is a dependence of the degree of positive polarization on the slope of the linear part of the polarization phase curve [28]. It should be noted that the Umov effect also applies to planets with dense atmospheres. It is confirmed by photometric measurements in the contours of molecular absorption bands of methane on Jupiter [33, 50].

In the 1920s, Lyot [36] discovered that the polarization plane of the reflected light from the Moon and different terrestrial materials rotates by 90° at a phase angle called the inversion angle α_i. A branch of negative polarization at phase angles smaller than α_i was also observed. Later it was discovered that this phase dependence is typical of all atmosphereless solar system bodies and of many particulate terrestrial materials in a broad wavelength interval. For different celestial bodies and for different terrestrial samples the value of the inversion angle varies from approx. 15° to approx. 27°. A quantitative relationship between the inversion angle and its wavelength dependence has not been established. At present, research results appear to be inconclusive: 1) Avramchuk [1] in 1962 carried out photoelectric measurements for 18 details of the lunar disk at 8 wavelengths within the spectral interval 355–600 nm. He discovered that at a phase angle of 20.3° the degree of polarization vanishes at all wavelengths studied. The same result was obtained in laboratory measurements of terrestrial materials (a detailed review of the laboratory measurements is available in [45]), in the observations of Mars in a long-wave spectral range [45], and in observations of some asteroids [35]. 2) Dolfus and Bowel [15] examined 14 details of the lunar disk in the spectral interval 327–1050 nm and came to the conclusion that the value of the inversion angle decreases with decreasing wavelength from 26° at 840–1050 nm down to 21° at 327–379 nm. 3) Kvartshelia [34] examined 190 lunar details and investigated samples of lunar soil delivered to Earth in the spectral interval 419–783 nm. For mares, embayments, regions similar to sea craters, and for the lunar soil samples, a decrease of the inversion angle took place in the short-wave spectral region. However, the decrease was not as big as 5°, but rather 1.5–2.0°. For bright details the inversion angle does not change with wavelength. The values of the degree of polarization for all details are given in [34, 55] for phase angles $\alpha < 25°$ at two extreme wavelengths. It can be seen that even if a spectral dependence of the inversion angle were to exist, the total variation would not exceed 2° in the spectral regions 327–600 and 419–704 nm. A quantitative relationship still has not been established.

For many years it was presumed that at the inversion point, the rotation of the polarization plane by 90° occurs abruptly. This was confirmed by observations of lunar details [34]. However, Kohan [31] and Morozhenko [41, 42] have shown that the rotation of the polarization plane of light reflected by terrestrial samples takes place within some interval of phase angles $\Delta\alpha$. The width of this interval depends on the sample morphology and composition [11]. This effect can be used in remote-sensing studies of atmosphereless celestial bodies.

In the early 1920s, Lyot [36] discovered a large value of negative polarization in radiation scattered by a magnesium particulate sample at extremely small phase angles. No special attention had been paid to this fact for many years. However, when astrophysical polarimetric observations were performed in 1986, Morozhenko found large values of negative polarization for all four Galilean satellites of Jupiter at a phase angle of 0.4°. The absolute value of the degree of polarization was 0.3–0.4% and the polarization plane rotated by 45° relative to the polarization plane of the Jovian Polar Region [54, 59]. In 1993 theoretical modeling by Mishchenko [39] showed that surfaces composed of fine particles can generate a polarization opposition effect at phase angles less than 1°. The optical mechanism of this phenomenon is coherent backscattering, otherwise known as weak photon localization. Subsequent observations of the Galilean satellites at small phase angles (see, for example, [64]) have corroborated the existence of this phenomenon.

Because this effect appears at phase angles less than 1°, the most useful observations would be those of giant-planet satellites and asteroids located beyond the Jovian orbit. It is typical of many observational papers to present phase angle values calculated without regard of the angular size of the Sun. However, the observed data fall into a phase angle interval from $\alpha + \Delta A$ to $\alpha - \Delta A$, where ΔA is the angular radius of the Sun as observed from a planet or an asteroid (see Table 1).

Table 1: Values of the angular diameter of the Sun as seen from planets

Planet	Mercury	Venus	Earth	Mars	Jupiter	Saturn	Uranus
$2\Delta A$ (deg)	1.29	0.69	0.50	0.33	0.10	0.05	0.03

In the study of the nature of atmosphereless solar system bodies it is also necessary to pay attention to the presence of circular polarization at large (> 120°) phase angles. Non-zero circular polarization is typical of surface layers having small amounts of metallic particles [12]. Circular polarization can be observed even when the surface is illuminated with unpolarized light.

2.2. Bodies with atmospheres

The Martian atmosphere has a very small optical thickness. Therefore, it occupies an intermediate place between atmosphereless bodies and planets with thick atmospheres. The Martian atmosphere has polarization properties similar to atmosphereless bodies. However, the spectral dependence of the inversion angle and the degree of negative are noticeably different. In accordance with the Umov effect, the absolute value of negative polarization increases with decreasing wavelength in the long-wave range of the spectrum (>500 nm) and decreases with decreasing wavelength at shorter wavelengths (<500 nm).

For planets with optically thick atmospheres, the first measurements in the visual spectral region were performed by Lyot [36] in 1922–1925. These observations revealed a series of new effects. Polarization measurements of details of the Jovian disk showed the following: 1) The polarization plane makes a 0° angle with the equatorial plane of intensity at the western and eastern limbs and a 90° angle at the polar regions. 2) At the polar regions, the degree of polarization weakly depends on the phase angle. These properties are the result of multiply scattered radiation and are confirmed by observations in a broad wavelength interval.

Polarization measurements of the entire disk of Jupiter at phase angles close to zero have shown unexpected results. In long-wave spectral regions at $\alpha < 3°$, the sign of linear polarization changes, and the value of the degree of polarization increases as α approaches zero [43, 46, 69]. Polarization originates from multiple light scattering in an optically nonuniform gas-aerosol atmosphere. One may expect that similar effects should be observable for all celestial bodies (including those without atmospheres) if their visible surfaces are optically inhomogeneous. The diffusely reflected radiation from subauroral regions of the disks of Jupiter and Saturn also appears to be elliptically polarized [27, 37]. The largest value of the degree of circular polarization (about 0.001%) decreases with decreasing phase angle. The sign of the degree of circular polarization is different for northern and southern regions. The sign also changes with the sign of the phase angle (i.e., before and after opposition). This effect has been attributed to multiple scattering by large spherically shaped, dielectric particles [21].

The polarimetric study of details of Saturn's disk indicates the existence of small oriented particles [7] in the upper atmospheric layers. It has been suggested that these particles are the results of the dropout of submicron particles from Saturn's rings that subsequently become crystallization nuclei for ammonia crystals. We highlight the following observations: 1) For eastern and western limbs at latitudes from −23° up to +23° and wavelengths < 450 nm, the position of the polarization plane corresponds neither to that modeled from single scattering (0° or 90° with the scattering plane), nor to that of multiple scattering (0° or 90° to the corresponding radius-vector) [2, 4, 19]. 2) The lowest values of polarization at the center of the planetary disk occur at phase angles ~2°, and the position of the polarization plane coincides neither with 0° nor with 90° (at $\lambda = 316$ nm) [48].

It is not surprising that for planets with dense atmospheres, it is impossible to identify common features in the phase and spectral dependences of the degree of linear polarization and in the spectral dependence of the inversion angle. Indeed, these quantities are determined by the morphology, chemical composition, size, and vertical distribution of cloud particles, which appear to be vastly different in various planetary atmospheres.

2.3. Comets

For comets, the spectral phase dependence of the degree of linear polarization is similar to that of atmosphereless solar system bodies. Moreover, for spectral regions within gaseous emission bands there is virtually no dependence of the inversion phase angle on wavelength [14]. At the same time, light from cometary halos (at least for comet Halley) is elliptically polarized in the continuum. At $\lambda = 484$ nm, $V = 0.40 \pm 0.09\%$ at phase angles ~30° [3, 10, 56]. Polarization measurements of comet Halley performed during five nights [56] at phase angles from 21° to 23.5° (i.e., in the vicinity of the inversion point) have shown that the polarization plane in continuum makes an angle of about 40°–50° with the scattering plane. For wavelengths corresponding to C_2 molecular emissions, this angle is 90°. Therefore, it is possible that at the inversion point, the orientation of the polarization plane changes and the degree of polarization does not vanish. It is likely that the above-mentioned polarization features of cometary atmospheres indicate that the cometary particles are aspherical and oriented, as was suggested previously in 1980 [49]. As a result, their polarization properties represent a peculiar combination of properties of small individual particles and those of an atmosphereless celestial body.

3. Physical and optical parameters of solar system bodies derived from polarimetric data

3.1. Atmosphereless bodies

At present, there is no general theory of the diffuse radiation field reflected from atmosphereless solar system bodies. Therefore polarization measurements are not typically used for a quantitative study of the physical properties of surface layers. However, the Umov effect can be used to derive qualitative conclusions about the surface microstructure and reflectivity of atmosphereless bodies. Such data allow quantitative estimates of the particle size.

Over the past few years, many papers have been dedicated to the development of the theory of formation of the polarization opposition effect. Since 1993, this theory allows for the determination of sizes of particles forming a surface layer [39]. Considerable attention in other chapters of this volume is paid to this problem. In this chapter we restrict ourselves to ideas developed from the Umov effect. There are two trends: 1) Investigations have indicated the following dependence between the maximum degree of positive polarization P_{max} and reflectivity $r(0)$ of atmosphereless celestial bodies (see, for example, [1, 15]):

$$\lg r(0) + n \lg P_{max} = \text{Constant} = C, \qquad (1)$$

where $n = 0.724 \pm 0.005$ [15] for lunar details. Relation (1) holds for the positive branch of polarization, but it is possible that it is valid for all values of the phase angle. In addition, observations of the degree of polarization at $\lambda =$ 720 nm and 620 nm agree with the following relationship [16]:

$$P(\alpha, \lambda) = P(\alpha, 504 \text{ nm})A(\alpha, 504 \text{ nm})/A(\alpha, \lambda). \qquad (2)$$

Here $A(\alpha, 504\text{nm})$ and $P(\alpha, 504 \text{ nm})$ are the phase dependences of the visual albedo and the degree of linear polarization at $\lambda = 504$ nm. 2) On the other hand, a logical corollary of the Umov effect is the existence of a dependence of the slope h of the linear part of the positive branch of the polarization phase curve on the surface reflectivity [13, 28]. Veverka and Liller [68] were the first to suggest that this dependence can be used to estimate asteroid reflectivities.

3.2. Cometary atmospheres

The dust-comet halo represents an extreme case of an extended and sparse planetary atmosphere. Its optical thickness is very small. Therefore, the role of multiple scattering is small too.

In the early 1980s, Morozhenko [49] attempted to fit all polarimetric and spectrophotometric results of cometary observations with calculations for spherical particles. This attempt was unsuccessful. The opinion expressed was that the cometary aerosols are very large, oriented, and rough (or aggregated) particles. Their sizes must be so large that their spectropolarimetric properties could be similar to those of atmosphereless celestial bodies. That is, a cometary atmosphere is, in fact, a collection of "microasteroids." Dollfus [13] arrived at the same conclusion. He observed that the phase dependence of the degree of polarization for comet Halley shows the best agreement with the observed phase dependence of 20-micrometer-radius fluffy aggregate particles.

Mukai et al. [60] presented a composite model consisting of two layers of small spherical particles. With such a model they achieved satisfactory agreement with observational data. It was suggested that the cometary particles were composed of water ice contaminated by silicates. The average values of the complex refractive indices for cometary particles were estimated to be 1.392–0.024i and 1.383–0.038i at wavelengths 365 and 730 nm, respectively.

Petrova et al. [62] used a model of small aggregated particles and showed that the phase dependence of the degree of linear polarization of light scattered by such particles is similar to that observed for comets. However, their results show a very large increase in the inversion angle with decreasing wavelength, which does not agree with observational data for comets.

3.3. Planetary atmospheres

3.3.1. Venus

In his seminal paper [36], Lyot demonstrated the efficiency of polarimetric methods for determining physical characteristics of cloud particles in planetary atmospheres. He estimated the average radius of particles in the Venus atmosphere to be 1.25 microns. This estimate was derived through a comparison of the observed phase dependence of the degree of linear polarization for Venus and that measured for water drops with varying size. It was not surprising that Lyot was able to reproduce only the general trend in the observed polarization phase dependence, whereas the specific values of the degree of polarization for Venus and the water drops could differ by a factor exceeding 3. The latter was explained by the effect of multiple scattering.

Sobolev [66] hypothesized that the formation of polarization properties of planetary atmospheres takes place in the highest atmospheric layers (where the optical depth does not exceed values of about 1). It is then sufficient to consider only the first-order scattering in the calculation of the second Stokes parameter, Q. Sobolev considered two models of the atmospheric vertical structure: 1) a single-layer model in the form of a semi-infinite cloud layer; and 2) a two-layer model, in which an optically thin gas layer is put on top of the semi-infinite cloud layer. Comparisons of the phase dependence of the degree of linear polarization observed in the visible with calculations for two values of the real part of the refractive index ($n_r = 1.33$ and 1.50) and for varying particle radii showed the best agreement for monodisperse particles with $n_r = 1.5$ and a radius of 1 micrometer.

At about the same time, Coffen [8] analyzed long-term observational data in the spectral range 340–990 nm. He also used the first-order-scattering approximation, but compared the observational data with results of calculations performed for a range of values of the real part of the refractive index from 1.335 to 2.5. The best fit to the observations was obtained for $1.43 < n_r < 1.55$ and $r = 1.25 \pm 0.25$ micrometers.

The inclusion of multiple scattering (with the so-called doubling method) [21, 22] to model a semi-infinite, homogeneous, gas-aerosol layer led to a further improvement in characterizing aerosol particles. It was assumed that the distribution of particle sizes can be approximated by a gamma distribution. It was concluded that $n_r = 1.43 \pm 0.01$ at the wavelength 990 nm and 1.46 ± 0.01 at 365 nm, the particle effective radius being $r_{eff} = 1.05 \pm 0.05$ micrometers and the effective variance being $v_{eff} = 0.07$. Similar aerosol parameters were derived through an analysis of the observed spectral dependence of the degree of linear polarization in a small interval of phase angles taking into account multiple scattering [58].

The great interest in this topic was further demonstrated by the paper of Kattawar *et al.* [23]. Several models of the vertical structure of the cloud layer were analyzed using the Monte-Carlo method and polarimetric observations. Using homogeneous and inhomogeneous atmosphere models yielded the same real part of the refractive index and the same particle size. This fact seems to confirm Sobolev's hypothesis that linear polarization is mostly formed by single scattering within an optically thin top atmospheric layer.

Ground-based polarimetric observations of some features on the planetary disk indicated the existence of a haze layer overlying the main cloud (especially in subauroral regions). This was confirmed by the results of photopolarimetric observations from Pioneer–Venus [24]. The value of the effective radius of the haze particles was about 0.23 micrometers; the value of the real part of the refractive index was the same as for the cloud particles.

There are noticeable differences in the degree of linear polarization for the Venus atmosphere measured in different years, which indicates long-term variations of atmospheric optical parameters, especially for upper layers. The most reliable data on such changes were obtained from the Pioneer–Venus spacecraft in 1978–1990 [29]. The changes in polarization can be explained by changes in the haze optical thickness. However, the same changes in polarization can be reproduced by lowering the height of the upper boundary of the haze layer from about 5 to about 20 mbar.

In addition to the long-term variations, there are also short-term variations with duration of about 4 days. It is interesting that the product of the amplitude of the polarization change ΔP and the visual albedo of the disk A shows little dependence on wavelength. The most probable explanation of the observed variations [51] is a variability of the optical thickness of the haze.

Inconsistency of the published estimates of the haze-particle effective radius are discussed in [24, 29, 51]. The cause of the differences is discussed in [53]. Using the assumption that the haze (for heights greater than 66 km) is a continuation of the underlying cloud layer [30], it was shown that stratification of particles should take place due to turbulent intermixing.

3.3.2. Mars

The Martian atmosphere is characterized by a small value of the optical thickness. Therefore polarization properties of the reflected light are formed both by the underlying surface and the gas-aerosol atmosphere. One fortunate exception occurs at phase angles near the inversion angle α_i^s of the surface, where the resulting polarization $P(\alpha_i^s)$ is formed only by the Martian atmosphere. In modeling the first-order scattering, the expression for the observed Q value becomes

$$Q(\alpha_i^s, \lambda) = -2\pi P(\alpha_i^s, \lambda) A(\alpha_i^s, \lambda)$$

$$= \{Q_a(\alpha_i^s, \lambda)[1 - \beta(\lambda)] - 0.75\beta(\lambda)\sin^2 \alpha_i^s\}\omega(\lambda)f(\tau_0, \alpha_i^s). \qquad (3)$$

Here $Q_a(\alpha_i^s, \lambda)$ is the second Stokes parameter due to the aerosol particles, $A(\alpha_i^s, \lambda)$ is the visual planetary albedo, $\tau_0 = \tau_g + \tau_a$, τ_g and τ_a are the gas and aerosol components of the total optical thickness, respectively, $\beta = \tau_g/(\tau_g + \tau_a)$, and $\omega(\lambda)f(\tau_0, \alpha_i^s)$ is the reflectivity of the atmospheric column. Morozhenko [58] used this expression to analyze polarization measurements in the spectral interval 355–450 nm and at $\lambda = 225$ nm [45]. He used the following assumptions: 1) the value of the inversion angle for the underlying surface does not depend on wavelength; 2) the aerosol particles are nonabsorbing spheres; 3) the value of the real part of the particle refractive index is 1.5 and does not depend on wavelength; 4) the particle size distribution is described by the Young power law. From this analysis, he inferred the following values of the model parameters: the optical thickness of the gas is $\tau_g = 0.016$ and that of aerosol is $\tau_a = 0.048$ at the wavelength 355 nm; the value of the Young parameter is ~ 4.25. A more unambiguous estimate of the real part of the aerosol refractive index was obtained from an analysis of spectral polarization data during global dust storms. The calculations were based on the assumption that the dust layer was semi-infinite, homogeneous, and conservatively scattering; furthermore, the following approximate expression [58] was used in order to take account of multiple scattering:

$$Q_a(\alpha, \lambda) \le Q(\alpha, \lambda) \le 2Q_a(\alpha, \lambda). \qquad (4)$$

It was concluded that $n_r = 1.59 \pm 0.01$ and $r_0 \ge 8$ micrometers at $\sigma^2 = 0.1$ [18]. Violation of even one of the four above-mentioned assumptions leads to erroneous results [16, 61, 67].

Great attention was paid to the Martian atmosphere after obtaining observational data of solar luminosity and the azimuthal dependence of the Martian sky radiance (at low altitudes) from Vikings 1 and 2. The value of the optical thickness ($\tau_0 \sim 0.4$) and effective radius (>1.5 micrometeres) were obtained [9, 63] for the case of high air transparency of the Martian atmosphere. A model of an optically homogeneous and stable atmosphere was considered in these studies, but during that time morning and evening haze was observed on Mars, leading to errors [17, 52]. Analyses of observational data and retrievals of the optical thickness from solar luminosity measurements at different altitudes [9] has shown that the value of τ_0 may be overstated ($\tau_0 \sim 0.35$), along with the value of solar luminosity outside the atmosphere by 1.7 times [17].

It is important to note that in analyzing photometric measurements, it is necessary to use relevant constraints on the value of the complex refractive index and on the particle shape and size distribution. In the case of optically thin atmospheres, the intensity of the reflected and transmitted radiation is directly proportional to the product of the scattering phase function $\chi(\alpha)$, the single-scattering albedo $\omega(\lambda)$, and a function of the atmospheric optical thickness, $f[\tau_0(\lambda)] - \omega(\lambda)\,\chi(\lambda)\,f[\tau_0(\lambda)]$. Imperfect knowledge of one of these factors yields errors [17]. Spectral values of the complex refractive index are given in Table 2.

It is impossible to analyze the observational data without some simplifications, but it is necessary to reduce to a minimum their number, and it is necessary that their uncertainty have little effect on the precision of the estimated parameters. So, for example, the effect of the assumption that the inversion angle of the underlying surface is wavelength independent can be diminished considerably if one uses observational data in the far ultraviolet region of the spectrum, where the contribution of the atmosphere to $Q(\alpha, \lambda)$ is large.

Table 2: Spectral values of the complex refractive index

λ, nm	$m(\lambda)$	λ, nm	$m(\lambda)$	λ, nm	$m(\lambda)$
260	1.62–0.0025i	366	1.58–0.0010i	654	1.57–0.00014i
308	1.60–0.0013i	433	1.57–0.00071i	717	1.57–0.00010i
336	1.59–0.0013i	536	1.57–0.00038i		

It is possible not to assume that the particles are non-absorbing [16]. In this case it is necessary to 1) calculate the spectral values of the phase function using the estimated values of the real part of the refractive index and the estimated particle size distribution as a first approximation; 2) calculate the spectral values of the single-scattering albedo $\omega(\lambda)$ for dust particles and the imaginary part of the refractive index $n_i(\lambda)$ [17] by comparing the theoretical results with the values of the visual albedo observed during global dust storms; 3) use the estimates of $n_i(\lambda)$ at the second stage of the analysis of polarimetric observations. In this case, the value of r_0 becomes approximately 1.4 times smaller and the value of the optical thickness increases slightly.

The problem of particle shape is much more complex. Undoubtedly, the assumption of spherical particles is a great idealization, but so is the assumption of any specific shape. Actually, aerosols can be represented by ensembles of randomly oriented particles of different shape. Moreover, all available polarimetric and photometric observational data do not allow us to make an unambiguous conclusion about the particle shape. Therefore, it is important to estimate the influence of the assumption about particle shape on the retrieved values of atmospheric optical parameters. In [16] it was shown that using particles of spherical shape and randomly oriented elongated ellipsoids (with an axis ratio of 1/2) causes differences up to a factor of 2 in the estimated values of the aerosol particle radius and optical thickness. The modeling calculations have indicated that it may be possible to select the most reliable shape of particles. It was shown in [16] that the phase dependence of the visual albedo for an optically thick cloudy layer at $0° < \alpha < 60°$ depends on the shape of particles at stationary values of the real part of the refractive index and on the size distribution parameters. This means that if in this interval the retrieved particle albedo or absorption do not depend on phase angle then they most closely resemble those of the model calculations.

3.3.3. Jupiter

The values $n_r = 1.36 \pm 0.01$, $r_0 = 0.19 \pm 0.01$ micrometers, and $\sigma^2 = 0.28 \pm 0.02$ [47, 58] were retrieved from analysis of ground-based polarimetric observations in the spectral interval 316–800 nm using the approximate multiple-scattering approach described above. The results agreed well with data from Pioneer-10 at a phase angle of 103°.

Mishchenko [38] retrieved essentially the same values from analyses of the same ground-based observations while taking multiple scattering into account rigorously. By assuming a gamma size distribution, he found that $n_r = 1.39 \pm 0.01$, $r_{eff} = 0.39 \pm 0.08$ micrometers, and $v_{eff} = 0.45$. For the log normal size distribution assumed in [58], one may use the following formulas:

$$r_{eff} = r_0 \exp(2.5\sigma^2), \tag{5a}$$

$$v_{eff} = \exp(\sigma^2/2) - 1, \tag{5b}$$

$$r_0' \exp(\sigma'^2/2) = r_0'' \exp(\sigma''^2/2). \tag{5c}$$

Applying these formulas to the results of [58] yields $r_{eff} = 0.40$ micrometers and $v_{eff} = 0.35$, which is indeed very close to the results of [38].

Kawata and Hansen [26] obtained different results from an analysis of circular polarization. They obtained very large values of particle sizes for a constant value of the real part of the refractive index and concluded that the particle parameters retrieved in [58] are incorrect. However, there is no inconsistency since the results of analyses of observational data on both linear and circular polarization indicate a stratification of the effective particle size with height [57] in the Jovian atmosphere. The circular polarization in subauroral regions is formed by multiple scattering in deeper atmospheric layers (with $\tau_0 > 1$) [20], whereas the linear polarization is formed primarily in the upper atmospheric layers (with $\tau_0 < 1$). Therefore, the analyses of linear and circular polarization data provide information on the presence (or absence) of vertical stratification of parameters for a distribution of particles and allow us to determine the values of the physical properties of cloud particles.

The results of spectropolarimetric measurements in methane absorption bands at the center of the planetary disk allow us to determine, with simplicity and high accuracy, the methane concentration along the line of sight [33]:

$$NL = -[\mu_0/(1+\mu_0)k_v]\ln[P_v(\mu_0)/P_c(\mu_0)R_v(\mu_0)]. \tag{6}$$

Here $P_v(\mu_0)$ and $P_c(\mu_0)$ are the degree of polarization at the center of the absorption band and in continuum, respectively, $R_v(\mu_0)$ is the residual intensity at the center of the absorption band, k_v is the monochromatic absorption constant of methane in the center of the absorption band, and μ_0 is the cosine of the incidence angle [33]. Later in [54] it was shown that Eq. (6) can be used in analyses of spectral observational data obtained with a slit oriented along the equator. It was shown in [50] that the upper boundary of the cloud layer above the Equatorial Zone is approximately $0.8 \cdot H_g$ higher than above the Northern and Southern Equatorial Bands (H_g is a scale of heights for the atmospheric gas component). The advantage of this method is that it does not require the assumption of horizontal homogeneity of the optical atmospheric properties over the planetary disk. As linearly polarized light is formed at small depths in the cloud layer, the errors due to the vertical structure are minimized.

3.3.4. Saturn

Among all planets except the Earth, only polarization measurements of light reflected by the Saturn atmosphere have indicated the presence of oriented particles in high layers. However, these indications are unequivocal only from the measurements at wavelengths $\lambda < 500$ nm. Therefore, for the analysis of the spectral dependence of linear polarization at wavelengths > 500 nm and at a phase angle of $6.4°$, the authors of [6, 48] used the model of spherical particles. For $\sigma^2 = 0.12$, two solutions were obtained: $1.35 < n_r < 1.42$, $r_0 = 1.0 \pm 0.1$ micrometers and $n_r = 1.93$, $r_0 = 0.4$ micrometers. The first solution agrees better with observational data of the brightness distribution over the planetary disk.

Data analyses of circular polarization also yielded an ambiguous solution: $n_r = 1.44$ and $r_{eff} = 1.4$ micrometers, and $n_r = 1.60$ and $2 < r_{eff} < 5$ micrometers, the effective variance being $v_{eff} = 0.07$ for both solutions [25]. Recalculation of the size distribution for the first solution using a log normal distribution gives an average geometrical particle radius of about 1.2 micrometers. Therefore, nearly the same results were obtained independently of the Stokes parameters used in the analyses. This means that the vertical stratification of particle sizes in the Saturn atmosphere is weak.

Despite difficulties and ambiguous solutions, the spectropolarimetric method is the only way of determining the spectral values of the real part of the refractive index. In addition, the particle size distribution and the optical thickness of the gas layer overlying the cloud layer are estimated more accurately. Furthermore, it is reasonable to expect that polarimetric methods will be more successful in the study of the physical characteristics of surface layers for atmosphereless celestial bodies. For this purpose it is necessary to obtain high-precision polarization measurements with an error smaller than 0.05%.

4. Conclusions

In summary, we have highlighted the problems associated with polarimetric observations of solar system bodies that warrant serious attention: 1) The use of CCD matrices in astrophysical observations does not allow the measurement of polarization with accuracy better than 0.1–0.2%, whereas it is necessary to increase the accuracy of the observational data for the calculation of the linear and circular polarization parameters. 2) A series of interesting effects indicates the necessity to improve (to 0.1–1°) the accuracy of determining the orientation of the polarization plane, especially at phase angles near the inversion point and near zero. 3) It is necessary to measure the degree of linear polarization at phase angles close to zero, and also in contours of strong methane absorption bands at large phase angles in order to identify different photometric details along the central meridian of the giant planets. 4) It is necessary to study the spectral dependence of the inversion angle for atmosphereless celestial bodies. 5) Circular polarization measurements can be very informative for a) atmosphereless celestial bodies at phase angles $> 90°$, b) planets with dense atmospheres, especially in the polar regions, and c) cometary comas at all phase angles. 6) It is necessary to study polarization properties of stellar light during star occultations by a cometary coma, because the presence of oriented cometary particles can cause elliptically polarized light.

References

[1]   V. V. Avramchuk, in "Physics of The Moon and Planets" (Naukova Dumka, Kyiv, 1964).

[2]   L. A. Bugaenko, O. I. Bugaenko, A. L. Guralchuk, et al., in A. Morozhenko (Ed.), "Photometric and Polarimetric Studies of Celestial Bodies," p. 169 (Naukova Dumka, Kyiv, 1985).

[3]   L. A. Bugaenko, M. A. Melnikov, L. E. Rogozina and V. S. Samoilov, in A. Morozhenko (Ed.), "Photometric and Polarimetric Studies of Celestial Bodies," p. 164 (Naukova Dumka, Kyiv, 1985).

[4]   O. I. Bugaenko, L. S. Galkin and A. V. Morozhenko, Astron. Zh. **48**, No. 2 (1972).

[5]   O. I. Bugaenko and A. L. Guralchuk, in A. Morozhenko (Ed.), "Photometric and Polarimetric Studies of Celestial Bodies," p. 160 (Naukova Dumka, Kyiv, 1985).

[6]   O. I. Bugaenko, Zh. M. Dlugach, A. V. Morozhenko, and E. G. Yanovitskij, Solar Syst. Res. **9**, No. 1 (1975).

[7]   O. I. Bugaenko and A. V. Morozhenko, in A. Morozhenko (Ed.), "Photometric and Polarimetric Studies of Celestial Bodies," p. 108 (Naukova Dumka, Kyiv, 1985).

[8]   D. L. Coffeen, PhD thesis (University of Arizona Tucson, AZ, 1968).

[9]   D. S. Colburn, J. B. Pollack and R. M. Haberle, Icarus **79**, No. 1 (1989).

[10]   A. L. Guralchuk, N. N. Kiselev, and A. V. Morozhenko, Kinem. Phys. Celest. Bodies **3**, No. 1 (1987).

[11]   V. S. Degtiarev, L. O. Koklokolova and A. V. Morozhenko, Astron. Tsirk. 1545 (1990).

[12]   V. S. Degtiarev, L. O. Koklokolova, A. V. Morozhenko and V. F. Tsirul, Lett. Astron. Zh. **18**, No. 3 (1992).

[13]   A. Dollfus, Astron. Astrophys **213**, No. 1-1 (1989).

[14]   A. Dollfus, P. Bastien, and J.-F. le Borgne et al., Astron. Astrophys **206**, No 2 (1988).

[15]   A. Dollfus and E. Bowell, Astron. Astrophys. **10**, No. 1 (1971).

[16]   Zh. M. Dlugach, M. I. Mishchenko and A. V. Morozhenko, Kinem. Phys. Celest. Bodies **18**, No. 1, 33–42 (2002).

[17]   Zh. M. Dlugach and A. V. Morozhenko, Kinem. Phys. Celest. Bodies **16**, No. 5 (2000).

[18]   A. Dolfus, Zh. M. Dlugach, A. V. Morozhenko and E. G. Yanovitskij, Solar Syst. Res. **8**, No. 4 (1974).

[19]   J. S. Hall and L. A. Riley, J. Atmos. Sci. **26**, No. 5 (1969).

[20]   J. E. Hansen, J. Atmos. Sci. **28**, No. 5 (1971).

[21]   J. E. Hansen and A. Arking, Science **171**, No. 3972 (1971).

[22]   J. E. Hansen and J. W. Hovenier, J. Atmos. Sci. **31**, No. 4 (1974).

[23]   G. M. Kattawar, G. N. Plass and C. N. Adams, Astrophys. J. **170**, No. 2 (1971).

[24]   K. Kawabata, D. L. Coffeen, J. E. Hansen, et al., J. Geophys. Res. **85**, No. A13 (1980).

[25]   I. Kawata, Icarus **33**, No. 1 (1978).

[26]   I. Kawata and J. E. Hansen, in T. Gehrels (Ed.), "Jupiter II" (Mir, Moscow, 1979).

[27]   J. S. Kemp, R. D. Wolstencroft and J. B. Swedlund, Nature **232**, No. 5307 (1971).

[28]   C. E. Kenknight, D. L. Rosenberg and G. K. Vehner, J. Geophys. Res. **72**, No. 12 (1967).

[29]   W. J. J. Knibble, J. F. de Haan and J. W. Hovenier, J. Geophys. Res. **10**, No. E4 (1998).

[30]   R. G. Knollenberg and D. M. Hunten, J. Geophys. Res. **85**, No. A13 (1980).

[31]   E. K. Kohan, Izvestiya GAO AN USSR, **5**, No. 17 (1964).

[32]   V. A. Kucherov and V. S. Samoilov, Opt. Mech. Industry No. 9, (1987).

[33]   V. A. Kucherov, M. I. Mishchenko and A. V. Morozhenko, Lett. Astron. Zh. **14**, 835–839 (1990).

[34]   O. I. Kvartshelia, Bull. Abastumani Astrophys. Observ. **64** (1988).

[35]   D. F. Lupishko, S. V. Vasilyev, Yu. S. Yefimov and N. M. Shahovskoj Icarus **119**, No. 1 (1995).

[36]   B. Lyot, Ann. Observ. Meudon **8** (1929).

[37]   J. J. Michalsky and R. A. Stokes, Publ. Astron. Soc. Pacific **86**, No. 514 (1977).

[38]   M. I. Mishchenko, Icarus **84**, 296–304 (1990).

[39]   M. I. Mishchenko, Astrophys. J. **411**, 351–361 (1993).

[40]   M. I. Mishchenko, J.-M. Luck and T. M. Nieuwenhuizen, J. Opt. Soc. Am. A. **17**, 888–891 (2000).

[41]   A. V. Morozhenko, in "Physics of The Moon and planets" (Naukova Dumka, Kyiv, 1966).

[42]   A. V. Morozhenko, Astron. Zh. **46**, No. 5 (1969).

[43]   A. V. Morozhenko, Astron. Zh. **50**, No. 1 (1973).

[44]   A. V. Morozhenko, Astron. Zh., **50**, No. 5 (1973).

[45]   A. V. Morozhenko, Solar Syst. Res. **8**, No. 3 (1974).

[46]   A. V. Morozhenko, Astrometry Astrophys. **30** (1975).

[47]   A. V. Morozhenko, Astrometry Astrophys. **31** (1976).

[48]   A. V. Morozhenko, Astrometry Astrophys. **33** (1977).

[49]   A. V. Morozhenko, in "Proc. Int. Conf. Cometary Explor." (Budapest, 1982).

[50]   A. V. Morozhenko, Astronomer. Bull. **24**, No. 3 (1990).

[51]   A. V. Morozhenko, Kinem. Phys. Celest. Bodies **8**, No. 4 (1992).

[52]   A. V. Morozhenko, Solar Syst. Res. **26**, No. 1 (1992).
[53]   A. V. Morozhenko, Kinem. Phys. Celest. Bodies **16**, No. 4 (2000).
[54]   A. V. Morozhenko, Kinem. Phys. Celest. Bodies **17**, No. 1 (2001).
[55]   A. V. Morozhenko, Kinem. Phys. Celest. Bodies **18**, No. 3 (2002).
[56]   A. V. Morozhenko, A. L. Guralchuk and N. N. Kiselev, Kinem. Phys. Celest. Bodies **3**, No. 2 (1987).
[57]   A. V. Morozhenko, A. S. Ovsak and P. P. Korsun, Kinem. Phys. Celest. Bodies **11**, No. 4 (1995).
[58]   A. V. Morozhenko and E. G. Yanovitskij, Icarus **18**, No. 4 (1973).
[59]   A. V. Morozhenko and E. G. Yanovitskij, in "50 Years of The Main Astronomical Observatory" (Naukova Dumka, Kyiv, 1994).
[60]   T. Mukai, S. Mukai and S. Kikuchi, Astron. Astrophys **187**, Nos. 1–2 (1987).
[61]   E. V. Petrova, J. Quant. Spectrosc. Radiat. Transfer **63**, Nos. 2–6 (1999).
[62]   E. V. Petrova, K. Jockers and N. N. Kiselev, Icarus **148**, No. 3 (2000).
[63]   J. B. Pollack, D. S. Colburn, R. Kahn, *et al*., J. Geophys. Res. **82**, No. 28 (1977).
[64]   V. K. Rosenbush, V. V. Avramchuk, A. E. Rosenbush and M. I. Mishchenko, Astrophys. J. **487**, 402–414 (1997).
[65]   G. V. Rozenberg, "Collected Papers in Memory of P. P. Lasarev" (1956).
[66]   V. V. Sobolev, Astron. Zh. **45**, No. 1 (1968).
[67]   V. P. Tishkovets and Yu. G. Shkuratov, Astron. Zh. **59**, No. 5 (1982).
[68]   J. Veverka and W. Liller, Icarus **10** (1969).
[69]   A. P. Vid'machenko, Kinem. Phys. Celest. Bodies **5**, No. 4 (1989).

Anatoliy Vid'machenko   *Alexander Morozhenko*

DISK-INTEGRATED POLARIMETRY OF MERCURY IN 2000-2002

DMITRIJ LUPISHKO AND NIKOLAI KISELEV
Institute of Astronomy of Kharkiv National University, 35 Sumska str., 61022 Kharkiv, Ukraine
lupishko@astron.kharkov.ua

Abstract. The orbital and rotation characteristics of Mercury lead to distinctly different solar irradiation intensities (photons, solar wind and cosmic rays) and radiative heating of the Mercury hemispheres with central meridians 0°, 180° on the one hand and 90°, 270° on the other. These differences allow us to hypothesize that the intensity of Mercury regolith maturation may depend noticeably on the planetocentric longitude, which would result in a corresponding variation of optical properties over the surface. The purpose of this study is to explore the variation of polarization over the Mercury surface. Polarimetric observations of Mercury were carried out during its three apparitions in 2000-2002 with a 70-cm reflector (Cassegrain configuration, f/16) and a single-channel photoelectric polarimeter. The polarization phase curve of Mercury obtained in the phase-angle interval of 39.1° → 135.5° shows measurements that approximate a curve with a maximal polarization difference $\Delta P = 1\%$, which is about one order of magnitude larger than the measurement accuracy. The dependence of differences between measured and approximated polarization ΔP on the planetocentric longitude of the center of the illuminated part of the visible disk revealed polarization variations with amplitude of approximately 1.5% over the range of observed planetocentric longitudes 265° → 330°. In order to obtain such variations in the whole interval of longitudes and to clarify their nature, the authors intend to carry out new observations of Mercury during its nearest apparitions.

Keywords: Mercury, polarimetry, observations, phase dependence, longitude variations of polarization

385

G. Videen et al. (eds.), Photopolarimetry in Remote Sensing, 385-392.
© 2004 *Kluwer Academic Publishers. Printed in the Netherlands.*

1.   Introduction

Mercury is a peculiar planet of the Solar system from many points of view. It is the smallest planet of the terrestrial group (D = 4878 km), it is also the closest to the Sun (a = 0,387 AU), and it has the most elongated (e = 0.205) and inclined (i = 7°) orbit among the terrestrial and giant planets. Mercury's rotation period (58.646^d) and its orbital period (87.969^d) are in exact commensurability 2:3. As a consequence, the solar day on Mercury's surface lasts two Mercurian years (176^d). The second important consequence is that at perihelion and aphelion Mercury always faces the Sun alternately by both hemispheres. In particular, the 0° and 180° longitude meridians always alternately face the Sun at perihelion, and the subsolar points of these meridians receive the greatest surface heating (they are the so called "hot poles" of Mercury). In addition, the 90° and 270° meridians always face the Sun at aphelion and their subsolar points receive less heating (they are called "warm poles"). Due to the close proximity to the Sun, the lack of a significantly dense atmosphere and relatively long day and night, Mercury experiences the largest temperature differences on its surface of all the planets and satellites. During the day the temperature can rise to 425°C and at night it can fall down to −183°C. Such periodic temperature variations of the surface can affect its structure.

Besides traditional ground-based observations of Mercury in optical wavelengths, radiometric and radar observations, a large amount of qualitatively new information has been obtained by the space mission Mariner-10 during its three approaches to the planet in 1974-1975. These are the data on the magnetic field of Mercury [1,2], on the surface structure [3-5], on the thin Mercurian atmosphere [6,7], on the possible presence of water ice (or sulfur deposits) in the polar regions [8,9] and so on. Nevertheless, among the terrestrial planets Mercury remains the least studied planet. At the same time this planet is of great interest primarily because, together with asteroids, Mercury can be viewed as the best conserved relic of the stage of large planet formation [10]. However, Mercury's regolith can be the most weathered due to irradiation by solar high-energy photons and charged particles. Peculiarities of Mercury's composition and interior, its magnetic field, surface relief and so on are undoubtedly linked with the features of the processes of planet formation and with its evolution during the post-accretion period of the late heavy bombardment which took place about 3.9 MY ago. That means that the study of Mercury can give us essential information on the origin and evolution of other planets, as well as of the Solar system as a whole. This is one of the main reasons for planning and preparation of the next space missions to Mercury – Messenger mission (NASA project for 2004) and BepiColombo mission (ESA project for 2009).

2.   Properties of Mercury regolith from previous observations

Most of the information on the presence and properties of Mercury's regolith was obtained from ground-based polarimetric and radiometric observations. The first polarimetric observations of Mercury were undertaken by Lyot [11] at Meudon Observatory (France) in 1922-24. In 1930 Lyot carried out observations of Mercury at Pic-du-Midi Observatory during daylight hours and obtained the polarization degree at small phase angles $\alpha = 5° \rightarrow 22°$ [12]. Comparing the properties of the Moon and Mercury with laboratory samples, Lyot suggested that their surfaces are covered by very porous material that structurally resemble volcanic ashes.

A. Dollfus continued polarimetric observations of Mercury in 1950 at Pic-du-Mi (France) with a Lyot-type visual polarimeter [13] and in 1966 and 1972 with a new photoelectric polarimeter [14]. The disk-integrated polarization at six wavelengths between 0.350 and 0.630 µm over a large interval of phase angles $\alpha \approx 4° \rightarrow 130°$, which includes the maximal negative ($\alpha_{min} \approx 10°$) and maximal positive ($\alpha_{max} \approx 110°$), was obtained and analyzed using Lyot's data and Dollfus' additional observations [14]. The principal result of this analysis was the following: the polarization phase dependence of Mercury shows a considerable similarity with that of the Moon (Table 1), thus confirming the work of Lyot that suggested a lunar-like regolith of Mercury consisting of dark and absorbing material altered by micrometeoritic bombardment.

Table 1. Mean parameters of polarization-phase curves of Mercury and Moon (λ550 mµ) [15].

Object	P_{min}, %	Polarimetric slope h	α_{inv}, deg	P_{max}, %
Mercury	1.4±0.1	0.147±0.002	25±2	8.2
Moon	1.2±0.0	0.148±0.005	23.6±0.5	8.6

But in spite of the considerable similarity of the optical properties of the Moon and Mercury regoliths, there should also exist essential differences between them due to the differences in conditions of forming and subsequent evolution (alteration) of their surfaces. These conditions are as follows:
1. Crater formation.
2. Micrometeorite bombardment.
3. Cosmic rays and solar wind.
4. Radiation heating of the surface.
It is worth examining the 3rd and 4th items in more detail.

Cosmic rays and solar wind. Two factors play a significant role in determining Mercury's surface morphology: the orbital resonance 2:3 and the screening by the Mercury magnetosphere. The result of the first factor is that the irradiation by the Sun (photons, cosmic rays and solar wind) of the surface element greatly depends on its longitude and is 2.3 times higher at perihelion (that is, at the Mercury 0° and 180° meridians) than at aphelion (90° and 270°). Because of the magnetospheric screening, it does not seem possible to estimate the strength of the solar wind flux reaching Mercury's surface. This flux results in the amorphization of an approximately 500-micron deep layer of outer particles of the surface that are directly irradiated. In the mature lunar regolith, up to 50% of the particles show such amorphous coatings, which decrease the albedo by up to 30% [16]. If the screening by Mercury's magnetosphere is absent, then we could expect that most particles of the Mercury regolith would be covered by such coatings because the strength of the solar wind flux would be one order of magnitude greater than on the Moon, and the Mercury surface albedo would be less than the Moon albedo. However this is not observed in reality. This can be explained by the fact that saturation occurs after some period of time, and the upper layer cannot be changed. This effect can take place for the young craters of the same age on the Moon and on Mercury.

Radiation heating of the surface. The irregular heating of the surface and the large night and day temperature differences result in accumulated radiation effects in the uppermost surface layers that include agglomeration of particles, losses of volatiles, rock annealing, fracturing, etc. The result of this can be surface structure differences between the hemispheres with central meridians 0° and 180° and the hemispheres with central meridians 90° and 270°.

But what do the observations tell about the possible global variations of the optical properties of the Mercury regoliths or surface microstructure? It might be significant that mapping the Mercury surface in 1995-1999 by the Swedish solar vacuum telescope (D = 0.5 m, La Palma) did not reveal any noticeable photometric differences between the hemispheres [17].

However, recently published results of CCD-photometry of Mercury obtained by using the large angle spectrometric coronograph of the SOHO-spacecraft together with ground-based observations [18] show brightness variations as a function of Mercurian longitude with an amplitude equal to about 0.05 mag. The maximum and minimum cannot be determined exactly, but it cannot be ruled out that they are near 120° and 300°, respectively. This effect is not large; the brightness difference is only about 5%. If this difference is the result of variations in the surface albedo, then differences in the surface polarization can be approximately 0.5%; that is, large enough to be detected.

Gehrels *et al.* [19] presented results of polarimetric observations of Mercury at $\alpha = 53° \rightarrow 130°$ and in the spectral range 0.340 - 0.960 μm obtained in 1964-70 using different telescopes but the same polarimeter. They reveal a time variation of the polarization with an amplitude of about $\Delta P = 0.7\%$. However, the minimal and maximal polarizations are distant from each other by only 40° of longitude. The authors interpret this as a longitudinal dependence of polarization caused by variations of composition and/or surface structure.

Taking all observations together, there appears to be a dependence of surface albedo and polarization degree on the central meridian longitude of Mercury. The aim of this work is to characterize the value and variations seen.

3.   Observations

Polarimetric observations of Mercury were carried out during its three apparitions in 2000-2002 at the Observation Station of the Institute of Astronomy of Kharkiv National University (75 km from Kharkiv city) with a 70-cm reflector (Cassegrain configuration, f/16) and a single-channel photoelectric photometer-polarimeter. Polarimetric measurements are based on a fast polaroid rotation (~33 *rev/sec*) and a quasi-simultaneous signal registration in the four channels corresponding to the polaroid positions of 0°, 45°, 90° and 135°. Such a method allows us to carry out polarimetric observations under the condition of large and rapidly changing air masses and under poor photometric conditions. The instrumental polarization and zero-point of the position angles of the polarization plane are determined by observations of standard stars with zero and large polarization degrees during each night. The parameters of instrumental polarization in the V-band were rather small and equal to $u_i = 0.015 \pm 0.025\%$ and $q_i = 0.130 \pm 0.031\%$. Because of the limited time of these observations, most of the measurements were obtained in one visual band of wavelengths (narrow-band filter GC 5256/57Å or in the standard V-band). Since the observations of Mercury were carried out at large air masses and therefore, relatively poor conditions, the focal diaphragm of 33 arcsec was used. The duration of the single exposure was equal to 15 *sec*.

4.   Results of observations

The results of observations are presented in Table 2, which contains the date of the observations, spectral band (BC – 4450/67Å, GC – 5256/57Å, RC – 7129/62Å, V – the standard band of UBV-system), phase angle, measured values of polarization degree, position angle, position angle of scattering plane χ and planetocentric longitude L of the center of the illuminated and visible part of Mercurian disk.

As one can see from Table 2, the polarization phase dependence of Mercury is obtained in the phase angle interval of $39.1° \rightarrow 135.5°$. These observations plotted in Figure 1 show the polarization measurements to be within $\Delta P=1\%$ of the curve approximating the polarization. This is about one order of magnitude larger than the accuracy of the measurements. These differences are significant and most likely are due to longitudinal variations of planet polarization. The differences ΔP between the measured and the approximated polarization degrees are plotted as a function of the planetocentric longitude of the center of the illuminated part of the visible disk in Figure 2. This figure shows polarization variations with amplitude of approximately 1.5% over the range of observed planetocentric longitudes 265° - 330°. Unfortunately, studies of the Mercurian surface are hampered by the sparsity of data currently in existence. While this dataset demonstrates a planetocentric longitudinal dependence, demonstrating that they might be useful in mapping surface characteristics, what is clearly needed is additional data, which the authors will be acquiring.

Table 2. The results of polarimetric measurements

Date, UT	Spectr band	P. A. Deg	$P \pm \sigma$ %	$\theta \pm \sigma$ deg	χ deg	L deg
2000, July 29.055	GC	97.0	7.65±0.07	179.5±0.6	272.3	292
30.053	BC	93.1	8.92±0.10	3.7±0.5	272.3	295
30.067	GC	93.0	7.72±0.06	5.5±0.5	273.4	295
August 01.054	RC	85.0	6.62±0.05	5.2±0.4	275.8	300
01.070	GC	84.9	7.52±0.17	6.6±0.8	275.8	300
07.056	RC	59.0	4.19±0.03	14.2±0.4	282.2	315
07.067	GC	59.0	5.06±0.03	15.6±0.5	282.2	315
08.063	RC	54.5	4.03±0.04	15.4±0.5	285.1	318
08.074	GC	54.5	3.94±0.05	14.4±0.6	285.1	318
09.066	RC	50.1	3.44±0.04	17.2±0.5	286.7	320
09.077	GC	50.0	4.05±0.06	22.5±0.6	286.7	320
10.065	RC	45.7	2.64±0.03	20.8±0.5	288.3	323
10.075	GC	45.7	3.42±0.04	24.9±0.6	288.3	323
2001, July 07.049	GC	113.6	7.03±0.23	170.0±0.7	283.4	274
09.051	GC	107.4	8.13±0.18	183.4±0.5	281.5	279
16.046	GC	83.8	7.05±0.08	177.0±0.3	273.6	298
17.047	GC	80.2	7.28±0.06	178.6±0.3	272.3	301
18.048	GC	76.4	6.02±0.06	0.0±0.3	269.7	304
20.049	GC	68.6	5.10±0.06	4.3±0.3	268.5	310
21.051	GC	64.5	4.86±0.06	6.3±0.3	267.4	312
26.065	GC	43.4	3.86±0.08	27.3±0.5	260.3	326
27.068	GC	39.1	2.31±0.14	29.9±1.2	258.5	330
2002, May 06.731	V	113.2	6.32±0.03	160.8±0.1	73.2	265
08.735	V	119.6	5.99±0.03	161.8±0.2	73.9	272
09.732	V	122.7	6.05±0.03	162.8±0.2	74.2	275
10.733	V	125.9	6.05±0.07	162.3±0.3	74.6	279
11.730	V	129.1	5.44±0.06	161.4±0.3	74.9	282
12.727	V	132.3	5.31±0.09	160.6±0.5	75.3	286
13.735	V	135.5	5.23±0.08	161.5±0.4	75.7	290

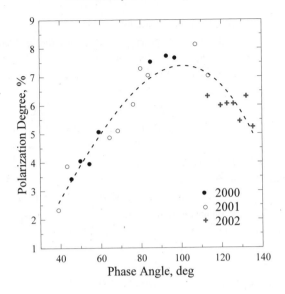

Fig. 1. Polarization-phase dependence of Mercury.

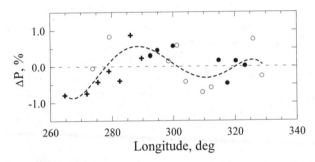

Fig. 2. Polarization as a function of longitude (symbols are the same as in Fig. 1).

5.   Acknowledgments

The authors thank Gorden Videen for his helpful comments and text editing. This work was supported by INTAS grant No. 99-403.

References

1.   N. F. Ness, K. W. Behannon, R. P. Lepping, and Y. C. Whang, J. Geophys Research 80, 2708 (1975).

2.   D. J. Jackson, and D. B. Beard, J. Geophys Research 82, 2828 (1977).

3.   B. Murray, M. Belton, G. Danielson, M. Davies, D. Gault, B. Hapke, B. O'Leary, R.

Strom, V. Suomi, and N. Trask, Science 185, 170 (1974).

4.   R. J. Pike, in F. Vilas, C. R. Chapman, M. S. Matthews, (Eds.), «Mercury», p. 165, (University of Arizona Press, Tucson, AZ, 1988).

5.   P. H. Schultz, in F. Vilas, C. R. Chapman, M. S. Matthews, (Eds.), «Mercury», p. 274, (University of Arizona Press, Tucson, AZ, 1988).

6.   D.M. Hunten, T.H. Morgan, and D. E. Shemansky, in F. Vilas, C. R. Chapman, M. S. Matthews, (Eds.), «Mercury», p. 562, (University of Arizona Press, Tucson, AZ, 1988).

7.   D. M. Hunten, and A. L. Sprague, Adv. Space Research 19, 1551 (1997).

8.   M.A. Slade, B.J. Butler, and D. O. Muhleman, Science 258, 635 (1992).

9.   B. Butler, D. Muhleman, and M. Slade, J. Geophys. Research 98, 15,003 (1993).

10. M. Ya. Marov, Solar System Planets, Moscow: Nauka, 99 (1986) (Russian).

11. B. Lyot, Ann Observatoire Meudon VIII, These. Paris, fasc.1, (1929).

12. B. Lyot, Comptes Rendus Acad. Science, 20 Oct., (1930).

       A.   Dollfus, L'Astronomie 67, 61 (1953).

13. Dollfus, and M. Auriete, Icarus 23, 465 (1974).

14. E. Bowell, and B. Zellner, in T. Gehrels (Ed.), «Planets Stars and Nebulae Studied with Photopolarimetry», p. 381, (University of Arizona Press, Tucson, AZ, 1974).

15. Y. Langevin, Planet. Space Science 45, 31 (1997).

16. J. Warell, and S. S. Limaye, Planet. Space Science 49, No. 14-15 Dec., 1531 (2001).

       A.   Mallama, D. Wang, and R. Howard, Icarus 155, 253 (2002).

17. T. Gehrels, R. Landau, and G. V. Coyne, Icarus 71, 386 (1987).

Nikolai Kiselev and Dmitrij Lupishko at Kiev workshop.

POLARIMETRY OF DUST IN THE SOLAR SYSTEM: REMOTE OBSERVATIONS, IN-SITU MEASUREMENTS AND EXPERIMENTAL SIMULATIONS

A. CHANTAL LEVASSEUR-REGOURD

Université Paris VI / Aéronomie CNRS-IPSL
BP 3 91371 Verrières, France

Abstract. With only a few *in-situ* studies and sample returns, information about the physical properties of dust in the solar system is obtained mostly through remote light-scattering observations. Amongst them, polarimetric observations are of special interest, since they only depend upon the physical properties at a given phase angle and wavelength, and do not vary with the changing dust concentration and geometry of the observations. We first present results on the polarimetry of dust in the solar system, with emphasis on cometary dust (for which numerous observations, leading to a better knowledge of the proto-solar dust, are already available). The spatial changes in the polarization are analyzed, together with the phase curves (which all present the same trend that is typical of irregular particles) and the wavelength dependence of the polarization. Such observations need to be interpreted through numerical or experimental simulations, the uniqueness of a proposed solution being always a difficult question. We thus present a modular series of laboratory experiments, which have been initiated in the mid-nineties and should give us, with the ICAPS multi-users facility on board the ISS, the opportunity of building particles and aggregates under conditions representative of those which had prevailed in the solar system. The analysis of light-scattering properties of such samples should provide the missing links between optical properties and physical properties (e.g. morphology, size distribution, albedo, porosity) of cosmic dust particles.

1. Dust in the solar system

Dust is almost everywhere in the solar system, in cometary comae and tails (dust released by solar heating from comets nuclei), on small atmosphere-less bodies (successive impacts, with electrostatic charging building up a regolith),

G. Videen et al. (eds.), Photopolarimetry in Remote Sensing, 393-410.
© 2004 *Kluwer Academic Publishers. Printed in the Netherlands.*

in the interplanetary dust cloud, in planetary atmospheres and in circumplanetary rings. Cometary dust particles, likely to be formed by aggregation of interstellar grains, represent material left from the protosolar nebula that could have played a key role in bringing organic compounds to telluric planets. The interplanetary dust cloud is a complex of dust coming from comets, from asteroidal collisions and, to a lesser extent, from the interstellar medium. The cosmic dust thus represents some remnant material from the proto-solar nebula, and from bodies at different stages in solar system evolution.

After more than 40 years of space exploration, *in-situ* studies of solar system dust are restricted to collections of particles in the near-Earth environment, to impact studies on space probes, and to a few missions to the Moon, planets and smaller bodies. Collection of interplanetary dust particles in the stratosphere (so-called IDPs) provides information on their composition and morphology; it is of interest to point out that particles of probably cometary origin - although modified by high velocity encounter with the Earth's atmosphere - are aggregates of submicron-sized grains in a size range of a few tens of micrometers. Impact studies mostly provide information on the concentration, direction and energy of particles encountered along the space-probe trajectory. As far as samples are concerned, hundreds of kilograms of lunar dust and rocks have been brought to Earth, and a few milligrams of cometary dust retrieved during the flyby of comet 81P/Wild 2 should arrive in early 2006. Amongst the space-mission discoveries, the diversity in chemical composition of cometary dust in comet 1P/Halley coma may be mentioned [1], together with the existence of regolith on low-gravity bodies, such as asteroid 433 Eros [2].

While only a few *in-situ* observations of dust are available, numerous remote observations of the above-mentioned bodies have been performed. Observation of emitted light, mostly in the infrared for solar system dust, provides information about temperature and composition, with the silicate emission feature near 11 μm. Observation of solar light scattered by dust, mostly in the optical and near infrared domains, provides clues to dust physical properties.

2.   Advantages of polarimetric observations

While solar-system objects orbit the Sun or a Sun-orbiting planet, their geometry relative to the observer (and thus the light they scatter) varies drastically as a function of time. The geometry of the observations (Figure 1) actually depends upon the distance R to the Sun, the distance Δ to the observer and the phase angle α (or the scattering angle θ equal to $\pi - \alpha$). Since the incident solar light is unpolarized, the light emerging from low concentration media (i.e. cometary comae, interplanetary dust cloud, aerosols clouds) is partially linearly polarized. The intensity I is then the sum of the polarized and unpolarized intensities; it is also the sum of the components perpendicular and parallel to the scattering plane (defined by the Sun, the scattering dust and the observer). The degree of linear polarization P, hereafter called polarization, is the ratio of the difference to the sum of these two components. It has the opposite sign, in comparison with the ratio of the second to the first component of the Stokes vector. Basics on polarization by scattering can be found in e.g. Gehrels [3], Bohren and Huffman [4], and Hapke [5].

$$I = I_{polarized} + I_{unpolarized} = (I_\perp + I_{//})$$

$$P = (I_\perp - I_{//}) / I = (I_\perp - I_{//}) / (I_\perp + I_{//}).$$

Because of these definitions, the polarization only depends upon the physical properties of the scattering particles, the phase angle α, and the wavelength λ. On the contrary, the intensity also depends upon the illumination, the distance R to the Sun, the distance Δ to the observer and the dust concentration. Indeed, polarization measurements are technically more difficult than intensity measurements; also, they require a longer integration time for a satisfactory signal to noise ratio. However, because P is a dimensionless ratio, it immediately allows a comparison of data from the same object at different times, or of different objects. Polarimetric phase curves are then retrieved, without needing to cope with light curves corrections, and to disentangle changes in dust physical properties from changes in dust concentration.

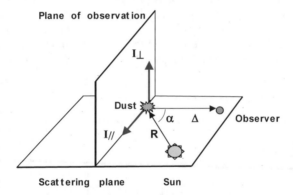

Figure 1. Geometry of solar-system observations, for the solar light scattered by dust under phase angle α at a distance R to the Sun and Δ to the observer.

The next part of this chapter provides a summary of our present understanding of the polarization properties of dust in the solar system. It is mostly illustrated with cometary dust data, although the polarization properties of asteroidal regoliths, of the interplanetary dust cloud and of Mars aerosols are also presented. The last part of the chapter provides a discussion of the need for experimental simulations to interpret the observations in terms of physical properties of the dust particles, e.g. morphology, size distribution, albedo, porosity. It then gives an overview of microgravity experiments already performed or under development; such experiments avoid sedimentation and multiple scattering on gravity packed layers, and should allow light-scattering measurements to be performed on dust particles and aggregates representative of solid particles in the proto-solar nebula.

3. Polarization properties of dust in the solar system

The polarization of the light scattered by lunar regolith has been studied as early as in 1929 by Lyot [6]. Data show significant spatial changes between the first and second quarter of the Moon and allow the study of smooth phase curves: a small negative branch appears near backscattering, followed by a large positive branch with a maximum near 90°. More recently, numerous data sets obtained from images and point observations allowed Shkuratov *et al.* [7, 8] to document accurately spatial changes in the polarization, to characterize the phase curves, and to point out a common origin of the backscattering phenomena for brightness and polarization. Spatial changes and phase-angle dependence are actually, together with the wavelength dependence, the main characteristics of the polarization properties of cosmic dust, which we now review.

3.1 Spatial changes in polarization of comets

Numerous data on the linear polarization in cometary comae are available, corresponding either to different regions of a given coma or to average values on a projected coma (see e.g. [9, 10] for reviews). The plane of polarization is almost always perpendicular or parallel to the scattering plane. The circular polarization (possibly produced by non-spherical aligned particles) is significantly below 2%, as estimated for 1P/Halley and C/1996 O1 Hale-Bopp. Spatial changes have been documented through local and remote observations.

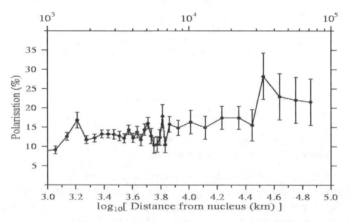

Figure 2. Variation of the polarization in comet 1P/Halley coma, as measured at 73° phase angle in the blue domain along the trajectory of Giotto spacecraft, for distances to the nucleus in the 1000-80000 km range. Local changes could be attributed to the crossing of swarms of particles with different scattering properties (from [11]).

A spatial change in polarization was first noticed with the Optical Probe Experiment on board Giotto during its 1986 flyby of 1P/Halley. At a constant phase angle (equal to 73°) and for a given wavelength along the trajectory of the spacecraft from the outer coma to the innermost coma, the polarization was found to present some local increases and a general trend to a decrease in the inner coma (Figure 2). Some of these local changes, also visible in the intensity data and in the flux recorded by the dust impact detector, could be attributed to swarms of particles with different scattering properties [11]. The use of a dust dynamical model to fit the optical probe and dust impact detector experiments allows the determination of the bulk density to albedo ratio. The best fit is obtained for particles with an albedo of 0.04, in excellent agreement with [12] and a density of 100 kg m^{-3}, i.e. for extremely absorbing and porous (or fluffy) particles [13]. We have, in the early nineties, developed a program of imaging polarimetry of bright distant comets to study extensively such changes in polarization in cometary comae.

The use of CCDs, initiated by Eaton *et al.* [14] with 1P/Halley, has allowed major progress in remote imaging polarimetry. Various teams are now performing CCD cometary polarimetry, either on dedicated instruments, or on large instruments in polarimetric mode. The techniques are typical of imaging polarimetry on extended sources [see e.g. 15-19]. Sequences of four polarized brightness images, corresponding to four orientations of the fast axis of the polarizer at 45° from one image to the next one, are usually required to obtain a polarization image. Calling I_i, I_{i+1}, I_{i+2} and I_{i+3} the corresponding components of the intensity and χ the angle between the direction of polarization and the direction of the fast axis of filter (i) in the plane of observation:

$$I = I_i + I_{i+2} = I_{i+1} + I_{i+3}$$

$$|P| = 2\ [(I_i - I_{i+2})^2 + (I_{i+1} - I_{i+3})^2]^{1/2}\ /\ [I_i + I_{i+1} + I_{i+2} + I_{i+3}]$$

$$\chi = 0.5\ \arctan\ [(I_{i+1} - I_{i+3})\ /\ [(I_i - I_{i+2})].$$

Observations are made through spectral filters, narrow enough to avoid contamination by cometary spectral lines, and spectral data obtained at the same time are used to verify this assumption. Reference stars are used for calibration; the sky background polarization and the alignment of the four images need to be taken into account in the data reduction.

The polarization images may reveal elongated fan-shaped structures where the polarization is higher than the average. They generally correspond to bright jet-like features on the brightness images. This effect has been noticed for quite a few so-called active comets, the case of the C/1995 O1 Hale-Bopp being the most documented [see e.g. 17, 18, 20]. Such features could be produced by freshly ejected dust particles with different physical properties. As first noticed for 1P/Halley [11, 21], the images (as well as polarimetry with a variable aperture) also reveal a region corresponding to a lower polarization in the innermost coma. This depolarization, noticed on numerous comets for distances to the nucleus below about 2000 km (see e.g. [22, 23]), could be attributed to dust particles with different physical properties (e.g. larger particles hovering around the nucleus, particles with volatile mantles). It nevertheless may be pointed out that the polarization images of a few comets do not show this lower polarization. It could possibly not always exist or not be detectable, either because of a low spatial resolution or because of the superimposition of highly polarized dust jets. Figures 3a and 3b provide examples of polarization images for the active comet C/1996 O1 Hale-Bopp and for the less-active comet 81P/Wild 2.

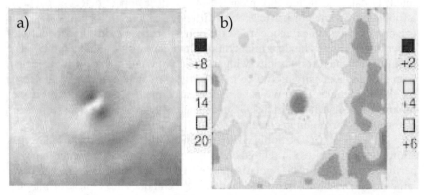

Figure 3. Polarimetric images of cometary coma in the red domain shown in percents from [19, 20]: a) 1999 O1 Hale-Bopp, field-of-view ≈ 80000 km, $\alpha = 44°$; b) 81P/Wild 2, field-of-view ≈ 25000 km, $\alpha = 37°$.

Finally, imaging polarimetry and aperture polarimetry (with a variation on the location of its center on the coma) may be used to study the variation with the distance to the nucleus along a given axis. Besides the above mentioned jets and halos, some differences are visible between the sunward and anti-sunward sides, and depending on the object, the polarization may increase (as illustrated with 1P/Halley, [11]) or decrease (as illustrated for C/1996 B2 Hyakutake, [24]) with the distance to the nucleus. Such variations indicate a temporal evolution of the physical properties of the particles ejected from the nucleus. More drastic temporal changes also can be observed after an outburst. These events, which usually take place while some fragments of the nucleus are released or before a complete disruption takes place (as illustrated for C/1999 S4 LINEAR), are immediately followed by an increase in polarization of 2 to 3 percents [see e.g. [21, 25], which indicates the presence of dust particles with different properties.

3.2 Phase angle and wavelength dependence of cometary polarization

While only few *in-situ* observations of dust are available, numerous remote observations of the above-mentioned bodies have been performed. Observation of emitted light, mostly in the infrared for solar system dust, provides information about temperature and composition, with the silicate emission feature near 11 μm. Observation of solar light scattered by dust, mostly in the optical and near infrared domains, provides clues to the dust physical properties. The analysis of the phase dependence of polarization data obtained within a given wavelength range $P_\lambda(\alpha)$ leads to smooth phase curves, although some local increases may be noticed during an outburst. The phase curves are similar to the lunar regolith phase curves. They all present a small negative branch in the backscattering region, an inversion for a phase angle of the order of 20°-23°, and a wide positive branch with a maximum in the 90° to 100° range, as illustrated in Figure 4a for the red domain. Such phase curves can be easily fitted by appropriate polynomial or trigonometric functions, which actually lead to similar fits within phase angle ranges where enough well distributed data points are available.

For α greater than about 30°-35°, such phase curves allow us to distinguish different cometary classes, corresponding to comets with a low maximum in polarization (10% to 15%, depending upon the wavelength), comets with a higher maximum in polarization (25% to 30%), and comet Hale-Bopp, the polarization of which was even higher [26, 27]. Such classes correspond to different properties of the dust; as noticed by Levasseur-Regourd *et al.* [26] and stressed by Hanner [28], comets with a high maximum in polarization seem to exhibit a silicate emission feature. Comets with a low maximum in polarization do not present such a feature. Comets with a high maximum in polarization actually seem to be dust-rich while comets with a lower maximum seem to be dust-poor and gas-rich [29].

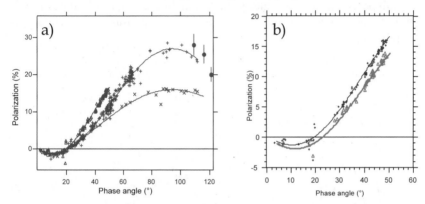

Figure 4. Phase angle dependence of the whole coma polarization: a) Cometary data in the red domain with (x) low maximum comets, (+) higher maximum comets and (Δ) C/1996 O1 Hale-Bopp. Data points above 120° are from C/1999 S4 (LINEAR) at its disruption [30]; b) C/1996 O1 Hale-Bopp data in the green domain (lower fit) and the red domain (upper fit) [20].

The polarization at a given phase angle, $P_\alpha(\lambda)$, usually increases with wavelength in the 30° to 90° range, at least in the optical domain. This trend, already noticed for 1P/Halley [21] and later confirmed for other comets [24, 31], is illustrated for C/1995 O1 in Fig. 4b. The increase appears to be linear, and the rate of increase can thus be obtained from observations at two wavelengths λ_1 and λ_2 from

$$dP_\alpha(\lambda)/ d\lambda = [P_\alpha(\lambda_2) - P_\alpha \lambda_1)] / [\lambda_1 - \lambda_2].$$

This rate of increase appears to be a linear function of the phase angle for Halley and Hale-Bopp, at least between 30° and 50° (illustrated in Fig. 5a for the latter). For a phase angle of 50°, it is of about 6% per μm for Halley and 14% per μm for Hale-Bopp [27]. In the innermost coma, the polarization can decrease with increasing wavelength, as revealed by *in-situ* measurements of Halley and suspected in remote observations of Hale-Bopp [27].

3.3 Polarization of asteroids and Kuiper Belt Objects

The scattering properties of regoliths on asteroidal surfaces have been reviewed by Muinonen *et al.* [32]. The polarization phase curves are smooth, as are those of the Moon and of cometary dust. The slope at inversion increases with decreasing albedo of the asteroid, and this parameter can be used to derive an asteroidal taxonomy [33, 34]. The maximum in polarization only can be determined for a few bright enough Near Earth Objects (NEOs), which are observable at large phase angles. The numerous observations obtained by different teams on an S-type asteroid, 4179 Toutatis (e.g. [35]), suggest that the phase curves obtained in different colors have the same inversion point, that the maximum is reached near 100°, and that the polarization decreases linearly with increasing wavelength, the decrease being more important for phase angles in the maximum region, as illustrated in Fig. 5b. For a phase angle of 50°, the above-defined rate of increase is about -4% per μm [27].

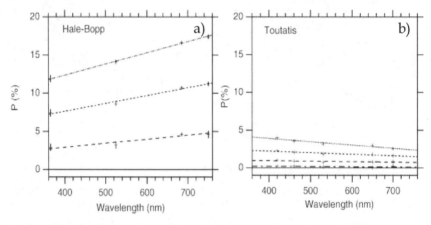

Figure 5. Wavelength dependence of the polarization at a given phase angle (from [26]): a) Comet Hale-Bopp, for α equal 30°, 40°, 50° (from bottom to top); b) Asteroid Toutatis, for α equal to 20°, 30°, 40°, 50° (from bottom to top).

The unique observations obtained on a C-type asteroid, 2100 Ra-Shalom, at almost 60° and in two colors [36] lead to a rate of increase of about -0.15 % per μm at 60°. Thus it may be estimated that the polarization does not vary significantly with wavelength, with a possible trend to a very low blue gradient. Of major interest are asteroidal observations at small phase angles, which reveal a significant opposition surge in the near-backscattering intensity together with a sharp increase of the negative polarization towards small angles. These effects originate in a mutual shadowing mechanism and/or in a coherent backscattering mechanism, as detailed in [37].

Centaurs and Kuiper Belt Objects may be rich in volatile organic compounds and ices. Their scattering properties are starting to be assessed (see e.g. [38]). Studying their polarization is a real challenge, because of the faintness of these remote objects and of limitations in the phase angles that can be explored near the backscattering region. However, deriving phase curves, without needing to cope with the difficult problem of the variations due to the rotation of an irregular object recently has triggered a series of observational campaigns at the VLT. Polarimetric observations of 28978 Ixion establish that the polarization is negative near inversion, with a value of -1.3% at $1.3°$ and a slope of about -1% per degree [39].

3.4 Polarization of interplanetary dust

The previous cometary and asteroidal polarization data need to be compared with those of the solar light scattered by interplanetary dust particles (mainly of cometary and asteroidal origin). The so-called zodiacal light observations are integrated along the line-of-sight from the observer to the outer fringe of asteroid belt (see e.g. [40] for a review). The polarization, as observed from the Earth, is mostly a function of the ecliptic latitude β and helio-ecliptic longitude $\lambda-\lambda_0$ (Fig. 6).

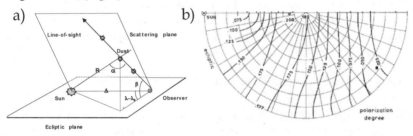

Figure 6. Zodiacal light observations: a) Geometry of the observations; b) Zodiacal polarization, as seen from the Earth. The outer circle corresponds to the ecliptic and the upper line to the Sun-anti-Sun line, with ecliptic pole at the center (from [41]).

An inversion is thus required to retrieve local information and derive polarization phase curves. Unfortunately, the wavelength dependence is estimated only from line-of-sight integrated data. A trend to a blue spectral gradient is found [41] with a slope tentatively estimated to be of about -6 % per μm at 90° elongation, corresponding to phase angles decreasing from 90° to 0° along the line-of-sight.

Inversion techniques have provided significant results in the vicinity of the ecliptic plane for distances to the Sun in the 0.3-1.5 AU range [40, 42]. As far as spatial changes are concerned, the local polarization at 90° phase angle in the ecliptic plane decreases with decreasing solar distance in the symmetry plane of the interplanetary dust cloud [43]. This trend remains true for large enough phase angles, at least between 0.3 and 1.5 AU. Conversely, the local albedo (as derived from a combination of infrared and visible observations) increases with decreasing distance to the Sun; such a result suggests a temporal evolution of the particles, which suffer some weathering (collisions, evaporation, sputtering) while they spiral towards the Sun under the Poynting-Robertson effect [40]. Up to 90° the phase-angle dependence is found to be smooth, with an inversion angle of about 15°.

3.5 Polarization for Mars surface and aerosols

Polarimetry of Mars, as studied from remote observations, provides information for phase angles up to 45°. Observations obtained through a clean atmosphere [44] reveal the scattering properties of the Martian surface. As expected, the phase curves are smooth, with a negative branch below 20°-25°. The polarization at a given phase angle (above 30°) decreases with increasing wavelength, but the effect of some Rayleigh scattering from the atmosphere of Mars cannot be ruled out.

In-situ observations, together with extensive remote surveys have revealed the scattering properties of the dust that may be present in the atmosphere. Three types of effects are observed: a layer of small (with respect to the wavelength) particles at an altitude of 40-60 km, clouds of CO_2 dominated crystals, and dense clouds of irregular dust particles [45-47]. It is of interest that although the data are noisy and restricted to 45° phase angle, the latter phase curves seem to present a deeper negative branch, with an inversion angle of about 40°. In-situ polarimetric observations are expected to be obtained with the infrared channel of the SPICAM experiment on-board the Mars Express orbiter [48].

3.6 Polarization properties of dust in the solar system, a summary

The analysis of cometary observations suggests the existence of very fluffy dust aggregates. Differences are observed in the light-scattering properties, e.g. structure of the comae, polarization phase curves maxima and minima, polarization wavelength dependence. They could be a clue to the temporal evolution of the physical properties of the dust particles, with collisional processes as well as evaporation of icy mantles and organic compounds. Table 1 presents some polarization properties of dust particles in comets, asteroids, in the interplanetary dust cloud, and on Mars, as retrieved by remote sensing.

All the phase curves correspond to the same smooth trend in the visible and near-infrared domains. Significant differences are noticed in the values of the minimum, in the inversion angle and slope at inversion, in the value of the maximum; besides, the negative branch may in some cases be asymmetric, with a significant opposition effect. The wavelength dependence of the polarization at a given angle (typically in the 30°-90° range) may correspond to an increase or a decrease of the polarization with the wavelength, so-called red or blue spectral gradient. In both cases, the rate of increase or decrease may be quite variable. These differences provide numerous constraints, to be taken into account to infer the physical properties of the medium (e.g. shape and morphology of the particles, size and size distribution, albedo, porosity and microstructure). Such an inverse problem requires careful approaches, which we now consider.

Table 1. Comparison between parameters characterizing the polarization properties of cosmic dust clouds and regoliths (precision of about ± 1% and ± 2°). Updated from [49]

Object	Phase curve P(α) Min. Inversion Max.			Polarization spectral gradient	Rate of increase dP/dλ]$_{\alpha=50°}$ (% per μm)
Comets				(Whole coma)	
Low P class	-2%	21°	10-15%	Red	
High P class	-2%	21°	25-30%	Red	6 (Halley)
Hale-Bopp	-2%	21°	> 30%	Red	14
Innermost coma				Blue?	
Asteroids					
(S type)	-1%?	20°	6-8%	Blue	-4 (Toutatis)
(C-type)	-2%?	20°?	15-25%?	Blue?	-0.15 or less neg.
Interplan. dust					
(ecliptic, ≈1 AU)	-1%?	15°?	25-30%	Blue?	–6 or less neg.
Mars					
Surface	-1%	20-25°	?	Blue	
Dense clouds	>-4%	40°?	?	?	

4. Laboratory simulations of light scattering in microgravity

4.1 Rationale for laboratory measurements under microgravity conditions

Very few significant results on the physical properties of cosmic dust, as revealed by *in-situ* experiments, may be expected in the coming years. The samples collected by the Stardust (interplanetary dust and comet 81P/Wild 2 dust) and Hayabusa (asteroid Itokawa regolith) missions should provide information primarily about chemical properties. The ground truth expected from space mission is thus quite far away in the future, with e.g. the Rosetta rendezvous with comet 67P/Churyumov-Gerasimenko and landing on its nucleus in 2014.

Most significant progresses are already taking place and will continue to take place with numerical simulations dealing with light scattering by irregular particles and aggregates [50]. However, one needs to keep in mind that such elaborate techniques are limited: some approximations need to be made, the contribution of multiple scattering needs to be taken into account, the value and wavelength dependence of the complex refractive index need to be estimated (a tricky task for the imaginary part, as well as for the index of a mixture). Even worse, the uniqueness of the solution can never be assessed, since the physical properties of the particles for which computations are performed are assumed.

A complementary approach may be provided by laboratory simulations. Measurements on particles illuminated by a laser had been initiated by the group of Bochum [51]. Successful measurements, avoiding multiple scattering on gravity packed layers and providing the whole Mueller matrix, are being made on ensembles of small particles levitating in an airflow by the group of Amsterdam [see e.g. 52, 53]. Most significant results also are obtained on large analogues illuminated in the microwave domain by the group of Gainesville [see e.g. 50, 54]. Finally, since the mid-nineties, we have initiated in the Paris area a modular series of light scattering experiments, which use nephelometer-type instruments to measure the scattering properties of dust particles of various sizes and shapes, illuminated by a collimated light source in the optical and near-infrared domains [55]. These instruments mainly operate under microgravity conditions.

Microgravity can be obtained for various durations by different techniques: up to 10 seconds with drop towers, up to 25 seconds in an aircraft flying a parabolic trajectory (and continuously compensating the deceleration resulting from the resistance of the air), about 10 minutes for a sounding rocket, and finally weeks, months and years on automatic satellites and space stations. In such conditions, dust particles are expected to be floating instead of suffering sedimentation, while the multiple scattering only takes inside large irregular particles.

Besides, if the duration is long enough, the particles may agglomerate under a variety of conditions, from Brownian motion to ballistic collisions and build up realistic aggregates under conditions somehow representative of the proto-solar nebula. Dust-cloud experiments actually have shown that low-velocity collisions in ensembles of particles lead to a rapid growth of fractal dust agglomerates. For Brownian-motion-driven agglomeration at the earliest stage of planetesimal formation, the fractal dimension is of about 1.3; for slightly larger aggregates, differential sedimentation and gas turbulence lead to higher fractal dimensions up to 1.9 (see e.g. [56, 57]).

4.2 The PROGRA^2 experiments in the laboratory and during parabolic flights

The PROGRA^2 (Propriétés Optiques des Grains Astronomiques et Atmosphériques) experiments have been designed and built at Service d'Aéronomie and at Laboratoire de Physique et Chimie de l'Environnement. As illustrated in Figure 7a, the instrument measures the linear polarization of the light scattered by a levitating dust sample illuminated by a laser beam and contained in a glass vial through a beam-splitter polarizing cube at a given moveable phase angle (tentatively down to 5°) with one measurement at a time. In an upgraded version, polarization images of the two components are obtained in two wavelengths [58, 59]. PROGRA^2 operates both in the laboratory with an airflow for very porous samples and during aircraft parabolic flight campaigns. With a total of 17 CNES flight campaigns from 1994 to 2003, the feasibility of the concept of light-scattering measurements during microgravity conditions has been demonstrated, with validation of the data obtained on micron-sized spheres through Mie computations. Also, a comparison with numerical simulations of the scattering properties of stochastic polyhedra has shown that this experimental approach allows an accurate estimation of the physical properties of dust particles [60].

A significant database is being built on various particles (e.g. natural and industrial samples, ashes, lunar analogues, Martian analogues, meteoritic powdered samples). Differences have been found in the phase curves obtained for deposited and levitating particles, and in the polarization spectral gradient for grey, compact and fluffy particles, as well as for mono-disperse media and mixtures [61]. Interestingly enough, although the uniqueness of the result is not demonstrated, mixtures of low-density aggregates of submicron-sized grains provide an excellent fit with cometary observations in terms of phase angle and wavelength dependence [62]. The blue polarization spectral gradient suspected in the vicinity of some cometary nuclei could be a clue to the presence of compact particles, possibly built up of grains glued together with ices and organic compounds, rapidly evaporating and forming fluffy aggregates as they move away from the nucleus.

4.3 CODAG-LSU experiments during parabolic flights and on-board a rocket

The CODAG-LSU (COsmic Dust Aggregation - Light Scattering Unit) experiment has been designed and built at Service d'Aéronomie. It has been operating during an ESA rocket flight in 1999. As schematically illustrated in Figure 7b, the instrument provides simultaneous determinations of the intensity and polarization phase curves of the light scattered by dust, progressively agglomerating under microgravity.

The illumination is provided by a laser diode in the near infrared at 830 nm. To avoid any specific orientation of the polarization of the light, the 4.8 mm-wide beam is directed by an optical fiber to a collimator and then to an optical device made of a Glan-Foucault prism and quarter-wave plate. An equatorial ring (inner diameter 118 mm) with vacuum tight optical windows is inserted in the wall of the low-pressure chamber in which the agglomeration of the dust particles takes place. A set of analyzers allows measurements to be obtained at 22 phase angles on a series of different phase angles, almost uniformly distributed from 10° to 165° [63].

CNES parabolic-flight campaigns devoted to calibrations took place before and after the rocket flight, with samples of spherical particles in a vial, and video monitoring of the light beam. The particles injected during the ESA Maser-8 rocket flight were micron-sized SiO_2 spheres, the precise size distribution and complex refractive index of which could be accurately determined through Mie-theory fits before the aggregation processes started to take place. Although the dust cloud was drifting in the experiment chamber, the data analysis indicates that the agglomeration of the particles was effective after about 300 s under microgravity conditions [64]. The result, compatible with a Ballistic Cluster-Cluster Agglomeration growth process, has demonstrated the feasibility of light-scattering measurements while aggregation processes take place under microgravity conditions.

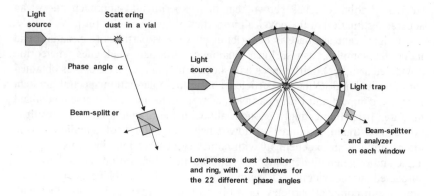

Figure 7. Schematic of light-scattering devices under micro-gravity conditions: a) PROGRA^2, with moveable arm around the dust sample enclosed in a glass vial; b) CODAG, with a ring of detectors around the dust, enclosed in a low-pressure chamber where agglomeration takes place.

4.4 ICAPS-LSU experiments on Space Station

Both PROGRA^2 and CODAG have contributed a semi-autonomous multi-user facility for research on Interactions in Cosmic and Atmospheric Particle Systems on board the International Space Station [65]. Amongst other objectives, the experiments should simulate aggregation processes from micron-sized particles under conditions representative of the proto-solar nebula, allow the formation of centimeter-sized regoliths, interpret the light-scattering observations in terms of physical properties of the dust, and document phase changes on icy particles. The brightness and polarization phase curves of the light scattered by the particles documented by 3D microscopic imaging will be monitored, together with their wavelength dependence, while they aggregate, break-off, and evolve under various physical processes, including condensation and evaporation. ICAPS is now in phase B at ESA and will be accommodated on board the Columbus laboratory on a common rack with the IMPF facility for dusty plasmas. The ICAPS facility could make use of two low-pressure chambers, with different diagnostics tools, including microscopes and Light-Scattering Units around each of the chambers. It will then be possible to monitor the formation of aggregates with different packing densities and sizes, as well as condensation (down to $-20°C$, possibly $-50°C$) and evaporation of water ice on particles that are representative of cosmic dust and atmospheric solid aerosols.

4.5 ICAPS-LSU precursor experiment

To prepare ICAPS efficiently, a precursor experiment, ICAPS-MSG (Microgravity Science Glovebox) should be launched in the 2006 time frame with a Soyuz/Progress delivery. It will test critical subsystems and monitor the evolution of the light-scattering properties of bi-disperse aggregates, built up from micron-sized SiO_2 grains of high and low albedos. Measurements with the LSU are to be performed at three different wavelengths (near 410, 630 and

830 nm) in order to study the wavelength dependence of the scattered light and assess its linearity. An electronic control of the light sources should allow high frequency measurements below 50 Hz. For the three above-mentioned wavelengths, an achromatic quarter-wave plate (broadband Fresnel rhomb) can be mounted behind the polarizer. Following the CODAG approach, the equatorial part of the experiment chamber consists of a horizontal ring with pressure-tight windows. The scattered light is monitored on 23, and tentatively 24, well-distributed phase angles from 5° to 175°, with possibly an exploration of smaller phase angles. Beam-splitter cubes are used to separate in each direction the two polarized components of the scattered light, together with Foster prisms, which provide a satisfactory achromaticity on the spectral range. Standard photodiodes, with a broadband spectrum of detection and a very high sensitivity, are to be used. Figure 8 compares the initial polarization phase curve for grains with a 0.75 ± 10 % radius illuminated at 830 nm, as computed by Mie theory and measured at 23 phase angles. It demonstrates the possibility of retrieving the initial phase curves for the available set of analyzers. Finally, the possibilities of obtaining polarization images at one phase angle (near 90°) and retrieving the whole Mueller matrix for one given phase angle are under study.

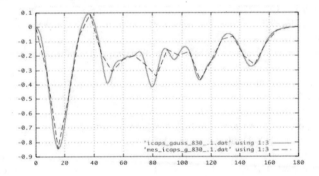

Figure 8. Percent polarization (from −0.9 to 0.1 %) of 0.75 ± 10 % μm spherical silicate particles illuminated at 830 nm, as a function of the phase angle (in degrees). Comparison of the Mie curve (continuous line) with the curve retrieved from measurements at 23 uniformly distributed phase angles (dashed line) demonstrates the possibility of retrieving phase curves [66].

4.6 Laboratory simulations of light scattering under microgravity, a summary

Numerical simulations, as well as laboratory and microgravity light-scattering measurements are needed to infer the properties of low-density dust clouds. An extensive series of light-scattering measurements on clouds of particles and aggregates should be performed with the ICAPS multi-users facility on board the ISS. They will provide a unique approach to deduce without ambiguity the physical properties of cosmic-dust particles from their optical properties, as well as their evolution when they aggregate, break-off, condense or sublimate.

Some significant constraints on the characteristics of some cosmic-dust particles, i.e. the existence of mixtures of low-density aggregates of submicron-sized grains in cometary comae and the existence of more compact particles around the nucleus can be deduced already from precursor laboratory simulations.

ICAPS laboratory simulations also will be used to validate some light-scattering codes. Hopefully, in a later stage [67], measurements at lower temperatures will be performed with a modified version of the experiment, in order to answer questions related to the formation and evolution of ices in interstellar and protoplanetary dust clouds.

Acknowledgements

This lecture could not have been presented without research efforts developed for quite a few years with various colleagues and former students. Special thanks to M. Cabane, A. Dollfus, R. Dumont, M. Fulle, P. Gaulme, B. Goidet-Devel, E. Hadamcik, M. Hanner, V. Haudebourg, L. Kolokolova, J. Lasue, N. McBride, P. Rannou, J.B. Renard and J.C. Worms.

References

1. M. Greving, F. Praderie and R. Reinhard (Eds.), "Exploration of Halley's comet" (Springer-Verlag, Berlin, 1987).

2. J. Veverka, Icarus **155**, 1 (2002).

3. T. Gehrels (Ed.), "Planets, stars and nebulae studied with photopolarimetry" (University of Arizona Press, Tucson, 1974).

4. C.B. Bohren and D.R. Huffman, "Absorption and scattering of light by small particles" (John Wiley & Sons, New-York, 1983)

5. B. Hapke, "Theory of reflectance and emittance spectroscopy" (Cambridge Univ. Press, Cambridge,1993).

6. Lyot, B., Ann. Obs. Paris **VIII**, 1, 1929.

7. Y.G. Shkutatov, N.V. Opanasenko and M.A. Kreslavsky, Icarus **95**, 283 (1992).

8. Y.G. Shkutatov and N.V. Opanasenko, Icarus **99**, 468 (1992).

9. A.C. Levasseur-Regourd, Space Science Rev. **90**, 163 (1999).

10. L. Kolokolova, M. S. Hanner, A.C. Levasseur-Regourd and B.Å.S. Gustafson, in "Comets II", Eds. M. Festou, U. Keller and H. Weaver (University of Arizona Press, Tucson, 2004), in press.

11. A.C. Levasseur-Regourd, N. McBride, E. Hadamcik and M. Fulle, Astron. Astrophys. **348**, 636 (1999).

12. M. Hanner and R.L. Newburn, Astrophys. J. **97**, 254 (1989).

13. M. Fulle, A.C. Levasseur-Regourd, N. McBride and E. Hadamcik, Astron. J. **119**, 1968 (2000).

14. N. Eaton, S.M. Scarrott and R.F. Warren-Smith, Icarus **76**, 270 (1988).

15. E.H. Geyer, K. Jockers, K., N.N. Kiselev and G.P. Chernova, Astrophys. Space. Sci. **239**, 259 (1996).

16. J.B. Renard, E. Hadamcik and A.C. Levasseur-Regourd, Astron. Astrophys. **316**, 263 (1996).

17. K. Jockers, V.K. Rosenbush, T. Bonev and T. Credner, Earth Moon Planets **78**, 373 (1997).

18. T.J. Jones and R.D. Gehrz, Icarus **143**, 338 (2000).

19. E. Hadamcik and A.C. Levasseur-Regourd, J. Quant. Spectrosc. Radiat. Transfer **79**, 661 (2003).

20. E. Hadamcik and A.C. Levasseur-Regourd, Astron. Astrophys. **403**, 757 (2003).

21. A. Dollfus, P. Bastien, J.L. Le Borgne, A.C. Levasseur-Regourd and T. Mukai, Astron. Astrophys. **206**, 348 (1988).

22. J.B. Renard, A.C. Levasseur-Regourd and A. Dollfus, Ann. Geophys. **10**, 288 (1992).

23. N. Eaton, S.M. Scarrott and P.W. Draper, Mon. Not. R. Astron. Soc. **273**, L59 (1995).

24. L. Kolokova, K. Jockers, B.Å.S. Gustafson and G.Lichtenberg, J. Geophys. Res. **106**, 10113 (2001).

25. E. Hadamcik and A.C. Levasseur-Regourd, Icarus **166**, 188 (2003).

26. A.C. Levasseur-Regourd, E. Hadamcik, and J.B. Renard, Astron. Astrophys. **313**, 327 (1996).

27. A.C. Levasseur-Regourd and E. Hadamcik, J. Quant. Spectros. Radiat. Transfer **79**, 903 (2003).

28. M.S. Hanner, J. Quant. Spectros. Radiat. Transfer **79**, 695 (2003).

29. E.P. Chernova, N.N. Kiselev and K. Jockers, Icarus **103** (1993).

30. E. Hadamcik and A.C. Levasseur-Regourd, Icarus, **166** (2003).

31. E. Hadamcik and A.C. Levasseur-Regourd, and J.C. Worms, ASP Conf. Ser. **104**, 391 (1996).

32. K. Muinonen, J. Piironen, Y.G. Shkuratov, A. Ovcharenko and B.E. Clark, in "Asteroids III", Eds. W.F. Bottke, A.Cellino, P. Paolicchi and R.P. Binzel (University of Arizona Press, Tucson, 2002), p.123.

33. D. and B. Zellner, in "Asteroids", Ed. T. Gehrels (University of Arizona Press, Tucson, 1979).

34. B. Goidet-Devel, J.B. Renard and A.C. Levasseur-Regourd, Planet. Space Sci. **43**, 779 (1995).

35. M. Kogachi, T. Mukai, R. Nakamura, R. Hirata , A. Okasaki, M. Ishiguro and H. Nakayama, Publ Astron Soc Japan **49**, L31 (1997).

36. N.N. Kiselev, V.K. Rosenbush and K. Jockers, Icarus **140**, 464 (1999).

37. Yu. Shkuratov, *et al.*, these proceedings.

38. P. Rousselot, J.M. Petit, F. Poulet, P. Lacerda and J. Ortiz, Astron. Astrophys. **407**, 149 (2003).

39. H. Boehnhardt, S. Bagnulo, K. Muinonen, M.A. Barucci, L. Kolokolova, E. Dotto, and G.P. Tozzi, Astron. Astrophys. Letters **415**, L21, 2004.

40. A.C. Levasseur-Regourd, I. Mann, R. Dumont and M.S. Hanner, in "Interplanetary dust", Eds. E. Grün, B. Gustafson, S. Dermott, and H. Fechtig (Springer-Verlag, Berlin, 2001), p.57.

41. C. Leinert, S. Bowyer, L. Haikala, M. Hanner, M.G. Hauser, A.C. Levasseur-Regourd, I. Mann, K. Mattila, W.T. Reach, W. Schlosser, J. Staude, G.N. Toller, J.L. Weiland, J.L. Weinberg and A.N. Witt, Astron. Astrophys. Suppl. Ser. **127**, 1 (1998).

42. R. Dumont and A.C. Levasseur-Regourd, Astron. Astrophys. **191**, 154 (1988).

43. A.C. Levasseur-Regourd, R. Dumont and J.B. Renard, Icarus **86**, 264 (1990).

44. A. Dollfus and J.H. Focas, Astron. Astrophys. **2**, 63 (1969).

45. R. Santer, M. Deschamps, L.V. Ksanfomaliti and A. Dollfus, Astron. Astrophys. **158**, 247 (1986).

46. P. Lee, S. Ebizawa and A. Dollfus, Astron. Astrophys. **240**, 520 (1990).

47. S. Ebisawa and A. Dollfus, Astron. Astrophys. **272**, 671 (1993).

48. J.L. Bertaux, D. Fonteyn, O. Korablev, E. Chassefière, E. Dimarellis, J.P. Dubois, A. Hauchecorne, M. Cabane, P. Rannou, A.C. Levasseur-Regourd, G. Cernogora, E. Quemerais, C. Hermans, G. Kockarts, C. Lippens, M. De Mazière, D. Moreau, C. Muller, B. Neefs, P.C. Simon, F. Forget, F. Hourdin, O. Talagrand, V.I. Moroz, A.V. Rodin, B. Sandel and A. Stern, Planet. Space Sci. **48**, 1303 (2000).

49. A.C. Levasseur-Regourd, Adv. Space Res. **31**, 2599 (2003).

50. L. Kolokolova, H. Kimura and I. Mann, These proceedings.

51. K. Weiss-Wrana, Astron. Astrophys. **126**, 240 (1983).

52. Muñoz O., H. Volten, J. F. de Haan, W. Vassen and J. W. Hovenier, Astron. Astrophys. **360**, 777 (2000).

53. H. Volten and O. Muñoz, These proceedings.

54. B.Å.S. Gustafson and L. Kolokolova, *J. Geophys. Res.,* **104**, 31711 (1999).

55. A.C. Levasseur-Regourd, M. Cabane, V. Haudebourg and J.C. Worms, Earth, Moon, Planets **80**, 343 (1998).

56. J. Blum et al., Phys. Rev. Lett. **85**, 2426 (2000).

57. G. Wurm and J. Blum, Icarus **132**, 125 (1998).

58. J.C. Worms, J. B. Renard, E. Hadamcik, A. C. Levasseur-Regourd and J. F. Gayet, Icarus **142** (1999).

59. J.B. Renard, J. C. Worms, T. Lemaire, E. Hadamcik and N. Huret, Appl. Opt. **91**, 609 (2002).

60. A. Penttilä, K. Lumme, J.C. Worms, E. Hadamcik, J.B. Renard and A.C. Levasseur-Regourd, J. Quant. Spectros. Radiat. Transfer 79 1043 (2003).

61. E. Hadamcik, J. B. Renard, J. C. Worms, A. C. Levasseur-Regourd and M. Masson, Icarus **155**, 497 (2002).

62. E. Hadamcik, J. B. Renard, A. C. Levasseur-Regourd and J. C. Worms, J. Quant. Spectros. Radiat. Transfer **79**, 679 (2003).

63. A.C. Levasseur-Regourd, M. Cabane, E. Chassefière, V. Haudebourg and J. C. Worms, Adv. Space Res. **23**, 1271 (1999).

64. V. Haudebourg, Ph. D. thesis, Univ. P. et M. Curie, Paris (2000).

65. J. Blum, M. Cabane, T. Henning, W. Holländer, A.C. Levasseur-Regourd, K. Lumme, J. Marijnissen, K. Muinonen, T. Poppe, F. Prodi, P. Wagner and J.C. Worms, ESA SP **433**, 285 (1999).

66. J. Lasue, Master thesis, Univ. P. et M. Curie, Paris (2003).

67. P. Ehrenfreund, H.J. Fraser, J. Blum, J.H.E. Cartwright, J. M. Garcia-Ruiz, E. Hadamcik, A.C. Levasseur-Regourd, S. Price, F. Prodi and A. Sarkissian, Planet. Space Sci. **51**, 473 (2003).

Artist's rendering of Anny-Chantal Levasseur-Regourd after leaping off the Rosetta spacecraft onto Comet 67P/Churyumov-Gerasimenko to collect dust samples. Courtesy of F. Castel.

POLARIMETRY OF COMETS: PROGRESS AND PROBLEMS

NIKOLAI KISELEV[1] AND VERA ROSENBUSH[2]

[1]*Institute of Astronomy, Kharkiv   National University,
35 Sumskaya Str., 61077 Kharkiv, Ukraine*
[2]*Main Astronomical Observatory of the National Academy of
Sciences of Ukraine, 27 Zabolotnoho Str., 03680 Kyiv, Ukraine*

Abstract. In this chapter we review the history and progress in the study of scattered light from comets using polarimetry along with its significance for understanding the nature of comets. We present observed characteristics of linear and circular polarized light and their angular, spectral, spatial and temporal dependencies. We discuss the main problems of taxonomy of comets based on direct observations of the scattered light. It is shown that at least a part of gas-rich comets has a low polarization at large phase angles and blue color mainly due to a low spectral and spatial resolution of the measurements. The existence of the two taxonomical classes of comets is still an open question.

1. Introduction

Solar light scattered by cometary solid particles is generally elliptically polarized. This polarization is an important parameter for defining both optical and physical properties (size, refractive index, shape, structure) of the dust grains. The observed phase-angle and wavelength dependence of polarization, its spatial and temporal variations depend on the processes of light scattering, properties of dust particles, distribution of dust and gas in the coma, and grain alignment. At present, the inverse problem of light scattering by arbitrary dust particles is still unsolvable and any interpretation of polarimetric data is uncertain. However, analyses of many observed characteristics of scattered light can yield strong constraints on physical properties of cometary dust, such as size distribution, composition, and particle structure. Besides, one of the aims of polarimetric observations of comets certainly consists in the attempt to establish taxonomy of comets and to connect properties of cometary dust with the origin and evolution of comets. It is hoped

411

G. Videen et al. (eds.), Photopolarimetry in Remote Sensing, 411-430.
© 2004 *Kluwer Academic Publishers. Printed in the Netherlands.*

that this problem can be solved through the direct observations of scattered light from comets.

In this paper, we provide both a brief historical review of the polarimetric researches of comets and analysis of the observed angular and spectral characteristics of linear and circular polarization of the scattered solar light by dust particles in order to reveal principal regularities in the polarimetric properties of cometary dust.

2. Historical background

The history of astronomical polarimetric investigations started from the invention of the visual polariscope by Arago with help of which he made the first observations of the Moon in 1811. In 1819, Arago discovered traces of polarization in radiation of the Great Comet 1819 II [1]. Later observations of comet Halley made in 1835 confirmed the existence of polarization in comets. Visual and thereafter photographic observations of many bright comets during the 19th and the beginning of the 20th century allowed for the establishment of some regular properties of the polarization of cometary light. It was found that the polarization plane of cometary radiation usually is perpendicular to the scattering plane. Variations in the polarization degree of isolated parts of a comet (coma, tail) as well as in different comets were detected. The polarization effects have been attributed to diffuse reflection and scattering of solar light.

The modern stage of polarimetric studies of comets begins with researches by Öhman. He found the degree of polarization changes with phase angle [2]. Polarization of the molecular emissions was explained as being due to resonance fluorescence [3]. Later a negative polarization branch at small phase angles was discovered for comets [4]. In 1986, the circular polarization was found in comets [5,6,7]. A brief history of progress in the cometary polarimetry is presented in Table 1.

Table 1: Chronology of cometary polarimetry

Years	Observational facts	Consequences
1819	The first visual polarimetric observations of comets: Arago[1] detected traces of polarization in Great comet 1819 II.	The first evidence of dust in comets.
1835	Arago detected polarized light coming from comet 1P/Halley.	The existence of polarization in comets was established as fact. Cometary grains scatter the solar light.
1861	Secchi [8] revealed different polarization of light in the comet's tail and nucleus.	Direct evidence of polarization variations over the comet.

1939-41	The first spectropolarimetric observations of comets by Öhman [2,3] at various phase angles.	The origin of the polarization of cometary light was established: scattering of sunlight by the cometary grains and resonance fluorescence emission. The phase-angle dependence of polarization was discovered.
1957	The first photoelectric polarimetric observations of comets C/1956 R1 (Arend-Roland) and C/1957 P1 (Mrkos) by Boyko and Kharitonov [9], Lipsky [10], Hoag [11], and Bappu and Sinvhal [12].	The dependence of polarization on the size of diaphragm. Detection of high polarization for the Na molecule.
1958-60	Elvius [13], Mirsoyan and Hachikyan [14], and Martel [15, 16] found local regions in the coma with different planes of polarization.	The idea is proposed of local alignment of non-spherical dust grains.
1965	Multicolor observations of polarization in the continuum throughout the tail of comet C/1965 S1 (Ikeya-Seki) by Weinberg and Beeson [17].	The presence of positive and negative polarization (polarization reversal) in the tail of the comet indicating changes in particle properties.
1973	The first near-infrared polarimetric observations of comets by Noguchi *et al.* [18].	The degree of polarization is similar to that for the visible domain, indicating particle sizes ≈ 1 μm or larger.
1976	Discovery of a negative polarization branch at small phase angles for comets by Kiselev and Chernova [4].	Because the negative polarization branch is observed for various objects of Solar system suggests that the mechanisms producing it have fundamental physical origin.
1981	Kiselev *et al.* [19–21] found differences in the polarization maximum for two groups of comets: dust-rich and gas-rich comets.	A problem of similarity and diversity of the polarization of comets was raised.
1986	Measurements of four Stokes parameters over the coma in comet Halley by Dollfus and Suchail [6].	There is evolution of dust physical properties with distance from the nucleus.
1986	The first successful detection of circular polarization in comet 1P/Halley by Metz and Haefner [5], Dollfus and Suchail [6], Morozhenko *et al.* [7].	There is strong evidence that aligned non-spherical dust particles are present in cometary atmospheres.
1986	The first imaging polarimetry of comets by Eaton *et al.* [22].	Detection of morphological features, dust jets and shells in the cometary coma with a high polarization degree.
1986, 1992	The first *in situ* polarimetric observations of comets P/Halley and P/Grigg-Skjellerup by Levasseur-Regourd *et al.* [23,24].	Evidence of dust particles of different physical nature near the nucleus.

1994, 1997	A non-zero polarization and deviations in the plane of polarization observed at forward-scattering direction during stellar occultation by Rosenbush et al. [25,26]	Direct evidence of aligned non-spherical dust particles in the cometary atmospheres.
1997	Optical linear and circular polarimetry of the dustiest comet, C/1995 O1 (Hale-Bopp) (Rosenbush et al. [26], Kiselev and Velichko [27], Hadamcik et al. [28], Jockers et al. [29], Manset and Bastien [30]).	The highest linear polarization among comets at corresponding phase angles was detected.
2000	Polarimetry of completely disrupted comet D/1999 S4 (LINEAR) by Kiselev et al. [31], and Hadamcik and Levasseur-Regourd [32]	Sharp increase in the degree of polarization of scattered light after total disintegration of the cometary nucleus.
2002	The SOHO polarimetric observations of comet 96P/Machholtz at largest phase angles (up to 167°) by Jockers et al. [33].	Useful information for understanding dust particle properties.

3. Linear polarization

The scattered light from comets is the superposition of radiation from dust particles and gas components. Scattering of dust and resonance fluorescence of the main cometary molecules C_2 and CN generally have different phase-angle dependences of linear polarization. The phase-angle dependence of polarization for other gas species is as yet unknown. Because emission from molecules, ions, and atoms are present in all ranges of the cometary spectrum, it is difficult to select the cometary continuum without contamination by these species. Therefore there are some difficulties in determining the intrinsic polarization of the cometary continuum. In addition, because of the scattering geometry and brightness limitations for comets, it is practically impossible to obtain the whole phase-angle and spectral dependence of polarization for a single comet in the phase-angle range $0° \leq \alpha \leq 130°$ accessible for ground-based observations. In practice, one usually receives the combined phase-angle dependence of polarization for many comets.

3.1   Phase-angle dependence of polarization

For obtaining and studying the phase-angle dependence of linear polarization for comets, we used our own homogeneous data sets received with narrowband filters [31,34-40]. Only comet C/1982 M1 (Austin) was observed through the wideband V filter. These data were supplemented with data for comets C/1982 M1 (Austin), 67P/Churyumov-Gerasimenko [41], C/1989 X1 (Austin) [42-45], C/1983 H1 (IRAS-Araki-Alcock) [46], and 1P/Halley [47]. All data for the blue and red domains of the continuum spectra are shown on the left and right panels in Fig. 1(a). Most measurements of polarization were made for the whole coma (several thousands or tens of thousands of kilometers). There are clear distinctions between polarization of the near nucleus area (small symbols) and polarization obtained for large areas of the coma (large symbols). Furthermore, the polarization degree measured in the wideband filter (large crosses) for comet C/1982 M1 (Austin) is significantly less than that in the narrowband filter (small crosses) because of the emission contribution.

The comets presented in Fig. 1a are differed by the strength of continuum and emission spectrum. To illustrate the degree of dustiness of the comet, Chernova *et al.* [34] and Kiselev [40] used the quantity $W(C_2)$, the so-called "equivalent width" that describes the ratio of radiation fluxes in the C_2 $\lambda 5140$ Å and $\lambda 4845$ Å (blue continuum) filters (see [48]). Comets with strong continuum (dust-rich comets) have equivalent width $W(C_2) \leq 500$ Å, while comets with faint continuum (gas-rich comets) have $W(C_2) \geq 1000$ Å. It was shown (see e.g. [34,49,50]) that the dust-rich comets also are characterized by higher infrared color temperature and stronger infrared silicate emissions near 10 and 18 μm. In contrast, for the gas-rich comets, usually only small color excess is observed and silicate features may be lacking completely or very weak. In the visible spectral region, the light scattered by the dust in dust-rich comets is generally redder relative to the Sun; whereas, light scattered by gas-rich comets (Fig. 2) generally appears to be bluer.

As follows from Fig.1a, the significant scatter of polarization data points may at first sight suggest that each comet has its proper behavior of polarization [51]. However, if the comets are divided according to the $W(C_2)$ criteria, then comets may be classed into two groups depending on the maximum degree of polarization at large phase angles (Fig 1b, c). The first group (Fig.1b) consists of 16 dust-rich comets: C/1975 VI (West), 67P/Churyumov-Gerasimenko, 22P/Kopff, P/1983 VI (Hartley-IRAS), 21P/Giacobini-Zinner, 1P/Halley, C/1987 P1 (Bradfield), C/1988 A1 (Liller), C/1990 K1 (Levy), 4P/Faye, C/1991 T2 (Shoemaker-Levy), 47P/Ashbrook-Jackson, C/1995 O1 (Hale-Bopp), C/1996 B2 (Hyakutake), C/1998 U5 (LINEAR), and C/1999 S4 (LINEAR). The scatter of data points for dust-rich comets may be caused by the molecular emissions contribution in the continuum filters (up to 2%), the polarization distribution over the coma that results in different results at different diaphragms (up to 2%), accuracy of measurements of polarization degree (in the range 0.1-2%), or by possible real variations in scattering properties of dust particles [40]. Nevertheless, most of the dust-rich comets show similar phase-angle dependence

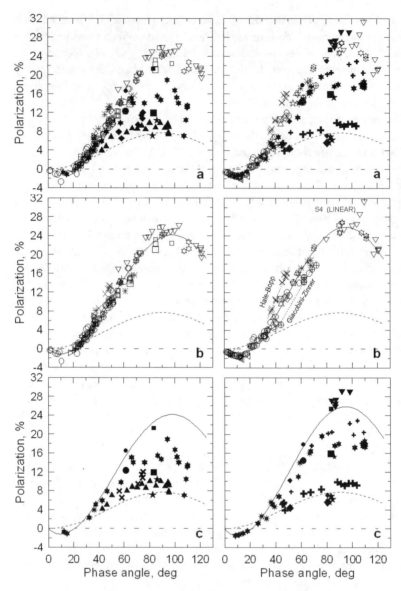

Figure 1. The phase-angle dependence of polarization for comets in the blue (left panel) and red (right panel) continuum: a – all comets; b – dust-rich comets; c – gas-rich comets. Dust-rich comets: □ - C/1975 V1 (West); ◁ - 67P/Churyumov-Gerasimenko; ▷ - 22P/Kopff; ✳ - P/1983 V1 (Hartley-IRAS); ⊕ - 21P/Giacobini-Zinner; ○ - 1P/Halley; △ - C/1987 P1 (Bradfield); ◇ - C/1988 A1 (Liller); + - C/1990 K1 (Levy); ✿ - 4P/Faye; ✖ - C/1991 T2 (Shoemaker-Levy); ⊞ - P/Ashbrook-Jackson; ✕ - C/1995 O1 (Hale-Bopp); ✪ - C/1996 B2 (Hyakutake); ⊠ - C/1998 U5 (LINEAR); ▽ - C/1999 S4 (LINEAR); Gas comets: ▲ - C/1975 N1 (Kobayashi-Berger-Milon); ✖ - 27P/Crommelin; ★ - 23P/Brorsen-Metcalf; ✚ - C/1982 M1 (Austin); ▼ - C/1983 H1 (IRAS-Araki-Alcock); ✺ - C/1989 X1 (Austin); ◆ - 2P/Encke; ■ - C/1996 Q1 (Tabur); ● - C/1999 J3 (LINEAR); ✱ - C/2001 A2 (LINEAR).

of polarization. In general, the phase-angle dependence of polarization for dust-rich comets consists of a negative polarization branch with minimum polarization $P_{min} \approx -1.5\%$ at phase angle $\alpha_{min} \approx 10°$, and a positive polarization branch with a maximum of $P_{max} \approx 25-30\%$ at $\alpha_{max} \approx 95°$ (Fig. 1b). The phase angle α_{inv} at which polarization changes sign is close to 21° and the slope of the linear part of the positive branch (usually from α_{inv} to $\alpha \approx 30°$) is $\Delta P / \Delta \alpha \approx 0.3\%/degree$.

As one can see in Fig. 1b (right panel), the comets Giacobini-Zinner, Hale-Bopp, and S4 (LINEAR) are an exception. Polarization degree measured in comet Hale–Bopp was largest among the comets at the same phase angles [27-29]. In contrast, Giacobini-Zinner has a very low polarization at corresponding phase angles. A sharp increase of polarization during a total disruption of nucleus was detected in comet S4 (LINEAR).

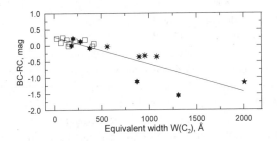

Figure 2. Correlation between the BC-RC color and equivalent width for dust-rich comets (□) and gas-rich comets: ✱ - C/1989 X1 (Austin); ✳- C/2001 A2 (LINEAR); ★ - 23P/Brorsen-Metcalf.

Gas-rich comets C/1975 N1 (Kobayashi-Berger-Milon), 27P/Crommelin, 23P/Brorsen–Metcalf, 2P/Encke, C/1982 M1 (Austin), C/1983 H1 (IRAS–Araki–Alcock), C/1989 X1 (Austin), C/1996 Q1 (Tabur), C/1999 J3 (LINEAR) and C/2001 A2 (LINEAR) have significantly larger scatter of data points than that for dusty comets: P_{max} is within the range 5% to 20% (Fig.1c). Such large scatter of data points precludes the construction of a single, composite, phase-angle dependence of polarization for this group of comets. It is important to note that the distinctions in polarization for both groups begin at phase angles $\alpha \geq 35°$. It is at this point that significant differences between continuum polarization and emission-band polarization begin (see the curves in Fig. 1c).

3.2   Taxonomy of comets

On the basis of wideband-filter polarimetry, Kiselev divided comets into two groups depending on the maximum degree of polarization at large phase angles [19-21]. He noted that such a distinction between comets is caused by an influence of the molecular emissions that fall within the wideband filters. The narrowband [34] and intermediateband [44] polarimetric observations of comets confirmed the existence of the two groups. Chernova *et al.* [34] indicated that a large part of the difference between the two groups can be ascribed to a contamination of the narrowband filters by molecular emissions, but a real difference of comets seems to be present. Levasseur-Regourd *et al.* [49] concluded that the existence of these two polarimetric classes of comets is most likely a clue to some significant differences in the bulk properties (albedo, size distribution, porosity) of dust in different comets. Later Levasseur-Regourd and Hadamcik [52] suggested that there are three classes comets: (i) comets with a low maximum of polarization, of about 10-15%; (ii) comets with a higher maximum, of about 25-30%; (iii) comet C/1995 O1 (Hale-Bopp). At present, there is widespread opinion that comets tend to form groups based on differing maximum polarization [49,50,52]. Correlation between a high maximum of polarization, infrared color temperature and silicate emissions in the dust-rich comets and absence of this correlation in the gas-rich comets (see section 3.1) is a strong argument for real differences of dust properties in these two groups of comets. Nevertheless, there are several observational facts that contradict this: (i) not all comets show a correlation between the polarization degree and infrared color temperature; (ii) there is a real depolarization of scattered light by the cometary grains due to the contribution of molecular emissions into both the wideband and narrowband filters. As pointed out by Li and Greenberg [53], high porosity of large dust particles can also provide a high infrared color temperature and silicate emissions. Moreover, the blue colors of gas-rich comets are in contradiction with the scattering of light by large dust particles, the sizes of which are defined from IR measurements. In addition, A'Hearn *et al.* [54] noted that bluer colors in comets with a very low dust/gas ratio are due primarily to instrumental artifact, namely, contamination by gas. Actually, polarimetric measurements in continuum are more sensitive to the gas influence because the maximum polarization due to resonance fluorescence is less than that for light scattered by the dust.

Jockers [55] noted that with respect to the contamination problems of the measurements, it seems premature to speak about a dichotomy (or "trichotomy") of comets. Kiselev [56] also has expressed a similar standpoint. A contribution of molecular emissions into continuum radiation and its influence on the observed degree of polarization was carefully examined in [34, 40]. It was shown that the polarization measured in the near-nuclear region of gas-rich comets Tabur, IRAS-Araki-Alcock, J3 (LINEAR), and Encke when corrected for the gas contribution is close to that for dust-rich comets. Nevertheless, even after this correction, the polarization measured near the nucleus of gas-rich comets through large diaphragms remains significantly smaller than that for the dust-rich ones. Fig. 3 demonstrates a decrease of the polarization continuum with an increase of the projected aperture for both groups of comets. For gas-rich comets this dependence is considerably steeper. Evidently, the total amount of polarization in gas-rich comets is influenced by the fact that dust in these comets is limited to the near-nucleus area that is caused by the low dust production of these nuclei. Fig. 4

shows a sharp decrease of polarization with distance from the nucleus of comet Tabur [38]. Because polarization does not depend on the number of dust particles suggests the following:

(i)   Correction for gas contamination is not sufficient, especially in the outer coma;

(ii)   Distinctions exist in the dust properties of dust-rich and gas-rich comets;

(iii)   The evolution of dust particles in gas-rich comets with distance from the nucleus differs strongly from the evolution of dust in dust-rich comets.

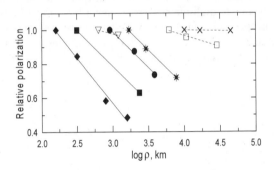

Figure 3. The diaphragm dependence of polarization for comets: ◆ - Encke, ■ - Tabur, ● - J3 (LINEAR), ✳ - A2 (LINEAR), ▽ - S4 (LINEAR), □ - West, ✕ - Hale-Bopp.

The separation of the continuum from emission features is a difficult problem. There are numerous unidentified emissions [57] that can form a pseudo-continuum of gas-rich comets. It is not clear whether the continuum is reached in the continuum bands; therefore, we can conclude that at least a part of the gas-rich comets have a low polarization at large phase angles and blue color mainly due to a low spectral and spatial resolution of observations. Thus, the existence of two taxonomical classes of comets with different physical properties of their dust particles still remains an open question.

3.3   Spatial distribution of polarization

Figure 3 displays a trend typically seen, that of decreasing polarization with increasing distance from the nucleus, which is obtained from aperture measurements with real (photoelectric polarimetry) or virtual (imaging polarimetry) diaphragms. In reality the distribution of polarization over the coma is more complicated. Maps of the direction and degree of polarization show considerable spatial structure, with polarization ranging from a few percents to 20-30%. There are also regions of negative polarization. Imaging polarimetry of comets Halley [58,59], Swift-Tuttle [60], Hale-Bopp [29,61], Encke [62], and others demonstrated higher polarization in jets, arcs, envelops, fans and other bright structures. It was shown [61,63,64] that the phase dependence of polarization for different areas of the coma differs from that obtained for the whole coma. This may indicate an evolution in the physical properties of dust particles ejected from the nucleus [63]. Evolution of dust in the near-nucleus coma caused by sublimation and fragmentation of core-mantle particles was studied by Kolokolova *et al.* [65]. The distribution of polarization and dust color over the coma also may be explained by spatial variations in particle size resulting from solar radiation pressure [66].

A polarization decrease was detected in the innermost coma in comets Halley [6,67] and Hale-Bopp [28,29] at distances ≤ 1000 km from the nucleus, which led to the hypothesis that this is due to fresh dust recently ejected from nucleus [6]. However, we believe that a decrease of the polarization in the vicinity of the nucleus also may be due to multiple scattering. According to Dollfus and Suchail [6] the optical thickness of comet Halley's atmosphere was about 0.9 at the cometocentric distance ≈ 500 km. Fernóndez [68] also indicated a significant optical thickness >1 within roughly 100 km of the nucleus's center of comet Hale-Bopp.

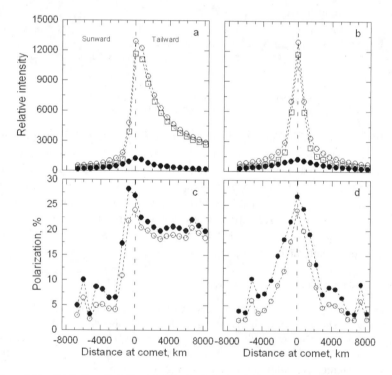

Figure 4. The distribution of brightness (a, b) and polarization (c, d) over the coma of comet Tabur. The left and right panels show the cuts along and perpendicular to the radial vector respectively, and directions are given in the upper panels: $\bigcirc$ - dust+NH$_2$, $\square$ - dust, $\bullet$ - NH$_2$; in bottom panels: $\bigcirc$ - the observed polarization, $\bullet$ - polarization corrected for the NH$_2$ contamination.

3.4   Heliocentric dependence of polarization

The study of the heliocentric dependence of polarization for comets is extremely difficult because of a lack of data for the same areas of the coma and at the same phase angles but at different distances from the Sun. Dollfus [69] noted that polarization of comets is higher at heliocentric distances larger than 2 AU. Figure 5 shows the negative polarization branch for comets obtained at different intervals of heliocentric distances [40]. These data show that the polarization in this phase-angle range and measured at similar wavelengths does not vary with heliocentric distance. However, imaging polarimetry [32,64] shows the negative polarization to vary over different regions of the coma. One possible explanation of the absence of a heliocentric dependence on polarization is that aperture polarimetry averages polarization over a large area of coma.

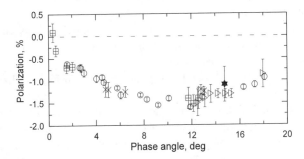

Figure 5. Negative polarization branch for comets at different heliocentric distances: ○ - 1P/Halley, 1.46<r<1.87 AU; ▷ - 22P/Kopff, r=1.70 AU; ✱ - C/1989 X1 (Austin), r=1.55 AU; ⊞ - P/Ashbrook-Jackson, 2.36<r<2.54 AU; ✕ - C/1995 O1 (Hale-Bopp), 3.47<r<4.05 AU.

3.5   Spectral dependence of polarization

The spectral dependence of polarization is also an important observational characteristic of comets that is an indicator of the composition and size of the scattering particles [70]. Prior to comet-Halley observations, two trends were discussed: polarization of comets increases with wavelength [4] and polarization of comets is wavelength independent [71,72]. Observations of comets West [4,73], Halley [34,47,74,75], and Hale-Bopp [27,30,76] confirmed that within α = 30° - 80° the polarization of dust-rich comets usually increases with increasing wavelength in the wide spectral range (0.36 - 2.2 μm) (Fig. 6).

Parameters $|P_{min}|$ and slope $\Delta P/\Delta\alpha$ at the inversion angle $\alpha_{inv} \cong 21°$ show a tendency to increase slightly in the visible domain for dust-rich comets. This result was obtained by comparison of trigonometric fits for the phase-angle dependence of polarization for dust-rich comets in the blue and red domain [40]. This tendency was recently confirmed by observations of comet C/2000 WM1 (LINEAR) [77]. It should be noted that the negative polarization branch of comet

Hale-Bopp was less deep at 2.2 µm than that in R band [76]. Therefore the spectral dependence of polarization at small phase angles requires further study.

For three comets, peculiarities in the spectral dependence of polarization were found. A negative gradient of polarization ($\Delta P/\Delta\lambda$ = -1.46%/1000 Å) was revealed for the dust-rich comet Giacobini-Zinner [37]. Most of the time comet S4 (LINEAR) also showed a decrease of polarization with the wavelength: the degree of polarization in the red filter was lower than that in the green and the blue spectral range [27]. Both comets belong to the group of comets with low abundance of C_2 [78]. The spectral gradient of the dustiest comet Hale-Bopp was larger ($\Delta P/\Delta\lambda$ = 1.13%/1000 Å) than the mean value for dust-rich comets ($\Delta P/\Delta\lambda$ = 0.80%/1000 Å) at the corresponding phase angle of about 45° [27]. We have already noted that these three comets, Giacobini-Zinner, Hale–Bopp, and S4 (LINEAR), differ distinctly from others also by the phase-angle dependencies of polarization (see Fig. 1b).

The spectral dependence of polarization for dust-rich comets changes with phase angle (Fig. 6b). This fact first was found for comet Halley [34,47]. A gradient $(\Delta P/\Delta\lambda)/\Delta\alpha$ is constant within the phase angle range 30 - 80° and equal to 0.025 ± 0.001 %/1000 Å per 1° [40]. One can see in Fig. 6b that for phase angles smaller than 20°, the gradient $(\Delta P/\Delta\lambda)/\Delta\alpha$ is close to zero.

We can say nothing about the average spectral dependence of polarization for gas-rich comets because of different gas contributions in the blue and red domain of spectra. It was shown only for comet Tabur [38] that correcting for the gas contamination changed the value of spectral gradient from 0.2%/1000 Å to 0.84%/1000 Å, which is close to the mean value for dust-rich comets.

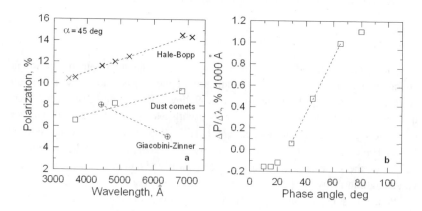

Figure 6. Variations of the degree of linear polarization with wavelength for dust-rich comets, Hale-Bopp, and Giacobini-Zinner (a). Spectral gradient of polarization versus phase angle for dust-rich comets (b).

3.6   Comparison of polarization in normal and disintegrated comets

Outbursts, partial fragmentation or complete disintegration of comets should be accompanied by the ejection of fresh internal material and therefore provide a good opportunity to study the inner composition of the nucleus and the physical processes occurring in the atmosphere of a splitting comet. Non-steady processes in comets can produce changes of polarization parameters. Such variations of the polarization degree and plane were registered during the outburst activity in comets Schwassmann–Wachmann 1 [79] and Levy [25]. Dollfus [80] also has observed variations of polarization degree for Comet Halley. He interpreted the temporary deviations of polarization in terms of an increased contribution of small-sized dust particles in the cometary atmosphere.

Kiselev et al. [81] have performed a comparative analysis of the dust polarimetric properties in split and normal (non-split) comets. Comets C/1975 V1 (West), 16P/Brooks 2, C/1988 A1 (Liller), D/1996 Q1 (Tabur), C/1999 S4 (LINEAR), and C/2001 A2 (LINEAR) can be related to the group of split comets. Comets West, S4 (LINEAR) and A2 (LINEAR) were observed directly during splitting. The partial fragmentation of comets has not resulted in a global change of its phase dependence on polarization. There is no significant evidence for differences of polarization between tidally split comets (e.g. Brooks 2), dissipating comets (e.g. Tabur), non-tidally split comets (e.g. West) and normal comets. Disintegration of comet S4 (LINEAR) represents an example of total disintegration of a cometary nucleus. After its final fragmentation, a dramatic increase in polarization was observed (see Fig. 1b). Apparently during the final disintegration of the comet, the properties of the dust particles changed. In particular, fresh dust in the form of small particles could have been ejected and this may have resulted in a considerable increase in polarization [66]. As we indicated above, the spectral dependence of polarization of comet S4 (LINEAR) was very unusual.

The split Comet A2 (LINEAR) showed a short-term variability of the polarization parameters on a time scale of a day. Inexplicable cases of polarization-plane rotation and a decrease in degree of polarization have been found for the comet [82]. The cause of such changes is unclear. It might be interpreted as the result of solar light scattering by aligned particles. Comet A2 (LINEAR) was observed in the active phase and displayed a series of sporadic outbursts due to disruption of its nucleus. It is impossible to eliminate the short-time changes of dust properties in the comet due to separation of small short-lived splinters that may not be directly observed.

4.   Circular polarization

Measurements of circular polarization are rare and conditions for producing circular polarization in comets are poorly understood. According to theoretical predictions [83], the degree of circular polarization arising from single scattering on aligned grains may be as high as 4% in comets. However, attempts to detect circular polarization in comets Tago-Sato-Kosaka, Kohoutek, Bradfield, and West were unsuccessful [84,85]. This run of bad luck may be explained by the usage of large diaphragms.

The first comet, for which circular polarization has been found, was comet Halley. Left handed (negative) circular polarization was detected on March 16-20, 1986 [6], but on April 8-28 it had variable sign [5,7]. The observed circular polarization was highly variable over the coma and showed temporal variations, from 0.04% up to 2.3%. The degree of circular polarization was found to be strongly dependent on the aperture size. It should be noted that these observations had a very low accuracy.

High-precision measurements of circular polarization across the coma including the nucleus, shells, regions between shells, and tail, were obtained for comet Hale-Bopp by Rosenbush *et al.* [26]. Only left circularly polarized light with a maximum value $-0.26 \pm 0.02\%$ was detected for all measured areas of the coma on March 11, 1997. The polarization was less than 0.08% at the cometary nucleus. Variations up to 0.1% in the degree of circular polarization were found across the coma (Fig. 7). The region of higher polarization coincided with shells. The smallest values of polarization were measured between the shells. Manset and Bastien [30] also measured circular polarization in this comet. The absolute value of the circular polarization was found at the same level up to 0.3%, however both left and right circularly polarized light was detected during their period of observations on April 2-16, 1997.

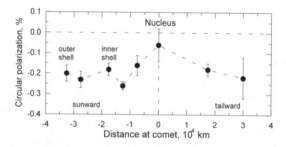

Figure 7. Variations of the degree of circular polarization with the cometocentric distance for comet C/1995 O1 (Hale-Bopp).

The spatial distribution of circular polarization along the cut through the coma and nucleus in the solar and antisolar directions was investigated for comet S4 (LINEAR) [86]. Its maximum value reached up to 1%. At most times, the degree of circular polarization at the cometary nucleus was close to zero. The left handed as well as right handed polarization was observed over the coma although left-circularly polarized light was systematically observed in the sunward part of the coma. It should be noted that on July 21, 2000, immediately after the complete disintegration of the nucleus, polarization in both directions was mainly left handed, on average -0.41 ± 0.07% (Fig. 8).

The phase angle of the comet at the time of observations varied from 61° up to 122° and it allowed for the study of variations of circular polarization with phase angle. The averaged degree of polarization along each cut in the sunward part of the coma clearly has a phase-angle dependence; i.e., the absolute value of the circular polarization increases with phase angle, as seen in Fig. 9. Using all available data one can construct the common phase dependence of circular polarization for comets Halley, Hale-Bopp, and S4 (LINEAR). The theoretical curve computed by Beskrovnaja *et al.* [87] for Rayleigh aligned particles has the same trend but much steeper.

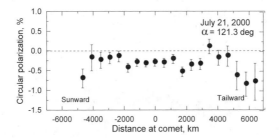

Figure 8. Variations of the degree of circular polarization along the cut through the coma and nucleus of comet C/1999 S4 (LINEAR). The position angle of the cut is about 151 deg.

The trend of the averaged degree of circular polarization on the tailward side (Fig. 10) differs considerably from that of the sunward side. It is remarkable that the variations of circular polarization coincide in time with the outbursts, which were caused by fragmentation of the nucleus. During observations from June 28 to July 22, 2000, the comet was in a very active state. The nucleus underwent multiple fragmentations culminating in complete disintegration around July 20.

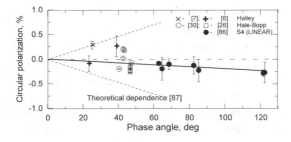

Figure 9. Composite phase-angle dependence of circular polarization for comets 1P/Halley [6,7], C/1995 O1 (Hale-Bopp) [26,30], and C/1999 S4 (LINEAR) [86].

A comparison of the circular polarization with nightly averaged visual magnitudes compiled from [88], which are sensitive to the dust and gas production rates, and the water production rate [89–91] for this observational period shows that the degree of circular polarization in the tail is strongly correlated with cometary activity (Fig. 10).

A qualitative understanding of the circular polarization of cometary radiation and its variability still remains to be solved. The observations of the same sign of circular polarization in different parts over the coma and the phase-angle dependence testify to the presence of global circular polarization. At the same time the variable sign across the coma may indicate the presence of local circular polarization.

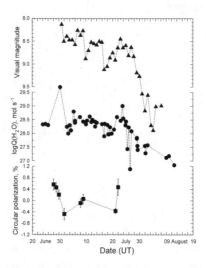

Figure 10. Comparison of the degree of circular polarization, the visual lightcurve, and the water production [89–92] of C/1999 S4 (LINEAR) during its final fragmentation in July 2000.

5. Polarimetry of comets during stellar occultations

Polarimetry of a transmitted star's radiation through the cometary coma supports the theoretical prediction that the dust grains are partly aligned in the cometary atmospheres. For the first time the existence of oriented particles was directly observed from polarimetric observations of a star occulted by the coma of comet C/1990 K1 (Levy), in which changes were detected in both the degree of linear polarization as well as the plane of polarization of the starlight transmitted through the coma [25]. Polarimetric observations of a stellar occultation by comet Hale-Bopp [26] confirmed these results. Since the stellar radiation was forward-scattered, this transformation points to the alignment of dust grains in the atmospheres of these comets.

6. Future prospects

One of the crucial problems of cometary physics is the compositional taxonomy of comets and its relation to their place of origin and evolution. One of the tools to analyze the difference in composition and structure of cometary material is polarimetry. We review the main observed characteristics of polarized light from comets obtained mainly through aperture polarimetry. Among almost 1000 comets known up to now [92] only several tens of comets have been studied using polarimetry. This represents a very small sample to obtain characteristics of comets as a whole. On the other hand, the characteristics and dependencies considered are related to the integrated properties of the entire coma. This is a limitation for the determination of intrinsic properties of dust particles as well as for the establishment of diversity and similarity of comets. Future progress in cometary science undoubtedly will be connected with a wide application of imaging polarimetry. Obvious tasks for future cometary polarimetry are as follows: (i) obtaining high resolution spectropolarimetric observations or simultaneous spectral, photometric, and polarimetric observations of gas-rich comets at large phase angles to resolve issues of diversity and similarity among comets; (ii) obtaining polarization and color maps with high spatial resolution to establish distributions of parameters of linear and circular polarization and color over the coma; (iii) obtaining time-dependent measurements to study the evolution of physical properties of dust particles ejected from the nuclei of dusty and gas-rich comets.

Acknowledgements

The authors are thankful to L. Kolokolova and G. Videen for many suggestions that greatly improved the manuscript. We also acknowledge the NATO ASI grant.

References

1. D.F.J. Arago, "Astronomie Populaire" (Paris, 1858).
2. Y. Öhman, Stockholm Obs. Ann. **13**, 3 (1941).
3. Y. Öhman, MNRAS **99**, 631 (1939).
4. N.N. Kiselev, G.P. Chernova, Astron. Zh. **55**, 1064 (1978).
5. K. Metz, R. Haefner, Astron. Astrophys. **187**, 539 (1987).
6. A. Dollfus, J. Suchail, Astron. Astrophys. **187**, 669 (1987).
7. A. Morozhenko, N. Kiselev, and A. Gural'chuk, Kinem. Phys. Sel. Bodies **2**, 89 (1987).
8. A. Secchi, Comptus Rendus **53**, 85 (1881).
9. P. Boyko and A. Kharitonov, Astron. Circ. 181, 6 (1957).
10. Ju. Lipsky, Astron. Circ. **185**, 3 (1957).
11. A. Hoag, PASP **413**, 203 (1958).
12. M. Bappu and S. Sinvhal, MNRAS **120**,152 (1960).
13. A. Elvius, Arkiv. Astron. **2**, 195 (1958).
14. L. Mirsoyan and E. Hachikyan, Byurokan Obs. **26**, 35 (1959).
15. M.T. Martel, C. R. Acad. Sci. **249**, 908 (1959).
16. M.T. Martel, Ann. d'Astrophys. **23**, 480 (1960).
17. J.L. Weinberg and D.F. Beeson, Astron. Astrophys. **48**, 151 (1976).

18. K. Noguchi, S. Sato, T. Naihara, et al., Icarus **23**, 545 (1974).

19. N. Kiselev, PhD Thesis, (Dushanbe, 1981).

20. O. Dobrovolsky, N. Kiselev, and G. Chernova, Earth, Moon, and Planets **34**, 189 (1986).

21. O. Dobrovolsky, N. Kiselev, and G. Chernova, "Photometric and polarimetric investigation of celestial bodies" (Naukova dumka, Kiev,1987).

22. N. Eaton, S. Scarrott, and R. Wolstencroft, MNRAS **250**, 654 (1988).

23. A.Ch. Levasseur-Regourd, J.L. Betraux, R. Dumont, et al., Adv. Space Res. **5**, 197 (1985).

24. A.Ch. Levasseur-Regourd, Adv. Space Res. **17**, 117 (1995).

25. V.K. Rosenbush, A.E. Rosenbush, and M.S. Dement'ev, Icarus **108**, 81(1994).

26. V.K. Rosenbush, N.M. Shakhovskoj, and A.E. Rosenbush, Earth, Moon, and Planets **78**, 373 (1999).

27. N.N. Kiselev and F.P. Velichko, Earth, Moon, and Planets **78**, 347 (1999).

28. E. Hadamcik, A.C. Levasseur-Regourd, and J.B. Renard, Earth, Moon, and Planets **78**, 365 (1999).

29. K. Jockers, V.K. Rosenbush, T. Bonev, and T. Gredner, Earth, Moon, and Planets **78**, 373 (1999).

30. N. Manset and P. Bastien, Icarus **145,** 203 (2000).

31. N.N. Kiselev and F.P. Velichko, S.F. Velichko, in K.I. Churyumov (Ed.)., "Fourth Vsekhsvyatsky readings. Modern Problems of Physics and Dynamics of the Solar System", p. 127 (Kyiv University, Ukraine, 2001).

32. E. Hadamcik and A.Ch. Levasseur-Regourd, JQRST **79-80**, 661 (2003).

33. K. Jockers, Y. Grynko, R. Schwenn, and D. Biesecker, In Book of Abstracts ACM-2002, July 29-August 2002, Berlin, Germany, #07-04 (2002)

34. G. P. Chernova, N. N. Kiselev, and K. Jockers, Icarus **103**, 144 (1993).

35. V.K. Rosenbush, V.P. Taraschuk, N.N. Kiselev, et al., Astron. Vestnik **31**, 448 (1997).

36. N. Kiselev and F. Velichko, Icarus **133**, 286 (1998).

37. N. Kiselev, K. Jockers, V. Rosenbush, et al., Planet. Space Sci. **48**, 1005 (2000).

38. N.N. Kiselev, K. Jockers, V.K. Rosenbush, and P.P. Korsun, Astron. Vestnik **35**, 1 (2001).

39. V.K. Rosenbush, N.N. Kiselev, S.F. Velichko, Earth, Moon, and Planets **90**, 423 (2002).

40. N.N. Kiselev, Doctoral Thesis, (Kharkov, 2003).

41. R. Myers and K. Nordsieck, Icarus **58**, 431 (1984).

42. N. Eaton, S. Scarrott, and T. Gledhill, MNRAS **258**, 384 (1992).

43. U. Joshi, A. Sen, M. Deshpande, and J. Chauhan, J. Astrophys. Astr. **13**, 267 (1992).

44. S. Kikuchi, A. Okazaki, N. Kondo, et al., Proc. 23rd ISAS Lunar and Planetary Symp., 39 (1990).

45. A. Sen, U. Joshi, and M. Deshpande, MNRAS **253**, 738 (1991).

46. S. Kikuchi, Proc. 16th Lunar and Planetary Symp. **16**, 36 (1983).

47. S. Kikuchi, Y. Mukai, T. Mukai, et al., Astron. Astrophys. **187**, 689 (1987).

48. Krishna Swamy., "Physics of Comets" (World Scientific, Singapore, 1986).

49. A. Levasseur-Regourd, E. Hadamcik, and J. Renard, Astron. Astrophys. **313**, 327 (1996).

50. M.S. Hanner, JQRST **79-80**, 695 (2003).

51. J.-M. Perrin, P.L. Lamy, ESA SP- **278**, 411 (1987).

52. A.C. Levasseur-Regourd and E. Hadamcik, J. Quant. Spectrosc. Radiat. Transfer **79-80**, 903 (2003).

53. A. Li, J.M. Greenberg, Astrophys. J. **498**, L.83 (1998).

54. M.F. A'Hearn, R.L. Millis, D.J. Schleicher, et al., Icarus **118**, 223 (1995).

55. K. Jockers, Earth, Moon, and Planets **79**, 221 (1997).

56. N.N. Kiselev, Proc. 173rd Colloquium of IAU, 223 (1999).

57. C. Arpigny, ASP Conf. Ser. **81**, 362 (1995).

58. N. Eaton, S. Scarrot, and R.F. Warren-Smith, Icarus **76**, 270 (1988).

59. A.K. Sen, U.C. Joshi, M.R. Deshpande, and C.D. Prasad, Icarus **86**, 248 (1990).

60. N. Eaton, S.M. Scarrott, and P.W. Draper, MNRAS **273**, L.59 (1995).

61. E. Hadamcik and A.C. Levasseur-Regourd, Astron. Astrophys. **403**, 757 (2003).

62. N. Kiselev, K. Jockers, and T. Bonev, Icasus, in press, (2004).
63. A. Dollfus, P. Bastien, J.-F. Le Borgne, et al., Astron. Astrophys. **206**, 348 (1988).
64. J.B. Renard, A.C. Levasseur-Regourd, and A. Dollfus, Ann. Geophys. **10**, 288 (1992).
65. L. Kolokolova, K. Jockers, B.S. Gustafson, and G. Lichtenberg, J. Geophys. Res. **106**, 10113 (2001).
66. K. Jockers, T. Bonev, M. Delva, et al., Astronomishe Gesellschaft **18**, 139 (2001).
67. J.B. Renard, E. Hadamcik, and A.-C. Levasseur-Regourd, Astron. Astrophys. **316**, 263 (1996).
68. Y.R. Fernández, Earth, Moon, and Planets **89**, 3 (2002).
69. A. Dollfus, Astron. Astrophys. **213**, 469 (1989).
70. L. Kolokolova and K. Jockers, Planet. Space Sci. **45**, 1543 (1997).
71. B. Kneissel, G.H. Schwehm, and C. Leinert, In T.I. Gombosi (Ed.), "Cometary Exploration II", p. 177 (Central Res. Inst. Physics Hungarian Acad. Science, Hungaria, 1983).
72. J. J. Michalsky, Icarus 47, 388 (1981).
73. M. Oishi, K. Kawara, Y. Kobayashi, et al., Publ. Astron. Soc. Jap. **30**, 149 (1978).
74. P. Bastien, F. Menard, and R. Nadeau, MNRAS **223**, 827 (1986).
75. T.Y. Brooke, R.F. Knacke, and R.R. Joyee, Astron. Astrophys. **187**, 621 (1987).
76. T. Jones and R.D. Gehrz, Icarus **143**, 338 (2000).
77. K. Jockers and N.N. Kiselev, Proccedings of Asteroids, Comets, Meteors ACM 2002, ESA **SP-500**, 567 (2002).
78. T.L. Farnham, D.G. Schleiher, L.M. Woodney, et al., Science 292, 1348 (2001).
79. N.N. Kiselev and G.P. Chernova, Sov. Astron. Lett. **5**, 294 (1979).
80. A. Dollfus, P. Bastien, and J.-F. Le-Borgne, Astron. Astrophys. **206**, 348 (1988).
81. N.N. Kiselev, K. Jockers, and V.K. Rosenbush, Earth, Moon, and Planets **90**, 167 (2002).
82. V.K. Rosenbush, N.N. Kiselev, and S.F. Velichko, Earth, Moon, and Planets **90**, 423 (2002).
83. A. Z. Dolginov, and I. G. Mytrophanov, Astron. Zh. **52**, 1268 (1975).
84. G. W. Wolf, Astron. J. **77**, 576 (1972).
85. J. J. Michalsky, Icarus **47**, 388 (1981).
86. V.K. Rosenbush and N.M. Shakhovskoj, In B.A.S. Gustafson, L. Kolokolova, and G. Videen (Eds.), "Electromagnetic and Light Scattering by Nonspherical Particles", p. 283 (Army Research Laboratory, Adelphi Maryland, 2002).
87. N.G. Beskrovnaja, N.A. Silant'ev, N.N. Kiselev, and G.P. Chernova, In "The Diversity and Similarity of Comets", p. 681 (Belgium, ESA SP-278, 1987).
88. International Comet Quarterly **23**, (2001).
89. J.T.T. Mäkinen, J-L. Bertaux, M.R. Combi, and E. Quémerais, Science **292**, 1326 (2001).
90. D. Bockelée-Morvan, N. Biver, R. Moreno, et al., Science **292**, 1339 (2001).
91. T.L. Farnham, D.G. Scleicher, L.M. Woodney, et al. Science **292**, 1348 (2001).
92. A.N. Cox (Ed.) "Allen's astrophysical quantities. Fourth edition" (Springer-Verlag, New, Inc., USA, 1999).

Vera Rosenbush and Nikolai Kiselev during reception of NATO ASI

Valeri Loiko, Ludmilla Kolokolova, and Anny-Chantal Levasseur-Regourd

CHARACTERIZATION OF DUST PARTICLES USING PHOTOPOLARIMETRIC DATA: EXAMPLE OF COMETARY DUST

LUDMILLA KOLOKOLOVA,[1] HIROSHI KIMURA,[2]
AND INGRID MANN[2]
[1]*University of Florida, Department of Astronomy,*
211 SSRB, Gainesville, FL, 32611, USA
[2]*Institut für Planetologie, Westfälische Wilhelms-*
Universität, Wilhelm-Klemm-Straße 10, D-48149,
Germany

Abstract. We present an approach to constrain characteristics of an ensemble of dust/aerosol particles using integrated photopolarimetric data. We test the approach with one of the best-studied and most intriguing ensembles – cometary dust. We analyze available cometary photopolarimetric data to identify characteristic light-scattering properties typical for cometary dust. We review attempts to interpret these properties using theoretical and laboratory simulations of light scattering by cometary dust. Particular attention is given to the most successful models: (1) aggregates of submicron particles and (2) ensembles of multishaped polydisperse particles. We explore the capabilities of these two models to describe the light scattering by cometary dust. We demonstrate that the model of multishaped polydisperse particles cannot reproduce cometary photopolarimetric data by using a single set of dust characteristics; whereas, the aggregate model can. Our calculations allow for further constraint of the cometary dust properties as large aggregates of absorbing material that is consistent with general ideas about the origin of comets.

1. Introduction

Studying physical properties of small particles through their interaction with radiation is a typical example of remote sensing that is usually associated with

G. Videen et al. (eds.), Photopolarimetry in Remote Sensing, 431-454.

the scattering inverse problem. This means that light-scattering data are used to determine or constrain particle size, shape, and composition (refractive index). Remote sensing of natural dusts and aerosols is based on studying the scattered sunlight that limits the experimental data to the first column of the scattering matrix [1]. As a result the dependencies of the intensity (I) and polarization (P) on the geometry of observations (mainly determined by the phase angle Sun-object-observer, α) and wavelength (λ) become the information that is used to solve this problem. In this paper we show how these characteristics can be used to constrain the properties of natural ensembles of particles.

We consider cometary dust as a test object for remote-sensing photopolarimetry. Our choice is based on these observational specifics: (1) the low concentration of the dust in cometary atmospheres, in the majority of cases, excludes multiple scattering; (2) the data for cometary dust have been obtained within a broad range of phase angles ($\alpha = 1.5-167°$ for intensity and within $\alpha = 0.3-122°$ for polarization); whereas, for the majority of cosmic bodies ground-based observations are limited to small phase angles, covering often only the backscattering region. Even though the spectral data for cometary dust are limited by wavelength bands free of gas contamination, so called continuum bands, reliable spectral trends in intensity and polarization have been obtained for a broad range of wavelengths, within which the dust interacts with light in the scattering regime (optical and near-infrared wavelengths).

This paper focuses on the general regularities in the light scattering by cometary dust. They are typical for the majority of comets regardless of differences in their properties, such as nucleus size, activity, and dust-to-gas ratio; for detailed reviews of the observational data see, for example, [2–5]. We review common interpretations of these regularities and show how constraints on the size, composition, and structure of cometary dust can be obtained from the observed angular and spectral characteristics of the brightness and polarization of the scattered light.

2. Observational data

2.1 Angular dependence of brightness

One direct way to characterize cometary dust is to study how much light it scatters and how this depends on the geometry of observations, determined by the phase angle. The scattering efficiency is usually characterized by the albedo; its variation with the phase angle is called phase function.

The geometric albedo A of a particle is defined as the ratio of the energy scattered at $\alpha = 0°$ to that scattered by a white Lambert disk of the same geometric cross section [6]. Since comets are usually observed at $\alpha \neq 0°$, it is convenient to define $A_p(\alpha)$ as the product of the geometric albedo and the normalized phase function at angle α. Hanner and Newburn [7] presented a plot of $A_p(\alpha)$ in the J band-pass (1.2 μm) for ten comets. The resulting albedos were found very low, close to 0.025 at $\alpha = 35–80°$ and about 0.05 at α near zero. Although there may be several components of the dust, with differing albedos, the low average albedo rules out a large population of cold, bright grains that contribute to the scattered light but not to the thermal emission.

Gehrz and Ney [8] compared the scattered energy to the absorbed energy reradiated in the infrared to derive the so-called bolometric albedo at the phase angle of observation. The method has been applied to a number of comets providing the albedo within the phase angle 35–80° [7–9]. Two comets have been observed at large phase angles $120° < \alpha < 150°$ [8, 10, 11]. Recently observations of comet 69P/Machholz were performed at phase angles $\alpha > 150°$ using SOHO/LASCO coronograph [12, 13]. Millis *et al.* [14] derived phase function data for small phase angles for comet 38P/Stephan-Oterma. These data are summarized in Fig. 1. One can see that a typical phase function for cometary dust possesses a distinct forward-scattering surge, a rather gentle back-scattering peak and is flat at medium phase angles.

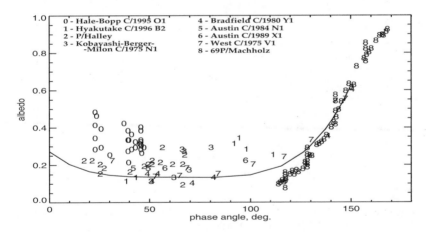

Figure 1. Dependence of bolometric albedo on phase angle. The data are from (0) [15], (1) [16], (2–7) [8], (8) [13]. The solid line represents the least-square fit to the data for comets C/1975 VI (West) and C/1980 Y1 (Bradfield) and is interpolated to connect smoothly to the backscattering curve from [13] normalized at α = 30°. The data from [13] are normalized to the data for comet Bradfield at α = 127°.

2.2 Spectral dependence of brightness

The wavelength dependence of the light scattered by dust can be presented as the color. In the case of comets this is determined traditionally as the difference of the comet apparent magnitude m in two continuum filters, e.g., blue (B ~ 0.4–0.45 μm) and red (R ~ 0.64–0.68 μm), i.e., $C_{B-R} = m_{blue} - m_{red}$. Thus, the color is a unitless characteristic, the logarithm of the ratio of intensities in two filters, that makes it independent of the concentration of the dust particles and distances to the Sun and observer. The determination of cometary dust colors requires use of continuum bands that are truly free of gas contamination. This is easier to achieve for near-infrared wavebands (I, J, H, K) rather than for the visible wavebands, but thermal emission from the warm dust contributes to the K (2.2 μm) band-pass for comets within 1 AU of the Sun. We summarize the data for the B-R visible colors and J-H and H-K near-infrared colors in Fig. 2. It can be seen that the scattered light is generally red (intensity increases with wavelength). Blue colors often are associated with a comet activity (see, e.g., [17]) or emission contamination of the continuum filters [3]. No tendencies of the color change with the heliocentric distance or phase angle have been observed.

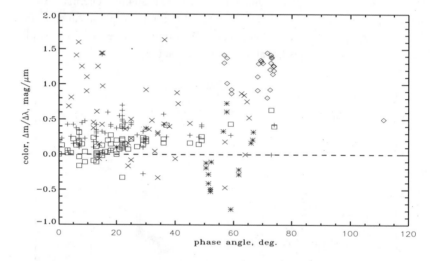

Figure 2. Color as a function of phase angle: J-H (+), H-K (□), and B-R (x) colors are from [18]; B-R colors (◊) are from [19] (* for comet 1P/Halley). Color is presented as $\Delta m/\Delta \lambda$, the solar color is subtracted. The dashed line indicates zero value.

2.3 Angular dependence of cometary polarization

As well as color, linear polarization presents a ratio of intensities of the scattered light, $P = (I_\perp - I_{/\!/}) / (I_\perp + I_{/\!/})$, where $I_\perp$ and $I_{/\!/}$ are the intensity components perpendicular and parallel to the scattering plane. Thus, polarization characterizes properties of the particles, and its values are not directly influenced by the distances to the Sun and observer, and the spatial distribution of the dust particles. For a given type of cometary dust at a given wavelength, the polarization mainly depends on the phase angle.

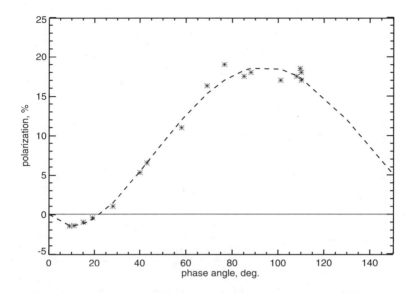

Figure 3. The dashed line shows a typical shape of the polarization phase dependence. Polarization measured in the narrow-band blue filter for comet C/1989 X1 (Austin) is shown by (∗) [20].

All comets show a shallow branch of negative polarization at the backscattering region, that inverts to positive polarization at $\alpha_0 \sim 21°$, and a bell-shape positive branch with a broad maximum near 90–100° (Fig. 3). The minimum polarization is typically -(1.5–2)% [21–23], while the polarization maximum usually stays within 10–30% [23, 24]. Only comet C/1995 O1 (Hale-Bopp) has been found to have maximum polarization that is distinctively higher than other comets as it reaches polarization equal to 18% already at $\alpha = 48°$ [3, 25].

2.4 Spectral dependence of polarization

First discovered in 1P/Halley [26] and later confirmed for other comets [22, 27, and references therein] that the polarization usually increases with increasing wavelength, at least in the visible domain. The polarization spectral gradient can be defined as polarimetric color, $\Delta P / \Delta \lambda = [P(\lambda_2) - P(\lambda_1)] / [\lambda_2 - \lambda_1]$. The polarimetric color changes with phase angle [22, 27] from zero and even negative values to gradually increasing positive values at $\alpha > 50°$ that is opposite to the tendency observed for asteroids and other atmosphereless bodies, which usually have negative polarimetric color [32]. Figure 4 summarizes the polarimetric color data for comets 1P/Halley, C/1996 B2

(Hyakutake), and C/1995 O1 (Hale-Bopp), measured within the range of wavelengths 0.4–0.8 µm. More details can be found in [2, 5].

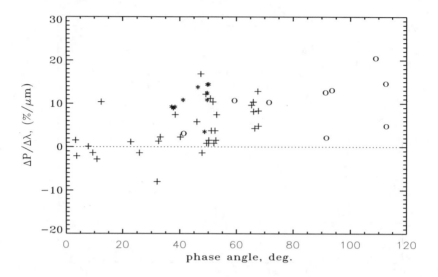

Figure 4. Polarimetric color vs. phase angle for comet (+) 1P/Halley, (o) C/1996 B2 (Hyakutake), and (*) C/1995 O1 (Hale-Bopp). The data are from [22, 28–31].

3. Theoretical and experimental simulations

Early models described cometary dust as homogeneous spherical grains. They could reproduce some observational data but failed to obtain smooth and stable angular dependencies of intensity and polarization within a broad range of wavelengths for realistic refractive indices [33]. Also, the non-zero circular polarization and variations of the polarization plane within the coma [34, 35] could not be explained with homogeneous spherical grains. It is reasonable to assume that real dust particles are not perfect spheres and vary in size, shape, and composition. Below we review the results of modeling cometary dust grains as non-spherical inhomogeneous particles and show how they managed to explain observational results presented in section 2 and what constraints on the dust properties could be formulated based on these studies.

3.1 Theoretical simulations

The main results for non-spherical grains in theoretical modeling have been obtained using two popular numerical techniques: the T-matrix approach and the coupled-dipole approximation.

The T-matrix approach is based on a harmonic solution to the boundary condition problem on the particle surface [36]. Popularity of the T-matrix approach is explained by its fast performance, especially for random oriented particles, and the online availability of the codes developed by M. Mishchenko (http://www.giss.nasa.gov/~crmim/t_matrix.html). The codes allow for the calculation of light scattering by mono- and polydisperse ensembles of a variety of particles, among them spheroids, cylinders, and Chebyshev particles. Results of T-matrix calculations for a variety of refractive indices, size distributions, and shapes of particles are presented in [1]. Mishchenko and Travis [37] undertook a survey of polarization for randomly oriented non-spherical particles with a power-law size distribution that is characterized by the effective radius and variance, weighted by the cross-sectional area. The polarization of such ensembles was found to significantly differ from that for similar ensembles of equal-volume or equal-area spheres. For example, the polarization for silicate spheroids at intermediate phase angles changes from negative, typical for spheres, to positive, leaving a negative branch at $\alpha < 30°$, i.e., it becomes similar to that of comets. The intensity phase function would be reminiscent of the cometary one, however at $\alpha \approx 10°-60°$ it reaches a deep minimum instead of being flat. The color of the spheroids changes both its value and even sign rapidly with phase angle. Extending the model by calculating over a shape distribution, i.e., using spheroids with a range of aspect ratios, can provide a good fit to the cometary phase function for polydisperse silicate spheroids whose effective size parameter ($x_{eff} = 2\pi r_{eff}/\lambda$) ranges within $6 < x_{eff} < 24$ [38, 39]. For the size distribution with $x_{eff} > 8$ and a constant refractive index $m = 1.53 + i0.008$ the color becomes neutral and then red with a very weak angular change. However, the polyshaped silicate spheroids [38] could not provide a good fit to the cometary polarization.

The T-matrix code by Mackowski and Mishchenko [40] was developed to calculate light scattering by multisphere clusters; it can be found online at ftp://ftp.eng.auburn.edu/pub/dmckwski/scatcodes. Petrova et al. [41] used this code for a comparison of cometary observations with the calculations for aggregated particles. The aggregates of highly ordered structures were built from particles of size parameters $1 < x < 3.5$, for a set of refractive indices with the real part equal to 1.65 and the imaginary part varying within the range 0.002–0.1. The maximum number of constituent particles (CP) in each aggregate was limited to 43 by the code convergence within the capabilities of their computing facilities. "Cometary" type polarization with low maximum at $\alpha \sim 90°$ and negative branch at small phase angles was obtained, but required a power-law size distribution of aggregates built of 8–43 CPs of $x = 1.3$, 1.5, and 1.65. The smooth curve and quantitative fit to the observational polarization were not achieved even for polydisperse aggregates. The characteristics of the negative polarization depend on the composition (it gets less pronounced with the absorption increasing from 0.01 to 0.05) and porosity (it is more pronounced for more compact aggregates). Rather good qualitative fit to the cometary phase function was achieved for the polydisperse aggregates with a size parameter of the CP x = 1.3 and an imaginary part of the refractive index κ = 0.01. Petrova and Jockers [42] showed that the same regularly structured aggregate model successfully reproduces the observed positive polarimetric color. However, the color of the aggregates was found to be blue, unlike the observed red color. In contrast, Kimura [43] showed that a randomly structured single aggregate consisting of silicate CPs of x = 1.57 reproduces the cometary polarization with a very good fit. It also reproduces the "cometary" shape of the brightness phase function. This model provides the best (among those available to date) fit to the observed phase curves of the brightness and polarization, although the spectral characteristics of the brightness and polarization were not verified. The number of CPs within 64 < N < 256, as well as the aggregate porosity, was found not to be important. A change in composition (carbon instead of silicate), however, could dramatically affect the value of maximum polarization, eliminate the negative branch, and flatten the phase function at medium phase angles.

A different group of light-scattering theories relies on solutions for the internal field from which the scattered field is calculated. The most popular among such methods is the coupled dipole approximation often called the Discrete Dipole Approximation (DDA), since it represents a particle as an ensemble of polarizable units, called dipoles [44]. Draine and Flatau [45] and, independently, Lumme and Rahola [46] improved the computational efficiency of the DDA, which allowed studying larger particles, among them aggregates of larger size. The most popular has been DDSCAT code by B. Draine and P. Flatau (http://www.astro.princeton.edu/~draine/DDSCAT.html).

West [47] used the DDA for a systematic study of light scattering by aggregates. He demonstrated that for compact aggregates both the intensity and polarization behave as it is typical for the spherical equivalent of the aggregate. For loose aggregates the projected area of the whole aggregate primarily defines the intensity in the forward scattering domain; whereas, the polarization is defined by the size of the constituent particles. Hence, even aggregates of thousand CPs of x < 1 behave like Rayleigh particles in polarization regardless of their composition [43, 48]. There is, however, evidence that for CPs of x > 0.2 a negative polarization at small phase angles can appear if the aggregate is sufficiently large, at least larger than the wavelength [43, 46, 49]. Xing and Hanner [50] confirmed that polarization is more sensitive to the properties of CPs whose size and refractive index determine most of the polarization; whereas, shape, number, and packing generate scattering effects of the second order. They achieved a reasonable fit to the observed cometary polarization and intensity for the mixture of carbon and silicate aggregates of CPs of x > 1 with an intermediate ("touching-particle") porosity.

The DDA was used in [51] for calculations of the complete scattering matrices of icy and silicate aggregates made of 20–200 CPs of x = 1.2–1.9 for comparison with results of some simplified theoretical methods. It was found that scattering in the forward domain could be calculated using the first-order scattering approximation (also known as coherent scattering) with the same accuracy as using the DDA, i.e., interaction between CPs is not manifested at large phase angles. It was also shown that excluding the case of Rayleigh constituent particles, the phase dependence of polarization differs more from that of an individual CP as the size parameter and the refractive index of the CPs increase. Some configurations of particles resulted in a cometary-type phase dependence of polarization, however, the phase trend in intensity was very different from the observed one for all aggregates considered in [51].

A study [52] of the light scattering by non-spherical, regularly shaped particles of size parameter 1 < x < 5 showed that particles consisting of transparent material (silicate) of different shapes produced very different polarization phase curves but the effect of particle shape was reduced for absorbing particles (carbon). The negative polarization at small phase angles could be reproduced by non-spherical silicate particles of x ≥ 2. Although the use of regular shapes caused oscillations that would be washed out in the scattering by irregular particles, the authors showed that mixed silicate and carbon particles with a distribution of size parameters 1 < x < 5 could produce phase curves of polarization and intensity qualitatively similar to the cometary curves. A similar conclusion was reached in [50], based on the scattering by small aggregates.

Lumme and Rahola [53] considered cometary particles as stochastically shaped, i.e., particles whose shape can be described by a mean radius and the covariance function of the radius given as a series of Legendre polynomials. They made computations for a variety of particle shapes and size parameters (x = 1–6) using the refractive index $m = 1.5 + i0.005$. They found that the particles should have size parameters x > 1 to provide the negative polarization and low maximum polarization. Ensembles of particles with a power-law size distribution showed phase functions of intensity and polarization similar to the cometary ones. No information of the spectral characteristics was presented.

3.2 Experimental simulations

Complexity of cometary and other natural dusts and aerosols stimulated several approaches to solve the scattering problem through experimental simulation.

In the most popular experimental techniques streams or suspensions of small particles, e.g., powdered terrestrial rocks, are illuminated by a source of light, recently mainly by a laser. By measuring the light scattered by the particles, the intensity, polarization, Stokes parameters, and even the complete Mueller scattering matrix can be obtained for a range of phase angles.

An early experimental work [54] showed significant differences between light scattering by large (20–40 μm) terrestrial/meteoritic particles and Mie calculations. The work also showed that the "cometary" polarization was not typical for transparent irregular particles (which had the polarization maximum located at $\alpha \approx 45°$ and negative polarization at large phase angles) as well as for absorbing compact particles (which had a bell-shaped positive polarization curve with high polarization maximum). However, slightly absorbing loose aggregates (e.g., fly ash) had the phase dependence of polarization similar to that of comets.

Extensive measurements of all 16 elements of the scattering matrix have been performed by Hovenier's group in Amsterdam [55–57]. They provide a public database at http://www.astro.uva.nl/scatter/. The scattering matrix of polydisperse, heterogeneous, irregular particles, presented by terrestrial and meteoritic powders, was studied in [57] at wavelengths 0.442 and 0.633 μm. The effective radius of the particles was approximately a few microns so that the size parameter of the particles peaked within x = 10–20. Both terrestrial and meteoritic samples showed polarization curves of shape similar to the cometary one and positive polarimetric color. However, the intensity at 0.442 μm exceeded the intensity at 0.633 μm, i.e., the color of the particles was blue. Also, the values and angle of the minimum polarization almost twice exceeded those observed; whereas, the maximum polarization was almost half of the observed values. The study shows that polydisperse heterogeneous irregular particles cannot be ruled out as candidates for cometary dust, although, at least for the size distributions that peak at x = 10–20, not all the observed characteristics are reproduced correctly.

Light scattering experiments under reduced gravity (microgravity) have been proposed to avoid sedimentation of the studied dust as well as particle sorting or orientation that can occur for particles dropped or suspended in airflow (see [2] for details). A systematic microgravity study of aggregated particles [58] showed that at small phase angles the samples made of single-type CPs demonstrated positive polarization, which could become negative for mixtures of small and large or dark and light particles. This can be evidence of a heterogeneous structure of cometary grains. A mixture of fluffy aggregates of submicron silica and carbon grains demonstrated a positive polarimetric color whereas a negative polarimetric color was typical for gray compact particles of size greater than the wavelength. This can explain the difference between the polarimetric color of comets and asteroids.

Another technique, called the microwave analog method, simulates the light scattering using microwave radiation [60, 61]. It takes advantage of the same scaling used in all theoretical approaches where the particle dimensions are given through its size parameter, i.e., as a ratio to the wavelength. This allows light scattering by a single submicron or micron particle to be simulated using a manageable millimeter or centimeter analog model as long as the size parameter of the particle and its refractive index are preserved. Systematic studies of light scattering by complex particles have been performed using microwave facilities designed and built at the University of Florida [62]. This facility works across a waveband 2.7–4 mm to simulate the visible region 0.4– 0.65 μm and thus to study not only angular but also spectral characteristics of light scattering (colors). A systematic study of microwave analog scattering by aggregates [63] confirmed the facts that the polarization is mainly determined by the size and composition of constituent particles and that the size parameter of cometary CPs must be $1 < x < 10$ to produce low maximum of polarization and negative polarization at small phase angles. The color and polarimetric color of aggregates also depend mostly on the size and composition of the CPs. For aggregates made of CPs of $x = 0.5$–20 and $m = 1.74 + i0.005$, an increase in the size of CPs results in larger (more red) color and smaller polarimetric color. Combination of positive polarimetric color and blue color is typical of aggregates consisting of non-absorbing particles of $0.5 < x < 10$, whereas red color and positive polarimetric color are indicative of dark, absorbing aggregates. Consequently, the aggregates, which scatter light similarly to the cometary dust, consist of absorbing constituent particles of size parameter $x > 1$. The shape of CPs could not be seen to influence the aggregates' color and just slightly changes the position and degree of maximum polarization. Notice that the results were obtained for a refractive index that does not change with wavelength. The intensity and polarization data for a variety of particles, including aggregates, are collected at http://www.astro.ufl.edu/~aplab/.

3.3 Summary

Table 1 summarizes the observational data and results of theoretical and laboratory simulations of light scattering by cometary dust. As mentioned in section 1 we indicate here only results related to the main regularities in cometary photopolarimetric data. More detailed analysis can be found in [5].

One can see that the theoretical and laboratory simulations of light scattering point to heterogeneous (silicates and some absorbing material) particles that are either aggregates of submicron particles or irregular/multishaped polydisperse grains as plausible models of cometary dust. In the next section we will check the capabilities of both models to reproduce the whole scope of cometary photopolarimetric data.

Table 1. Summary of observational facts and their interpretations.

Observational facts	Dust models with successful interpretation
1. Low albedo of the dust.	1. Absorbing particles.
2. Prominent forward-scattering and gentle back- scattering peaks in the phase dependence of intensity, "flat" behavior at medium phase angles.	2. Aggregates of particles of x = 1–5; silicate polydisperse (power law) elongated particles (e.g., spheroids) with a distribution of the aspect ratios; silicate polydisperse (power law) irregular particles.
3. Negative branch of polarization for $\alpha < 20°$ with the minimum $P \approx -$ (1.5–2)% and bell-shaped positive branch with low maximum of value $P \approx 10$–30% at $\alpha \approx 90$–100°.	3. Polydisperse (power law) irregular submicron particles; aggregates of particles of x = 1–5.
4. Usually red or neutral color for a broad range of wavelength that does not change with the phase angle	4. Slightly absorbing particles of x > 6; absorbing particles of x >1; particles containing material with a spectrally dependent refractive index.
5. Polarization at a given α above $\approx$ 30° usually increases with λ (positive polarimetric color).	5. Aggregates of particles of x = 1–5 or polydisperse non-spherical particles of x > 1 (but not x >> 1) for a broad range of refractive indices independent of wavelength; particles made of materials with spectrally dependent refractive index.

4. Survey of inversion capabilities of the successful models.

4.1 Aggregate model

To calculate light scattering by aggregates we use the double-precision superposition T-matrix code [40] (see section 3.1). We represent cometary grains as fractal aggregates of identical spherical particles. We build fractal aggregates as described in [64, 65], using the BCCA (Ballistic Cluster-Cluster Aggregate) model to present highly porous aggregates and the BPCA (Ballistic Particle-Cluster Aggregate) model for more compact aggregates. BCCAs grow in a system where clusters of the same size follow linear trajectories and stick upon collision, while BPCAs grow in a system where a cluster produced by previous collisions and single particles follow linear trajectories.

We set the radius of the constituent particles equal to $a = 0.1$ μm, the same radius as was inferred in [66] from the arguments for cometary dust temperature and has long been used for modeling cometary dust [67]. We refer the reader to [68] for a discussion of the CP's size as well as for details of the computational techniques. The number N of the CPs is $N = 64$, 128, or 256; the larger numbers of N fall outside of the limitation of our computational resources for the selected refractive index, radius, and configuration of CPs. As a result, the aggregate with $a = 0.1$ μm has a radius of a volume-equivalent sphere $a_V = 0.400$, 0.504, or 0.635 μm.

The time-consuming computations require a careful selection of the particle material based on previous light-scattering simulations and other than light-scattering information. We reject that the composition is dominated by silicates, since we have found that even the best-fit model [43] cannot provide the correct spectral characteristics of the scattered light. We assume that the composition of CPs is consistent with the composition of the dust in comet 1P/Halley [69]; i.e., it is a mixture of silicates, metals, and carbonaceous materials with the volume filling factors of silicate 31.76%, iron 2.56%, and carbonaceous materials 65.68% (for more details see [68, 70]). Using the Maxwell-Garnett mixing rule [71] we obtain the complex refractive index equal to $m = 1.88 + i0.47$ at $\lambda = 0.45$ μm and $m = 1.98 + i0.48$ at $\lambda = 0.6$ μm. The chosen wavelengths are close to the effective wavelengths of blue- and red-band filters. For both wavelengths we calculate the intensity and polarization of the scattered light for randomly oriented BCCAs and BPCAs (Fig.5) particles as a function of phase angle. We present the intensity using the geometric albedo defined as $A_p = (S_{11}/k^2) (\pi/G)$ where S_{11} denotes the (1,1) element of the scattering matrix S_{ij}, k is the wave number, $k=2\pi/\lambda$, and G denotes the geometric cross section [6]. The degree of linear polarization is given by $P = -S_{12}/S_{11}$ where S_{12} is the (1,2) element of the scattering matrix.

Figure 5 illustrates that the results for the BPCA are qualitatively identical to, and quantitatively only slightly different from those for the BCCA. This implies that, in accord with [43], the morphology of cometary grains is of minor importance for their light-scattering properties in the visible. The results for $N = 64$, 128, and 256 clearly demonstrate the similarity in the light-scattering properties for aggregates of different sizes. As a result, a size distribution of aggregates, even if taken into account, does not drastically change the light-scattering characteristics of an ensemble of aggregates. This contrasts with the results from the majority of previous modeling, which often had to apply averaging over various sizes or compositions to produce smooth phase functions. The weak size dependence most likely results from the fact that we use a large number of CPs and highly irregular configurations.

The geometric albedo is a smooth function of phase angle with a slight increase toward small α and a significant enhancement toward large phase angles. The value for backscattered radiation increases gradually with N, but at small phase angles is almost independent of N and structure. Consequently, the low geometric albedo of cometary dust at small phase angles does not provide information on the overall size nor structure of aggregates but indicates the predominance of optically dark material.

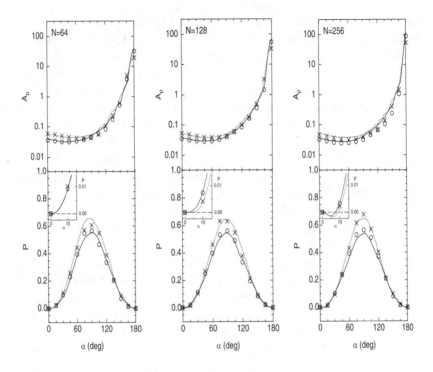

Figure 5. The geometric albedo A_p and polarization P for randomly oriented BCCA (shown by lines) and BPCA (shown by symbols) aggregates consisting of optically dark submicron grains. Solid lines and crosses are the numerical results for λ=0.45 µm and dotted curves and circles are for λ=0.60 µm. Dashed horizontal straight lines in the upper left corner of each lower panel illustrate P=0. The number of constituent particles in the aggregate is N=64 (left panel), 128 (middle panel), or 256 (right panel).

The polarization has an overall bell-shaped curve with a maximum at $\alpha \sim$ 90°. A close look at the curves shows a shallow negative branch at small phase angles whose depth and width gradually grow with N. In previous studies, the negative polarization was often attributed to optically bright grains (large dielectric particles), although such grains contradict the low albedo of cometary dust. The new results show that the negative polarization at small phase angles is also characteristic of large aggregates built of absorbing materials. We also notice a tendency of decreasing the maximum polarization with increasing N. This tendency does not show up clearly in Fig. 5, but becomes more evident in [72], where the DDA is applied.

The important achievement of the model is that it can reproduce the observed spectral trends in the intensity and polarization. Fig. 5 indicates that both the intensity and polarization of aggregates increase with the wavelength (have red color) over wide ranges of phase angle. Within the phase-angle range of 0–90°, the color is approximately constant, tending to decrease with the phase angle, which corresponds to the tendency seen in Fig. 2 for the B-R color. The polarimetric color increases as the phase angle changes from 0 to 90° as was observed for cometary dust.

The spectral variations of refractive indices significantly influence the color of intensity and polarization [27]. The color as well as the shape of intensity and polarization curves is also sensitive to the refractive index. The refractive index used for these calculations is typical for optically dark materials: it is high and has a positive gradient with wavelength. Thus, we conclude that the use of optically bright materials prevented previous researchers from reproducing the observed spectral characteristics of the intensity and polarization as well as the observed low albedo.

The radius, number, and refractive index of the CPs in the aggregates presented above were obtained through a comprehensive survey of parameters. Future calculations with improved computer capabilities hopefully can confirm the tendencies that we have discussed in this section.

4.2 Multishaped polydisperse model

To perform the calculations of light scattering by ensembles of multishaped polydisperse particles, we use the extended-precision T-matrix code for randomly oriented non-spherical particles by Mishchenko discussed in section 3.1. Following the successful approach described in [38], we build our ensemble from a mixture of prolate and oblate spheroidal particles whose aspect ratio varies from 1.4 to 2.6. The particles of each aspect ratio have a power-law size distribution characterized, as it is defined in [37], by variance 0.05 and effective radius; we have performed the calculations with effective radii from 0.01 to 1.5 μm. Figures 6–8 show results of our calculations for different refractive indices of the material. The x-axis in the figures is the phase angle and the y-axis is the effective radius of particles in the ensemble. Thus, each horizontal scan through each figure shows the phase dependence and the vertical scan shows the size dependence of a light-scattering characteristics. Intensity (a) is shown as a logarithm of the quantity S_{11}/k^2 defined in section 4.1, polarization (b) is defined as $-S_{12}/S_{11}$, color (c) is defined as the difference of logarithms of intensity in two wavelengths (we consider the wavelengths 0.45 and 0.6 μm), i.e., the color is $log\ I_{0.6} - log\ I_{0.45}$, and polarimetric color (d) is shown as the difference of polarization in two wavelengths, i.e., $P_{0.6} - P_{0.45}$. The value of the characteristics shown is indicated by its darkness: darker colors correspond to smaller values.

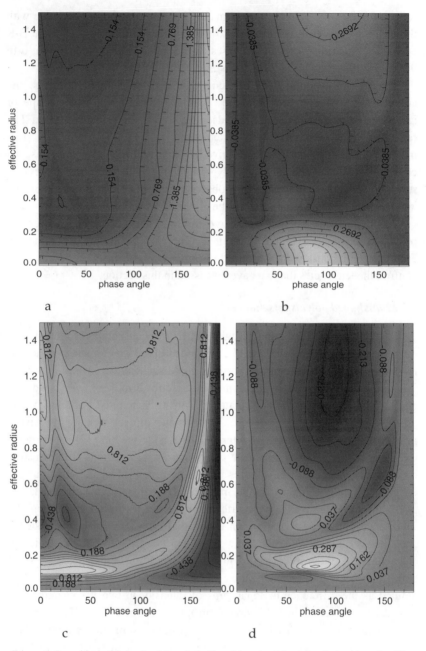

Figure 6. Logarithm of intensity (a), polarization (b), color (c), and polarimetric color (d) as functions of the phase angle in degrees (x-axis) and effective radius of the particle size distribution in microns (y-axis) for silicates. Smaller values are shown darker (see the labels on contours for more details).

Figure 6 shows the results for the most accepted component of cometary dust [73], silicates, characterized by the refractive index 1.5 + i0.05 at both wavelengths. Analysis of Fig. 6 shows that the size distributions characterized by small effective radii $r_{eff} < 0.2$ μm do not reproduce the cometary data since they show no increase in brightness and negative polarization at small phase angles. For large effective radii, $r_{eff} > 1$ μm, we see both effects at small phase angles: the negative polarization and backscattering peak in the brightness. However, unlike what is observed in comets, the color of particles of this size is rapidly changing with phase angle and the polarimetric color is negative and becomes smaller with the phase angle. Negative polarimetric color, phase variations of color together with the low albedo of cometary dust require more absorbing materials than silicates.

Figure 7 presents the results for the so-called "cosmic organic refractory" whose refractive index is $m = 1.91 + i0.317$ at 0.45 μm, and $m = 1.98 + i0.268$ at 0.6 μm [74]. The results shown in Fig. 7 are also typical of comet 1P/Halley composition [69] since the average refractive indices of the comet 1P/Halley material (see section 4.1) are close to those for the cosmic organic refractory.

Absorbing particles of radius less than 0.2 μm also cannot reproduce the observed behavior of brightness and polarization at small phase angles. Moreover, in this case, the opposition peak and negative polarization do not appear for large particles. Only the color of this dust is consistent with the cometary properties: it is red and does not change with the phase angle, at least within phase angles 0–90°, that are typical for observations of comets. However, the polarimetric color of such dust does not fit the one observed for comets: it decreases with the phase angle reaching negative values at the angles around 90°.

Figure 8 shows results of calculations using a multishaped polydisperse model to simulate porous aggregates. For this purpose the refractive index was calculated using the Maxwell-Garnett mixing rule for a mixture of 20% Halley-like material and 80% of voids; this gives us the refractive index equal to 1.158 + i0.07 at 0.45 μm and 1.176 + i0.06 at 0.6 μm.

One can see that such a method to mimic aggregates fails to reproduce the backscattering peak in brightness and shows unrealistic values and behavior of the polarimetric color. This can be expected since the Maxwell-Garnett rule, as well as many other effective medium theories, were derived for a medium built of small, Rayleigh-type inhomogeneities. They cannot be applied to aggregates made of non-Rayleigh constituent particles and therefore fail to reproduce their light scattering correctly.

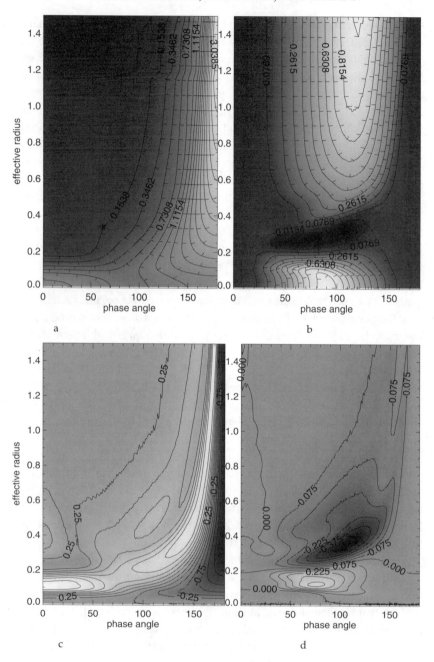

Figure 7. Same as Fig. 6 but for cosmic organics [74]. Similar plots have been obtained for the particles that have the composition of the dust of comet 1P/Halley.

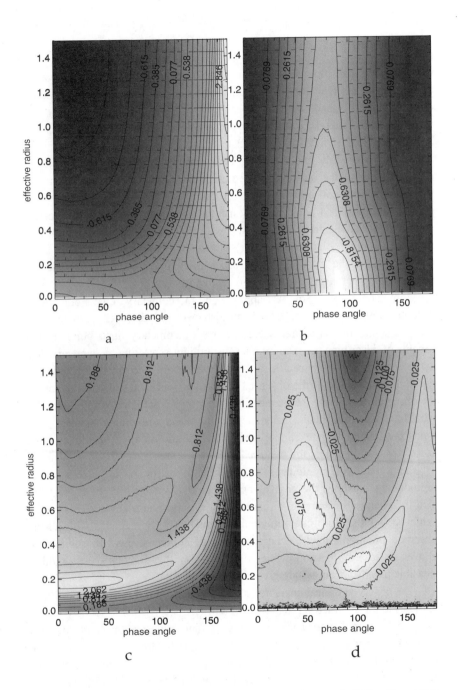

Figure 8. The same as Figs. 6 and 7 but for porous particles that contain 80% voids and 20% material with optical properties of the dust of comet 1P/Halley [69].

5. Conclusions

Our study has provided us with a set of conclusions of general value, applicable to the inversion of the light-scattering data for a variety of natural dusts and aerosols, and with some specific conclusions for cometary dust.

Among the general conclusions the most important one is that the inversion of a limited set of light-scattering data (only intensity or only polarization; only angular or only spectral data, etc.) may lead to a completely wrong conclusion about the physical properties of the particles. The correct model should simulate not only the correct angular behavior of intensity and polarization but also their correct spectral characteristics and the correct variations of the spectral characteristics with phase angle. The spectral change in polarization (polarimetric color), even just its sign, provides strict constraints on the particle properties, particularly, it can be crucial at distinguishing between solid and aggregated particles.

More specifically, in its application to cometary dust, our study demonstrates that the model of multishaped polydisperse particles can simulate *some* observational characteristics of cometary dust, but it is not capable of describing the main regularities in light-scattering by cometary dust using a single set of particles, although we cannot necessarily exclude the possibility that some of the observed optical features stem only from a part of the observed dust. Note that simulation of aggregate particles as solid particles with low refractive index, calculated using effective-medium theories with voids considered as one of the components of the medium, cannot provide correct characteristics of the scattered light.

The model of aggregates of submicron particles is successful at reproducing not only the angular and spectral characteristics of comets but also the change of the spectral characteristics with phase angle. This model achieved, for the first time, simultaneous qualitative agreements with all the observed photopolarimetric characteristics of cometary dust. In addition, it allows us to put the following constraints on the cometary dust models (feasible parameter ranges are given in parentheses):

− the observed optical characteristics altogether limit the size of constituent particles to the submicron range ($a \sim 0.1$ μm);

− the spectral variations of intensity and polarization, and the low geometric albedo require optically dark materials, the refractive index of which commonly increases with wavelength, the best results have been achieved at $\mathrm{Re}(m) > 1.9$, $\mathrm{Im}(m) > 0.1$, $d\mathrm{Re}(m)/d\lambda > 0$, $d\mathrm{Im}(m)/d\lambda > 0$;

– the presence of negative polarization at small phase angles indicates the large overall sizes of aggregates compared to the wavelength of the incident radiation (for the visible, $a_V > 0.6$ μm).

These results were obtained with mineralogically and morphologically reasonable parameters of cometary dust. A mixture of silicates, metals, and carbonaceous materials, random aggregate structures, and submicron CPs are in agreement with characteristics not only the dust in comet 1P/Halley, but also interplanetary dust particles (IDPs), collected in the Earth's atmosphere [75].

Further constraints on the properties of cometary dust require including other light-scattering information. Among them the most obvious is the consistency with the thermal emission properties of particles (temperature, spectral energy distribution, shape and strength of thermal emission spectral features). In the case of cometary dust, it is also necessary for the model to be consistent with the dynamic properties of particles (their spatial distribution resulting from the influence of gravitation, radiation pressure, etc.), evolutionary theories (processes in the early solar system, interstellar medium, etc.), the *in-situ* (Giotto, Vega, DS3, Stardust missions) results, and properties of collected IDPs.

Acknowledgements

We are grateful to D. W. Mackowski, K. A. Fuller, and M. I. Mishchenko for providing the T-matrix codes. L. Kolokolova acknowledges the NATO ASI travel grant and support from the NASA Planetary Atmospheres program. H. Kimura and I. Mann appreciate the support by the German Aerospace Center DLR (Deutschen Zentrum für Luft- und Raumfahrt) under the project "Kosmischer Staub: Der Kreislauf interstellarer und interplanetarer Materie" (RD–RX–50 OO 0101–ZA).

References

1. M. I. Mishchenko, L. D. Travis, and A. A. Lacis, Scattering, Absorption, and Emission of Light by Small Particles, (Cambridge Univ. Press, Cambridge, 2002).
2. A. -C. Levasseur-Regourd, this issue.
3. N. Kiselev and V. Rosenbush, this issue.
4. K. Jockers, Earth, Moon, Planets, **79**, 221 (1999).
5. L. Kolokolova, M. S. Hanner, A.-C. Levasseur-Regourd, and B. Å. S. Gustafson in Comets II, Eds. M. Festou, U. Keller, and H. Weaver, (Arizona Press, Tucson, 2004).
6. M. S. Hanner, R. H. Giese, K. Weiss, and R. Zerull, Astron. Astrophys. **104**, 42 (1981).
7. M. S. Hanner and R. L. Newburn, Astrophys. J., **97**, 254 (1989).
8. R. D. Gehrz and E. P. Ney, Icarus, **100**, 162 (1992).
9. A. T. Tokunaga, W. F. Golisch, D. M. Griep, C. D. Kaminski, and M. S. Hanner, Astron. J., **92**, 1183 (1986).
10. E. P. Ney and K. M. Merrill, Science, **194**, 1051 (1976).

11. E. P. Ney, in Comets, 323, (Univ. of Ariz. Press, Tucson, 1982).
12. K. Jockers, K., Ye. Grynko, R. Schwenn, and D. Biesecker, In *Book of abstracts ACM-2002, July 29-August 2002, Berlin, Germany,* #07-04 (2002).
13. Ye. Grynko, K. Jockers, and R. Schwenn, Astron. Astrophys., in press, (2004).
14. R. L. Millis, M. F. A'Hearn, and D. T. Thompson, Astron. J., **87**, 1310 (1982).
15. C. G. Mason, R. D. Gerhz, T. J. Jones, C. E. Woodward, M. S. Hanner, and D. M. Williams, Astrophys. J. **549**, 635 (2001).
16. C. G. Mason, R. D. Gerhz, E. P. Ney, D. M. Williams, and C. E. Woodward, Astrophys. J., **507**, 398 (1998).
17. T. Bonev, K. Jockers, E. Petrova, M. Delva, G. Borisov, and A. Ivanova, Icarus, **160**, 419 (2002).
18. D. Jewitt and K. Meech, Astrophys. J., **310**, 937 (1986).
19. L. Kolokolova, K. Jockers, G. Chernova, and N. Kiselev, Icarus, **126**, 351 (1997).
20. S. Kikuchi, A. Okazaki, M. Kondo *et al.* in Proc. 23th ISAS Lunar and Planet. Symp., 39 (1990)
21. S. Mukai, T. Mukai, and S. Kikuchi, in Origin and Evolution of Interplanetary Dust, Eds. A. C. Levasseur-Regourd and H. Hasegawa, 249, (Kluwer, Dordrecht, 1991).
22. G. Chernova, N. Kiselev, and K. Jockers, Icarus, **103**, 144 (1993).
23. A.-C. Levasseur-Regourd, E. Hadamcik, and J. B. Renard, Astron. Astrophys., **313**, 327 (1996).
24. O. V. Dobrovolsky, N. N. Kiselev, and G. P. Chernova, Earth, Moon, Planets, **34**, 189, (1986).
25. A.-C. Levasseur-Regourd and E. Hadamcik, J. Quant. Spectrosc. Radiat. Transfer, **79**, 903 (2003).
26. A. Dollfus, P. Bastien, J.-F. Le Borgne, A. C. Levasseur-Regourd, and T. Mukai, Astron. Astrophys., **206**, 348 (1988).
27. L. Kolokolova and K. Jockers, Planet. Space. Sci., **45**, 1543 (1997).
28. K. Jockers, V. K. Rosenbush, T. Bonev, and T. Credner, Earth, Moon, Planets, **78**, 373 (1999).
29. N. N. Kiselev and F. P. Velichko, Earth, Moon, Planets, **78**, 347 (1999).
30. A. Dollfus and J.-L. Suchail, Astron. Astrophys., **187**, 669 (1987).
31. S. Kikuchi, Y. Mikami, T. Mukai, S. Mukai, and J. Hough, Astron. Astrophys., **187**, 689 (1987).
32. T. Mukai, T. Iwata, S. Kikuchi, R. Hirata, M. Matsumura, Y. Nakamura, S. Narusawa, A. Okazaki, M. Seki, and K. Hayashi, Icarus, **127**, 452 (1997).
33. T. Mukai, S. Mukai, and S. Kikuchi, Astron. Astrophys., **187**, 650 (1987).
34. V. K. Rosenbush, A. E. Rosenbush, and M. S. Dement'ev, Icarus, **108**, 81 (1994).
35. V. K. Rosenbush, N. M. Shakhovskoj, and A. E. Rosenbush, Earth, Moon, Planets, **78**, 381 (1999).
36. P. C. Waterman, Phys. Rev. D, **3**, 825 (1971).
37. M. I. Mishchenko and L. Travis, J. Quant. Spesctrosc. Radiat. Transfer, **51**, 759 (1994).
38. M. I. Mishchenko, Appl. Opt. **32**, 4652 (1993).
39. M. I. Mishchenko, L. Travis, R. Kahn, and R. West, J. Geophys. Res., **102**, 16831 (1997).
40. D. W. Mackowski and M. I. Mishchenko, J. Opt. Soc. Am. A, **13**, 2266 (1996).
41. E. V. Petrova, K. Jockers, and N. Kiselev, Icarus, **148**, 526 (2000).
42. E. V. Petrova and K. Jockers, in Electromagnetic and Light Scattering by Nonspherical Particles, Eds. B. Gustafson, L. Kolokolova, and G. Videen, 263 (ARL, 2002).
43. H. Kimura, J. Quant. Spectrosc. Radiat. Transfer, **70**, 581 (2001).
44. E. M. Purcell and C. R. Pennypacker, Astrophys. J., **186**, 705 (1973).
45. B. T. Draine and P. J. Flatau, J. Opt. Soc. Am. A, **11**, 1491 (1994).
46. K. Lumme and J. Rahola, Astrophys. J., **425**, 653 (1994).
47. R. West, Appl. Opt., **30**, 5316 (1991).
48. T. Kozasa, J. Blum, H. Okamoto, and T. Mukai, Astron. Astrophys., **276**, 278 (1993).

49. V. Haudebourg, M. Cabane, and A.-C. Levasseur-Regourd, Phys. Chem. Earth C, **24**, 603 (1999).

50. Z. Xing and M. S. Hanner, Astron. Astrophys., **324**, 805 (1997).

51. K. Lumme, J. Rahola, and J. Hovenier, Icarus, **126**, 455 (1997).

52. P. A. Yanamandra-Fisher and M. S. Hanner, Icarus, **138**, 107 (1998).

53. K. Lumme and J. Rahola, J. Quant. Spectrosc. Radiat. Transfer, **60**, 439 (1998).

54. K. Weiss-Wrana, Astron. Astrophys., **126**, 240 (1983).

55. J. W. Hovenier, in Light Scattering by Nonspherical Particles: Theory, Measurements, and Applications, Eds. M. I. Mishchenko, J. W. Hovenier, and L. D. Travis, 355, (Acad. Press, San Diego, 2000).

56. J. W. Hovenier, H. Volten, O. Muñoz, W. J. van der Zande, and L. B. Waters, J. Quant. Spectrosc. Radiat. Transfer, **79-80**, 741 (2003).

57. O. Muñoz, H. Volten, J. F. de Haan, W. Vassen, and J. Hovenier, Astron. Astrophys., **360**, 777 (2000).

58. E. Hadamcik, J.-B. Renard, J.-C. Worms, A-C. Levasseur-Regourd, and M. Masson, Icarus, **155**, 497 (2002).

59. E. Hadamcik and A.-C. Levasseur-Regourd, J. Quant. Spectrosc. Radiat. Transfer, **79-80**, 661 (2003).

60. J. M. Greenberg, N. E. Pedersen, and J. C. Pedersen, J. Appl. Phys., **32(2)**, 233 (1961).

61. R. H. Zerull and R. H. Giese, in Cometary Exploration; Proceedings of the International Conference, Budapest, Hungary, November 15-19, 1982 143; (Akademiai Kiado, Budapest, 1983).

62. B. Å. S. Gustafson, in Light Scattering by Nonspherical Particles: Theory, Measurements, and Applications, Eds. M. I. Mishchenko, J. W. Hovenier, and L. D. Travis, 367, (Acad. Press, San Diego, 2000).

63. B. Å. S. Gustafson and L. Kolokolova, J. Geophys. Res., **104**, 31711 (1999).

64. T. Mukai, H. Ishimoto, T. Kozasa, J. Blum, and J. M. Greenberg, Astron. Astrophys., **262**, 315 (1992).

65. Y. Kitada, R. Nakamura, and T. Mukai, in The Third International Congress on Optical Particle Sizing, Ed. M. Maeda, 121, (Keio University, Yokohama, 1993).

66. J. M. Greenberg and J. I. Hage, Astrophys. J., **361**, 260 (1990).

67. A. Li and J. M. Greenberg, Astrophys. J., **498**, L83 (1998).

68. H. Kimura, L. Kolokolova, and I. Mann, Astron. Astrophys., **407**, L5 (2003).

69. E. K. Jessberger, A. Christoforidis, and J. Kissel, Nature, **332**, 691 (1988).

70. I. Mann, H. Kimura, and L. Kolokolova, J. Quant. Spectrosc. Radiat. Transfer, submitted.

71. C. Bohren and D. Huffman, Absorption and Scattering of Light by Small Particles, (John Wiley & Sons, New York, 1983).

72. H. Kimura and I. Mann, J. Quant. Spectrosc. Radiat. Transfer, submitted.

73. M. S. Hanner and J. P. Bradley, in Comets II, Eds. M. Festou, U. Keller, and H. Weaver, (Arizona Press, Tucson, 2004).

74. A. Li and J. M. Greenberg, Astron. Astrophys. **323**, 566 (1997).

75. D. E. Brownlee, L. Pilachowski, E. Olszewski, and P. W. Hodge, in Solid Particles in the Solar System; Proceedings of the Symposium, Ottawa, Canada, August 27-30, 1979; 333, (D. Reidel Publishing Co., Dordrecht, 1980).

Ludmilla Kolokolova (top left),
Hiroshi Kimura (top right),
Ingrid Mann (left),
and others at Chufut-Kale (bottom).

INVITATION TO SPECTROPOLARIMETRY

YU. S. YEFIMOV
Crimean Astrophysical Observatory,
Nauchny, Crimea, 98409, Ukraine

Abstract. Recent research has made clear the exceptional importance of spectropolarimetry in the study of physical conditions in atmospheres and circumstellar envelopes of different kinds of stars, in galactic nuclei and to obtain knowledge about interstellar matter. The goal of this brief review is to illustrate the field with several examples and to present new revelations made using ground-based spectropolarimetry with large telescopes. I conclude with a short description of a newly proposed spectropolarimeter for the WSO/UV mission.

1. Introduction

The main purpose of this presentation is to draw attention to the rather new and powerful tool of investigation that combines the advantages of spectroscopy and polarimetry, i.e., spectropolarimetry. The use of this method dates to the end of the 1970s. At first it was applied only to study polarization variations across the spectral lines of bright stars. The impressive results obtained shone new light on the physical processes in galaxies, stellar atmospheres and circumstellar envelopes. This field expanded greatly in the 1990s when a new generation of large telescopes and high-sensitivity devices as CCD cameras and high-speed computers became accessible. The quick look on the number of the publications on spectropolarimetry since 1989 through 2000 shows rapid growth (gray boxes in Fig.1) while the number of publications on active galaxies demonstrates relatively constant attention focused on these objects (dark grey boxes in Fig.1).

G. Videen et al. (eds.), Photopolarimetry in Remote Sensing, 455-478.

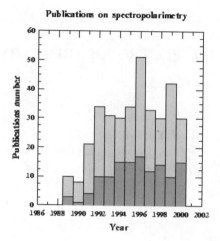

Figure 1. The dynamics of publications on spectropolarimetric observations. Gray boxes represent the total number of publications, dark grey boxes are for number of publications on active galaxies in 1989 - 2000 (from Astronomy and Astrophysics Abstracts).

The contents of this presentation consist of the following sections:

1. Scientific problems;
2. Spectropolarimetry versus polarimetry;
3. Brief review of polarizing mechanisms;
4. Some scientific tasks;
5. Some ground-based instrumentation;
6. Some examples;
7. Spectropolarimeter for WSO/UV mission;
8. Some scientific tasks for the space spectropolarimeter at WSO/UV mission.

2. Scientific problems

Polarimetric measurements provide essential information in space physics that cannot be found through other methods of observations. Obtained and analysed polarimetric results have shown that such measurements provide important information about a) diagnostics of non-thermal radiation mechanisms; b) determination of spatial structure of matter, a field of radiation and magnetic fields of small angular-size objects; c) determination of optical, geometrical, physical and chemical properties of dust particles in space - circumstellar, interstellar, and intergalactic.

3. Spectropolarimetry versus polarimetry

Up to now the measurements of polarization of light from stars and other objects has been carried out by either polarimeters, selecting a necessary range of a spectrum by filters, or with spectropolarimeters on the basis of slit spectrographs. To each of them disadvantages are inherent, lowering efficiency of devices and enlarging time of measurements. Polarimetry performed through consecutive filter changes to study spectral dependence of polarization parameters is time consuming. To reduce this time multichannel devices with selective reflecting interference filters are used. But with an increasing number of channels, light losses increase too, limiting the possible number of channels, usually not more than 5 - 8. It also is difficult to use the most effective modern detectors CCDs in multichannel devices.

4. Brief review of some polarizing mechanisms

There are several mechanisms producing polarization. Some of these mechanisms are briefly discussed by Angel [1], Dolginov et al. [2], and Hillier [3]. Unpolarized or linearly polarized light incident on free electrons is partially linearly polarized after scattering with no wavelength dependence of the degree of polarization. This mechanism plays an important role in atmospheres and envelopes of hot stars like close binary stars, Be, planetary nebulae, Wolf-Rayet, novae, supernovae, and also in galaxies with active nuclei (AGN) and quasars (QSS).

4.1 Rayleigh scattering

The light scattered by atoms, molecules and small dust particles has strong wavelength dependence that rises toward ultraviolet as λ^{-4}. This mechanism is important for stars with circumstellar dust envelopes like R CrB type, T Tauri type, Ae/Be stars, miras, slow supernovae, AGN.

4.2 Dust scattering

Dust is responsible for interstellar polarization and polarization in dust circumstellar envelopes and stars embedded in dark nebulae. As a rule, the wavelength dependence in these cases has a maximum whose position depends on the size and matter of particles. The wavelength dependence of interstellar polarization in optics is well represented by the relation: $p(\lambda)/p_{max}=\exp[-K\ln^2(\lambda_{max}/\lambda)]$, where K is accepted equal to 1.15 and λ_{max} is the wavelength of the maximum of polarization p_{max} (Serkowski et al. [4]). For this dependence the term "the Serkowski law" was established. Soon after publication of this dependence it was found that the factor K depends on the location of the maximum of the degree of polarization in the spectrum. However, three modifications of the Serkowski law have been proposed to better fit observed data: Wilking et al. [5], Whittet et al. [6], and Martin et al. [7].

4.3 Raman scattering

This mechanism may be used to explain the appearance of two broad emission lines near 6830 and 7088 Å seen in the near infrared region in symbiotic stars. In these stars Raman scattering of OVI photons by hydrogen is possible. The details of this mechanism have been considered by (Schmid [8]; Schmid and Schild [9]).

4.4   Coherent background effect (opposition effect)

Recently Mishchenko *et al.* [10] predicted that at small phase angles near opposition, asteroids produce a sharp narrow spike of the measurable linear polarization due to coherent backscattering of reciprocal rays (named Opposition Effect). This mechanism operates in planets, asteroids and comets. The polarimetric observations made recently by Rosenbush *et al.* [11] with 125cm telescope at the Crimean Astrophysical Observatory confirm this prediction.

4.5   Synchrotron emission

This type of emission is well known in astrophysical objects. It arises when relativistic electrons propagate through matter in the presence of a magnetic field. Linear polarization is produced in regions of uniform magnetic field. The wavelength dependence of the degree of polarization is flat. The maximum degree of linear polarization may be very high, up to 70% in a well ordered magnetic field. This mechanism is known for blazers, quasars, gamma-ray bursts and supernovae remnants, like Crab nebula.

4.6   Cyclotron emission

Optical cyclotron radiation may be generated in hot plasma when nonrelativistic electrons spiral in very strong magnetic fields with $H > 10^8$ Gauss. This mechanism is responsible for large circular polarization (up to several dozen percents) seen in AM Her type stars (polars).

4.7   Zeeman effect

This effect is well known for stars with moderate magnetic fields ($H > 100$ Gauss) and is used to measure the strength of magnetic fields from the spectral lines sensitive to magnetic fields.

4.8   Hanle effect

This effect arises in areas of weak magnetic field strengths with $H < 100$ Gauss and like the Zeeman effect, it may be useful to measure stellar magnetic fields. The comparison of the applications of Zeeman and Hanle effects is demonstrated by the diagnostic diagram for circumstellar magnetic fields in Fig.2 taken from the paper by Nordsieck [12].

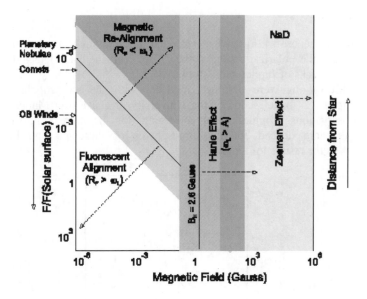

Figure 2. Diagnostic diagram of circumstellar magnetic fields for NaD. Vertical axis: illuminating flux, or distance from illuminator. Light grey: unsaturated diagnostics, where both field magnitude and angle may be recovered. Dark grey: saturated diagnostics, where the geometry, plus a lower limit of the field magnitude may be recovered [12].

4.9   Faraday effect

This effect is connected with the polarization plane rotation in magneto-active matter. In astronomy it is observed mostly in the radio region. The rotation is wavelength dependent as λ^2 and allows an estimate of magnetic field strength from observations at different wavelength.

All these mechanisms may operate simultaneously in different regions of the spectrum. To distinguish between various radiation mechanisms one needs to analyze the wavelength dependencies of all polarization parameters and their variation with time.

5.   Some ground-based instrumentation

There are a number of detailed descriptions of ground-based spectro-polarimeters [13-25]. Providing a detailed description of such devices is beyond the scope of this chapter. The reader can consult for many details in these papers.

6.   Some examples of the observations of different objects

I provide a few examples to demonstrate the power of spectropolarimetry to study features of different kinds of astronomical objects.

6.1   Saturn

Smith and Wostencroft [26] used spectropolarimetry to study the variation of linear and circular polarization across the 7270Å methane band detected from the integrated disks of Jupiter and Saturn.

The observations were obtained in 1980 with the spectropolarimeter attached to the 24-in telescope of the University of Hawaii on Mauna Kea. These observations provide useful constraints for the more sophisticated models that are needed for the atmospheres of these planets. Fig. 3 demonstrates the results for Saturn.

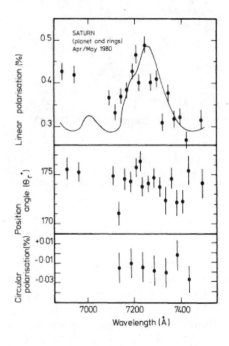

Figure 3. The linear and circular polarization of Saturn during 1980, April and May. The solid curve is the inverted intensity profile across the 7270 Å methane band scaled to the polarization measurements. The Earth was in the plane of the rings at that time so that the contribution of the rings to the integrated intensity and polarization should be negligible [26].

6.2   Jupiter

Several years later the spectropolarimetry was used by Kucherov *et al.* [27] to study the vertical structure of planetary atmospheres from observations of polarization within absorption bands. A method was developed to estimate the abundance of absorbing gas in a layer of planetary atmosphere above the cloud deck by analyzing the variation of the polarization across an absorption band profile (Fig.4). The observations were performed at the center of Jupiter's disk in two methane absorption bands in 1986 with 60 cm telescopes at the soviet field station in Bolivia and on Mt Maidanak in Uzbekistan. Using the wavelength dependence of polarization parameters of aerosol particles they found the abundance of methane in the upper atmosphere of Jupiter.

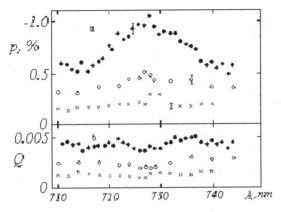

Figure 4. Average values of (a) the percentage polarization *p* and (b) the Stokes parameter *Q* measured at the center of the Jovian disk in the 727 nm methane absorption band. Dots correspond to measurements on 26/27 June and 10/11 July 1986 (phase angle α ≈ 11°); circles are for data on October 8/9, 13/14, 14/15 (α ≈ 6°.9); crosses are for October 1/2, 2/3 (α ≈ 4°.6). The wavelength resolution is 2.5nm. The error of measurements remains constant within the band, and is indicated by a vertical bar for each phase angle [27].

6.3   Cool stars

It is well known that many variable cool giants exhibit intrinsic linear polarization. Both the degree and position angle of polarization tend to vary with time. The polarization usually increases toward short wavelength. It seems quite likely that the polarization originates by scattering from asymmetric clouds of circumstellar dust. However, due to huge optical amplitude of variability, these objects were studied mostly in the bright stage. Landstreet and Angel [28] studied the wavelength dependence of polarization on several Mira-type stars and one RV Tauri star in the range 3300 – 11000 Å with the 5m telescope at Mt Palomar Observatory in 1973 - 1975, some near to brightness minima. All polarization spectra show significant structure on a scale of a few hundred angstroms in some molecular or atomic absorption features, indicating that at least part of the polarization of these stars arises in the stellar atmosphere rather than in a circumstellar dust shell. An example of such a study is given in Fig. 5 for the mira-type star R And (zirconium star), observed near a brightness maximum.

The most striking variations in the polarization spectrum occur in the region 5000-7000 Å, where strong peaks in the polarization degree and marked variations of the position angle coincide with strong ZrO and NaI absorption. A large polarization change also occurs across the ZrO band at 9299 Å, whose behavior changed with time.

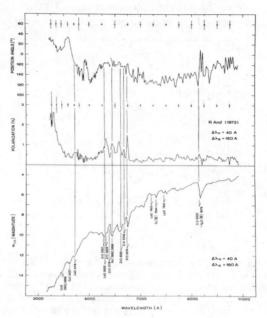

Figure 5. Spectrophotometric and polarimetric observations of the S-type Mira variable R And on 1973 June 22 [28].

The recently discovered unusual star V854 Cen relates to the R Coronae Borealis type star. It is very active, and had shown three major declines since 1987. It is of particular interest to know where the dust forms with respect to the star and emission region. The wavelength dependence of the polarization provides data on how much stellar and emission line radiation is seen directly and how much is scattered from dust. Observations were obtained in 1991 with a spectropolarimeter at the 3.0 m Anglo-Australian Telescope. The data are shown in Fig. 6. The spectropolarimetry data show that the line emission is unpolarized, and unscattered. Thus, the emission line region is not eclipsed by the dust cloud causing the decline. This region may be associated with several dust clouds expelled in many directions. The polarization of the scattered continuum steeply rises into the blue. This could be the result of scattering from small grains or wavelength-dependent dilution effects of direct photospheric light seen through the dust cloud.

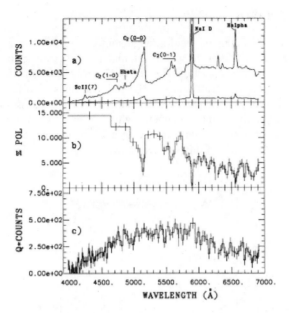

Figure 6. The spectrum (*a*), degree of linear polarization (*b*), and polarized flux of V854 Centauri (*c*) as a function of wavelength. The fact that the polarized flux does not vary across the lines indicates that the lines are unpolarized, and, therefore, arise in a region unobscured by dust [29].

6.4 Hot stars

It is well known that electron scattering plays an important role in hot stars. An example is recent work on Be star ζ Tauri. Fig. 7 shows the optical and UV polarization for this star obtained by Wood *et al.* [30]. Note the large change in polarization across the Balmer jump. While the model (thick line) explains the change in polarization across the Balmer jump, it does not match the level of polarization in the UV.

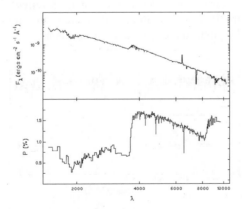

Figure 7. Plot of the optical and UV polarization for the Be star ζ Tauri [30].

6.5. White dwarfs

The distribution of magnetic field strengths among white dwarfs may reveal critical processes in the origin of stellar magnetism. Among the two-dozen single magnetic white dwarfs, WD 0637+477 (V=14.8) in 1992 revealed strong polarization reversals across H_α and H_β lines, indicating a mean longitudinal field of $B_{eff} = +349 \pm 19$ kG. WD 0637+477 possesses the weakest magnetic field yet found on a magnetic white dwarf. The observations were made with a spectropolarimeter attached to a 2.3 m telescope. Results of the spectropolarimetric study of this star are shown in Fig. 8.

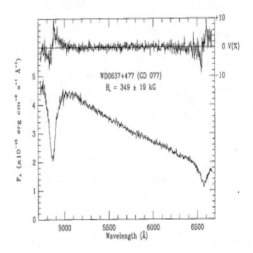

Figure 8. Flux and circular polarization spectra of the white dwarf WD 0637+477 showing strong polarization reversals across both H_α and H_β. From Schmidt et al. [24].

In 1995 a new magnetic white dwarf PG2329+267 was found with magnetic-field strength $B_{eff} = 1.58 \pm 0.08$ MG. The observations were performed with the 1.2 m William Hershel Telescope at La Palma. The circular polarization in the H_α line is about 10%. Fig. 9 shows the structure of the circular polarization across H_α. The two displaced σ components are evident.

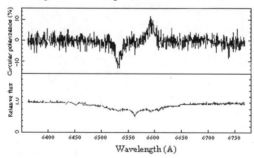

Figure 9. The top panel shows the percentage of circularly polarized light in the spectrum of white dwarf PG 2329+267. The lower panel shows the normalized spectrum of H_α line [31].

The results of phase-resolved spectropolarimetry of the magnetic cataclysmic variable starV884 Her (V~16, period 1.88h), observed in 1998 - 2000 with the 2.3 m Bok telescope at Kitt Peak, reveal a circular polarization spectrum far more structured than any yet seen from a magnetic cataclysmic variable. Very broad polarization humps near 7150 Å and below 4000 Å are interpreted as emission in the cyclotron fundamental and first harmonic in a magnetic field B ~ 115 – 130 MG (see Fig.10). This star is the second AM Her - type system with a magnetic field exceeding 100 MG and the first case in which the cyclotron fundamental has been directly observed from a magnetic white dwarf.

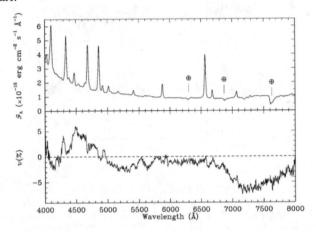

Figure 10. Orbit-averaged total flux and circular polarization spectra of the high-field magnetic variable V884 Her from 1998 September 14. The weak level of circular polarization for a magnetic cataclysmic variable and the highly structured appearance are seen, particularly for λ< 5000 Å. Terrestrial absorption features affecting the spectral flux are indicated [32].

6.6. Supernovae

We describe the structure of polarizing spectra in some Supernovae stars. Understanding spectropolarimetry may be critical to reveal the complicated nature of supernovae. One group of models assumes that the polarization arises from a combination of Thomson and line scattering through the aspherical supernova envelope. Another group of models assumes that the supernovae are associated with an aspherical dusty circumstellar environment.

Figures 11and 12 show the evolution of the polarizing spectra in the region of the H$_\alpha$ line for one of these stars, SN 1998S (Type IIn), obtained with 10m at 5 days after discovery, and with 2.1m telescope in two nights at 10 and 41 days after the optical maximum of the supernova. The data show that the polarization varied with time, meaning an intrinsic origin. The existence of intrinsic polarization supports the idea that the ejection is highly aspherical. However, due to a lack of observational data it is difficult to determine which mechanism is best. In many cases it likely a combination of mechanisms.

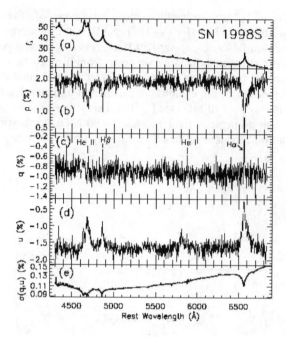

Figure 11. Polarization of SN1998S, obtained on 1998 March, 7.5 days after discovery. (a) Total flux, in units of 10-15 ergs s^{-1} cm^{-2} A^{-1}. (b) Observed degree of polarization. (c,d). The normalized Q and U Stokes parameters, with prominent narrow-line features identified. (e) Average of the (nearly identical) 1σ uncertainties (statistical) in the Stokes Q and U parameters [33].

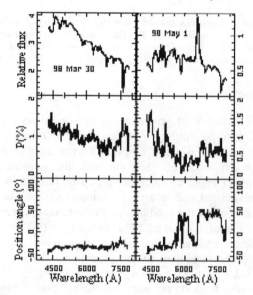

Figure 12. Spectropolarimetry of SN1998S. The left column gives the total flux spectrum, the percent polarization and the polarization angle for the data of 1998 March 30, respectively, and the right column gives the corresponding data for 1998 May 1 [34].

Spectropolarimetry allows us to compare the detailed structure of polarization spectra in supernovae. Fig. 13 provides a comparison of the data on two supernovae: SN 1993J and SN 1996cb of type IIb, obtained with different telescopes at different times. Both sets of data were taken near the maximum brightness of the stars. The real surprise is that two stars are not only spectroscopically similar but also spectropolarimetrically similar. Sharp polarization changes are seen across the H_α and He 5876 Å lines. The variation of the polarization is as large as 1.5% in both supernovae. The fact that these stars show similar spectropolarimetry suggests strongly that the two supernovae have a similar geometry with similar orientation to the observer. This similarity means that they may belong to the sub-group of supernovae of more common phenomena.

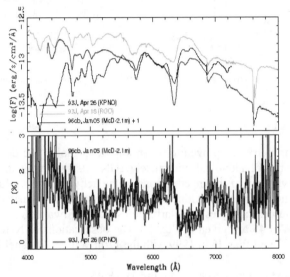

Figure 13. The total flux and polarization spectra of SN 1996cb are compared to those of SN 1993J [34].

6.7. Gamma-ray afterglow

In the last few decades, attention has focused on the investigation of gamma-ray-burst afterglows. An example of this study is given in Fig. 14 of the afterglow of GRB 021004. The observations were made October 5, 2002 with the Very Large Telescopes (UT3) of ESO and show the polarizing spectrum of this object across its strongest absorption lines in Ly_α (left panel) and C IV and Si IV lines (right panel). The authors (Wang et al. [35]) proposed a simple model for the observed polarization variations across spectral lines based on the hypothesis of patchy structure of the circum-burst matter.

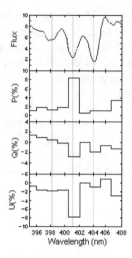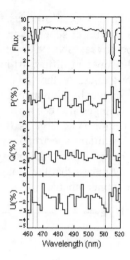

Figure 14. Polarization around the strongest narrow lines of Ly_α (left) and C IV and Si IV (right) [35].

6.7.   Galaxies

Spectropolarimetry has proven to be an important tool in the development of unified theories of active galactic nuclei (AGN). Its strength is that it provides an alternative view of the inner regions of the active nucleus. This allows for the study the structure and kinematics of both polarizing materials and the emission source. Examples of polarization spectra of two active galaxies NGC 5548 and Was 45 are given in Fig. 15. The observations of NGC 5548 were made in 1997 and Was 45 in 1999 with the 1.2m William Hershel Telescope.

6.7.1   Seyfert galaxies NGC 5548 and WAS 45

Spectropolarimetric data of NGC 5548 show that the broad H_α line is polarized at much lower level than the adjacent continuum. There is a sharp drop of polarization on the blue side of the line and gradual increase over the core and red wing back to the continuum level. The position angle of polarization displays unusual structure across the line. In Was 45 the picture is the opposite. The $H\alpha$ line has a significantly higher polarization than the adjoining continuum and there is small change of position angle. In both galaxies at least some of the observed polarization is intrinsic to the sources. Understanding polarization features in these and other AGN is not without difficulties.

Another interesting problem is the determination of the nature of low ionization nuclear emission-line regions (LINERs) found in 38% of nearby emission-line galactic nuclei. The results of the observations of one such galaxy NGC 1052 are shown in Fig. 16. The observations were obtained in 1997 with the 10m Keck-II telescope. It shows a rise of polarization flux in the wings of the H_α line (panel c). This is the first detection of a polarized broad emission line in LINER, demonstrating that unified models of AGN are applicable to at least some LINERs.

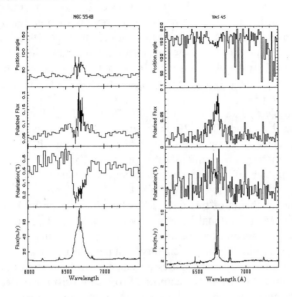

Figure 15. Spectropolarimetric data for NGC 5548 and Was 45. In each frame the panels show, from the bottom, the total flux density, an total flux density, the percentage polarization, the polarized flux density and the position angle of polarization [36].

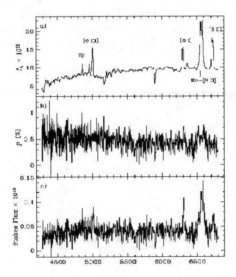

Figure 16. Polarization data for NGC 1052 (liner). (a) Total flux in units of 10-15 erg s^{-1} cm^{-2} Å^{-1}. (b) Degree of polarization. (c) Stokes flux equal to $p \times f_\lambda$ [37].

6.8.   Spectropolarimeter for WSO/UV mission

Polarimetric observations in the far ultraviolet can be performed only from outside the Earth atmosphere. An excellent review of polarimetry in the far ultraviolet from balloons, aircrafts and rockets through 1981 was given by Coffeen [38]. The brief review of information about the equipment and results obtained in the Wisconsin Ultraviolet Photo-Polarimetric Experiment (WUPPE) and with the Hubble Space Telescope (HST) has been published by Kucherov *et al.* [39].

The complete data obtained by WUPPE and HST can be found at http://www.sal.wisc.edu/WUPPE/ and http://archive.stsci.edu/. Due to the prime importance of the polarimetric data from the far ultraviolet region the spectropolarimetric observations are proposed to continue with a new spectropolarimeter of special design which is planned to be on-board the World Space Observatory for UV (WSO/UV).

The optical schema of this device has been developed by Kucherov *et al.* [39] from the Main National Observatory of Academy of Science of Ukraine. It is slitless low-resolution spectropolarimeter. A schema of this device is based on two original optical elements: a superachromatic revolving retarder containing from 5 to 7 plastic plates to measure all four Stokes parameters in the range from 0.3 to 1.5 μm, a complex Wollaston prism with decentered and deformed surfaces which allow us to obtain the two spectra with mutual orthogonal polarization. This unit replaces an ordinary two-block unit containing a normal Wollaston prism to split the light into two (ordinary and extraordinary) beams and a dispersion element to produce the two orthogonal polarized spectra. One other advantage is its ability to measure simultaneously both linear and circular polarization in a very wide spectral range. The general layout and arrangement drawing of this device are shown in Figs. 17 and 18 accordingly.

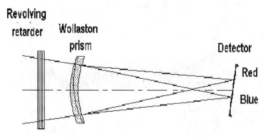

Figure 17. The optical lay-out of the spectropolarimeter with a deformed Wollaston prism.

6.9.   Some scientific tasks for the spectropolarimeter on the WSO/UV mission

In conclusion, I present a number of astrophysical tasks for which spectropolarimetry would be useful.

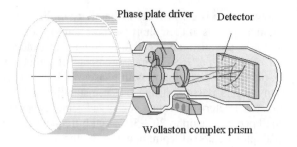

Figure 18. The diagram of the spectropolarimeter.

6.9.1 Extragalactic objects

The numerous ground-based observations of a number of extragalactic objects - blazars, quasars and galaxies with active nuclei - have advanced understanding of their nature considerably. At the same time the necessity of expanding such observations for more detailed study of physical processes in their nuclei is clear. In some blazars, these processes manifest themselves in variability on long (years and months) and short (days and hours) timescales, changing degree and position angle of polarization with wavelength. To study radiation mechanisms, it often is important to determine the degree of circular polarization. Data available for several objects indicate its value is less than one percent. However, measurement accuracy is insufficient for their unequivocal interpretation. Some blazars show bursts of radiation (BL Lac, 3C 279). Statistics of observations of polarized radiation of such objects during flares is rather small, and no total clarity exists in understanding the difference between XR- and radio-selected objects. Thus, the above-mentioned features define scientific tasks for research of these objects. The most perspective objects for ground-based observations are OJ 287, 3C 66A, S5 0716+71, PKS 0735+178, Mkn 421, 3C 279, 3C 345, PG 1551+11, H 1722+11, BL Lac, as well as all bright Seyfert galaxies with stellar-like nuclei and some others active in gamma, X-ray and radio ranges. The linear polarization of a majority of these objects is rather high (not less than several per cents, but sometimes up to 30 %!) and can be measured with error of the order 0.5%.

6.9.2 Stars

The variety of objects in the stellar population of our Galaxy is rather large. I select only the most interesting from the point of view of spectropolarimetric observations.

6.9.2.1 Magnetic interacting binary systems -- polars

This small class of unique stars presents the opportunity to study the structure and behavior of matter in extremely strong magnetic fields. It is well known that these systems consist of compact objects and red dwarfs. The general feature of all known polars is exclusively fast activity in different parts of the optical spectrum in all phases of their short (several hours) photometric period.

In addition, there are also appreciable individual distinctions between different objects of this type. Recent measurements of circular polarization paint a complex picture of its evolution with phase of the light curve and with wavelength. Many polars possess strong phase instability in both brightness and polarization. Characteristic values observed circular polarization are limited from small percentage up to 20 – 30 %, with frequent sign changes. To estimate physical parameters of components and the nature of observable instabilities of their characteristics it is necessary to increase statistics of polars with detailed and repeated polarimetric monitoring with the highest possible time resolution, about a few minutes.

Polarimetric accuracy of measurements of circular polarization for objects 15–16 mag may be about 1%. Linear polarization is less by a factor of ten than circular polarization, so the accuracy of such measurements must be about 0.1%.

6.9.2.2 Magnetic white dwarfs

Currently, white dwarfs are considered the final product of the stellar evolution. Among single white dwarfs stands a small group of objects with very strong magnetic fields of tens and hundreds MG. Usually these objects show rather large (some per cents), strongly wavelength dependent circular and linear polarization. This polarization is the main carrier of information on the features of geometry and of magnetic field intensity on white dwarfs. Some of them show changes of polarization parameters that can indicate a drift of their magnetic axis in space. However, the estimations of magnetic field intensity are field-averaged over a surface of a star. More detailed data on magnetic fields – their topology, inclination of a magnetic axis to an axis of rotation, their time dependence, etc – is necessary to measure polarization parameters in continuum as well as in wide absorption lines and bands. The proposed spectropolarimeter will enable us to study the spectral dependence of polarization changes of this type of star up to 14 mag with an error of about 0.1% .

6.9.2.3 Hot stars

Among hot stars Ae/Be stars with non-periodic brightness minima are of especial interest. Ground-based observations indicate the existence of circumstellar protoplanet disks. Research of such objects provides a unique opportunity to understand the origin of planetary systems around stars, and, in particular, the planets of the Solar system. To study chemical and physical properties of circumstellar dust polarimetric observations of a number of these objects are necessary in the whole spectral range accessible for observations, from far ultraviolet up to far infrared region. Changes of the degree of linear polarization in an optical range of these stars are of several per cents. The accuracy of polarimetric measurements should be about 0.1%.

Other perspective spectropolarimetric observations are from a group of massive Be and WR stars – single as well as components of close binary systems. Data on polarization of their radiation and its changes with time provides valuable information on geometry of extended envelopes of these stars. Spectropolarimetry also enables us to compare the polarization in continuum and in bright lines that help to distinguish between different components – intrinsic and interstellar – in observed polarization and to understand the mechanism of their generation.

6.9.2.4 Cool stars

Among cool stars, giants and supergiants are of especial interest in spectropolarimetry. The presence of temperature variations on the surfaces of stars, spottiness, extended and nonstable gas-dust envelopes, outflow of matter – all strongly complicates the study of these stars. Many aspects of these stars, like their atmospheric structures and dynamics are as yet insufficiently investigated.

Ground–based observations have established that complex changes of linear-polarization parameters take place along the spectra of different type stars, from T Tauri type up to miras. One of the main scientific tasks is the research for process of matter outflow in jets or part of an envelope with dust formation, as in the R CrB type stars, and also accretion, as in the T Tauri type stars. Spectropolarimetry may be extremely useful in a broad spectral range, especially during deep – up to 8 mag – minima, when the linear polarization increases from parts of a per cent up to 5 – 15 % and an appreciable rotation of the polarization plane occurs. For spectropolarimetric research, it is necessary that measurement errors do not exceed 0.5%.

One specific case is the T Tauri type stars. The intrinsic polarization of their radiation arises in processes of scattering in their environments. The main task of research of this star type is the study of processes in envelope, leading to time-dependent the spectral changes of intrinsic polarization. These stars, as a rule, are connected with gas-dust nebulae, which introduces significant – up to 2 % and more – additional polarization in partly polarized stellar radiation passing through it and make it very difficult to find an intrinsic polarization of the object under consideration. Therefore a reliable study of time-dependent spectral changes of intrinsic polarization of T Tauri stars and related objects is necessary to carry out a careful investigation of their environment. This in itself provides valuable information on physical and optical parameters of nebulae, connected with T Tauri type stars. In the majority of cases their polarization can be measured with an accuracy of about 0.1%.

6.9.3 Interstellar polarization

Despite extensive research of characteristics of interstellar polarization in previous years, many problems remain. The research of interstellar polarization is important not only to study the properties of interstellar dust, but also to extract the interstellar polarization from what is observed.

6.9.3.1 Deviations from the Serkowski law

Numerous broad-spectral observations of stars have led researchers to the general behavior of polarization with wavelength of the interstellar polarization:

$$p(\lambda)/p_{max} = \exp[- K\ln^2 (\lambda_{max} /\lambda)],$$

where $K = 1.15$ and λ_{max} is a wavelength of the maximum of polarization p_{max} (Serkowski et al.,1975) [4]. For this dependence the term "the Serkowski law" was established. Soon after publication of this dependence it was found that the factor K depends on the location of the maximum of the degree of polarization in the spectrum. Three modification of the Serkowski law were proposed, with

$$K = (-0.10 \pm 0.05) + (1.86 \pm 0.09) \lambda_{mzx} \qquad \text{(Wilking et al. [5]),}$$
$$K = (0.01 \pm 0.05) + (1.66 \pm 0.09) \lambda_{max} \qquad \text{(Whittet et al. [6]),}$$

and much more complicated formula

$$p(\lambda) = p_{IR}S + p_{UV} (1 - S) \quad \text{(Martin et al. [7]),}$$

where

$$S = 1/[1 + (\lambda_s /\lambda)^\gamma ,$$

$$pir(\lambda) = b\exp(d^\delta) (\lambda_{max} /\lambda)^\beta \exp[-(d\lambda_{max}/\lambda)^\delta],$$

$$p_{UV} (\lambda) = a \exp[-\alpha \ln^2 (\lambda_a /\lambda)],$$

$$d^\delta = \beta/\delta,$$

where λ is in microns, and the coefficients are obtained by fitting the available wavelength-dependent polarization.

However, despite a good average fit to the available spectrum of interstellar polarization, there are rather many essential deviations from the Serkowski law. As the parameters in the Serkowski law depend on structure, size, form and orientation of interstellar particles, the reason for the deviations may be real distinctions of characteristics of the interstellar environment in various localities of the Galaxy. The systematic analysis of these deviations has never been made. To carry out such analyses it is necessary to increase the statistics of the known deviations from the Serkowski law substantially. Values of this polarization in some areas close to the galactic plane can reach up to several percent. For reliable localization of a maximum of polarization the measurements should be carried out not less than in 10 spectral regions. The error of measurements in all accessible spectral range should be less than 0.1 %.

6.9.3.2 Polarization maps of the Galaxy

To find localities in the Galaxy with well marked characteristic features of interstellar matter, wavelength-dependence interstellar polarization measurements are necessary. The ideal goal is the production of detailed maps of interstellar polarization on galactic coordinates, similarly to maps of interstellar absorption. Such a work was carried out early by a number of researchers. However, results of this work can be used only for very rough estimation of parameters λ_{max} and p_{max} and their connection with interstellar reddening close to the galactic plane, and they are quite unsuitable for evaluation of interstellar polarization of particular objects.

7. Conclusions

The goal of this brief review is to demonstrate the power of the spectropolarimetry. We enter now a golden age of polarimetry. The combination of new a generation of giant telescopes, high sensitive detectors, powerful computers and fast communication open the possibility to investigate tiny structures of the spectra of cosmic objects from far ultraviolet to the radio domain for extremely faint objects, to discriminate between various physical mechanisms of radiation and to look into the early stages of the Universe. Spectropolarimetry is now in the point of intersection of microcosm and macrocosm.

References

1. J. R. P. Angel, Mechanisms that produce linear and circular polarization. Planets, Stars and Nebulae studied with Photopolarimetry. Ed. T. Gehrels, Univ. Arizona Press, Tucson, Arizona, 54-63 (1974).

2. A. Z. Dolginov, Yu. N. Gnedin, and N. A. Silant'ev, Propagation and Polarization of Radiation through Cosmic Matter. Moskva, Nauka, 423 (1979).

3. D. H. Hillier, Spectropolarimetry and Imaging Polarimetry. ASP Conference Series, **164**, 90-101 (1999).

4. K. Serkowski, D. S. Mathewson, and V. L. Ford, *Astrophys. J.* **196**, 261 (1975).

5. B. A. Wilking, M. J. Lebofsky, and G. H. Rieke, *Astron. J.* **87**, 695 (1982).

6. D. C. B. Whittet, P. G. Martin, J. H. Hough, M. F. Rouse, J. A. Bailey, and D. J. Axon, *Astrophys. J.* **386**, 562 (1992).

7. P. G. Martin, G. C. Clayton, and M. J. Wolff, *Astrophys. J.* **510**, 905 (1999).

8. H. M. Schmid, *Astron. Astrophys.* **211**, L31-L34 (1989).

9. M. Schmid, and H. Schild, *Astron. Astrophys.* **281**, 145 (1994).

10. M.I. Mishchenko, J-M. Luck, and Th. M. Nieuwenhuizen, *J. Opt. Soc. Am.* A, **17**, 888 (2000).

11. V. K. Rosenbush, N. N. Kiselev, V. V. Avramchuk, N. M. Shakhovskoj, and Yu. S. Efimov, in *"NATO Advanced Research Workshop on the Optics of Cosmic Dust"* Bratislava, Chech Republic (2001).

12. K. H. Nordsieck, New circumstellar magnetic field diagnostics. *ASP Conf. Ser.* **248**, 607-616. Eds. G. Mathys, S. V. Solanki, and D. T. Wickramasinghe (2001).

13. C. Aspin, Polarimetry at the Royal Observatory Edinburgh. *Polarized Radiation of Circumstellar Origin*. Eds. G. V. Coyne, S. J., A. M. Magalhães, A. F. J. Moffat *et al.*, Vatican Observatory-Vatican City State, 693-704 (1988).

14. J. A. Bailey, Polarimetric Instrumentation at the Anglo-Australian Observatory. *Polarized Radiation of Circumstellar Origin*. Eds. G. V. Coyne, S. J., A. M. Magalhães, A. F. J., Moffat, *et al.*, Vatican Observatory-Vatican City State, 747-757 (1988).

15. Th. Eversberg, A. F. J. Moffat, M. Debruyne, J. B. Rice, N. Piskunov, P. Bastien, W. H. Wehlau, and O. Chesneau, *PASP* **110**, 1356 (1998).

16. S. Kawabata, A. Okazaki, H. Akitaya, N. Hirakatas, R. Hirata, Y. Ikeda, M. Kondoh, S.Masuda, and M. Seki, *PASP* **111**, 898 (1999).

17. V. G. Klochkova, V. E. Panchuk, and V. P. Romanenko, Stellar spectropolarimeters (in Russian). *SAO Russian Akademie of Science, Preprint No* **156** (2001).

18. O. L. Lupie, and H. S. Stockman, Calibration of the Hubble Space Telescope (HST) Polarimetric Modes. *Polarized Radiation of Circumstellar Origin*, Eds. G. V. Coyne, S. J., A. M. Magalhães, A. F. J. Moffat, *et al.*, Vatican Observatory-Vatican City State, 705-726 (1988).

19. A.M. Magalhães, and W. F. Velloso, Optical Polarimetry at the University of Saõ Paolo. *Polarized Radiation of Circumstella Origin*, Eds. G. V. Coyne, S. J., A. M. Magalhães, A. F. J. Moffat, *et al.*, Vatican Observatory-Vatican City State, 727- 733 (1988).

20. K. H. Nordsieck, K. P. Jaehnig, E. B. Burgh, H. A. Kobulnicky, J. W. Percival, and M. P. Smith, *Proc. SPIE*, vol. **4843**, Polarimetry in Astronomy, rep.170 (2003).

21. S. McLean, S. R. Heathcote, M. J. Paterson, J. Fordham, and K. Shortridge, *Mon. Not. Roy Astr. Soc.* **209**, 655 (1984).

22. S. Miller, L. B. Robinson, and G. D. Schmidt, *PASP* **92**, 702 (1980).

23. S. Miller, L. B. Robinson, and R. W. Goodrich, A CCD Spectropolarimeter for the Lick Observatory 3-Meter Telescope. Instrumentation for Ground-Based Optical Astronomy. Present and Future. *The Ninth Santa Cruz Summer Workshop in Astronomy and Astrophysics*, Ed. L. B. Robinson. (Springer-Verlag, New York, 1988) 157-171 (1988).

24. G. D. Schmidt, H. S. Stockman, and P. S. Smith, *Astrophys. J.* **398**, L57 (1992).

25. R. D. Wolstencroft, W. A. Cormack, J. W. Campbell, and R. J. Smith, *Mon. Not. Roy Astr. Soc.* **205**, 23 (1983).

26. R. J. Smith, and R. D. Wolstencroft, *Mon. Not. Roy Astr. Soc.* **205**, 39 (1983).

27. V. A. Kucherov, M. I. Mishchenko, A. V. Morozhenko, Sov. Astron. Lett. **14**, 354 (1988).

28. J. D. Landstreet, and J. R. P. Angel, *Astrophys. J.* **211**, 825 (1997).

29. B. A. Whitney, G. C. Clayton, R. E. Schulte-Ladbeck, and M. R. Meade, *Astron. J.* **103**, 1652 (1992).

30. K. Wood, K. S. Bjorkman, and J. E. Bjorkman, *Astrophys. J.* **477**, 926 (1997).

31. C.Moran, T. R. Marsh, and V. S. Dhillon, *Mon. Not. Roy Astr. Soc.* **299**, 218 (1998).

32. G. D. Schmidt, L. Ferrario, D.T. Wickramasinghe, and P. S. Smith, *Astrophys. J.* **553**, 823 (2001).

33. C. Leonard, A. V. Fillipenko, A. J. Barth, and Th. Matheson, *Astrophys.J* **536**, 239 (2000).

34. L. Wang, D. A. Howell, P. Höflich, and J. C. Wheeler, *Astrophys. J.* **550**, 1030 (2001).

35. L. Wang, D. Baade, P. Höflich, and J. C. Wheeler, Spectropolarimetry of GRB 021004 – Evidence for High Velocity Ly–α Absorptions. *astro-ph/* **0301266** (2003).

36. J. E. Smith, S. Young., A. Robinson, E. A. Corbett, M. E. Giannuzzo., D. J. Axon and J. H. Hough, *Mon. Not. Roy Astron. Soc.* **335**, 773 (2002).

37. A. J. Barth, A. V. Fillipenko, and E. C. Moran, *Astrophys. J.* **515**, L61 (1999).

38. D. L. Coffeen, Optical polarimeters in space. *Planets, Stars and Nebulae studied with photopolarimetry*. Ed. T. Gehrels, Univ. Arizona Press, Tucson Arizona, 189-217 (1974).

39. V. A. Kucherov, Yu. S. Ivanov, Yu. S. Efimov, A. V. Berdyugin, and N. M. Shakhovskoy, Ultraviolet low-resolution spectropolarimeter for the space mission SPECTRUM-UV (UVSPEPOL project). *Kosmichna nauka i technologiya (in Russian)*. **3**, No 5/6, 3 - 26 (1997).

*Yuri Yefimov (Efimov) surveying ruins of
ancient safety wall near the Simeiz
Observatory in Crimea*

*Entertainers Sasha Krysyuk,
Ivan Andronov
and Anatoliy Vid'machenko*

Uspensky Monastery on the road to CrAO.

Chufut-Kale

ASTRONOMICAL POLARIMETERS AND FEATURES OF POLARIMETRIC OBSERVATIONS

ALEXANDER MOROZHENKO AND ANATOLIY
VID'MACHENKO
*Main Astronomical Observatory of Ukrainian National
Academy of Sciences*
27 Akademika Zabolotnogo Street, Kyiv, 03680, Ukraine

Abstract. We present a general description of ground-based astronomical polarimeters, and provide a detailed description of the spectropolarimeter of the Main astronomical observatory (MAO) of a National Academy of Sciences of Ukraine (NASU). Using a polarization modulator of a rotating quarter-wave phase plate (FP) allows us to measure the parameters of linear and circular polarization simultaneously. In 1983 O. I. Bugaenko with the colleagues from MAO of NASU produced an automatic astronomical spectropolarimeter (ASP), which used a continuous rotation of polarizer with frequency of 61 Hz. Observations in two beam modes allowed it to accommodate changes of transparency of the Earth's atmosphere, air mass the of observational object, inexactness of guiding and displacement from an optical axis because of atmospheric turbulence. In 1995 the spectropolarimeter was upgraded and its spectral interval expanded to 1 micron. Sources of errors and methods of their elimination are described.

1. Principles of treatment of polarimetric measurement

Practical schemes for constructing polarimeters up to 1970 can be found in [11]. All detectors of optical radiation react to a change of intensity of luminous flux, but do not react to a state of its polarization. Therefore all methods for determining polarization are grounded in the transformation of the Stokes vector a scalar value. A polarization modulator installed prior to the detector is used for this purpose. The simplest modulator is used for the analysis of linearly polarized radiation. It may be a polaroid, prism, grating

G. Videen et al. (eds.), Photopolarimetry in Remote Sensing, 479-486.

etc. Changing the Stokes parameters of light passing through an ideal polarizer is described by the transformation matrix:

$$
\begin{vmatrix} I \\ Q \\ U \\ V \end{vmatrix} = 0.5 \begin{vmatrix} 1 & \cos 2\varphi_2 & \sin 2\varphi_2 & 0 \\ \cos 2\varphi_2 & \cos^2 2\varphi_2 & \cos 2\varphi_2 \sin 2\varphi_2 & 0 \\ \sin 2\varphi_2 & \cos 2\varphi_2 \sin 2\varphi_2 & \sin^2 2\varphi_2 & 0 \\ 0 & 0 & 0 & 0 \end{vmatrix} \cdot \begin{vmatrix} I_0 \\ Q_0 \\ U_0 \\ V_0 \end{vmatrix} \quad (1)
$$

The light intensity is modulated by a polarizer according to

$$I(\omega_1 t) = 0.5(I_0 + Q_0 \cos2(\varphi_2 - \omega_2 t) + U_0 \sin2(\varphi_2 - \omega_2 t)$$

$$= 0.5(I_0 + P\cos2\varphi\cos2\omega + P\sin2\varphi\sin2\omega) = 0.5[I_0 + P\cos2(\varphi - \omega)], \quad (2)$$

where ω_2 – speed of polarizer's rotation.

A two-element modulator is more complicated, as the phase-shifting plate (PSP) is rotating and the polarizer is fixed. The transformation matrix of a PSP with phase shift Δ is

$$
\begin{vmatrix} I \\ Q \\ U \\ V \end{vmatrix} = \begin{vmatrix} 1 & 0 & 0 & 0 \\ 0 & G + H\cos4\varphi_1 & H\sin4\varphi_1 & -\sin\Delta\sin2\varphi_1 \\ 0 & H\sin4\varphi_1 & G - H\cos4\varphi_1 & \sin\Delta\cos2\varphi_1 \\ 0 & \sin\Delta\sin2\varphi_1 & \sin\Delta\sin2\varphi_1 & \cos\Delta \end{vmatrix} \cdot \begin{vmatrix} I_0 \\ Q_0 \\ U_0 \\ V_0 \end{vmatrix} \quad (3)
$$

The light is modulated according to

$$I(\omega_2 t) = 0.5\{I_0 + GQ_0\cos2\varphi_2 + HQ_0\cos2(2\varphi_1 - \omega_1 t) + HU_0\sin2(2\varphi_1 - \omega_1 t)$$

$$- V_0\sin\Delta\sin2(\varphi_1 - \omega_1 t)\}, \quad (4)$$

where ω_1 – speed of PSP's rotation and

$$G = 0.5\,(1 + \cos\Delta) \quad (5a)$$

$$H = 0.5\,(1 - \cos\Delta) \quad (5b)$$

For analyzing linear-polarized light, a PSP with phase shift on 180^0 is used. In this case,

$$I(\omega_2 t) = 0.5\{I_0 + Q_0\cos2(2\varphi_1 - \omega_2 t) + U_0\sin2(2\varphi_1 - \omega_2 t)\}. \quad (6)$$

Here the modulating frequency is twice that in formula (2). The polarization modulator must be a fixed polarizer ($\omega_2 = 0$) and phaseshifter with a phase shift of 90^0 or the polarizer's angular rate must be twice the speed of angular rotation of the PSP ($\omega_2 = 2\omega_1$). Then formula (4) becomes

$$I(t, 90^0) = 0.5\{I_0 + 0.5Q_0(1 + \cos4\omega_1 t) + 0.5U_0\sin4\omega_1 t \pm V_0\sin2\omega_1 t\}. \quad (7)$$

"+" and "-" signs appear before the fourth Stokes parameter according to the state of the polarizer (fixed or rotating, accordingly).

When a quarter-wave rotating phase plate is used in a polarization modulator, it is possible to measure parameters of linear and circular polarization simultaneously, but the accuracy of determining Q and U parameters is half that of parameter V. When a PSP with a phase shift of $\Delta = 126.52^0$ is used, the accuracy of measurements of all three Stokes parameters is identical. When $\varphi_2 = 0$, the formula for light modulation becomes

$$I = 0.5\{I_o + Q_o[0.2 + 0.8\cos4\omega_1 t] + 0.8U_o\sin4\omega_1 t \pm 0.8V_o\sin2\omega_1 t\} \qquad (8)$$

The use of PSPs was restrained by the dependence of the phase shift of single-element plates upon wavelength. It was necessary to produce separate plates even for a narrow spectral range. Now these restrictions have been removed. In the middle of the nineteen eighties Kucherov [7, 8] (MAO of NASU) calculated multicomponent superachromatic phase-shifting plates (SPSP), which could be manufactured not only from one material (for example from a crystalline quartz), but also from expanded films and other plastic materials. Polarimeter construction is determined by the type of detector used, devices of spectral selection, etc. So for his pioneer polarimetric investigations of solar system bodies Lio [10] used a visual polarimeter based on Savar's interference polariscope. The error in determination of the degree of polarization of the light was about 0.1 %.

In the 1950s, panoramic detectors and photomultipliers tubes (PMT) came into use. Their principles of operation are grounded on the use of modulation's formula (2). For panoramic detectors a method of discontinuous variation of position of polarizer's optical axis (with step $\Delta\varphi$) was used. The degree of polarization P and plane position of polarization φ were determined from a system of three equations:

$$I_1 = 0.5(I_0 + Q_0\cos2\varphi_0 + U_0\sin2\varphi_0)$$

$$I_2 = 0.5(I_0 + Q_0\cos2(\varphi_0 + 60^0) + U_0\sin2(\varphi_0 + 60^0)$$

$$I_3 = 0.5(I_0 + Q_0\cos2(\varphi_0 + 120^0) + U_0\sin2(\varphi_0 + 120^0), \qquad (9)$$

or a system of four equations:

$$I_1 = 0.5(I_0 + Q_0\cos2\varphi_0 + U_0\sin2\varphi_0$$

$$I_2 = 0.5(I_0 + Q_0\cos2(\varphi_0 + 22.5^0) + U_0\sin2(\varphi_0 + 22.5^0)$$

$$I_3 = 0.5(I_0 + Q_0\cos2(\varphi_0 + 45^0) + U_0\sin2(\varphi_0 + 45^0)$$

$$I_4 = 0.5(I_0 + Q_0\cos2(\varphi_0 + 67.5^0) + U_0\sin2(\varphi_0 + 67.5^0). \qquad (10)$$

It is necessary to maintain step values of the polaroid rotation precisely with accuracy less than 1.0 arc minutes, otherwise systematic errors become greater than 0.1 %. For PMTs, Mehkur's method is used at observatories with single-channel radiation detectors. In this method the polarizer rotates continuously with a small angular rate ω, and the intensity $I(\varphi)$ is registered by the corresponding optical detector. All above mentioned polarimetric observational methods involve subsequent mathematical processing of observational data to calculate the degree of polarization of light from celestial objects. At the end of the 1950s, Ksanfomaliti [6] proposed a scheme for automated data processing. It was used through a system of automatic control of the PMT in analogue mode. In practice it does not have a deficiency photon-counting mode. Bugaenko and colleagues [1] elaborated and produced the first automatic polarimeter with using this operational mode. They used a continuous polarizer rotation rate of 75 Hz. For the first time they registered the modulated intensity (2) successively in four counters during a half period. In each counter the beginning of a signal recording was displaced by special electronic keys at a phase of $\pi/2$. The numerical data in these counters could be computed from

$$I_1 = I_0(\pi - 2U) \qquad \text{for } 0 \leq 2\omega t \leq \pi$$

$$I_2 = I_0(\pi + 2U) \qquad \text{for } \pi \leq 2\omega t \leq 2\pi$$

$$I_3 = I_0(\pi - 2Q) \qquad \text{for } (\pi/2) \leq 2\omega t \leq (3\pi/2)$$

$$I_4 = I_0(\pi + 2Q) \qquad \text{for } 3\pi \leq 2\omega t \leq 5\pi, \tag{11}$$

Values of Stokes parameters are determined by combinations:

$$I_1 + I_2 + I_3 + I_4 = 4\pi I_o$$

$$I_2 - I_1 = 4I_oU = 4u$$

$$I_4 - I_3 = 4I_o = 4q \tag{12}$$

In practice, three counters are used for recording these values: $4\pi I_o$, $4u$ and $4q$. Photon quantity in counter 1 ($4\pi I_o$) was selected according to the requirement that parameters Q and U are registered with some statistical error. The accuracy of is ~0.05 % using of this polarimeter.

The same idea was used in the development of a spectropolarimeter for simultaneous measurement of all four Stokes parameters [3-5]. A 127-degree SPSP is rotated with frequency 61 Hz and modulated intensity (4) was divided into 8 equal intervals with the use of special electronic keys:

$$I_1 = 0.5\{I_0\pi/8 + (Q_0/8)(0.25\pi + 1) + U_0/8 \pm 0.146V_0\} \text{ for } 0 \leq \pi/8$$

$$I_2 = 0.5\{I_0\pi/8 + (Q_0/8)(0.25\pi - 1) + U_0/8 \pm 0.353V_0\} \text{ for } \pi/8 \lesssim \pi/4$$

$$I_3 = 0.5\{I_0\pi8 + (Q_0/8)(0.25\pi - 1) - U_0/8 \pm 0.353V_0\} \text{ for } \pi/4 \leq 3\pi/8$$

$$I_4 = 0.5\{I_0\pi8 + (Q_0/8)(0.25\pi + 1) - U_0/8 \pm 0.146V_0\} \text{ for } 3\pi/8 \lesssim \pi/2$$

$$I_5 = 0.5\{I_0\pi8 + (Q_0/8)(0.25\pi + 1) + U_0/8 \pm 0.146V_0\} \text{ for } \pi/2 \leq 5\pi/8$$

$$I_6 = 0.5\{I_0\pi8 + (Q_0/8)(0.25\pi - 1) + U_0/8 \pm 0.353V_0\} \text{ for } 5\pi/8 \leq 3\pi/4$$

$$I_7 = 0.5\{I_0\pi8 + (Q_0/8)(0.25\pi - 1) - U_0/8 \pm 0.353V_0\} \text{ for } 3\pi/4 \leq 7\pi/8$$

$I_8 = 0.5\{I_0 \pi 8 + (Q_0/8)(0.25\pi + 1) - U_0/8 \pm 0.146V_0\}$ for $7\pi/8 \leq \pi$ (13)

The combination of such values allows us to acquire information about all four Stokes parameters in four counters:

$$I_1 + I_2 + I_3 + I_4 + I_5 + I_6 + I_7 + I_8 = \pi(I_0 + 0.2Q_0)/2 = I'/2$$

$$I_1 - I_2 - I_3 + I_4 + I_5 - I_6 - I_7 + I_8 = 0.8\ Q = q$$

$$I_1 + I_2 - I_3 - I_4 + I_5 + I_6 - I_7 - I_8 = 0.8\ U = u$$

$$I_1 + I_2 + I_3 + I_4 - I_5 - I_6 - I_7 - I_8 = 0.8\ V = v \qquad (14)$$

If in such a modulator one uses a 180-degree phase-shifting plate, equation (14) reduces to

$$I_1 + I_2 + I_3 + I_4 + I_5 + I_6 + I_7 + I_8 = \pi I_0/2 = I'/2$$

$$I_1 - I_2 - I_3 + I_4 + I_5 - I_6 - I_7 + I_8 = Q = q$$

$$I_1 + I_2 - I_3 - I_4 + I_5 + I_6 - I_7 - I_8 = U = u$$

$$I_1 + I_2 + I_3 + I_4 - I_5 - I_6 - I_7 - I_8 = 0, \qquad (15)$$

2.   Astronomical polarimeter of the Main Astronomical Observatory

There are few spectropolarimeters for simultaneous measurement of all four Stokes parameters of solar system bodies. One such device was developed at the Main Astronomical Observatory of National Academy of Sciences of Ukraine. The scheme of its optical-mechanical construction is shown in Fig. 1.

In its development, it was necessary to combine four devices into one housing: photometer, polarimeter with filters, spectrometer and spectropolarimeter. The change of operational modes was executed by the introduction on an optical axis or by removal of partially passing (~10 %) inclined mirror 5 and polarized modulator 4. In making observations, it is possible to set exposure time in seconds (from 1 up to 9000 s) or quantity of impulses (from 10^3 up to 10^7), which are accumulated in the channel $\pi I_0/2$. The necessary quantity of impulses accumulates in a photometer channel of the two-beam polarimeter. This automatically takes into account the change of Earth's atmosphere transparency, the air mass of the examined object, the inexactness of its guiding and the displacement from an optical axis because of atmospheric turbulence. Since the values of the first two factors depend on wavelength, light filters are used to decrease errors of non-dispersed light.

It is possible to take into account sky background and dark noise in the automatic mode of operation. For this purpose diaphragms are put in a permanent magnetic field. The management system may displace it from one position to another with frequency of 6 Hz. The diaphragm displacement value can be regulated from 0 up to 2 centimeters. The transition time of the diaphragm is about 8 ms.

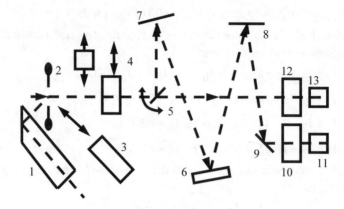

Figure 1. Block-scheme of Spectropolarimeter of MAO of NASU. 1, 3 – forward and back guiders; 2 – gang of diaphragms; 4 – block of polarized modulator; 5 – partially passing (~10%) mirror; 6 – replaceable diffraction gratings with step-changed inclination; 7, 8 – monochromator' mirrors; 9 – fixed oblique mirror; 10 – filters of order; 12 – filters of photometer; 11, 13 – photomultiplier tubes; 14 – source of 100% polarized radiation.

Observational experience with such a system has shown its suitability, but for polarization observations it is not possible to achieve the necessary accuracy to take into account sky background polarization. These observations have shown that the errors increase considerable with increasing amplitude of diaphragm "jump." It is possible to explain this by the instrumental polarization changes with deviations from the optical axis. Two such devices were manufactured in 1985 for a spectral range of 340-760 nm and were placed in the Cassegrain focus of 60-cm telescopes at observatories in Bolivia (near of the Tarija city) and in Uzbekistan (on the Maidanak mountain). In 1995 the spectropolarimeter was upgraded [12] on a spectral interval up to 1 micron by using photomultiplier tubes with InAsGa-photocathode.

3.   Sources of errors and methods for their elimination

All sources of error can be divided into systematic and casual. The instrumental polarization (IP) of a "telescope-polarimeter" system is of the first type. It is stipulated, mainly, by that fact that the polarization degree of the light (both reflected and transmitted) varies with incident angle on the optical elements. In astronomical observations such surfaces are the telescopic mirrors, monochromator mirror, diffraction grating, filters, polarizers, SPSP etc. Therefore, even unpolarized light that falls on the main telescope mirror becomes partially elliptically polarized. The contribution of the polarization attributable to the instrument (see [2], for example) is P_i, φ_i Q_i, U_i and V_i.

Additional sources of equipment error can be the following:
1. Poorly manufactured diaphragm (especially when its cone does not reach a forward surface).

2. Misalignment of filter surfaces from being perpendicular to the optical axis.

3. Wedge-form of polarized modulator or inclination of its optics to the optical axis. In this case there is a beam deflection from the optical axis and a "dancing" image appears when the Polaroid is rotated. This can result from imperfect focusing of a Fabri lens onto a photocathode, which has different spectral response at different locations.

4. Photomultiplier tubes with a lateral entrance have a sensitivity dependence of their photocathode upon polarization degree of the incident light.

5. Optical inhomogeneity of polarizers or phaseshifters.

6. Inexact registration of the rotation angle of the phaseshifter, or the time displacement at the beginning of intensity integration in different channels.

Methods of eliminating instrumental polarization can be divided into technological (which should be executed during device manufacturing) and astronomical (which should be executed during observation and data reduction). Technological methods are the following:

1. Minimization of the number of optical elements in construction.

2. High resolution in focusing and adjustment of telescope and polarimeter optics.

3. All elements that can polarize light should be placed after the polarized modulator. Polarizer or SPSP should be in the forefront, close to the telescope focal plan. The rotor length and diaphragm diameter should be minimal so the light of examined object does not interact with rotor walls.

4. When using a two-element polarized modulator in spectropolarimetry mode, the fixed polarizer should be installed in such a way that its polarization plane is parallel to the polarization plane of the inclined mirrors and to lines of the diffraction grating. Such installation minimizes errors when using polarization-sensitive detectors, like PMT with a lateral photocathode.

5. It is necessary to select carefully electronic filters and keys for maximal elimination of extraneous frequencies, and to maintain equal integration intervals $\Delta\omega t_i$ with a large accuracy (< 1 minute of arc).

6. Using a Wollaston prism as a polarizer, the light after the polarized modulator becomes partially elliptically polarized. In this case it is necessary to use depolarizers.

7. Polarizing properties of some elements, like prisms and SPSP, are very sensitive to the incident angle. This is why it is expedient to install an additional optical system after the diaphragm to form parallel bundle rays.

8. It is necessary to execute observations in the focus of telescopes.

9. It is necessary to use high-frequency (>30 Hz) modulation of luminous flux.

The value of instrumental polarization is determined by the observations of stars with zero polarization. Systematic errors are determined through observations of stars with large polarization. Reliability criterion is the comparison of observation data with statistical error:

$$\sigma = \Delta N/N \approx (N)^{-1/2} \qquad (16)$$

In [9] it was shown that instrumental polarization for the polarimeter installed at the 60-cm telescope of the observatory near Tarija (Bolivia), had statistical error 0.024 % and the mean square was 0.023%. Investigations have shown that instrumental polarization in Newton and Nesmith focus can reach 5%, whereas it is less than 0.05% in a well adjusted Cassegrain. In the former case, values of P_i increase with wavelength; whereas they decrease in the latter.

The change of the transparency of Earth's atmosphere, turbulent picture jitter, and inexactness of guidance are random errors. One large problem is the correct registration of the sky background. This is especially true during twilight and when there is a lunar presence in the sky. In the latter case the sky background polarization can reach 10% depending on lunar phase and the angular distance to observed object. Sky background can either increase or decrease the Stokes parameters of the observed objects.

There are also considerations in the use of polarimeters in panoramic detectors. Because of differences in spectral response for every pixel, the primary source of error is image "jumping" on the detector. Even small beam displacements onto adjacent pixels are registered with a different spectral response. Because of the large information content in a panoramic detector and the inertia of separate pixels the possibility of using an analyzer with high-frequency rotation is restricted considerably. A change of transparency of Earth's atmosphere, guiding errors, turbulent trembling and flickering all play an essential role. Therefore, actual errors of polarimeters with panoramic detectors are more than several times that for polarimeters with PMTs.

References

1. O.I. Bugaenko, L.A. Bugaenko, V.D. Krugov, V.G. Parusimov, Astrometry Astrophys. (1968).
2. O.I. Bugaenko, L.S. Galkin, A.V. Morozhenko, Astron. Zh., **48**, No2, (1971).
3. O.I. Bugaenko, L.A. Bugaenko, A.L. Guralchuk, *et al*, "Photometric and polarimetric investigations of celestial bodies," ed. A.V. Morozhenko, (Naukova dumka, Kyiv, 1985) 169.
4. O.I. Bugaenko, M.A. Melnikov, L.E. Rogozina, V.S. Samoylov, "Photometric and polarimetric investigations of celestial bodies," ed. A.V. Morozhenko, (Naukova dumka, Kyiv, 1985), 164.
5. O.I. Bugaenko, A.L. Guralchuk, "Photometric and polarimetric investigations of celestial bodies," ed. A.V. Morozhenko, (Naukova dumka, Kyiv, 1985), 160.
6. L.V. Ksanfomaliti, Sci. notes of Kharkiv University, **6**, (1962).
7. V.A. Kucherov, "Photometric and polarimetric investigations of celestial bodies," ed. A.V. Morozhenko, (Naukova dumka, Kyiv, 1985).
8. V.A. Kucherov, V.S. Samoylov, Optical and Mechanical Industry, No 9, (1987).
9. A.V. Morozhenko, Kinematics and physics of celestial bodies, **4**, No 1, (1988).
10. B. Lyot, Ann. Observatoire Meudon, **8**, (1929).
11. K. Serkowski, "Planets, Stars, and Nebular Studies with Photopolarimetry," (The University of Arizona Press, Tucson, 1974).
12. A. P. Vid'machenko, P.V. Nevodovskiy, Kinem. Phys. Celest. Bodies, **16**, 1 (2000).

Participants

Ilya Yu. Alekseev, Crimean Astrophysical Observatory, Nauchny, Crimea 98409, Ukraine, ilya@crao.crimea.ua

Ivan L. Andronov, Odessa National University, Department of Astronomy, Shevchenko Park, Odessa 65014, Ukraine, il-a@mail.od.ua

Kirril A. Antonyuk, Crimean Astrophysical Observatory, Nauchny, Crimea 98409, Ukraine, antoniuk@crao.crimea.ua

Ludmila G. Astafyeva, B.I. Stepanov Institute of Physics, National Academy of Sciences of Belarus, F. Scorina Ave. 68, Minsk 220072,Belarus, astafev@dragon.bas-net.by

Vladimir V. Barun, B.I. Stepanov Institute of Physics, National Academy of Sciences of Belarus, Skaryna Pr. 68, Minsk 220072, Belarus, barun@dragon.bas-net.by

Andriy Bilinsky, Ivan Franko National University, Astronomical Observatory, Kyrylo and Mefodiy Str. 8, Lviv, Ukraine, bilian@astro.franko.lviv.ua

Yaroslav Blagodyr, Ivan Franko National University, Astronomical Observatory, Kyrylo and Mefodiy Str. 8, Lviv, Ukraine, blagod@astro.franko.lviv.ua

Sergiy A. Borysenko, Main Astronomical Observatory, National Academy of Sciences of Ukraine, Zabolotnoho Str. 27, Kyiv 03680, Ukraine, borisenk@mao.kiev.ua

Tamara Bulba, Main Astronomical Observatory, National Academy of Sciences of Ukraine, Zabolotnoho Str. 27, Kyiv 03680, Ukraine, tamara@mao.kiev.ua

Varya V. Butkovskaya, Crimean Astrophysical Observatory, Nauchny, Crimea 98409, Ukraine, varya@crao.crimea.ua

Ludmila I. Chaikovskaya, B.I. Stepanov Institute of Physics, National Academy of Sciences of Belarus, F. Scorina Ave. 68, Minsk 220072,Belarus, lch@zege.bas-net.by

Karine Chamaillard, National University of Ireland, I.T Department and Physics Department, Galway, Ireland, Karine.Chamaillard@nuigalway.ie

Klim I. Churyumov, Kyiv Shevchenko National University,Astronomical Obseravory, Observatorna Str. 3, Kyiv 04053, Ukraine, klim.churyumov@observ.univ.kiev.ua

Vladimir Damgov, Space Research Instituite, Bulgarian Academy of Sciences, Moscowska Str. 6, POBox 799 1000 Sofia, Bulgaria, vdamgov@bas.bg

Vladimir P. Dick, Institute of Physics, Academy of Sciences of Belarus, F. Scaryna Ave. 68, 220072 Minsk, Belarus, dick@dragon.bas-net.by

G. Videen et al. (eds.), Photopolarimetry in Remote Sensing, 487-493.
© 2004 *Kluwer Academic Publishers. Printed in the Netherlands.*

Zhanna M.. Dlugach, Main Astronomical Observatory, National Academy of Sciences of Ukraine, Zabolotnoho Str. 27, Kyiv 03680, Ukraine, dl@mao.kiev.ua

Helmut Domke, Ludwig Richer Str. 31, D-14467 Potsdam, Germany, hdomke@t-online.de

Daria N. Dubkova, Sobolev Astronomical Institute, St. Petersburg State University, 5th line V.O.,46-22, St. Petersburg 199004,Russia, daria@dd8103.spb.edu

Oleg Dubovik, Laboratory for Terrestrial Physics, Code 923,NASA Goddard Space Flight Center, Greenbelt, MD 20771, USA, dubovik@ltpmailx.gsfc.nasa.gov

Matthew Easley, Rockwell Scientific, 1049 Camino Dos Rios, Thousand Oaks, CA 91360, USA, Matthew.Easley@usarc-emh2.army.mil

Jay Eversole, Code 5611, Naval Research Laboratory, 4555 Overlook Ave. SW, Washington DC, USA, eversole@nrl.navy.mil

Anatoliy B. Gavrilovich, B. I. Stepanov Institute of Physics, National Academy of Science of Belarus, F. Scorina Ave. 70, Minsk 220072, Belarus, gavril@dragon.bas-net.by

Igor V. Geogdzhayev, Columbia University, Department of Applied Physics and Applied Mathematics, 2880 Broadway, New York, NY 10025, USA, igor@giss.nasa.gov

Larissa Golubeva, Shemakha Astrophysical Observatory, Azerbaijan Academy of Science, Shemakha 373243, Azerbaijan, land@azdata.net

Francisco González, Grupo de Óptica, Dpto. Física Aplicada, Universidad de Cantabria, 39005 Santander, Spain, gonzaleff@unican.es

Oksana S. Goryunova, Astronomical Institute of Kharkiv National University, Sumskaya Str. 35, Kharkiv 61022, Ukraine, dslpp@astron.kharkov.ua

Vladimir Grinin, Crimean Astrophysical Observatory, Nauchny, Crimea 98409, Ukraine, grinin@VG1723.spb.edu

Vladimir P. Kuz'kov, Main Astronomical Observatory, National Academy of Sciences of Ukraine, Zabolotnoho Str. 27, Kyiv 03680, Ukraine, kuzkov@mao.kiev.ua

Yevgen Grynko, Max Planck Institut für Aeronomie, 37191 Katlenburg-Lindau, Germany, grynko@linmpi.mpg.de

John J. Hillman, University of Maryland, Department of Astronomy, College Park, Maryland 20742, USA, jhillman@astro.umd.edu

Keith Hopcraft, University of Nottingham, School of Mathematical Sciences, Applied Mathematics Division, University Park, Nottingham NG7 2RD, UK, keith.hopcraft@nottingham.ac.uk

James H. Hough, University of Hertfordshire, Department of Physics, Astronomy & Mathematics, Hatfield Al10 9AB, UK, jhh@star.herts.ac.uk

Vsevolod V. Ivanov, Astronomical Department and Sobolev Astronomical Institute, St. Petersburg University, Universitetskii Pr. 28, St. Petersburg 198504, Russia, viva@pobox.spbu.ru

Klaus Jockers, Max-Planck-Institut für Aeronomie, 37191 Katlenburg-Lindau, Germany, jockers@linmpi.mpg.de

Olga V. Kalashnikova, Jet Propulsion Laboratory, MS 169-237, 4800 Oak Grove Dr., Pasadena CA 91109, USA, Olga.V.Kalashnikova@jpl.nasa.gov

Vadim Kaydash, Institute of Astronomy of Kharkiv National University, Sumskaya str. 35, Kharkiv 61022, Ukraine, VKaydash@astron.kharkov.ua

Boris N. Khlebtsov, Institute of Biochemistry and Physiology of Plants and Microorganisms, Russian Academy of Sciences, Pr. Entuziastov 13, Saratov 410049, Russia khlebtsov@ibppm.sgu.ru

Nikolai G. Khlebtsov, Institute of Biochemistry and Physiology of Plants and Microorganisms, Russian Academy of Sciences, Entuziastov Pr. 13, Saratov 410049, Russia, khlebtsov@ibppm.sgu.ru

Nikolai N. Kiselev, Institute of Astronomy of Kharkiv National University, Sumskaya str. 35, Kharkiv 61022, Ukraine, kiselev@kharkov.ukrtel.net

Sergey Kolesnikov, Crimean Astrophysical Observatory, Nauchny, Crimea 98409, Ukraine, efimov@astro.crao.crimea.ua

Ludmilla Kolokolova, University of Florida, Department of Astronomy, Gainesville, FL, 32611, USA, ludmilla@astro.ufl.edu

Victor V. Korokhin, Astronomical Institute of Kharkiv University, Sumskaya Str. 35, Kharkiv 61022, Ukraine, dslpp@astron.kharkov.ua,

Theodor Kostiuk, NASA Goddard Space Flight Center, Greenbelt, Maryland 20771, USA, Kostiuk@gsfc.nasa.gov

Ayse Gulcin Kucukkaya, University Trakya, Department of Eng.& Architecture, Trakya Universitesi, Muh. Mimarlik Fakultesi, 22030, Edirne, Turkey, kucukkaya@mailcity.com

Irina V. Kulyk, Main Astronomical Observatory, National Academy of Sciences of Ukraine, Zabolotnoho Str. 27, Kyiv 03680, Ukraine, leda@mao.kiev.ua

Halina P. Ledneva, B. I. Stepanov Institute of Physics, National Academy of Sciences of Belarus, F. Scorina Ave. 68, Minsk 220072, Belarus, astafev@dragon.bas-net.by

A.Chantal Levasseur-Regourd, Aeronomie CNRS-IPSL/ Universite Paris VI, BP 3, 91371 Verrieres, France, aclr@aerov.jussieu.fr, Anny-Chantal.Levasseur@aerov.jussieu.fr

Pavel V. Litvinov, Institute of Radio Astronomy, National Academy of Sciences of Ukraine, Chervonopraporna Str. 4, Kharkiv 61002, Ukraine, Litvinov@ira.kharkov.ua

Valery A. Loiko, Institute of Physics, National Academy of Sciences of Belarus, F.Scaryna Ave. 68, Minsk 220072, Belarus, loiko@dragon.bas-net.by

Albert Lomach, Crimean Astrophysical Observatory, Nauchny, Crimea 98409, Ukraine, lomach@crao.crimea.ua

Dmitriy F. Lupishko, Institute of Astronomy of Kharkiv National University, Sumskaya Str. 35, Kharkiv 61022, Ukraine, lupishko@astron.kharkov.ua

Andreas Macke, Institute for Marine Research, Düsternbrooker Weg 20, D-24105 Kiel, Germany, amacke@ifm.uni-kiel.de

Oleg Makarenkov, Voronezh State University, Mathematical Department, Universiteskaja Pl. 1, Voronezh 394006, Russia, omakarenkov@kma.vsu.ru

Hal Maring, NASA Headquarters, 300 E Street SW, Washington, DC 20024-3210, USA, hal.maring@nasa.gov

Irina Martynova, Voronezh State University, Universiteskaja Pl. 1, Voronezh 394006,Russia, imakarenkova@st.vsu.ru

James McDonald, Scientific Computing Group, I.T. Department, NUI, Galway, Ireland, james@it.nuigalway.ie

Ivan A. Mishchenko, National Ukrainian Agricultural University, Geroiv Oborony Str. 15, Kyiv 03041, Ukraine, mishiv@ukr.net

Lidiya T. Mishchenko, Taras Shevchenko Kiev National University, Department of Biology, Radiophysics, Volodimirska Str. 64, Kyiv 01033, Ukraine, lmishchenko@ukr.net

Michael Mishchenko, NASA Goddard Institute for Space Studies, 2880 Broadway, New York, NY 10025, USA crmim@giss.nasa.gov

Alexandr V. Morozhenko, Main Astronomical Observatory, National Academy of Sciences of Ukraine, Zabolotnoho Str. 27, Kyiv 03680, Ukraine, mor@mao.kiev.ua

Karri Muinonen, Observatory, University of Helsinki, P.O. Box 14, FIN-00014 Univ, Helsinki, Finland, Karri.Muinonen@Helsinki.Fi

Olga Muñoz, Instituto de Astrofisica de Andalucia, c/Camino Bajo de Huetor 24, Apartado 3004, 18080 Granada, Spain, olga@iaa.es

Vitaliy V. Omelchenko, Astronomical institute of Kharkiv National University, Sumskaya Str. 35, Kharkiv 61022, Ukraine, omelchenko@astron.kharkov.ua

Nikolay V. Opanasenko, Astronomical Institute of Kharkiv National University, Sumskaya Str. 35, Kharkiv 61022, Ukraine, opanasenko@astron.kharkov.ua

Andrey A. Ovcharenko, Institute of Astronomy of Kharkiv National University, Sumskaya Str. 35, Kharkiv 61022, Ukraine, Ovcharenko@astron.kharkov.ua

Dmitriy V. Petrov, Institute of Astronomy of Kharkiv National University, Sumskaya Str, 35, Kharkiv 61022, Ukraine, petrov@astron.kharkov.ua

Elena V. Petrova, Space Research Institute, Profsoyuznaya Str. 84/32, Moscow 117997, Russia, epetrova@iki.rssi.ru

Sergey I. Plachinda, Crimean Astrophysical Observatory, Nauchny, Crimea 98409, Ukraine, plach@crao.crimea.ua

Alina N. Ponyavina, Institute of Molecular and Atomic Physics, National Academy of Sciences of Belarus, F. Skaryna Ave. 70, Minsk 220072, Belarus, ponyavin@imaph.bas-net.by

Oleg V. Postylyakov, A.M. Obukhov Institute of Atmospheric Physics, RAS, Pyzhevsky Per. 3, Moscow 117997, Russia, ovp@omega.ifaran.ru

Natalija Primak, Crimean Astrophysical Observatory, Nauchny, Crimea 98409, Ukraine, efimov@astro.crao.crimea.ua

Marina Prokopjeva, Sobolev Astronomical Institute, St. Petersburg University, Universitetskij Prosp. 28, St. Petersburg 198504, Russia, marina@dust.astro.spbu.ru

Vera K. Rosenbush, Main Astronomical Observatory, National Academy of Sciences of Ukraine, Zabolotnoho Str. 27, Kyiv 03680, Ukraine, rosevera@mao.kiev.ua

Alla N. Rostopchina-Shakhovskaya, Crimean Astrophysical Observatory, Nauchny, Crimea 98409, Ukraine, arost@crao.crimea.ua

Jose Maria Saiz, Grupo de Óptica, Dpto. Física Aplicada, Universidad de Cantabria, 39005 Santander, Spain, saizvj@unican.es

Anton V. Samoilov, Institute of Semiconductor Physics, of Ukrainian National Academy of Sciences, prospekt Nauki 41, Kyiv 03028, Ukraine, samoylov@isp.kiev.ua

Sergiy N. Savenkov, Radiophysics Department, Kyiv Taras Shevchenko University, Vladimirskaya Str., 64, Kyiv 01033, Ukraine, sns@mail.univ.kiev.ua

Ayhan Sayin, Turkish State Meteorological Service, Research Department, Cc 401 Ankara, Turkey asayin@meteor.gov.tr

Sazuman Sazak, Trakya Universitesi Muh. Mimarlık Fakultesi, 22030 Edirne, Turkey, Sadumans@yahoo.com

Dmitry N. Shakhovskoj, Crimean Astrophysical Observatory, Nauchny, Crimea 98409, Ukraine, dshakh@crao.crimea.ua

Nikolay M. Shakhovskoj, Crimean Astrophysical Observatory, Nauchny, Crimea 98409, Ukraine, dshakh@crao.crimea.ua

Evgen V. Shalygin, Institute of Astronomy of Kharkiv National University, Sumskaya Str. 35, Kharkiv 61022, Ukraine, evgen@astron.kharkov.ua

Dmitry Shestopalov, Shemakha Astrophysical Observatory, Azerbaijan Academy of Science, Shemakha 373243, Azerbaijan, shestopalov_d@mail.ru

Yuri G. Shkuratov, Institute of Astronomy of Kharkiv National University, Sumskaya Str. 35, Kharkiv 61022, Kharkov shkuratov@vk.kh.ua

Natalija G. Shchukina, Main Astronomical Observatory, National Academy of Sciences of Ukraine, Zabolotnoho Str. 27, Kyiv 03680, Ukraine, shchukin@mao.kiev.ua

Leonid M. Shulman, Main Astronomical Observatory, National Academy of Sciences of Ukraine, Zabolotnoho Str. 27, Kyiv 03680,Ukraine, shulman@mao.kiev.ua

Tracy Smith, Space Science Institute 2192 Smith Laboratory, 174 West 18th Avenue Columbus, OH 43210, tsmith@pacific.mps.ohio-state.edu

Dmitiy Stankevich, Institute of Astronomy of Kharkiv National University, Sumskaya Str. 35, Kharkiv 61022, Ukraine, Stankevich@astron.kharkov.ua

Mikhail A. Sviridenkov, A.M. Obukhov Institute of Atmospheric Physics RAS, Pyzhevsky Per. 3, Moscow 119017, Russia, misv@mail.ru

Larisa Tambovtseva, Central Astronomical Observatory, Pulkovo, St. Petersburg, Russia, grinin@VG1723.spb.edu

Edvard I. Terez, National Taurida Vernadsky University, Department of Astrophysics and Atmospheric Physics, Yaltinskaya Str. 4, Simferopol 95007, Ukraine, terez@ccssu.crimea.ua

Svetlana A. Terpugova, Institute of Atmospheric Optics, Academicheskii Ave. 1, Tomsk 634055, Russia, swet@iao.ru

Mark Thoreson, Purdue University, School of Electrical and Computer Engineering, West Lafayette, Indiana 47907, USA, mthoreso@purdue.edu

Victor P. Tishkovets, Astronomical Institute of Kharkiv National University, Sumskaya Str. 35, Kharkiv 61022, Ukraine, tishkovets@astron.kharkov.ua

Lyuba Trachuk, Institute of Biochemistry and Physiology of Plants and Microorganisms, Russian Academy of Sciences, Pr. Entuziastov 13, Saratov 410049,Russia, LyubaT2001@mail.ru

Oleg S. Ugolnikov, Space Research Institute, RAS, Profsoyuznaya Str. 84/32, Moscow 117997, Russia, ugol@tanatos.asc.rssi.ru

Ben Veihelmann, FOM-Institute, AMOLF, Kruislaan 407, 1098 SJ Amsterdam, Netherlands, veihelmann@amolf.nl

Alexander A. Veles, Main Astronomical Observatory, National Academy of Sciences of Ukraine, Zabolotnoho Str. 27, Kyiv 03680, Ukraine, veles@mao.kiev.ua

Sergei F. Velichko, Department of Astronomy, School of Physics of Kharkiv National University, Svobody Sqr. 4, Kharkiv 61022, Ukraine, velichko@astron.kharkov.ua

Yuri I. Velikodsky, Astronomical Institute of Kharkiv University, Sumskaya Str. 35, Kharkiv 61022, Ukraine dslpp@astron.kharkov.ua

Gorden Videen, US Army Research Laboratory, 2800 Powder Mill Road, Adelphi, Maryland 20783, USA, gvideen@arl.army.mil

Anatoliy P. Vid'machenko, Main Astronomical Observatory, National Academy of Sciences of Ukraine, Zabolotnoho Str. 27, Kyiv 03680, Ukraine, vida@mao.kiev.ua

Tõnu Viik, Tartu Observatory, Tõravere, Tartumaa 61602, Estonia viik@aai.ee

Rosario Isabel Vilaplana, Universidad Politécnica de Valencia, Departamento de Física Aplicada of Aplicada, Pl.Ferrándiz Carbonell, s/n, Alcoy, Spain rosario@iaa.es

Hester Volten, University of Amsterdam, Astronomical Institute 'Anton Pannekoek', Kruislaan 403, 1098 SJ Amsterdam, The Netherlands, volten@amolf.nl

Nikolai Voshchinnikov, Astronomical Department and Sobolev Astronomical Institute, St. Petersburg University, Universitetskii Pr. 28, St. Petersburg 198504, Russia, nvv@astro.spbu.ru

Yeva Vovchyk, Ivan Franko National University, Astronomical Observatory, Kyrylo and Mefodiy Str. 8, Lviv, Ukraine, eve@astro.franko.lviv.ua

Yaroslav Yatskiv S., Main Astronomical Observatory, National Academy of Sciences of Ukraine, Zabolotnoho Str. 27, Kyiv 03680, Ukraine, yatskiv@mao.kiev.ua

Yuri S. Yefimov, Crimean Astrophysical Observatory, Nauchny, Crimea 98409, Ukraine, efimov@astro.crao.crimea.ua

Tatiana B. Zhuravleva, Institute of Atmospheric Optics SB RAS, Akademicheskii Ave. 1, Tomsk 634055, Russia, ztb@iao.ru

Klaus Ziegler, Institut fur Physik, Universitaet Augsburg, D-86135 Augsburg, Germany, Klaus.Ziegler@physik.uni-augsburg.de

Dmitry A. Zimnyakov, Saratov State University, Optics Department, Astrakhanskaya Str. 83, Saratov 410026, Russia, zimnykov@sgu.ru

Evgenij S. Zubko, Institute of Astronomy of Kharkiv National University, Sumskaya Str. 35, Kharkiv 61022, Ukraine, zubko@astron.kharkov.ua